19th Chemnitz Seminar on Materials Engineering

19. Werkstofftechnisches Kolloquium

IOP Conference Series: Materials Science and Engineering
Volume 181

Chemnitz, Germany
16 – 17 March 2017

Editors:

Thomas Lampke
Guntram Wagner
Martin F.-X. Wagner

ISBN: 978-1-5108-3772-0
ISSN: 1757-8981

Printed from e-media with permission by:

Curran Associates, Inc.
57 Morehouse Lane
Red Hook, NY 12571

Printed by Curran Associates, Inc. (2017)

For permission requests, please contact the Institute of Physics at the address below.

Institute of Physics
Dirac House, Temple Back
Bristol BS1 6BE UK

Phone: 44 1 17 929 7481
Fax: 44 1 17 920 0979

techtracking@iop.org

Additional copies of this publication are available from:

Curran Associates, Inc.
57 Morehouse Lane
Red Hook, NY 12571 USA
Phone: 845-758-0400
Fax: 845-758-2633
Email: curran@proceedings.com
Web: www.proceedings.com

Table of contents

Volume 181

19th Chemnitz Seminar on Materials Engineering – 19. Werkstofftechnisches Kolloquium

16–17 March 2017, Chemnitz, Germany

Accepted papers received: 23 February 2017
Published online: 17 March 2017

Preface

Invited Talks

Joining and Solid State Processing

Thermal Coatings

Mechanical Testing Fatigue Damage

Composite Materials Material Compounds

Materials Engineering Heat Treatment

Light Metals Ultrafine-grained Materials

Galvanic Deposition Anodization

IOP Conf. Series: Materials Science and Engineering **181** (2017) 011001 doi:10.1088/1757-899X/181/1/011001

Preface

This year, the Institute of Materials Science and Engineering (Institut für Werkstoffwissenschaft und Werkstofftechnik, IWW) of Chemnitz University of Technology is proud to host the Chemnitz Seminar on Materials Engineering – Werkstofftechnisches Kolloquium – for the 19th time.

This traditional scientific meeting is an annual get-together of scientists and engineers both from academia and industry. Most recent research results and current issues are presented and discussed in various scientific sessions, such as materials engineering and testing, surface engineering with focus on thermal coating processes, composite materials, and joining technologies.

Like last year, we have provided our contributors with the opportunity of publishing their work not only in the printed Proceedings Volume but also in *IOP Conference Series: Materials Science and Engineering*. The vast majority of participants chose to disseminate their interesting research results in this international, peer-reviewed journal. Consequently, this issue provides an excellent overview of the main topics of our colloquium.

Chemnitz, March 2017

Prof. Thomas Lampke
Prof. Guntram Wagner
Prof. Martin F.-X. Wagner

IOP Conf. Series: Materials Science and Engineering **181** (2017) 011002 doi:10.1088/1757-899X/181/1/011002

Materials combining structure and chemical function – A perspective

M Armbrüster

Institute of Chemistry, Chemnitz University of Technology, 09111 Chemnitz, Germany

Abstract. Production efficiency as well as the reduction of energy consumption and waste are the main drivers for the improvement of large-scale production in chemical industry. In addition, the demand for new processes – these days especially related to the chemical energy conversion as basis for the Energiewende – are additional triggers for innovation. Catalysis, well recognized by the public e.g. in form of the 3-way automobile catalyst, plays an important role for about 90% of all processes in chemical industry. The development of new and improved catalysts is a challenge for chemists – as is the search for new structural materials for mechanical engineers. Due to the often-necessary noble and thus expensive metals (e.g. palladium) in many catalysts, structural and chemical function have been developed separately in the two communities. Recent progress in catalytically active materials allows replacing expensive noble metals in selected catalytic processes by significant cheaper elements [1]. This offers the new perspective to combine structural and chemical functionality by innovative materials. Within this area, intermetallic compounds with their peculiar combination of crystal and electronic structure [2], represent an interesting class of materials to address this vision as will be shown in the contribution.

[1] Armbrüster M, Kovnir K, Friedrich M, Teschner D, Wowsnick G, Hahne M, Gille P, Szentmiklósi L, Feuerbacher M, Heggen M, Girgsdies F, Rosenthal D, Schlögl R and Grin Y 2012 *Nature Mater.* **11** 690

[2] Armbrüster M, Schlögl R and Grin Y 2014 *Sci. Technol. Adv. Mater.* **15** 034803

IOP Conf. Series: Materials Science and Engineering **181** (2017) 011002 doi:10.1088/1757-899X/181/1/011002

Beside all euphoria for A.M. and Industry 4.0 thermal spray remains the burner of technology!

K Nassenstein

GTV Verschleißschutz GmbH, 57629 Luckenbach, Germany

Abstract. Within the last 2 to 3 years GTV has realized a lot of heavy duty projects in thermal spray as well as in laser cladding technology. There will be interesting projects and applications shown. Unfortunately, the most interesting are under NDA. Nevertheless, an idea what is possible by coating technology will be given in this presentation. Moreover, component handling and masking, data acquisition and MES is in most cases a bigger challenge for equipment manufacturing than coating development itself. Nothing more, to upscale prototype coatings into serial production is a far under rated duty. The second part will show what kind of results are possible using high power plasma and arc spray burner. As a high light of this presentation a new idea of a plasma burner will be introduced for the first time.

IOP Conf. Series: Materials Science and Engineering **181** (2017) 011002 doi:10.1088/1757-899X/181/1/011002

Cold spraying for additive manufacturing and repair

T Klassen, S Krebs, K Binder, M Villa-Vidaller, F Gärtner, T Gartner and H Assadi

Helmut Schmidt University, Hamburg, Germany
Helmholtz-Zentrum Geesthacht, Centre for Materials and Coastal Research GmbH
Lufthansa

Abstract. Materials with high specific strength are required for aerospace applications. This presentation focuses on deposition of Ti and Al alloys as well as bronze by cold spraying, in view of cost-effective near net shape production of bulk parts, as well as refurbishing of used parts. Comprehensive understanding of the cold spray process allows for systematic optimization of process parameters. A generalized parameter can relate process conditions and feedstock powders with coating properties such as porosity, mechanical strength and electrical conductivity [Assadi et al., J. Thermal Spray Techn. 20 (2011) 1161]. This unifying process variable considers even complex interdependencies between several process conditions, and thus reveals unexpected tuning potential. For example, for most high strength materials, higher impact temperatures and higher impact velocities are desired. However, detailed analysis shows that, for some materials, lower pressures actually yield better results. This obviously applies to materials with limited deformability at higher strain rates, but also to materials with low density and/or powder cuts with smaller particle sizes. The concept facilitates prediction of prospective coating performance, and a respective user software tool is available. In addition, for comparison of cold spraying with thermal spraying, a generalized concept based on impact energy will be presented.
Based on the scientific understanding, dense deposits with tensile strengths up to the level of respective bulk material are successfully achieved, demonstrating that cold spraying is a powerful and energy efficient tool for additive manufacturing and repair.

IOP Conf. Series: Materials Science and Engineering **181** (2017) 011002 doi:10.1088/1757-899X/181/1/011002

Rivetless aircraft structures: Possible application of solid-state joining processes

J dos Santos

Institute of Materials Research, Helmholtz-Zentrum Geesthacht GmbH, 21502 Geesthacht, Germany

Abstract. It is estimated that the market demand for new aircraft in all classes will reach 23.000 units by the year 2020. Moreover, it is expected that this increase in production should be coupled with improved structural safety and reduced costs by fulfilling present environmental requirements ("clean manufacturing"). One way to achieve such demands is to introduce new production processes, in replacement to riveting, currently the most common assembly technology in aircraft manufacturing (i.e. rivetless aircraft structures). In this context, solid-state joining process offer ideal basis for application in such structures. These processes have been shown to be suitable to join all structural Al alloys in similar or dissimilar configurations, producing joints with superior properties and surface quality, in comparison to riveting. Currently two solid-state joining processes are being considered for possible application in the aeronautical industry: friction stir welding (FSW) and refill friction spot welding (FSpW). Apart from the standard FSW, two variants of the process have shown great potential for a future application: stationary-shoulder bobbin tool (SS-BTFSW) and stationary shoulder FSW (SS-FSW).

In this work results obtained with these two FSW process variants are presented and discussed. Emphasis has been placed on a comparison between FSW and SS-FSW in terms of mechanical properties, including toughness behaviour. Main process characteristics and surface quality aspects complement this section. The state of art on the application of the FSpW to aircraft structures is described based on an example centred on the alloy 2024-T356. The results cover the process development based on a DoE methodology, mechanical properties including fatigue and the development of numerical models predicting the mechanical performance of the joints.

In summary the results present in this study show that FSW and FSpW have achieved a very high maturity level for a possible application in aircraft structures. The potential weight savings (butt joints instead of overlap joints, absence of rivets, etc.) associated with higher productivity (welding speeds above 1 m/min) and satisfactory damage tolerance behaviour fulfil the expectations of the industry. Further steps are required in order to implement these processes in production and to expand the database on damage tolerance allowables for relevant alloys and joint configurations.

IOP Conf. Series: Materials Science and Engineering **181** (2017) 011002 doi:10.1088/1757-899X/181/1/011002

Materials science for making gold

S Odenwald

Department of Sports Equipment and Technology, Chemnitz University of Technology, 09126 Chemnitz, Germany

Abstract. From 4th to 5th century B.C. material scientists sought for technologies to transmute base elements into gold [1]. In 1894, it was Pierre de Coubertin who discovered the formula: he inspired the invention of the first Olympic Games that were held in 1896 in Athens [2]. Since then researches frequently managed to convert latest material developments into Olympic gold medals [3].

Citius, altius, fortius – running or gliding faster, jumping higher, batting stronger – in the field of sports material science is of vital importance in fighting for fractions of seconds or inches. While optimized materials [4], specific modifications [5] or functionalised surfaces enable athletes to exploit their full physiological capabilities, other material inventions like energy absorbing polymers in Helmets and mats [6] help to maintain the athlete's health and to provide safety. For material science the primary objective when it comes to man-machine-interaction in a sportive environment is often not to strive for the maximum of certain properties. As the restrictions are many and each athlete has individual demands the researchers' challenges are legion [7]. The ultimate goal is to achieve an optimized adaption of material and equipment to the moving athlete. Besides the resulting solutions have stay within the tight limits of frequently altered technical regulations. Appropriate testing procedures should follow the sport-specific requirements regarding kinetics and kinematics. Therefore courageous steps have to be taken away from standard material testing towards application oriented mechanical simulations [8].

[1] Deshpande V 1984 Transmutation of base-metals into gold as described in the text RASARNAVAKALPA and its comparison with the parallel chinese methods *Indian J. Hist. Sci.* **19** 186–92

[2] MacAloon J J 2006 The olympic idea *Int. J. Hist. Sport* **23** 483–527

[3] Caine M, Blair K and Vasquez M 2012 Materials and technology in sport *Nat. Mater.* **11** 655–8

[4] Kaufmann J, Kroll L and Odenwald S 2010 Application-specific design of sports equipment from anisotropic fiber-reinforced plastics with stiffness related coupling effect *Procedia Eng.* **2** 2569–74

[5] Schwanitz S, Frint P and Wagner M F-X 2016 High-strength aluminum materials with customized damping features for sled in high-performance sports *Tagungsband zum 18. Werkstofftechnischen Kolloquium in Chemnitz* p 639

[6] Schwanitz S, Costabile G, Amodeo G, Odenwald S and Lanzotti A 2014 Modelling head impact safety performance of polymer-based foam protective devices *Procedia Eng.* **72** 581–6

[7] Nathan A M, Crisco J J, Greenwald R M, Russell D A and Smith L V 2011 A comparative study of baseball bat performance *Sports Eng.* **13** 153–62

[8] Brückner K, Odenwald S, Schwanitz S, Heidenfelder J, Milani T L 2010 Polyurethane-foam midsoles in running shoes – impact energy and damping *Procedia Eng.* **2** 2789–93

IOP Conf. Series: Materials Science and Engineering **181** (2017) 011002 doi:10.1088/1757-899X/181/1/011002

In-line surface treatment and diffusion bonding of oxide forming metals

S Habisch and P Mayr

Institute of Joining and Assembly, Chemnitz University of Technology, Germany

Abstract. The formation of oxide layers is a critical factor for the weldability of oxide forming metals in diffusion bonding. To improve diffusivity and thus the joint properties it is necessary to treat the joining surfaces directly before joining. Any regeneration of the oxide layer before joining must be avoided. Therefore, a unique concept for in-line surface treatment and diffusion bonding was developed. A diffusion bonding chamber coupled to a glove box, to treat specimens and transfer samples under inert atmosphere was established. Results have shown, that improving joint properties or decreasing bonding temperature is possible by this in-line technology. Joining strategies for various metals and material combinations, such as Al-Mg, Al-Cu, Cu-Nb and W-Cu are presented.

IOP Conf. Series: Materials Science and Engineering **181** (2017) 011002 doi:10.1088/1757-899X/181/1/011002

Interface microstructure characterization of joints fabricated by oblique impact welding

S Sharafiev[1], B Niessen[2], C Pabst[3], A Rebensdorf[4], M F-X Wagner[1], P Groche[2] and S Böhm[4]

[1]Institute of Materials Science and Engineering, Chemnitz University of Technology, 09125 Chemnitz, Germany

[2]Institute for Production Engineering and Forming Machines, Technische Universität Darmstadt, Otto-Berndt-Straße 2, Darmstadt 64287, Germany

[3]PSTproducts GmbH, Junkersstraße 1, Alzenau 63755, Germany

[4]Institute of Production Technology and Logistics, Universität Kassel, Kurt-Wolter-Str. 3, Kassel 34125, Germany

E-Mail: semen.sharafiev@mb-tu-chemnitz.de

Abstract. The oblique collision of two metallic plates, accelerated towards each other, can result in joining of plates. A cohesive metallic bond is formed at certain values of the collision velocity and the collision angle between joining partners. In the current work, conventional magnetic pulse welding (MPW) and joining in a novel test rig are analyzed and compared. Typically, joint interfaces exhibit a wavy morphology with a wavelength that directly depends on the collision angle. During the welding process in MPW, the collision angle changes because of the specific plate deformation caused by the electromagnetic impulse. In contrast, the test rig was designed with the aim of enforcing constant collision angles. To compare both methods of joining, morphologies and microstructures of several interfaces are examined using optical and scanning electron microscopy. We show that the wavelength in the MPW seam increases with increasing collision angle, whereas the wavy morphology in the test rig joint stays homogeneous throughout the whole interface. This confirms that the collision angle does remain constant during the joining process in the test rig. Consequently, it is possible to analyze the joining mechanisms of the oblique impact welding at discrete collision angles by setting up certain values in the test rig experiments – this novel experimental method provides the means for well-defined experiments that pave the way to a more detailed understanding of impact joining processes.

Gas nitriding as thermochemical post heat treatment of thermal spray stainless steel coatings

P Kutschmann, T Lindner and T Lampke

Institute of Materials Science and Engineering, Chemnitz University of Technology, 09125 Chemnitz, Germany

Abstract. HVOF coatings of austenitic stainless steel (AISI 316L) were post-treated by gas nitriding. A variation of the process temperature has a signific effect on microstructure development. At conventional temperature range of 520 °C a combined layer and a diffusion zone is formed. The formation of nitride phases can be successfully prevented by decreasing the process temperature to 420 °C. A solid solution with a high amount of nitrogen can be achieved. In comparison to the conventional treated coating a significant increase in corrosion and wear resistance was detected. The interstitially dissolved nitrogen causes the formation of strong compressive residual stresses and high surface hardness. Increases in the service life of existing applications or new material combinations with face-centred cubic friction partners are possible.

IOP Conf. Series: Materials Science and Engineering **181** (2017) 011002 doi:10.1088/1757-899X/181/1/011002

Zinc-doped hydroxyapatite coatings developed by plasma spraying of precursor solution

R Candidato[1], R Serge[1] and L Pawlowski[2]

[1]University of Modena and Reggio Emilia, 41125, Modena, Italy
[2]SPCTS-CEC, University of Limoges, 12 Rue Atlantis, Limoges, France

Abstract. Hydroxyapatite (HA) is a naturally occurring mineral form of apatite having similar mineral composition with human bones. In a structural point of view, hydroxyapatite is flexible and accommodates anionic and cationic substitutions. The cation sites can be replaced by various cations and the physico-chemical and biological properties might be modified due to this incorporation of foreign ions. Among cations, zinc is considered to have great potential since it accommodates bone growth, inhibit osteoclast and promote osteoblast activity. In addition zinc also exhibits antibacterial activity which might be useful for HA when employed in medical applications.HA coatings used as medical implants are commonly developed by atmospheric plasma spraying using powder feedstock. An emerging process called solution precursor plasma spraying (SPPS) is an interesting one as it allows for the deposition of coatings directly from liquid precursors avoiding drawbacks of powder preparation process.

The present work presents about the synthesis of hydroxyapatite coatings incorporated with zinc (HA-Zn) using SPPS process. Varying mol% of zinc was incorporated to HA structure by using also different zinc ion precursors. Thermal behaviour of precursor solutions with different zinc concentrations was studied first under equilibrium condition by DTA-TG analysis. The influence of varying concentrations of zinc to the microstructural characteristics of hydroxyapatite was examined by X-ray diffraction (XRD) analysis, Fourier Transform Infrared (FTIR) spectroscopy and Scanning Electron Microscopy-Energy Dispersive Spectroscopy (SEM-EDS). Suitable parameters for the development of HA-Zn hybrid coating with appreciable characteristics were determined. The results of this work could open new opportunities of producing various oxide materials via plasma spraying of solution.

IOP Conf. Series: Materials Science and Engineering **181** (2017) 011002 doi:10.1088/1757-899X/181/1/011002

Investigation of porosity and oxide contents in thermally sprayed coatings – application and discussion of different methods

I Ali[1], E Saborowski[1], H Liborius[2], R Börner[2], A Schubert[2] and T Lampke[1]

[1]Institute of Materials Science and Engineering, Chemnitz University of Technology, 09125 Chemnitz, Germany
[2]Professorship Micromanufacturing Technology, Chemnitz University of Technology, Germany

Abstract. The microstructure of atmospheric plasma sprayed (APS) coatings contains process-dependent features such as globular pores, inter-lamellar pores and oxides. These features lower the thermal conductivity of the coating. Furthermore, they negatively affect the corrosion and wear resistance. On the other hand, open pores provide a reservoir for lubricants in tribological applications, thereby avoiding metal-to-metal contact between the respective parts. In order to be able to predict the properties of a thermally sprayed coating, a reliable investigation method for the porosity and oxide contents is of great interest.

In the present study, thermally sprayed Fe16Cr2Ni0.2C coatings based on different atmospheric plasma spray parameters were applied on the surface of cylindrical substrates. The spraying parameters were adapted to generate coatings with high and low porosity and oxide contents. These contents were investigated by laser scanning microscopy and image analysis software, based on the micrographs detected by light microscopy and scanning electron microscopy. The investigations were performed on the cross section polishes in the longitudinal direction of the coating. The results show that the laser scanning microscopy is suitable to provide a reliable estimation for the porosity content meanwhile it is not suitable to investigate the oxide contents. The image analysis is appropriate for the investigation of the porosity and oxide contents in the coatings. However, it overestimates the porosity content. Consequently, both methods of analysis should be used for detailed characterisations.

IOP Conf. Series: Materials Science and Engineering **181** (2017) 011002 doi:10.1088/1757-899X/181/1/011002

3D Scanning Laser Doppler Vibrometer for non-contact defect recognition on composite materials

M Fritzsche

Polytec GmbH, Polytec Platz 1-7, 76337 Waldbronn, Germany

Abstract. 3D Scanning Laser Doppler Vibrometers (3D-SLDV) allow a very precise, three dimensional, full field and non-contact vibration measurement on structures from a few mm² up to a few m². Additional to the standard vibration application of the 3D-SLDV such as measuring Operation Deflection Shape (ODS) and Frequency Response Function (FRF) for high spatial resolution Experimental Modal Analysis (EMA), today more and more application in the field of Non-Destructive Testing (NDT) of composites establish. One of the well know methods for defect recognition in plate like composite structures is the Lamb-Wave method. By means of an interaction of the symmetric and antisymmetric wave with the crack or delamination, the animation of the 3D-SLDV Vibration measurement can clearly visualise the defect position. A quite new method for defect selective NDT on Composite structures is the evaluation of non-linear effects appearing at the defect. Utilizing a low frequency (LF) and high frequency (HF) ultrasonic acoustic excitation at the composite structure one can determine with the 3D-SLDV a non-linear frequency modulation of the HF with the LF at the defect. A visualiation of the RMS value distribution of the modulation frequencies clearly indicate the defect position. Based on a high spatial resolution non-contact 3D experimental modal analysis (EMA) it is shown, how to find a defect by adjusting a defect in the FE model to the results of the EMA. Changing the position and the size of a defect in the FE model will lead to a change of the resonance frequency of the modes in the model. The defect position and size is known when e.g. 5 resonance frequency of the FE model of higher order modes are the same as in the experiment.

IOP Conf. Series: Materials Science and Engineering **181** (2017) 011002 doi:10.1088/1757-899X/181/1/011002

Evolution of adiabatic shear bands in the beta titanium alloy Ti-10V-2Fe-3Al

S Winter, M Scholze and M F-X Wagner

Institute of Materials Science and Engineering, Chemnitz University of Technology, 09125 Chemnitz, Germany

Abstract. Because of their low thermal conductivity, titanium alloys tend to a pronounced formation of adiabatic shear bands at high strain rates. The mechanical behavior of β-titanium is strongly affected by microstructural parameters like β-grain size, morphology and volume fraction of primary and secondary α-phase precipitates, but the effect of these microstructural features on shear band morphology is not yet well understood. In this study we investigate the evolution of adiabatic shear bands of the β-titanium alloy Ti-10V-2Fe-3Al for three different types of microstructures: beta titanium without any precipitates, beta titanium with primary α-phase and β-titanium with primary and secondary α-phase precipitates. To study the influence of these microstructural characteristics, we carried out compression-shear tests in a Split Hopkinson pressure bar (initial strain rate: 10^3 s-1) for each condition. Different amounts of straining (2 %, 5 % and 10 %) were applied to the material. This testing regime made it possible to investigate different stages of nucleation and propagation of the shear bands. To reveal the characteristics of the bands, scanning electron microscopy and nanoindentation was performed. Our results show that the formation mechanisms of adiabatic shear bands strongly depend on the initial microstructure. Shear bands in pure β-titanium are dominated by the formation of stress induced martensite. In β-titanium with primary α-phase, solid bands with a thickness of 5 µm can be observed. In contrast, β-titanium with primary and secondary α-phase exhibits many branching shear bands.

IOP Conf. Series: Materials Science and Engineering **181** (2017) 011002 doi:10.1088/1757-899X/181/1/011002

An experimental and theoretical study on the effect of Cu interlayers on microstructures of NiTi thin films on graphene substrates

S Hahn[1], A Schulze[1], M Böhme[1], T Hahn[2] and M F-X Wagner[1]

[1]Institute of Materials Science and Engineering, Chemnitz University of Technology, 09125 Chemnitz, Germany
[2]Institute of Theoretical Physics, TU Freiberg, 09669 Freiberg

Abstract. The present study explores how thin NiTi films can be deposited on graphene substrates. We deposited NiTi with thicknesses in the range of 2 to 4 microns on graphene substrates. The physical vapour deposition (PVD) processed thin NiTi and Cu films were subsequently annealed at 450 °C for several minutes. Different microstructures, and particularly a pronounced change in terms of the predominant grain orientation, were observed when an additional Cu interlayer was added between graphene and NiTi. These microstructures were characterized by scanning electron microscopy, electron backscatter diffraction, and transmission electron microscopy. Our results show that grain sizes and grain orientations in the thin NiTi films are considerably affected by the Cu interlayer. Theoretical investigations within the density functional theory framework were used to calculate the binding energies for different orientations of NiTi on the substrates. These numerical results support the experimental findings and provide details on the interaction at the NiTi / substrate interface that can explain the different microstructures of the functional NiTi layers.

IOP Conf. Series: Materials Science and Engineering **181** (2017) 011002 doi:10.1088/1757-899X/181/1/011002

Deformation analysis of pseudoelastic NiTi shape memory alloys

C Elibol, M Pouya and M F-X Wagner

Institute of Materials Science and Engineering, Chemnitz University of Technology, 09125 Chemnitz, Germany

Abstract. The deformation behavior of pseudoelastic NiTi shape memory alloys (SMAs) has been extensively characterized under uniaxial tensile loading. Previous studies have shown that NiTi SMAs exhibit a distinct mode of localized deformation (nucleation and propagation of well-defined martensite bands) under uniaxial tension that leads to a constant stress plateau in the corresponding stress-strain curves. Using digital image correlation (DIC) for experimental investigations, and the finite element method (FEM) for numerical analysis, the present study investigates the effect of the position of various virtual extensometers (with respect to the nucleation side of the martensite band) on the macroscopic stress-strain curves, especially on the shape of the nucleation peak that is typically observed prior to the pseudoelastic plateau stress. We also discuss the strain-softening nature of the isotropic material model that allows simulating the mechanical behavior of NiTi SMAs, including the localization and growth of individual martensite bands. We show for the first time that the specific softening behavior of NiTi SMAs under uniaxial tension can to some extent be directly characterized experimentally by positioning an extensometer (in post-experimental DIC data analysis) directly on the region corresponding to the nucleating martensite band. With excellent agreement between experimental and numerical results, our investigations demonstrate that the positioning of extensometers has a strong influence on the macroscopic stress-strain curve, and that numerical modeling of the mechanical behavior of the NiTi SMAs in tension requires a consideration of softening phenomena.

IOP Conf. Series: Materials Science and Engineering **181** (2017) 011002 doi:10.1088/1757-899X/181/1/011002

Punctual magnesium-aluminum-joints

D Serafinski and R Winkelmann

Fertigungstechnik/Tribologie, Brandenburgische Universität Cottbus-Senftenberg, Universitätsplatz 1, 01968 Senftenberg

Abstract. Lightweight construction with multi material systems requires a load bearing interconnection between different materials. Selection of the joining technology bases on material and joint properties. Because of limited miscibility of aluminum and magnesium in melted state some technological problems for using a thermal joining technology exist. For this reason a thermo-mechanical technology for punctual joining of magnesium and aluminum has been developed. Based on the principle clinching required thermal conditions to generate a loadable punctual joint have been elaborated. Whereas most aluminum alloys are - based on their crystal structure - good formable and therefore well suited for clinching it is not so with magnesium. The hexagonal crystal structure of magnesium requires adequate heating of the contact point prior to the joining process. This alone is not, however, enough, because the high thermal conductivity of the joining tool dissipates the supplied energy quickly. Consequently preheating of the tools is mandatory. At first thermal conditions were simulated with the aim of reduction of the test effort. This model has been validated experimentally. The thermal conditions of the joining part and the tools have been varied with the aim to increase the joint strength. As an optimization criterion shear strength has been defined. To realize a reproducible heating of both parts - the materials and the tools - a combination of Joule and induction heating proved useful. The temperature of the joined parts was measured to utilize the temporal effect of the inductor power in a defined position to the component as a criterion of the applied thermal energy. An adjustment of tool geometry and movement parameters resulted in reproducible high strength joints.

IOP Conf. Series: Materials Science and Engineering **181** (2017) 011002 doi:10.1088/1757-899X/181/1/011002

Brazing of steel-plastic-steel-composites

A Kilian and R Winkelmann

Fertigungstechnik/Tribologie, Brandenburgische Technische Universität Cottbus-Senftenberg, Universitätsplatz 1, 01968 Senftenberg

Abstract. Consequent lightweight construction is an imperative requirement to meet the growing social expectations in the fields of power efficiency, climate protection, fuel consumption, comfort and sustainability. The developed steel-plastic-steel-composites represent new materials that show a great potential in lightweight construction. These materials allow its final user - the mobility producing industry - to work with preassembled lightweight compounds. Utilizing the new steel-plastic-steel-composites in an intelligent way requires adequate joining techniques. Sustainable joining of steel-plastic-steel-composites by perpetuation of the multi material concept becomes the central component in lightweight construction. The addition of goals like the speed of joining of 4m/min, direct further processing of the joined parts, no negative influence on the steel-plastic bond, tight and metallic joints, process reliability and highest aesthetic ambitions lead to the development of a new joining technology. Conventional thermal techniques have been tested and found to be limited in their applicability. Therefore it is necessary to reduce the currently utilized minimal thermal processes energy by about 2/3. That was achieved by energetic splitting of the melting of the filler metal from the process of component activation. Independently from that it is necessary to use solder with low melting point based on Sn-Zn. Surface activation of the substrates to be joined can be realized applying flux but also by laser radiation. Additionally arcs have been tested. These cannot be applied reliably in the intended speed range. Using the example of flange seams the usability of spray brazing with laser support is shown. As soldering the steel-plastic-steel-composites on to massive components one has to take the differences in thermal conduction into account. Both energy sources are placed with respect to this influence. Characteristics of the constructed joints were evaluated physically and metalographically. A method to evaluate delaminations has been developed.

IOP Conf. Series: Materials Science and Engineering **181** (2017) 011002 doi:10.1088/1757-899X/181/1/011002

Materials and their combinations for spot-welding connections of aluminum

D Krüger[1], P Schütte[2] and R Winkelmann[1]

[1]Fertigungstechnik/Tribologie, Brandenburgische Technische Universität Cottbus-Senftenberg, Universitätsplatz 1, 01968 Senftenberg
[2]GLAMAtronic Schweiß- und Anlagentechnik GmbH

Abstract. Connection of lightweight construction materials, like Aluminum with conventional resistance spot welding provokes intensive reactions between electrodes and substrate. Diffusion processes influence the surface of electrodes. This changes the properties of a welding point. After a few welding points, it is necessary to rework the electrodes. A possibility to reduce these reactions is to consider electrode materials with low affinity to aluminum. Additionally the thermal and electrical properties of these materials must be considered. Furthermore, a secure and effective connection to the electrode material copper has to be established. As a result of assessing and weighting electrical and thermal properties as well as the economical aspect, the technically interesting materials graphite, tungsten and molybdenum were considered in the tests. In addition, physical vapour deposition (PVD)-coated copper electrodes were investigated. Information about the suitability of the materials was gained via long-term studies. A problem that needed to be addressed was the creation of connections between materials showing no or just a minimum reaction with aluminum and copper of the welding electrodes. Capacitor discharge welding proved suitable. Parameters creating secure connections in an extremely short time were determined. In contrast to soldering or casting it is an effective process with high potential for mass production. The test results can be considered as a prerequisite for the multiplication of similarly produced welding points supporting lightweight constructions.

IOP Conf. Series: Materials Science and Engineering **181** (2017) 011002 doi:10.1088/1757-899X/181/1/011002

Punctual fiber composite-metal-joints

D Krüger and R Winkelmann

Fertigungstechnik/Tribologie, Brandenburgische Technische Universität Cottbus-Senftenberg, Universitätsplatz 1, 01968 Senftenberg

Abstract. Consequent lightweight construction requires the use of materials with differing properties. Therefore it is necessary to use multi material solutions. Implementation of these concepts requires strong connections. For this purpose mechanical joining technology offers good conditions, because different thermal properties of materials can be compensated. However, the application requires deformability of the parts at the point of joint which is limited in fiber composites. Hence possibilities for thermal and thermo-mechanical pretreatments of the joints were investigated, with the aim to produce high-strength connections. In addition to conventional energy sources, such as heated gas, induction and resistance heating, laser was considered during investigations. Also variants of the mechanical and thermal processes such as thermal assisted pre-punching were examined for achieving high bond strengths. By thermo-mechanical pretreatment of the joint, a doubling of the ultimate tensile strength compared to the conventional self-piercing rivet connections was achieved. Mechanical qualities of the joints were determined applying valuation of the used connecting technologies. Metallographic studies help to understand the results of the mechanical tests. Finally, corrosion tests were performed on the created joints and coated rivets. In addition to the visual examination of the corrosion resistance changes of mechanical properties of the connection were evaluated. The result of this investigation is a technology allowing the generation of reproducible connections.

IOP Conf. Series: Materials Science and Engineering **181** (2017) 011002 doi:10.1088/1757-899X/181/1/011002

Influence of laser beam hardening on the mechanical properties of ledeburitic chrome steel

P Landgraf[1], J Schubert[2], T Grund[1], E Haack[2] and T Lampke[1]

[1]Institute of Materials Science and Engineering, Chemnitz University of Technology, 09125 Chemnitz, Germany
[2]LHW GmbH, Am Winkel 11, 15528 Spreenhagen, Germany

Abstract. Heat treatment via laser beam hardening allows to heat up the edge layer of a component locally and therefore also to specifically change the properties. The advantages are, among other things, short treatment durations, a high degree of flexibility and a very good automation capability of the process. However, the fast heating rates and the short austenitization periods require an adjustment of the conventional heat treatment routines. Especially in the case of high-alloyed hypereutectoid steels, an adaptation of the laser treatment parameters is necessary, since insufficient heat treatment can, for example, lead to low hardness and increased corrosion susceptibility. Furthermore, the high carbon content causes the formation of residual austenite because the martensite finish temperature is below room temperature, so that subsequent freezing is necessary after a conventional self-quenching at room temperature. The various influences on the microstructure and the resulting properties are analysed in the present study. The influencing factors considered are the austenitizing temperature, the surface state of the steel and the feed rate of the laser, as well as the cooling conditions. These where varied and their effects on the heat treatment result are objects of investigation. Light- and electron-microscopic examination methods are used to analyze the microstructure and to draw conclusions on the microstructural transformations and laser beam depths. XRD-measurements enable a quantitative phase analysis. Hardening measurements allow to record hardness profiles and, in combination with materialography, to describe the individual laser traces. The effects of the different influencing factors on the heat treatment results are presented in a poster.

IOP Conf. Series: Materials Science and Engineering **181** (2017) 011002 doi:10.1088/1757-899X/181/1/011002

Localized shear deformation in different α-brass sheets during accumulative roll bonding

M Böhme, P Fritzsche, S Hahn and M F-X Wagner

Institute of Materials Science and Engineering, Chemnitz University of Technology, 09125 Chemnitz, Germany

Abstract. Accumulative roll bonding (ARB) is a promising method to produce metal/metal compound sheet materials. For many material combinations the thickness of the individual constituting metal sheets has to differ significantly to allow for repeated stacking and rolling. In this study we present investigations on the deformation behavior of two different α-brass alloys (CuZn15 and CuZn37) during accumulative roll bonding (ARB), and on the formation of symmetric compound sheets formed from these two alloys. Longitudinal sections (RD-ND plane) of the produced metal sheets were investigated using optical and scanning electron microscopy to estimate the homogeneity of grain refinement as a function of the number of ARB cycles. Tensile testing was performed on samples oriented parallel and perpendicular to the rolling direction in order to evaluate the associated changes in mechanical properties resulting from grain refinement, work hardening and texture changes. The CuZn37 ARB sheets suffer from a massive localization of deformation, and we observe shear bands passing through the entire cross section of the ARB samples already at low numbers of ARB cycles. The CuZn15 sheets, in contrast, exhibit similar localization zones only in the final investigated ARB pass (N = 5) – and to a much weaker extent. In the compound sheets, the localization zones in the CuZn37 layers are deflected and dissipated in a zone close to the CuZn15/CuZn37 interface.

IOP Conf. Series: Materials Science and Engineering **181** (2017) 011002 doi:10.1088/1757-899X/181/1/011002

Small scale mechanical testing of bulk ultrafine-grained laminated structures

P Frint and M F-X Wagner

Institute of Materials Science and Engineering, Chemnitz University of Technology, 09125 Chemnitz, Germany

Abstract. In this study, we report on bulk laminated microstructures after processing of AA6060 by extrusion and subsequent ECAP at room temperature. Macroscopic deformation bands, embedded in a matrix containing significantly less strain, are observed. A visio-plastic method, where local deformation is studied by analyzing the distortion of equi-axed grids of micro-indents, helps to understand the unstable plastic flow of the material and the formation of distinct shear bands. We analyze the local (anisotropic) mechanical behavior of individual bands by micro tensile- and micro shear-testing. We furthermore consider local vs. global behavior and the role of local strain rate sensitivity and load path-dependent behavior. Digital image correlation data obtained during compression testing clearly show that further straining exclusively takes place within the shear bands when their orientation is 45° to the applied load, even though the shear bands exhibit a higher strength. We discuss how this type of micromechanical behavior can be fully rationalized by considering local textures and strain rate sensitivities.

IOP Conf. Series: Materials Science and Engineering **181** (2017) 011002 doi:10.1088/1757-899X/181/1/011002

Effect of compression shear loading on the deformation behavior of a strongly textured AZ31 magnesium alloy

S Seipp[1], S Pfeiffer[1], P Frint[1], S Roy[2] and M F-X Wagner[1]

[1]Institute of Materials Science and Engineering, Chemnitz University of Technology, 09125 Chemnitz, Germany

[2]Materials Science Center, Indian Institute of Technology IIT-Kharagpur, India

Abstract. Magnesium has a low formability at room temperature that is directly related to its hexagonal close packed crystal structure. Its mechanical behavior is strongly anisotropic and significantly controlled by texture. In this study, we focus on the deformation behavior of such a strongly textured magnesium alloy under complex stress states during compression shear testing. The influence of texture on the deformation behavior is studied experimentally for different alignments of the dominant crystal orientations with respect to the loading direction. Moreover, we use an elastic finite element simulation to analyze the special stress state caused by the geometry of the compression shear specimen. Based on the simulated stress states at characteristic points in the specimen, and using a set of representative orientation data measured via electron backscatter diffraction (EBSD), we perform Schmid-Factor calculations for the $\{0001\} < 11\bar{2}0 >$ basal slip system. The results are compared to compression shear experiments at different strain rates, indicating additional strain rate effects on the deformation behavior. Our study contributes to a more detailed understanding of how the deformation of a strongly textured magnesium alloy is affected by an interplay between textures and (complex) stress states imposed by special sample geometries.

IOP Conf. Series: Materials Science and Engineering **181** (2017) 011002 doi:10.1088/1757-899X/181/1/011002

On the effect of natural aging prior to low temperature ECAP of a high-strength aluminum alloy

S Fritsch and M F-X Wagner

Institute of Materials Science and Engineering, Chemnitz University of Technology, 09125 Chemnitz, Germany

Abstract. High-strength aluminum alloys are hard to deform. Conventional severe plastic deformation methods that generate ultra-fine grained microstructures and thus increase strength lead to cracking of the billets. In this study, we investigate low temperature deformation in a cooled ECAP tool (internal angle 90°) as an alternative approach to strongly deform a 7075 aluminum alloy. By varying the natural aging time between 1 min and 24 h after solutionizing, we analyze how the formation of precipitates prior to low temperature ECAP affects the formability. We analyze the fracture surfaces of the as-ECAPed materials by scanning electron microscopy, and we document a transition from brittle to ductile fracture behavior with increasing natural aging times. Complementary transmission electron microscopy investigations further demonstrate that natural aging prior to ECAP leads to an increased formation of sub-grain structures and smaller grain sizes after ECAP. Finally, we characterize, by tensile and compression testing, how these microstructural changes influence the mechanical properties. Our results highlight the potential of low temperature forming to produce improved mechanical properties of high-strength aluminum alloys after severe plastic deformation.

IOP Conf. Series: Materials Science and Engineering **181** (2017) 011002 doi:10.1088/1757-899X/181/1/011002

On the interplay of strengthening mechanisms in low-temperature extruded AA6060

N Berndt, P Frint and M F-X Wagner

Institute of Materials Science and Engineering, Chemnitz University of Technology, 09125 Chemnitz, Germany

Abstract. Compared to conventional extrusion processes, extrusion below recrystallization temperature leads to considerably accelerated precipitation kinetics (including a technologically attractive reduction of aging time) and higher hardness, which offers a high potential for the integration in industrial process chains. In the present study, we investigate the influence of the extrusion temperature on the microstructures and precipitation behavior of the age hardening 6060 aluminum alloy. Cast billets are solid-solution annealed and extruded at different temperatures ranging from room temperature to 170 °C, which corresponds to the alloy's aging temperature. Electron back-scatter diffraction is performed in order to characterize the microstructure at certain locations (near the surface and in the center) within the cross-section of the extrudates. The center region exhibits a coarse-grained microstructure while fine- and ultrafine-grained pancake structures are observed near the outer surface. Despite this microstructural gradient, a homogeneously distributed hardness is observed after subsequent short-time aging of the material extruded at room-temperature (108 HB) and 120 °C (116 HB). In contrast, the material extruded at 170 °C exhibits a heterogeneous hardness distribution (mean value: 109 HB) in the as-extruded condition due to dynamic aging. These differences in hardness distribution likely result from a complex interplay of work hardening as well as static and/or dynamic recovery and precipitation. Studying the influence of extrusion temperature on these microstructural mechanisms may help to tailor mechanical properties as well as processing routines for future industrial applications.

IOP Conf. Series: Materials Science and Engineering **181** (2017) 011002 doi:10.1088/1757-899X/181/1/011002

Numerical design of electrolytes for electrodeposition of alloys

M Müller, I Scharf, E Schmidl and D Höhlich

Institute of Materials Science and Engineering, Chemnitz University of Technology, 09125 Chemnitz, Germany

Abstract. Over 500 binary alloys can be deposited from aqueous solution. But only a small amount of them has been used industrially so far. The problem lies mainly in the strong dependence of the alloy composition from electrolyte composition and the working conditions. The empirical approach in the development of new electrolytes depends extremely on the experimenter. The upscaling from laboratory to industrial scale is also very little implemented because of the local optimization of electrolyte properties.

In order to standardize the design of electrolytes, new mathematical optimization algorithms are required. They allow a prediction of the alloy composition as a function of essential process parameters.

A prediction of the composition and the workspace takes place entirely by simple experiments for the determination of thermodynamic and kinetic characteristics. The use of the Design of Experiments decreases the number of experiments additionally. The production of a basic electrolyte, without discard many electrolytes and a big amount of experiments, opens up new perspectives for the development of novel alloys.

Peer review statement

All papers published in this volume of *IOP Conference Series: Materials Science and Engineering* have been peer reviewed through processes administered by the proceedings Editors. Reviews were conducted by expert referees to the professional and scientific standards expected of a proceedings journal published by IOP Publishing.

Recent developments in plasma spray processes for applications in energy technology

G Mauer, M O Jarligo, D Marcano, S Rezanka, D Zhou and R Vaßen

Forschungszentrum Jülich GmbH, Institute of Energy and Climate Research, IEK-1: Materials Synthesis and Processing, 52425 Jülich, Germany

Abstract. This work focuses on recent developments of plasma spray processes with respect to specific demands in energy technology. High Velocity Atmospheric Plasma Spraying (HV-APS) is a novel variant of plasma spraying devoted to materials which are prone to oxidation or decomposition. It is shown how this process can be used for metallic bondcoats in thermal barrier coating systems. Furthermore, Suspension Plasma Spraying (SPS) is a new method to process submicron-sized feedstock powders which are not sufficiently flowable to feed them in dry state. SPS is presently promoted by the development of novel torch concepts with axial feedstock injection. An example for a columnar structured double layer thermal barrier coating is given. Finally, Plasma Spray-Physical Vapor Deposition (PS-PVD) is a novel technology operating in controlled atmosphere at low pressure and high plasma power. At such condition, vaporization even of high-melting oxide ceramics is possible enabling the formation of columnar structured, strain tolerant coatings with low thermal conductivity. Applying different conditions, the deposition is still dominated by liquid splats. Such process is termed Low Pressure Plasma Spraying-Thin Film (LPPS-TF). Two examples of applications are gas-tight and highly ionic and electronic conductive electrolyte and membrane layers which were deposited on porous metallic substrates.

1. Introduction

Material development covers material synthesis on the one hand and processing techniques on the other hand. These scopes are closely related to each other and must be both considered correspondingly [1]. With respect to energy technology, some present development goals are strain tolerant microstructures with low thermal conductivity, thin and gas tight functional layers, and processing of new materials with particular difficulties such as decomposition and inhomogeneous evaporation.

In this work, three recent developments of plasma spray processes are introduced with respect to some demands in energy technology. Processing of MCrAlY bondcoats for TBCs by High Velocity-Atmospheric Plasma Spraying (HV-APS) was established as an alternative to common processes such as Low Pressure Plasma Spraying (LPPS) or High Velocity Oxy-Fuel Spraying (HVOF). HV-APS manufactured microstructures, phase compositions and oxygen contents were investigated with respect to their impact on oxidation behaviour. Moreover, the processing of nano-sized powder feedstock by Suspension Plasma Spraying (SPS) is described as a processing method to obtain new microstructures such as dense vertically cracked or columnar structured ceramic thermal barrier coatings (TBCs). Finally, the processing of oxide ceramics by Plasma Spray-Physical Vapor Deposition (PS-PVD) is shown to obtain novel microstructures such as strain tolerant, columnar structured TBCs on the one hand as well as thin and dense gas separation membranes and electrolytes on the other hand.

2. High Velocity-Atmospheric Plasma Spraying

MCrAlY alloys are typically applied to protect e. g. gas turbine components from high temperature oxidation and to promote the bonding of ceramic thermal barrier coatings [2, 3]. The basic constituents are M=Ni, Co or a combination thereof, further alloying elements are 17–30% Cr, 8–15% Al and smaller amounts of reactive elements such as Y. MCrAlY coatings should exhibit a low porosity, a sufficient low oxygen content, and a controlled oxidation behaviour. Typical coating processes are LPPS (formerly often termed VPS), HVOF, and sometimes APS. HV-APS however, aims to offer an alternative method operated under atmospheric conditions but without oxygen as process gas. Furthermore, particle temperatures can be well adjusted as high as necessary but as low as possible. High particle velocities are obtained at justifiable costs as the gas consumption is much lower compared to other kinetic spray processes [4].

2.1 Process development

The HV-APS process was realized applying a TriplexPro™-210 torch (Oerlikon Metco AG, Wohlen, Switzerland) equipped with a 5 mm nozzle. The plasma gas flow was increased to the maximum possible values of 190 slpm Ar and 40 slpm He. Higher flows were not available due to limitations of the spray facility and gas supply. The current was increased to 450 A which is the highest value being allowed to be applied with the 5 mm nozzle according to the torch specifications. Figure 1 shows a plasma jet under HV-APS conditions. Since the flow is supersonic, the characteristic sequence of compression and expansion cells (shock diamonds) is generated.

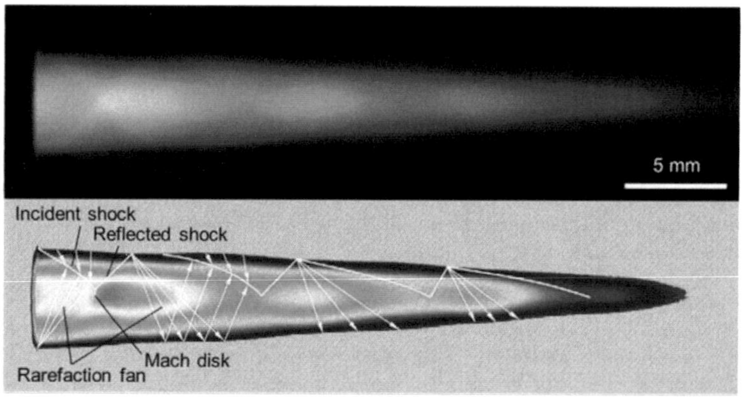

Figure 1. Shock diamonds in a supersonic HV-APS plasma jet.

Initially, the Co-32Ni-21Cr-8Al-0.5Y powder feedstock was optimized. As a result, a very fine cut (-22 / +5 µm) was selected. The carrier gas flow had to be increased up to 6.5 slpm in order to achieve an appropriate injection depth of the fine powder particles into the high-velocity plasma jet. The lowest coating porosities and best deposition efficiencies were observed at a spray distance of 100 mm.

IOP Conf. Series: Materials Science and Engineering **181** (2017) 012001 doi:10.1088/1757-899X/181/1/012001

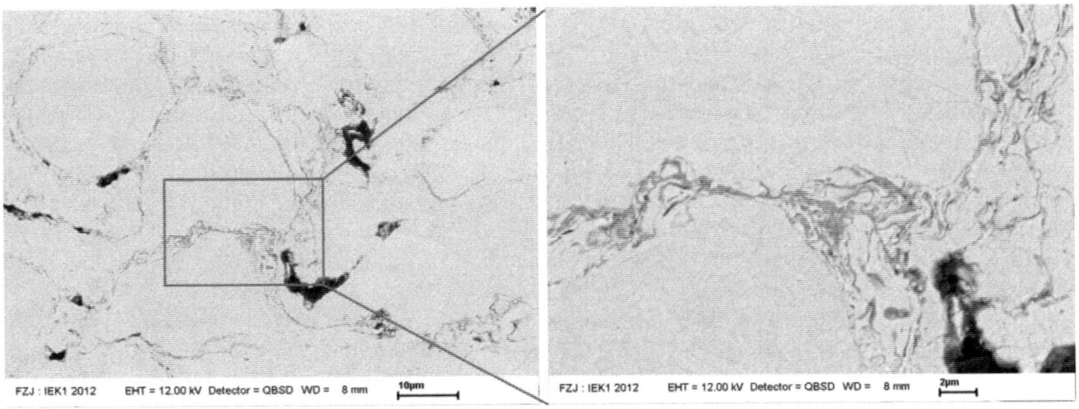

Figure 2. HV-APS as-sprayed microstructures [4].

Figure 2 gives insight into the microstructures obtained at these parameters. It is obvious that the powder particles were softened just so much that a dense structure could be achieved. On the other hand, particle temperature was low enough to avoid extensive oxidation; during the short flight time, only a thin oxide scale was formed on the particle surfaces. A chemical analysis revealed an oxygen content of 0.41 ± 0.04 wt% which is more than what is typically obtained by LPPS [5]. However, it is much less than obtained by APS [6] and lower compared to most of the HVOF coatings [7]. By HVOF, lower oxygen contents are still possible, but only using a much coarser powder feedstock [8]. However, allowances have to be made with respect to a small porosity.

2.2 Oxidation tests

Isothermal oxidation tests were performed with freestanding coating specimen at 1100°C in synthetic air for 72 h [9]. HV-APS samples were compared with LPPS reference samples. Despite their higher oxygen content, the HV-APS samples showed a distinctly lower oxidation rate, Figure 3. The aluminium-rich beta phase was less depleted and the thermally grown oxide layer (TGO) was thinner, more homogeneous, and consisted of virtually pure alumina.

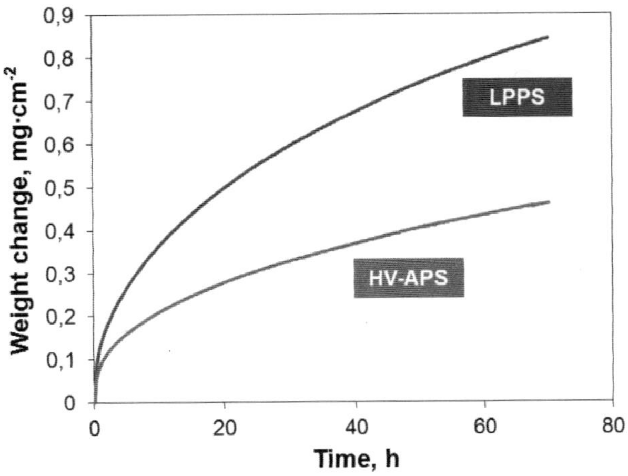

Figure 3. Isothermal oxidation rates of freestanding HV-APS and LPPS coating specimen at 1100°C in synthetic air for 72 h [9].

So called Y-Al oxide pegs, as they are typically formed in LPPS coatings, were missing in the HV-APS samples, Figure 4. Y was present in form of nano-sized oxide particles. No obvious depletion of Y-rich oxides occurred during oxidation. In contrast, in the LPPS reference samples, Y was present in Ni-Y intermetallic phases. It was used up during oxidation. This unique, new distribution of Y-rich nano-sized oxide precipitates in the HV-APS samples was able obviously to moderate the thermal growth of the oxide layer. Positive effects were also revealed for cyclic oxidation. Tests (2 h cycles) at 1100 °C in laboratory air for 300 h confirmed the better oxidation resistance of the HV-APS coatings. The LPPS specimen even showed indications of delaminated oxide scales at the edges. This was not the case for the HV-APS coatings which obviously did not reach a critical thickness of the TGO layer [10].

Hence, HV-APS appears to be a promising, cost effective alternative to established spraying techniques such as LPPS and HVOF. However, the formation mechanisms of the particular oxide distribution and the effect on the oxidation behavior still need further clarification.

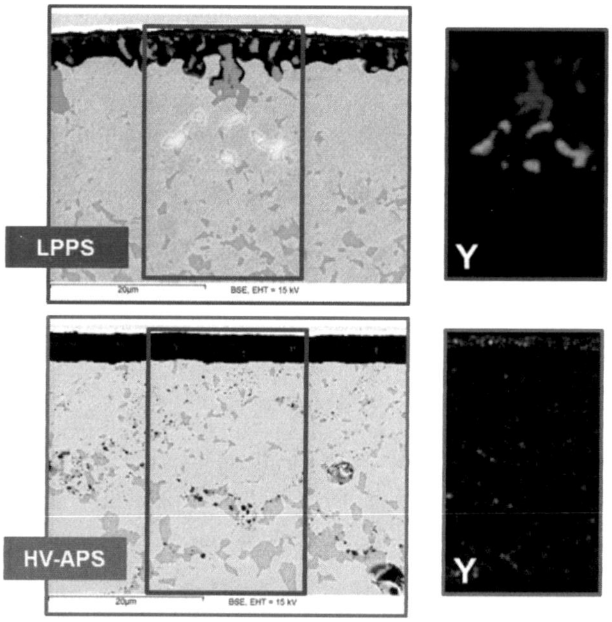

Figure 4. EDX mapping of yttrium distribution near surface zone after 72 h isothermal oxidation (1100 °C) [9].

3. Suspension plasma spraying

Suspension Plasma Spraying (SPS) is a new method to process submicron-sized feedstock powders which are not sufficiently flowable to feed them in dry state [11]. Novel microstructures such as dense vertically cracked or columnar structured ceramic thermal barrier coatings (TBCs) can be obtained [12]. SPS is presently promoted by the development of novel torch concepts with axial feedstock injection such as the Axial III™ plasma torch with three cathodes, anodes, and converging plasma jets, each (Northwest Mettech Corp., Vancouver, Canada).

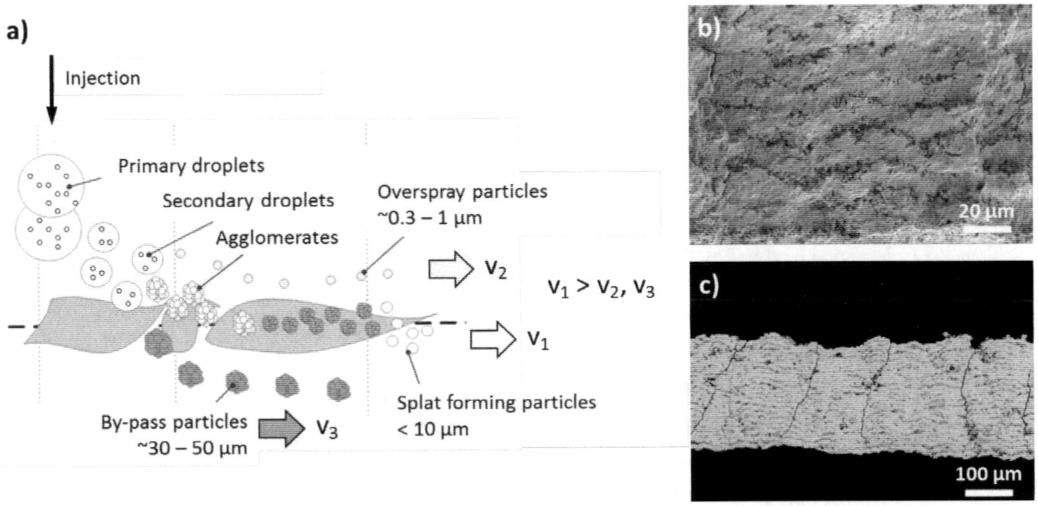

Figure 5. Formation of form highly porous, weakly bonded overspray layers due to radial injection of the suspension feedstock; schematic diagram [13] a), examples: fracture surface b) and polished cross-section c).

Figure 5 illustrates the problem which is associated to SPS with radial injection. After evaporation of the carrier fluid, the solid particles agglomerate. Their trajectories are widely fanned out in radial direction depending on their size. Due to their small inertia, the fine overspray particles do not penetrate deep enough into the hot plasma core so that they are not properly molten. However, they are partly deposited on the substrate on top of each pass and form highly porous, weakly bonded layers. Figure 5 gives two examples. For TBCs, such overspray layers were found to be disadvantageous as they can channel crack propagation and affect the cohesion of the coating. This can be avoided by axial injection of the suspension so that the advantages of nano-sized microstructures can also benefit for TBCs.

Figure 6 shows a columnar structured Y_2O_3-stabilized ZrO_2 (YSZ)/$Gd_2Zr_2O_7$ double layer TBC by SPS with axial injection. Coatings were sprayed at 105 kW torch input power on Inconel 738 LC substrates with a MCrAlY bondcoat. The YSZ solid load of the suspension for the TBC was 10 wt% and the feed rate 30 g/min. The solid load of the $Gd_2Zr_2O_7$suspension had to be reduced to 5 wt% to achieve columns. This might be related to the larger particle size in the suspension. The detail in Figure 6 shows that the interface between YSZ and $Gd_2Zr_2O_7$ appears free of any interference. The investigation of service life performance and failure modes is still pending.

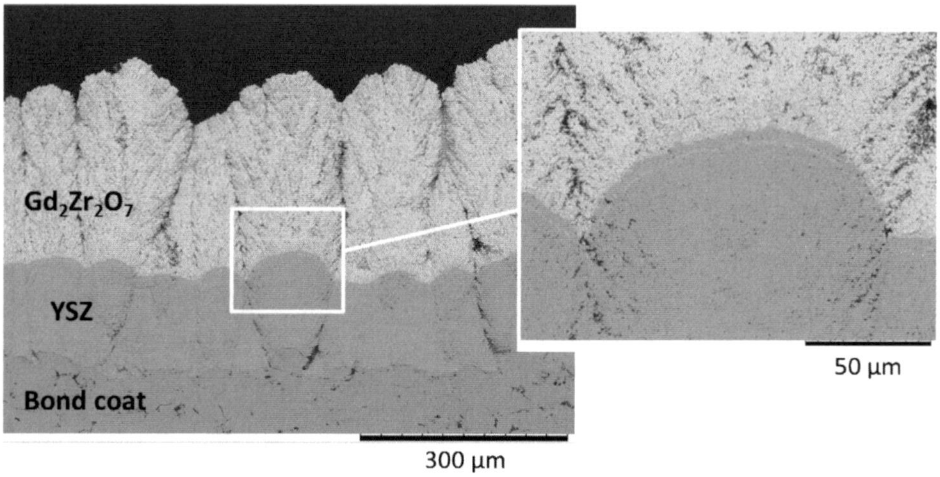

Figure 6. YSZ/$Gd_2Zr_2O_7$ double layer TBC by SPS with axial injection.

4. Very low pressure plasma spraying

The combination of plasma spraying at very low pressures with enhanced electrical input power led to the development of a process family which is termed Very Low Plasma Spraying (VLPPS) [14]. Applying electrical currents up to 3 kA, an input power level of 150 kW can be achieved. As long as the deposition is dominated by the liquid phase, particular thin and dense coatings can be manufactured. This specific process is termed Low Pressure Plasma Spraying-Thin Film (LPPS-TF). Beyond that, it is even possible to evaporate the feedstock material, if appropriate parameters and powders are applied. In this particular case, the process is termed Plasma Spray-Physical Vapor Deposition (PS-PVD) enabling the manufacture of columnar, strain tolerant microstructures for TBCs. The experiments described in the following were carried out on a Sulzer Metco VLPPS Multicoat system with an O3CP torch (Oerlikon Metco, Wohlen, Switzerland). It resulted from a comprehensive reconstruction of an existing conventional LPPS system. In particular, it was equipped with an additional vacuum pumping unit, a large vacuum blower to provide sufficient pumping capacity at low pressures, enlarged cooling capacity, and new control units.

4.1 Thermal Barrier Coatings by PS-PVD

Two different YSZ microstructures of TBCs manufactured by PS-PVD are given in Figure 7. The corresponding plasma parameters are indicated in the caption. The coating in Figure 7 a) was designed to achieve an increased erosion resistance. However, the thermal conductivity was higher due to the relatively small porosity. In contrast, the coating shown in Figure 7 b) was manufactured at hotter conditions. The highly porous, columnar structure is strain tolerant and exhibits a lower thermal conductivity [15]. However, this is inherently at the expense of erosion resistance.

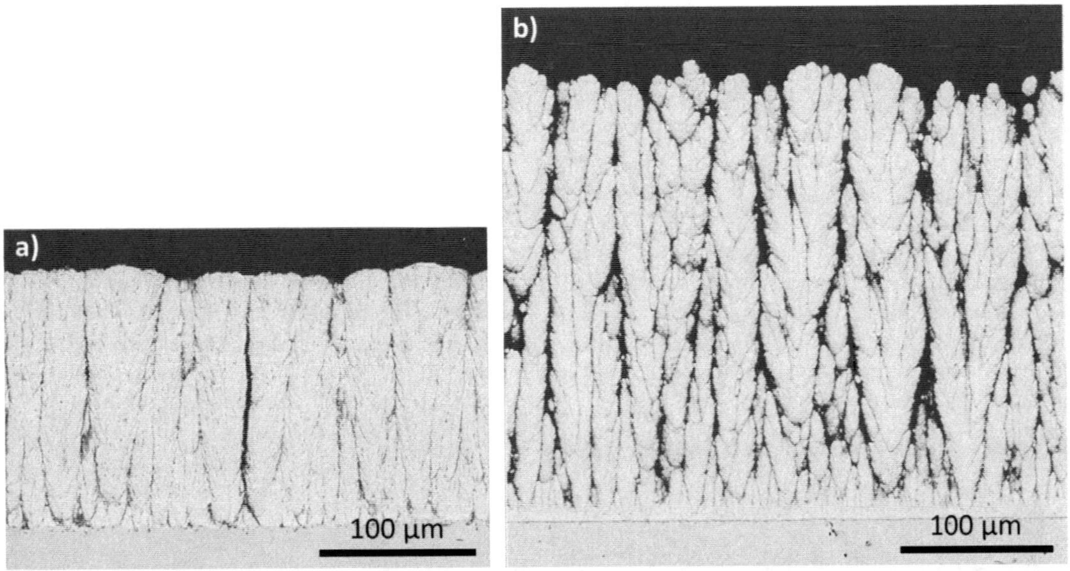

Figure 7. YSZ microstructures for TBCs by PS-PVD; a) plasma parameter 60 slpm He, 35 slpm Ar, 10 slpm H_2, current 2200 A; b) plasma parameter 60 slpm He, 35 slpm Ar, current 2200 Ar [15].

The failure analysis of a TBC system with a YSZ topcoat such as shown in Figure 7 b) is given in Figure 8 after a cyclic thermal gradient test. Details of the experimental setup and procedure can be found elsewhere [16]. The test conditions are indicated in the caption.

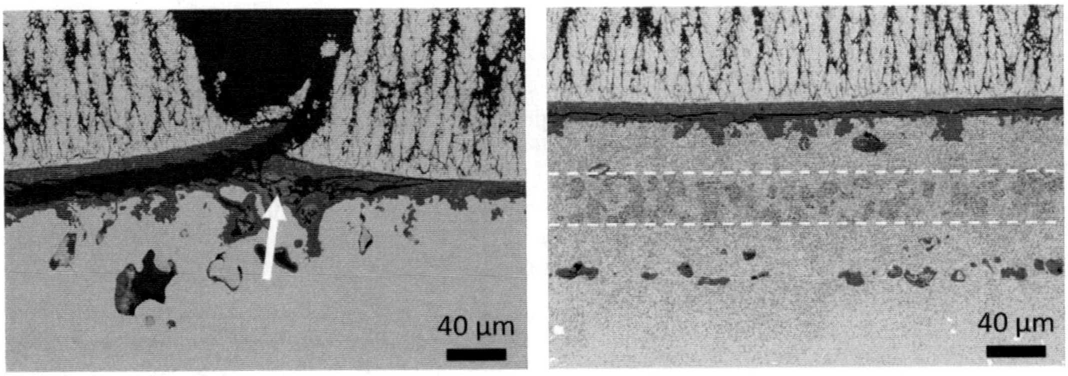

Figure 8. TBC failure analysis after 1185 cycles (~99 h) at cycling conditions $T_{surface} = 1382°C$, $T_{bondcoat}$ ~1113°C; the arrow indicates fast growing transient oxides, the β-phase is coloured to enhance contrast, the dotted lines demarcate the zones of ß-phase depletion.

A thick TGO layer was formed which is deep-rooted in the bondcoat by many "pegs". The ceramic topcoat was locally bulged and behaved highly deformable. No failure within the ceramic TBC was observed indicating a high strain tolerance. For conventional TBCs, such failure mode during cycling is unusual. Below the bulged areas, a non-uniform growth of TGO due to the formation of fast growing transient oxides (Cr_2O_3, spinel $(Co,Ni)Cr_2O_4$) occurred [17]. The bondcoat showed pronounced zones of ß-phase depletion. Obviously, the durability of ceramic top coat exceeded the service capability of the bondcoat.

4.2 Oxygen transport membranes by LPPS-TF

The perovskite material $La_{0.58}Sr_{0.4}Co_{0.2}Fe_{0.8}O_3$ (LSCF) is a mixed ion and electron conductor applied for oxygen transport membranes and cathodes of solid oxide fuel cells (SOFCs). Due to the large difference between the vapor pressures of its constituents, LSCF is prone to decomposition and inhomogeneous evaporation. Thus, single phase coatings could be obtained by APS only at moderate torch power and short dwell time of the feedstock particles in the plasma jet [18]. However, such coatings are not gastight as required for gas separation membranes.

An alternative approach is LPPS-TF. The feedstock was an agglomerated powder with d_{50} = 10 µm consisting of single phase rhombohedral perovskite; the feed rate was 20 g min^{-1}. The membranes were sprayed directly on a porous, sintered NiCoCrAlY substrate at a plasma torch input power of 90 kW and an Ar/He plasma gas mixture. The chamber pressure was 200 Pa and a small amount of extra O_2 was introduced into the chamber to maintain the LSCF stoichiometry. The example given in Figure 9 demonstrates that the large substrate porosity of approx. 40% could be closed effectively. The detail reveals that the splats were flattened widely so that voids and pores were levelled out [19].

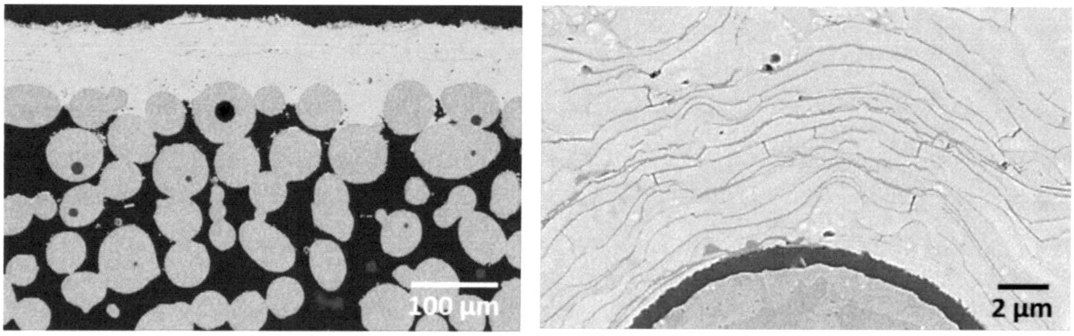

Figure 9. Thin and gastight LSCF membrane on a porous MCrAlY substrate by LPPS-TF [19].

Such LSCF coatings meet the requirements on oxygen separation membranes. They are dense (porosity 0.4%), gastight, and thin (approx. 50 µm). The metal substrate is a mechanical stable alternative to conventional ceramic supported cells. No subsequent sintering procedure is needed; hence, no shrinking occurs as on wet chemical routes [20]. An oxygen permeation test for 300 h at temperatures up to 1000°C revealed an oxygen permeation rate of 5.3 ml/min cm² which is amongst the highest fluxes measured up to now for LSCF [21]. The scalability of crack-free samples was preliminary demonstrated up to a diameter of 105 mm.

4.3 Electrolytes for solid oxide fuel cells

Metal supported SOFCs were manufactured by a combination of APS and LPPS-TF [22]. The substrate was a Fe-Cr alloy with a porosity of 40%. The anode was deposited by APS containing YSZ unmelted globules in a Ni/NiO grain matrix at a porosity of 9%. The YSZ electrolyte was manufactured by LPPS-TF with a thickness of 26 µm and 2% porosity. For the cell tests, a simple LSCF cathode was screen-printed which was kept non-sintered.

Figure 10 shows the metal supported fuel cell assembly and a cross-section at the working electrode site after testing. Air-tightness (anode & electrolyte) as well as a good contact at the metal support/anode and anode/electrolyte interfaces were achieved. The contact at the cathode/electrolyte interface however still needs improvement.

The cell test at 817°C resulted in a good open-circuit voltage (OCV) of 1.033 V. At the same temperature, the current density was 1.380 A cm^{-2} at 0.7 V corresponding to a power density of 0.976 W cm^{-2}. These values are higher than known from literature for solid oxide fuel cells with thermally sprayed functional layers. Details can be found elsewhere [22]. The excellent cell performance was attributed to the high ionic conductivity and good inter-splat contact in the electrolyte layer, even though it is not absolutely dense. Further optimization potential is suggested with respect to powder processing in order to reduce the still existing porosity and the thickness of the membrane.

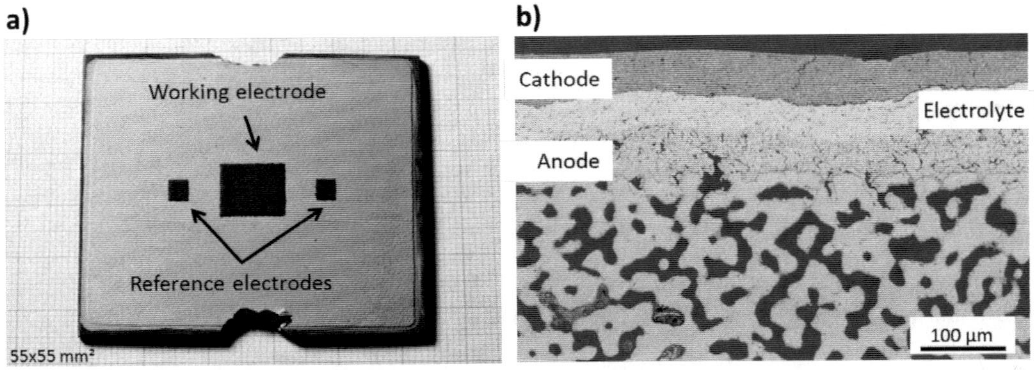

Figure 10. a) Metal supported solid oxide fuel cell by APS/LPPS-TF; b) cross section of anode, electrolyte, and cathode assembly at the working electrode site after testing.

5. Summary and conclusions

In this work, recent developments of plasma spraying techniques are described with respect to specific demands in energy technology. Three processes and some typical applications are given as examples. HV-APS parameters were developed for the TriplexPro™-210 torch with a 5 mm high velocity nozzle. The best MCrAlY microstructures were obtained using a fine powder with $d_{50} = 14$ µm. Although the spray conditions were generally cold, reasonable deposition efficiencies and rather dense coatings were achieved. The oxidation rates of free standing HV-APS coatings were lower than those of LPPS reference coatings. Far less Y incorporation in the oxide scale and a better oxide scale adherence were observed. The reason for the superior oxidation performance was suggested to be related to the unique distribution of Y-rich nano-sized oxide precipitates in the coatings.

TBCs with different structures, such as vertical cracked and columnar, can be manufactured by SPS axial suspension injection. A columnar structured YSZ/$Gd_2Zr_2O_7$ double layer system was given as an example. The investigation of service life performance and failure modes is still pending.

LPPS-TF and PS-PVD are significantly different from spraying under atmospheric conditions. Novel structural features can be achieved. One example are highly strain tolerant columnar structured TBCs by PS-PVD with excellent adhesion to the bondcoat. Other examples are dense and effective oxygen transport membranes by LPPS-TF. Thin and gastight LSCF thin films exhibiting high oxygen permeation rates were deposited directly on porous metallic supports. Finally, an air-tight metal supported SOFC assembly was obtained by a combination of APS and LPPS-TF. This cell showed a good OCV and an excellent power density.

Acknowledgments
The cooperation on HV-APS with A. Barth, M. Gindrat, and H.-M. Höhle (Oerlikon Metco, Wohlen) is gratefully acknowledged. The research on the oxygen transport membranes has partly received funding from the European Community's Seventh Framework Programme (FP7/2007-2013) under grant agreement no. 241309 (Project acronym: DEMOYS). The oxygen permeation measurements of the LSCF membranes were carried out by J. Serra and J. Garcia (CSIC, Valencia). The SOFC cell tests were performed by A. Weber (KIT, Karlsruhe).

References

[1] Mauer G, Jarligo M O, Mack D E and Vaßen R 2013 Plasma-Sprayed Thermal Barrier Coatings: New Materials, Processing Issues, and Solutions *J. Therm. Spray Technol.* **22** 646-58

[2] Toscano J, Vaβen R, Gil A, Subanovic M, Naumenko D, Singheiser L and Quadakkers W J 2006 Parameters affecting TGO growth and adherence on MCrAlY-bond coats for TBC's *Surf. Coat. Technol.* **201** 3906-10

[3] Naumenko D, Shemet V, Singheiser L and Quadakkers W J 2009 Failure mechanisms of thermal barrier coatings on MCrAlY-type bondcoats associated with the formation of the thermally grown oxide *J. Mater. Sci.* **44** 1687-703

[4] Mauer G, Sebold D and Vaßen R 2014 MCrAlY Bondcoats by High-Velocity Atmospheric Plasma Spraying *J. Therm. Spray Technol.* **23** 140-6

[5] Mauer G, Vaßen R and Stöver D 2007 Controlling the oxygen contents in vacuum plasma sprayed metal alloy coatings *Surf. Coat. Technol.* **201** 4796-9

[6] Patterson T, Leon A, Jayaraj B, Liu J and Sohn Y H 2008 Thermal cyclic lifetime and oxidation behavior of air plasma sprayed CoNiCrAlY bond coats for thermal barrier coatings *Surf. Coat. Technol.* **203** 437-41

[7] Shibata M, Kuroda S, Murakami H, Ode M, Watanabe M and Sakamoto Y 2006 Comparison of Microstructure and Oxidation Behavior of CoNiCrAlY Bond Coatings Prepared by Different Thermal Spray Processes *Mater. Trans.* **47** 1638-42

[8] Rajasekaran B, Mauer G and Vaßen R 2011 Enhanced Characteristics of HVOF-sprayed MCrAlY Bond Coats for TBC Applications *J. Therm. Spray Technol.* **20** 1209-16

[9] Mauer G, Sebold D, Vaßen R, Hejrani E, Naumenko D and Quadakkers W J 2016 Impact of processing conditions and feedstock characteristics on thermally sprayed MCrAlY bondcoat properties *Surf. Coat. Technol.*

[10] Evans H E 1995 Stress effects in high temperature oxidation of metals *Int. Mater. Rev.* **40** 1-40

[11] Killinger A, Gadow R, Mauer G, Guignard A, Vaßen R and Stöver D 2011 Review of New Developments in Suspension and Solution Precursor Thermal Spray Processes *J. Therm. Spray Technol.* **20** 677-95

[12] Guignard A, Mauer G, Vaßen R and Stöver D 2012 Deposition and Characteristics of Submicrometer-Structured Thermal Barrier Coatings by Suspension Plasma Spraying *J. Therm. Spray Technol.* **21** 416-24

[13] Siegert R 2006 A Novel Process for the Liquid Feedstock Plasma Spray of Ceramic Coatings with Nanostructural Features. PhD thesis Ruhr University Bochum. In: *Berichte des Forschungszentrums Jülich vol. 4205,* (Jülich: Forschungszentrum Jülich)

[14] Mauer G, Gindrat M and Smith M F 2016 Very Low Pressure Plasma Spraying (VLPPS), including PS-TF, PS-PVD, and PS-CVD. In: A. Vardelle et al., The 2016 Thermal Spray Roadmap *J. Therm. Spray Technol.* **25** 1383-8

[15] Mauer G 2014 Plasma Characteristics and Plasma-Feedstock Interaction Under PS-PVD Process Conditions *Plasma Chem. Plasma Process.* **34** 1171-86

[16] Traeger F, Rauwald K-H, Vaßen R and Stöver D 2003 Thermal Cycling Setup for Testing Thermal Barrier Coatings *Adv. Eng. Mater.* **5** 429-32

[17] Rezanka S, Mauer G and Vaßen R 2014 Improved Thermal Cycling Durability of Thermal Barrier Coatings Manufactured by PS-PVD *J. Therm. Spray Technol.* **23** 182-9

[18] Harris J and Kesler O 2009 Atmospheric Plasma Spraying Low-Temperature Cathode Materials for Solid Oxide Fuel Cells *J. Therm. Spray Technol.* **19** 328-35

[19] Jarligo M O, Mauer G, Bram M, Baumann S and Vaßen R 2014 Plasma Spray Physical Vapor Deposition of $La_{1-x}Sr_xCo_yFe_{1-y}O_{3-\delta}$ Thin-Film Oxygen Transport Membrane on Porous Metallic Supports *J. Therm. Spray Technol.* **23** 213-9

[20] Mauer G, Jarligo M O, Rezanka S, Hospach A and Vaßen R 2015 Novel opportunities for thermal spray by PS-PVD *Surf. Coat. Technol.* **268** 52-7

[21] Marcano D, Mauer G, Sohn Y J, Vaßen R, Garcia-Fayos J and Serra J M 2016 Controlling the stress state of $La_{1-x}Sr_xCo_yFe_{1-y}O_{3-\delta}$ oxygen transport membranes on porous metallic supports deposited by plasma spray–physical vapor process *J. Membr. Sci.* **503** 1-7

[22] Marcano D, Mauer G, Vaßen R and Weber A 2016 Manufacturing of high performance solid oxide fuel cells (SOFCs) with atmospheric plasma spraying (APS) and plasma spray-physical vapor deposition (PS-PVD) *Surf. Coat. Technol.*

Formation of a diffusion-based intermetallic interface layer in friction stir welded dissimilar Al-Cu lap joints

R Marstatt[1], M Krutzlinger[2], J Luderschmid[1], M F Zaeh[2] and F Haider[1]

[1]Chair for Experimental Physics 1, University of Augsburg, Universitaetsstr. 1, 86159 Augsburg, Germany
[2]Institute for Machine Tools and Industrial Management (*iwb*), Technical University of Munich, Boltzmannstr. 15, 85748 Garching, Germany

E-mail: roland.marstatt@physik.uni-augsburg.de

Abstract. The joining of dissimilar metals is an important issue in modern lightweight design. Friction Stir Welding (FSW) is suitable for this task since the solidus temperature is usually not exceeded during the process. As a consequence, dissimilar joints can be produced with a minimum of deteriorating intermetallic phases. The latest studies showed the formation of intermetallic layers at the bonding interface with no significant negative influence on the seam quality. In this study, those intermetallic nanolayers at the interface of aluminium / copper lap joints were analysed. For the experiments, the commercially pure alloys EN AW-1050 and CW008A were chosen. The process temperature changed with respect to the parameter setup and was measured at different locations of the seam. The intermetallic layers at the interface were analysed by scanning electron microscopy (SEM) and scanning transmission electron microscopy (STEM). The experiments show that the thickness of the interlayer clearly correlates with the process temperature using an Arrhenius equation. It is supposed, that the rotating probe removes the oxide layers of the metal surfaces and a metallic bonding between the Al- and the Cu-phase is formed. Due to the elevated temperature after the probe has passed, the intermetallic layer has emerged by interdiffusion.

1. Introduction

The use of materials adapted to the locally specified requirements of structures (e.g. load capability, thermal/electrical conductivity) is one of the most promising solutions in modern lightweight design to reach higher efficiencies in terms of energy, mass and cost reduction. This results in the necessity of combining dissimilar materials. The thermal and physical properties of these material combinations are often very different or even competing. Therefore, the joining of dissimilar materials remains challenging for traditional fusion welding technologies. However, innovative solid state welding technologies, such as Friction Stir Welding (FSW), are suitable for this task. The solidus temperature is usually not exceeded during this process, which reduces the formation of thin intermetallic phase layers. As a consequence, high quality joints can be produced.

1.1 Literature review

The use of FSW to join dissimilar materials was reported and confirmed in numerous studies [1]. Most studies focused on feasibility issues or the dependency between the welding conditions, the interface zone structure and the mechanical properties. Many studies also proved the existence of diffusion based intermetallic compounds (IMC) at the interface between the materials.

Giera et al. [2] reported on laser-assisted FSW of butt joints of aluminium (AA5182) and steel (ZStE340). The authors described a saw tooth shaped interface area, which provides mechanical interlocking on the one hand and enlarges the interface area for diffusion on the other hand. Marginal diffusion processes were also discussed by Uzun et al. [3] for butt joints of aluminium (Al 6013-T4) and stainless steel (X5CrNi18-10). Kostka et al. [4] assumed a positive influence of intermetallic phases (particularly Fe_2Al_5) on joints of lap-welded aluminium (Al6181) and steel (ZStE340).

Former studies analysed the interface of friction stir welded aluminium / titanium joints both in butt and in lap joint configuration. Aonuma & Nakata [5] observed mixed regions with the IMC $TiAl_3$ at the interface of butt welded aluminium (2024-T3 and 7075 T651) and titanium (pure and Ti6Al4V) joints. A continuous IMC, consisting of TiAl3 with a thickness of about 100 nm, was detected by Wu et al. [6] for butt joints of aluminium (Al6061) and titanium (Ti6Al4V). The authors stated a dependency between the layer thickness, which is also influenced by the rotational speed, and the tensile strength, since the joint strength was reduced for higher layer thicknesses. Dressler et al. [7] analysed butt welded joints of aluminium (AA2024) and titanium (TiAl6V4). It was suggested, that an IMC layer existing in small regions could positively influence the mechanical properties. The formation of $TiAl_3$ was also reported by Chen & Nakata [8] for lap joints of aluminium (ADC12) and titanium (commercially pure). The authors stated a strong dependency between the heat input during FSW and the diffusion based formation of $TiAl_3$, which influences the mechanical properties. Chen et al. [9] performed FSW on lap joints of aluminium (LF6) and titanium (TC1). The existence of IMCs at the interface zone was also indicated in their studies and decreasing feed rates lead to increasing shear strengths. Krutzlinger et al. [10] detected an interface layer with a constant thickness of about 20 nm at the bonding area of lap joints of aluminium (EN AW-1050) and titanium (Ti grade 1). It was assumed that this layer consists of $TiAl_3$. In some experiments, also possibilities for macroscopic form-fit were provided by a recess of the titanium surface. The tensile shear strengths of such joints were not significantly increased compared to joints without any recess and macroscopic form-fit, respectively. Similar observations were reported by Chen & Yazdanian [11] for lap joints of aluminium (6060) and titanium (Ti6Al4V). The authors achieved the highest tensile shear strength when the probe was positioned close to the interface without penetrating the titanium surface. They assumed, that a very thin interface layer leads to high tensile shear strengths, since no interface layer could be detected with the used resolution of 250 nm.

Besides that, the combination of aluminium and copper materials is of increasing interest, especially in the electrical industry. A detailed review on friction stir welded aluminium / copper joints is given by Galvão et al. [12]. The studies presented in the following also discuss the occurrence of IMCs (e.g. CuAl, $CuAl_2$, or Cu_9Al_4). Abdollah-Zadeh et al. [13] and Saeid et al. [14] performed FSW on joints of aluminium (1060) and copper (commercially pure). Both studies suggest that a sufficient joint strength results neither from high nor from low heat input conditions. While Abdollah-Zadeh et al. [13] observed an increasing amount of IMCs for an increasing rotational speed or a decreasing feed rate, Saeid et al. [14] reported on cavity defects for increased and on microcracks for reduced feed rates. Continuous and uniform IMC layers with thicknesses of about 1 μm were detected by Xue et al. [15; 16] for butt joints of aluminium (1060) and copper (commercially pure). Similar results were reported by Tan et al. [17], who welded butt joints of aluminium (5A02) and copper (T2), and by Muthu & Jayabalan [18], who welded butt joints of aluminium (AA1100-H14) and copper (commercially pure). It is concluded, that thin IMC layers provide high joint strengths. To maintain such thin layer thicknesses, a probe offset from the interface can be used as described by Genevois et al. [19] and Galvão et al. [20], who welded butt joints of aluminium and copper (1050-H16 / Cu-bl and AA6082-T6 / Cu-DHP, respectively). With the probe only stirring in the aluminium matrix, the thickness of the interface layer was reduced to 200 nm in the first study. It was not reported, if there was any influence on the joint strength. In Galvão et al. [20], the IMC build-up could be inhibited with

IOP Conf. Series: Materials Science and Engineering **181** (2017) 012002 doi:10.1088/1757-899X/181/1/012002

a probe offset. However, there was no improvement of the mechanical properties of the joints. FSW of aluminium (Al99.5) and copper (Cu-DHP) was performed with a stationary shoulder tool by Doehner et al. [21]. It was observed that a probe length causing a small distance between the probe tip and the interface leads to a minimal recess of the copper sheet and results in increased joint strengths. A deeper recess of the copper sheet by an increased probe length causes a drastic reduction of the joint strength due to the occurrence of cavities at the interface zone.

Summarizing the above studies, the build-up of IMC layers during FSW of aluminium and copper is evident. These layers have to be very thin to improve the mechanical properties, especially the strength of the joints. There is a strong dependency between the layer thickness and the heat input on the interface zone, which can be influenced by the welding conditions. However, a detailed description of the mechanisms during welding, which lead to the final joint properties, is still subject of research. In the following, these dependencies are discussed concerning process temperature, joint strength and a detailed analysis of the bonding interface.

2. Experimental and analysis setup

2.1 Experimental setup

A schematic diagram of a lap welded aluminium / copper joint is shown in Figure 1. The sheets had dimensions of 245 mm x 100 mm with a thickness of 4.0 mm for the commercially pure aluminium EN AW-1050 and 2.0 mm for the commercially pure copper CW008A. A constant lap width of 40 mm was set for all experiments with the aluminium sheet on the advancing side (AS) and the copper sheet on the retreating side (RS) of the tool. The tool dimensions were 14 mm for the shoulder diameter, 5.0 mm for the base and 3.6 mm for the tip diameter of the conical probe and 3.5 mm for the probe length. In combination with a shoulder plunge depth of 0.1 mm, this resulted in an effective probe-tip-to-interface distance of about 0.3 mm for a tool tilt angle of 2 °. The probe was threaded and furnished with three equally distributed flats. A CNC milling machine (Heller MCH250) was used in position controlled mode for the experiments.

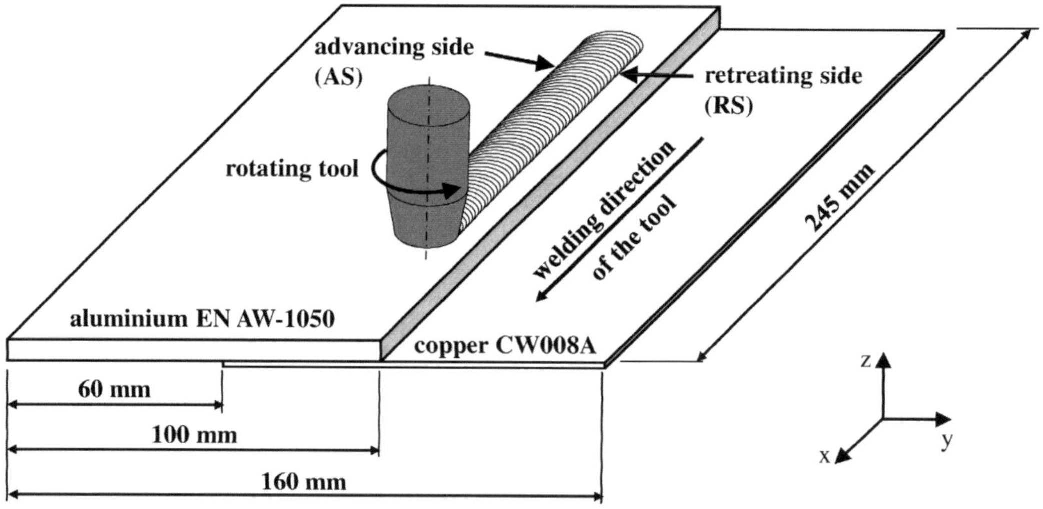

Figure 1. Schematic diagram of the sample dimensions and of the lap welded aluminium / copper joints.

In order to measure the process temperature, each of the sheets (aluminium and copper) was outfitted with three type K thermocouples at various positions along the welding path. The thermocouples were placed $60\,mm$, $122.5\,mm$ and $182.5\,mm$ away from the edge where the process began. The thermocouples with a diameter of $0.5\,mm$ were positioned in drill holes with a depth of $13\,mm$. The holes were placed at the face side of the RS for the aluminium sheets with a distance of $0.55\,mm$ to the interface and at the face side of the AS for the copper in the middle of the sheets (figure 2).

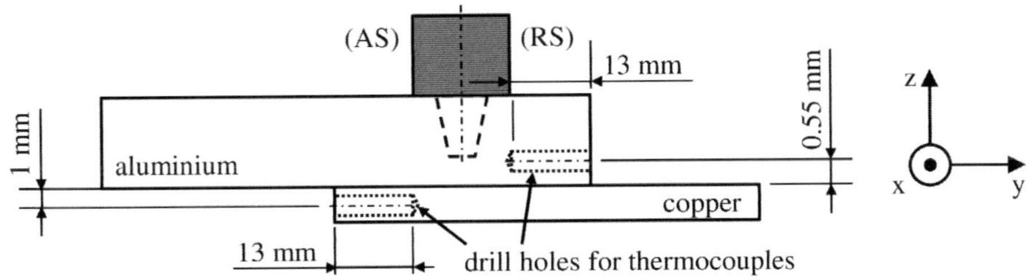

Figure 2. Positions of the drill holes for temperature measurements via thermocouples.

Three specimens were cut out of each joint perpendicular to the welding seam to test the shear strength, tensile tests were carried out on a Zwick Z020 testing machine. The extraction positions were at $62.5\,mm$, $102.5\,mm$ and $142.5\,mm$ from the starting edges of the sheets and the nominal width of the specimens was $20\,mm$.

The rotational speed n was varied during the welding study (see Table 1), while the feed rate remained constant at $v = 300\,mm/min$.

Table 1. Rotational speed settings for the experiments.

Sample	Rotational speed in min^{-1}
1	800
2	1300
3	1800
4	2300
5	2800

2.2 Analysis setup

Two specimens of each sample were extracted perpendicular to the welding seam at $87\,mm$ and $127\,mm$ from the starting edges of the sheets for electron microscopy. The cross sections were prepared metallographically for scanning electron microscopy (SEM) with a final thermomechanical polishing using $40\,nm$ silica particles with hydrogen peroxide. The specimens for transmission electron microscopy (TEM) were electrolytically thinned and finalised by a brief ion polishing. High resolution SEM was executed using a ZEISS MERLIN microscope. A JEOL JEM-2100F allowed TEM with a resolution up to atomic scale.

3. Results and discussion

3.1 Process temperature

Figure 3 illustrates the average of the measured peak temperatures in the aluminium and in the copper sheet for different rotational speeds. The error bars show the minimum and the maximum peak temperature for each rotational speed. The mean temperature in both materials increased with an increasing rotational speed and saturated at a temperature of about $390\,°C$. However, the difference between the minimum and the maximum peak temperatures in the aluminium sheet significantly rises for rotational speeds $n > 1300\,min^{-1}$. This was not observed for the temperatures in the copper sheet.

3.2 Tensile shear strength

The results of the tensile tests are shown in Figure 4. The symbols represent the average and the error bars of the minimum and maximum tensile shear strength of each joint. For a rotational speed of $n = 800\ min^{-1}$, the sheets were not bonded over the entire welding length and only one specimen could be cut out and tested. Therefore, the peak temperatures at $n = 800\ min^{-1}$ seem to be very low for a sufficient bonding. The minimum rotational speed for a sufficient bonding and joint quality should be above $n = 800\ min^{-1}$ for a feed rate of $v = 300\ mm/min$. Comparing the tensile shear strengths to the temperature data, three similar trends can be concluded. First, an increasing rotational speed basically leads to an increased joint strength. Second, the joint strength is reduced at a rotational speed of $n = 2800\ min^{-1}$. As a consequence, the maximum achievable joint strength seems to occur at a rotational speed between $n = 1800\ min^{-1}$ and $n = 2800\ min^{-1}$. Further experiments have to be carried out to prove this. Third, the variations between the maximum and the minimum shear strengths extend above a rotational speed of $n = 1300\ min^{-1}$.

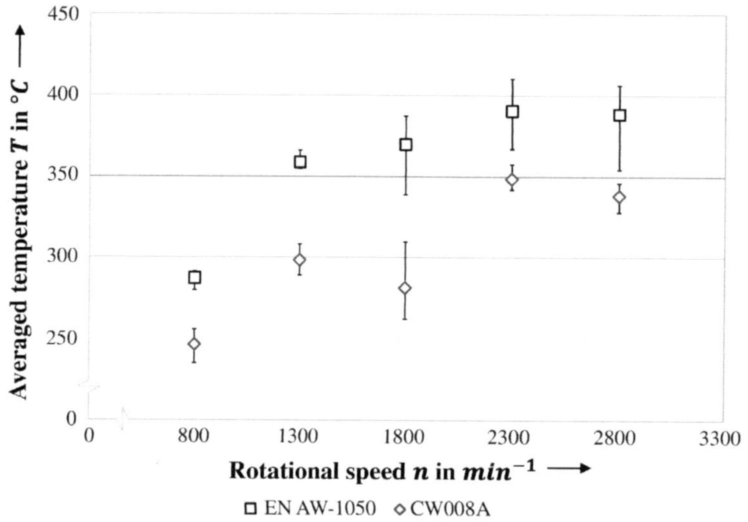

Figure 3. Averaged peak temperatures for different rotational speeds.

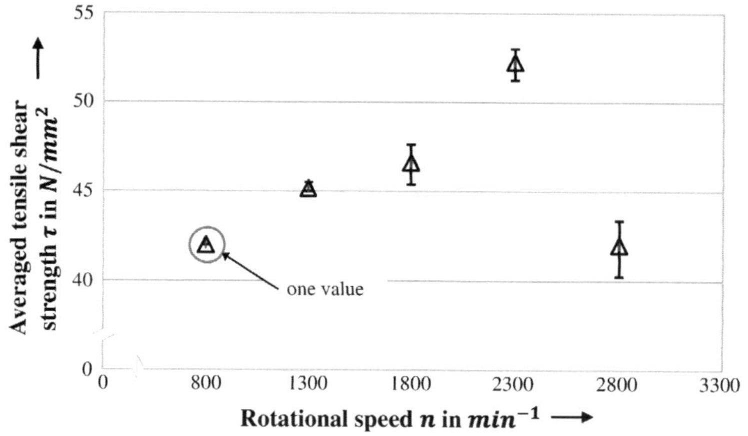

Figure 4. Averaged tensile shear strength for different rotational speeds.

The results seem to indicate a significant dependency between the temperature and the achievable tensile shear strengths of the joints. A further examination of the bonding area was carried out to determine how this dependency relates to the metallographic structure.

3.3 Interface analysis

For all analysed samples, the analysis of the interface area revealed an intermetallic compound (IMC) layer causing the bonding (see Figure 5 a). This layer discernibly consists of two sublayers. Xue et al. [15] observed similar double layers of Al_2Cu and Al_4Cu_9 for intermixed butt joints. In contrast, TEM-EDX analysis distinguished the stoichiometry of the phases as AlCu and $AlCu_3$ in this study. However, the shape of the interface layer, which is smooth and plane on the aluminium side but waved on the copper side, is in good agreement with Xue et al. [15]. The TEM image also shows that the crystals of both phases are columnar on the interface plane. Figure 5 b) shows a high resolution TEM (HRTEM) image of the interface between the aluminium matrix and the AlCu phase layer. It is clearly visible, that the AlCu sublayer consists of a long range crystalline structure with a semi coherence of the atomic plains at the imaged interface. This semi coherence is presumably crucial for the mechanical properties of this phase boundary and therefore of the joint strength. EDX did not detect residual oxides or the occurrence of any elements other than aluminium or copper at the interface, which could influence the bonding.

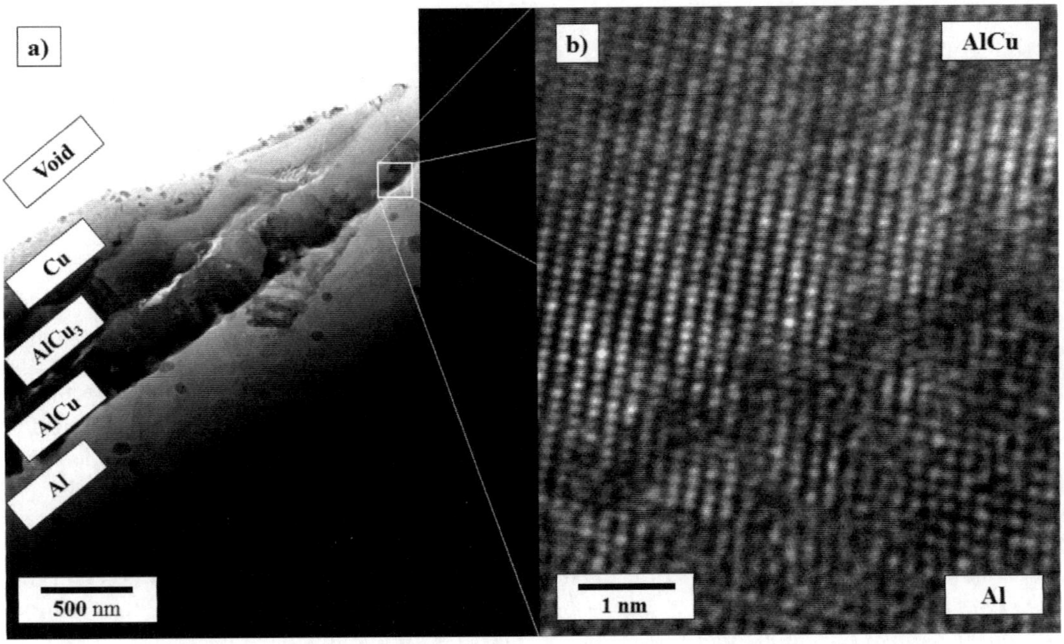

Figure 5. a) TEM image of the IMC interface layer b) HRTEM image of the semi coherence of the atomic plains at the Al to AlCu interface.

TEM and SEM were used to measure the thickness of the IMC interface layer of samples, which were welded with different rotational speeds. As shown in Figure 6, the layer thickness reaches a maximum of 704.6 nm at the rotational speed of $n = 2300\ min^{-1}$. At $n = 800\ min^{-1}$, the formed layer is not continuous and the fragments are only 47.4 nm thick. The error bars represent the standard deviation for at least 5 thickness measurements. The thickest IMC layer was detected for the sample welded at a rotational speed of $n = 2300\ min^{-1}$. The highest peak temperatures as well as the highest tensile shear strengths were also observed at this rotational speed. Lower temperatures lead to a thinner IMC layer and to lower joint strengths. Based on the observed results, the highest joint strengths can be obtained with a layer thickness of $600 - 700\,nm$.

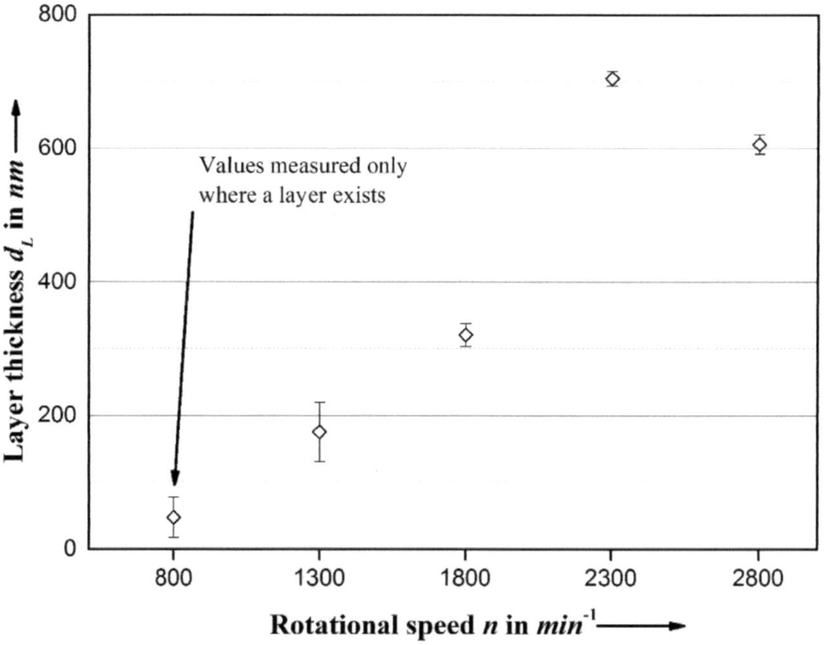

Figure 6. Thickness of the IMC layer for different rotational speeds.

3.4 Correlation between bonding double layer and process temperature

The measured temperature-time curves are equally shaped for all process parameters and differ only in the value of the peak temperatures. The short time interval at the peak temperature primarily controls the thermal activation of the interface region. Figure 7 shows an Arrhenius diagram where the thickness of the IMC interface layer is plotted logarithmically against the inverse peak temperature. The thickness of the barely bonded joint at a rotational speed of $\boldsymbol{n = 800\,min^{-1}}$ is also illustrated in Figure 7 but not used for the calculations since the layer is only partially formed. Hence, the measured thicknesses for this rotational speed only represent maximum values and the average cannot be compared quantitatively to the other thicknesses. The remaining valid layer thicknesses exhibit a linear dependency of the inverse temperatures. As a consequence, the layer thickness of friction stir welded aluminium / copper joints is a result of a thermal activation. Although not taken into account, this linear correlation would also be valid for the broad error margin of the weld at $\boldsymbol{n = 800\,min^{-1}}$. Assuming a thermal activation for diffusion the Arrhenius equation follows

$$\boldsymbol{D(T) = D_0\, e^{\frac{-E_{act}}{k_B T}}}$$

$\boldsymbol{D}$ is the diffusion coefficient, $\boldsymbol{T}$ the temperature, $\boldsymbol{D_0}$ is the prefactor of diffusion, $\boldsymbol{E_{act}}$ is the activation energy and $\boldsymbol{k_B}$ is the Boltzmann constant. Further assuming a diffusion controlled and time dependent growth of the IMC layer thickness $\boldsymbol{d(t)}$ as

$$\boldsymbol{d(t) = \sqrt{2Dt}}$$

a linear function for the fit in the Arrhenius plot can be defined as

$$\ln(d) = \frac{1}{2}\ln(2t_{eq}D_0) - \frac{E_{act}}{2k_B}\frac{1}{T_{peak}}\,.$$

t_{eq} is set to the equivalent time interval around the temperature peak T_{peak}, during which considerable thermal diffusion takes place. It is of the magnitude of a few seconds. Determined from the slope of the Arrhenius plot, the activation energy is $E_{act} = 2.33\ eV$ and the prefactor D_0 ranges from $10^{8}cm^2/s.$ to $10^9\ cm^2/s$. The activation energy is in reasonable agreement with data for Al-Cu interdiffusion in Funamizu & Watanabe [22]. Although the diffusion prefactor is higher than the values found by Funamizu & Watanabe [22], the value reported here is considered still plausible due to the complex diffusion system generated during the FSW process.

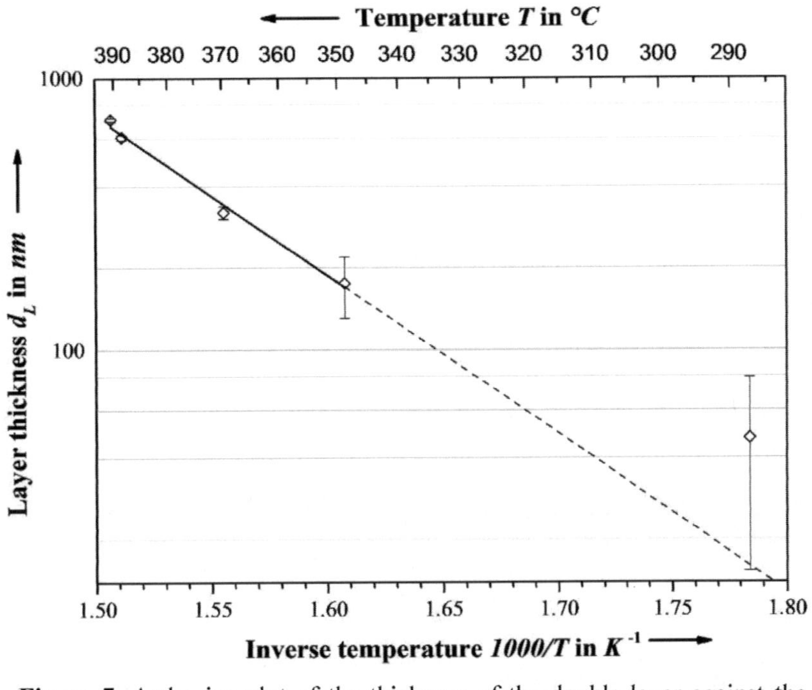

Figure 7. Arrhenius plot of the thickness of the double layer against the process peak temperature with a linear fit for calculations.

4. Conclusions and outlook

Since the material is heated by the friction during FSW, the process temperature is directly affected by the rotational speed. The temperature rises with an increasing rotational speed and saturates at about $390\ °C$ as a result of a decreasing friction based heat input caused by material softening. Similar tendencies could be observed for the dependencies between the rotational speed and the tensile shear strength of the joints. The ideal process conditions seem to exist shortly before the process temperature saturates. During FSW, a nanoscaled intermetallic double layer consisting of AlCu and $AlCu_3$ forms. The structure of the IMC layer at the bonding interface is columnar grown with a well-defined orientation. The layer thickness is controlled by thermal activation. Hence, this layer is formed by interdiffusion after the oxide layers at the interface are removed by the FSW process. This leads to a direct contact of the aluminium phase and the copper phase under the elevated thermal conditions.

The thickness of the double layer can be predicted by measuring the temperature during FSW. This predictability is the first step to a defined and reliable application of FSW for the commercial joining of aluminium and copper.

Acknowledgements

The research project "Mechanisms in joining of dissimilar materials by friction stir welding" is part of the priority programme 1640 (SPP1640) "Joining by plastic deformation", which is funded by the German Research Foundation (DFG). The authors would like to thank for the funding and the support.

References

[1] Murr L E 2010 A review of FSW research on dissimilar metal and alloy systems *J. Mat. Eng. Perf.* **19** 8 pp 1071–1089

[2] Giera A, Merklein M and Baumeister P 2007 Laser-assisted Friction Stir Welding of dissimilar steel and aluminum Alloys *Proc. 2nd Int. Conf. New Form. Techn.* 20.-21. Sept. 2007 Bremen, Germany (Bremen: BIAS Verl.) pp 421–430

[3] Uzun H, Dalle Donne C, Argagnotto A, Ghidini T and Gambaro C 2005 Friction Stir Welding of dissimilar Al 6013-T4 to X5CrNi18-10 stainless steel *Materials & Design* **26** pp 41–46

[4] Kostka A, Coelho R, Dos Santos J and Pyzalla A 2008 Microstructure and properties of Al to steel friction stir overlap welds *Proc. 7th Int. FSW Symp.* 20.-22. May 2008 Awaji Island, Japan

[5] Aonuma M and Nakata K 2011 Dissimilar metal joining of 2024 and 7075 aluminium alloys to titanium alloys by Friction Stir Welding *Mat. Trans.* **52** pp 948–952

[6] Wu A, Song Z, Nakata K, Liao J and Zhou L 2015 Interface and properties of the friction stir welded joints of titanium alloy Ti6Al4V with aluminum alloy 6061 *Mat. & Des.* **71** pp 85–92

[7] Dressler U, Biallas G and Alfaro Mercado U 2009 Friction Stir Welding of titanium alloy TiAl6V4 to aluminium alloy AA2024-T3 *Mat. Sci. Eng. A* **526** pp 113–117

[8] Chen Y C and Nakata K 2009 Microstructural characterization and mechanical properties in Friction Stir Welding of aluminum and titanium dissimilar alloys *Mat. & Des.* **30** pp 469–474

[9] Chen Y-h, Ni Q and Ke L-m 2012 Interface characteristic of Friction Stir Welding lap joints of Ti/Al dissimilar alloys *Trans. Nonf. Met. Soc. China* **22** pp 299–304

[10] Krutzlinger M, Marstatt R, Suenger S, Luderschmid J, Zaeh M F and Haider F 2014 Formation of joining mechanisms in friction stir welded dissimilar Al-Ti lap joints *Adv. Mat Res.* **966-967** pp 510–520

[11] Chen Z W and Yazdanian S 2015 Microstructures in interface region and mechanical behaviours of friction stir lap Al6060 to Ti–6Al–4V welds *Mat. Sci. Eng.: A* **634** pp 37–45

[12] Galvão I, Loureiro A and Rodrigues D M 2016 Critical review on Friction Stir Welding of aluminium to copper *Sci. Techn. Weld. Join.* **21** pp 523–546

[13] Abdollah-zadeh A, Saeid T and Sazgari B 2008 Microstructural and mechanical properties of friction stir welded aluminum/copper lap joints *J. All. Comp.* **460** pp 535–538

[14] Saeid T, Abdollah-zadeh A and Sazgari B 2010 Weldability and mechanical properties of dissimilar aluminum-copper lap joints made by friction stir welding *J. All. Comp.* **490** pp 652-655

[15] Xue P, Xiao B L, Ni D R and Ma Z Y 2010 Enhanced mechanical properties of friction stir welded dissimilar Al–Cu joint by intermetallic compounds *Mat. Sci. Eng.: A* **527** pp 5723–5727

[16] Xue P, Ni D, Wang D, Xiao B and Ma Z 2011 Effect of Friction Stir Welding parameters on the microstructure and mechanical properties of the dissimilar Al–Cu joints *Mat. Sci. Eng.: A* **528** pp 4683–4689

[17] Tan C W, Jiang Z G, Li L Q, Chen Y B and Chen X Y 2013 Microstructural evolution and mechanical properties of dissimilar Al–Cu joints produced by Friction Stir Welding *Mat. & Des.* **51** pp 466–473

[18] Muthu M F X and Jayabalan V 2015 Tool travel speed effects on the microstructure of friction stir welded aluminum-copper joints *J. Mat. Proc. Techn.* **217** pp 105–113

[19] Genevois C, Girard M, Huneau B, Sauvage X and Racineux G 2011 Interfacial reaction during Friction Stir Welding of Al and Cu *Met. Mat. Trans. A* **42** pp 2290–2295

[20] Galvão I, Loureiro A, Verdera D, Gesto D and Rodrigues D M 2012 Influence of tool offsetting on the structure and morphology of dissimilar aluminum to copper friction-stir welds *Met. Mat. Trans. A* **43** pp 5096–5105

[21] Doehner J, Weigl M, Bergmann J-P, Regensburg A and Schuerer R 2016 Rührreibschweißen mit stehender Schulter am Beispiel von Kupfer-Aluminium-Verbindungen (engl.: Friction Stir Welding of aluminium-copper joints with a stationary shoulder tool) *Proc. 7th FSW-Workshop. Rührreibschweißen und verwandte Verfahren* 24.-26.02.2016 Geesthacht, Germany (Geesthacht: Helmholtz-Zentrum Geesthacht)

[22] Funamizu Y and Watanabe K 1971 Interdiffusion in the Al-Cu system *Trans. Jap. Inst. Met.* **12** 3 pp 147–152

Hybrid Al/steel-joints manufactured by ultrasound enhanced friction stir welding (USE-FSW): Process comparison, nondestructive testing and microscopic analysis

M Thomä[1,4], G Wagner[1], B Straß[2], B Wolter[2], S Benfer[3] and W Fürbeth[3]

[1]Institute of Materials Science and Engineering, Chemnitz University of Technology, 09125 Chemnitz, Germany
[2]Fraunhofer Institute for Nondestructive Testing IZFP, Campus E3 1, 66123 Saarbrücken, Germany
[3]DECHEMA-Forschungsinstitut, Theodor-Heuss-Allee 25, 60486 Frankfurt am Main, Germany
[4]Corresponding author: marco.thomae@mb.tu-chemnitz.de

Abstract. The process of friction stir welding (FSW) is an innovative joining technique, which proved its potential in joining dissimilar metals that are poorly fusion weldable. This ability opens a wide range for applications in industrial fields, where weight reduction by partial substitution of conventional materials through lightweight materials is a current central aim. As a consequence of this, the realization of aluminum / steel-joints is of great interest. For this material compound, several friction stir welds were carried out by different researchers for varying Al/steel-joints, whereas the definition of optimal process parameters as well as the increase of mechanical properties was in the focus of the studies. To achieve further improved properties for this dissimilar joint a newly developed hybrid process named "ultrasound enhanced friction stir welding (USE-FSW)" was applied. In this paper the resulting properties of Al/steel-joints using FSW and USE-FSW will be presented and compared. Furthermore, first results by using the nondestructive testing method "computer laminography" to analyze the developed joining area will be shown supplemented by detailed light-microscopic investigations, scanning electron microscopic analysis, and EDX.

1. Introduction

Recent developments in many industrial sectors regarding lightweight constructions to save energy in different ways lead to the necessary realization of components out of dissimilar materials like aluminum or aluminum alloys and steel. But in this case, established fusion welding methods struggle with the development of brittle intermetallic (IM) phases in the welding area, which results in a significant reduction of the strength of the joint [1, 2]. In comparison to fusion welding friction stir welding (FSW) as a low temperature joining process is a more appropriate method, because of the reduced development of intermetallic phases. FSW is characterized by a rotating cylindrical tool, which gets plunged in a butt or overlap joint of two materials and then runs along contact area with a defined feed rate [3]. Several research work has been performed on the field of Al/steel-joints concerning the realization of high-strength welds [4–7]. Watanabe et al. reached a maximum tensile strength for their FSW-joints of 86 % of their aluminum base material [4]. But in the most cases in literature the presence of intermetallic phases in the nugget and at the interface of the joints after the friction stir welding process is described [4, 6–8]. This verified that IM-phases also cannot be completely reduced by using FSW. Furthermore, the intermetallics appear as brittle separating layers between the base materials. This weakens the joint strength additionally. To reduce these detrimental effects, different approaches were investigated for FSW of Al/steel-joints. Liu et al. and Ferrando worked with an electrically assisted friction stir welding system, which enables a local electrical current density field to run synchronous to the FSW tool. The occurring electro-plastic effect results in a material softening due to the elevated process temperatures and also in the formation of a thin layer of IM-phase, which was believed to increase the joint strength by micro-interlocking [8, 9]. This additional heating approach for the contact area was also applied by Merklein et al. who used laser assistance for their friction stir welding process. They further observed a reduced tool wear and a higher drawing ratio for their joints [10]. Another method for extra heating was the gas tungsten arc welding (GTAW) supported version of FSW introduced by Bang et al., where a GTAW torch runs in front of the friction stir welding tool and also leads to joints with higher tensile strength. This is attributed to an improved material plastic flow and partial annealing effects [11]. A further concept was realized by Klag et al. and Straß et al., who investigated Al/Mg-joints by using additional power ultrasound during the friction stir welding process [12, 13]. This hybrid joining method named "ultrasound enhanced friction stir welding (USE-FSW)" improves the joint strength by the mechanical breakup and fine dispersion of the continuous brittle intermetallic layers. With this method an increase of the tensile strength of 25 % for USE-FSW compared to conventional FSW can be achieved directly.

Based on this positive effects in the recent work the potential of the USE-FSW for dissimilar joints of aluminum with steel will be presented. Therefore, FSW- as well as USE-FSW-joints were realized and analyzed by light-microscopic and scanning electron microscopic methods. Additional nondestructive computer laminography examinations were carried out to find possible weld defects in the bonding area.

2. Experimental Procedure

For the investigations the commercially available aluminum wrought alloy EN AW-6061 T6 (AlMg1SiCu) and the deep-drawing steel DC04 (1.0338) are used in a sheet geometry of 280 mm length, 100 mm width and 3 mm thickness. The chemical composition of the materials according to the manufacturer specifications can be seen in table 1.

Table 1. Chemical composition of the investigated materials.

Material	Elements (wt%)								
	Si	Fe	Cu	Mn	Mg	Cr	Zn	Ti	Al
EN AW-6061 T6	0.64	0.51	0.21	0.14	0.89	0.15	0.04	0.05	bal.
	C	Si	Mn	P	S	Al	Ti	Nb	Fe
DC04	0.041	0.015	0.3	0.01	0.0077	0.49	0.0007	0.0037	Bal

For the friction stir welding process a universal four axis machining center DMU80T from DMG MORI was used extended with a pneumatic clamping for the metal sheets as well as four load cells from Kistler to realize a force controlled process (figure 1).

IOP Conf. Series: Materials Science and Engineering **181** (2017) 012003 doi:10.1088/1757-899X/181/1/012003

Figure 1. Machining center DMU80T extended for FSW.

For the investigations a butt-joint configuration was chosen, whereby the steel plate was placed on the advancing side and the aluminum plate on the retreating side. The welding direction was perpendicular to the rolling direction of EN AW-6061 and parallel to the rolling direction of DC04. For the FSW tool a hot-work steel 1.2344 (X40CrMoV51) with a shoulder diameter of 16 mm, a probe length of 2.8 mm and a metric thread of M4.5 was used. An axial force of 4 kN, a tool rotation speed of 1500 rpm, a travel of 20 mm/min, a tilt angle of 2°, and a lateral offset of the probe surface to the faying surface of the steel of 0.3 mm were chosen for conventional friction stir welding. For USE-FSW the machining center was additionally equipped with a Lab VIEW-controlled ultrasonic roll seam module of Schunk Sonosystems, which moves synchronously and parallel to the tool (figure 2).

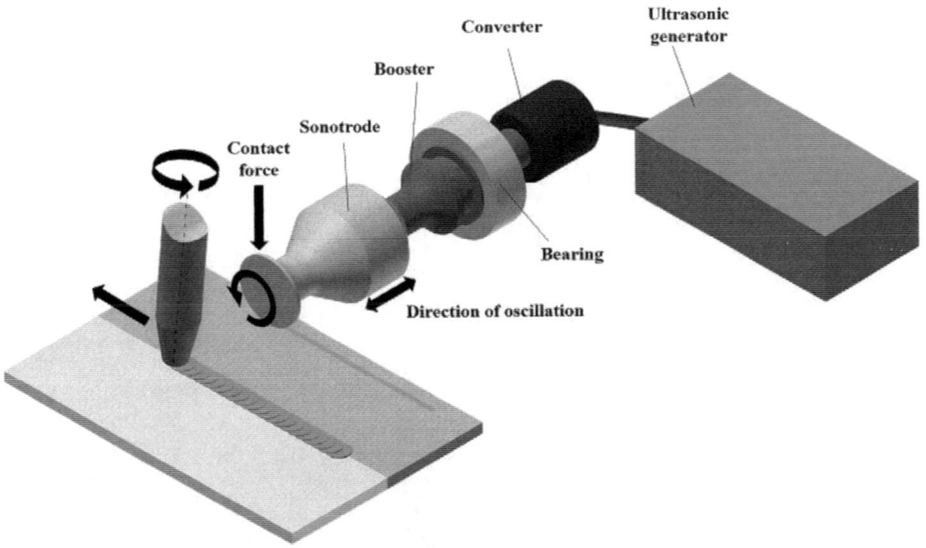

Figure 2. Schematic USE-FSW configuration.

The roll seam module works with a resonant frequency of 20 kHz, a generator maximum of 3000 W, and a constant amplitude of 18 µm. The transmission of the ultrasound takes place in the steel sheet located on the advancing side

3. Results and Discussion

The analysis on the microstructure of the developing joining area was carried out by light microscopic investigations first. The comparison of the two cross-sections clearly show the influence of additional ultrasound power on the joining area (figure 3).

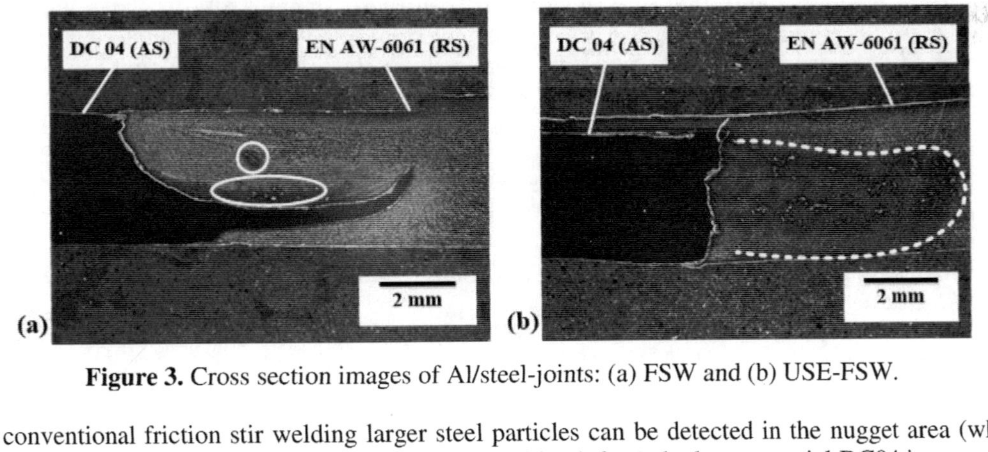

Figure 3. Cross section images of Al/steel-joints: (a) FSW and (b) USE-FSW.

For conventional friction stir welding larger steel particles can be detected in the nugget area (white circles in figure 3 (a)). Furthermore, an extensive steel hook from the base material DC04 is apparent. In comparison to this for USE-FSW Al/steel-joints the interface between EN AW-6061 and DC04 is nearly perpendicular and the nugget clearly defined (white dashed line figure 3 (b)). It is recognizable that simultaneous transmitted power ultrasound improves the development of the bonding zone during the FSW process.

Furthermore, a detailed analysis of the microstructure of the stirred zone for FSW and USE-FSW was carried out by scanning electron microscopy (SEM). Thereby different types of particles could be found in the welding area of FSW-joints. In figure 4 (a) larger particle within the nugget of an FSW Al/steel-joint is depicted and analyzed.

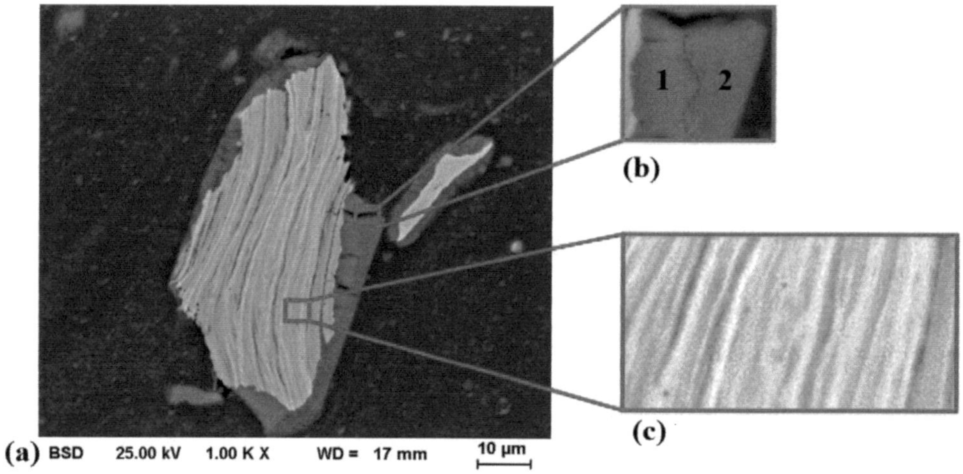

Figure 4. (a) Larger particle in the nugget of a FSW Al/steel-joint consisting of IM-phases, (b) detail of the partially present shell, (c) detail of the layer structure in the particle.

The particle consists of a partially present shell and a fine stripped layer structure inside. For the shell two different intermetallics were identified by energy dispersive X-ray spectroscopy (EDX) spot measurements (figure 4 (b) 1 and 2). The particle facing layer (1) was determined as a Fe_2Al_5-phase and the nugget facing IM-phase (2) contents out of the aluminum-rich $FeAl_3$. Due to the small width of the intermetallic layers in the inner structure of the particle (figure 4 (c)) only an average EDX analysis could be performed. With that, a quantity proportion of 83 % iron and 17 % aluminum was detected. The structure and morphology of this particle leads to the assumption that it was formed at the contact area of the base materials due to the FSW process and afterwards, as a result of the stirring, transferred into the nugget. Similar observation are also described by different authors in the literature [4, 6, 7, 11].

The same procedure can be assumed for another type of particles consisting of nearly pure steel and found in the nugget, too. But, in this case it can be presumed that the steel particles are from the DC04 joining partner or from the FSW-tool.

In comparison to the FSW-joints, for USE-FSW-joints in the nugget only steel particles were found by SEM-investigations (figure 5 (a)).

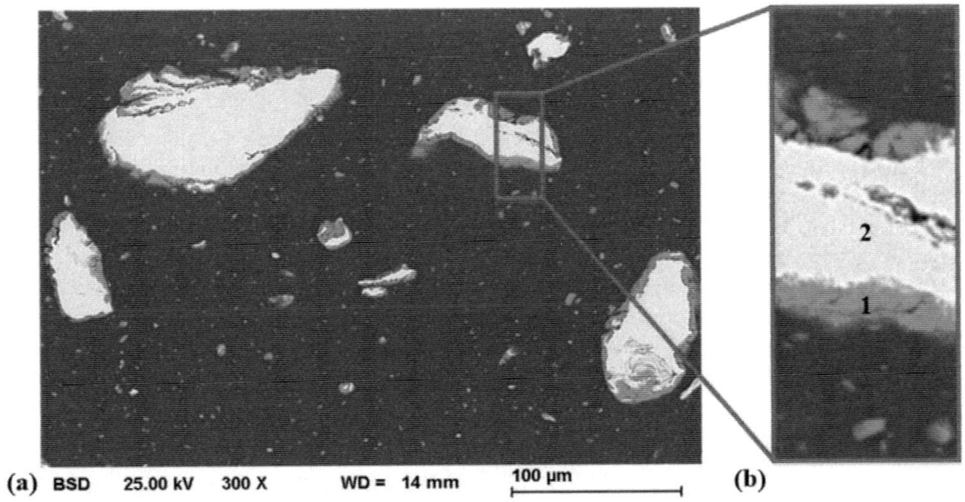

Figure 5. (a) Steel particles in the stirred zone of USE-FSW Al/steel-joint, (b) detail of a steel particle.

Unlike the fine stripped intermetallic layer particle for FSW-joints the particles of USE-FSW-joints are nearly completely surrounded by an IM-phase shell. Again, EDX-spot measurements were performed for the shell and the inner structure (figure 5 (b)). It revealed that the intermetallic layer (1) consists of the aluminum-rich $FeAl_3$ whereas the particle itself (2) contains of 98 % iron. The formation can be explained by the transfer of steel particles from the DC04 through ablation caused by the material flow on the surface of the contact area between the base materials. Moreover, the process temperature is high enough to enable the development of the $FeAl_3$-phase on the surface of the particle.

Furthermore nondestructive computer laminography (CL) investigations were carried out for both process variants (figure 6).

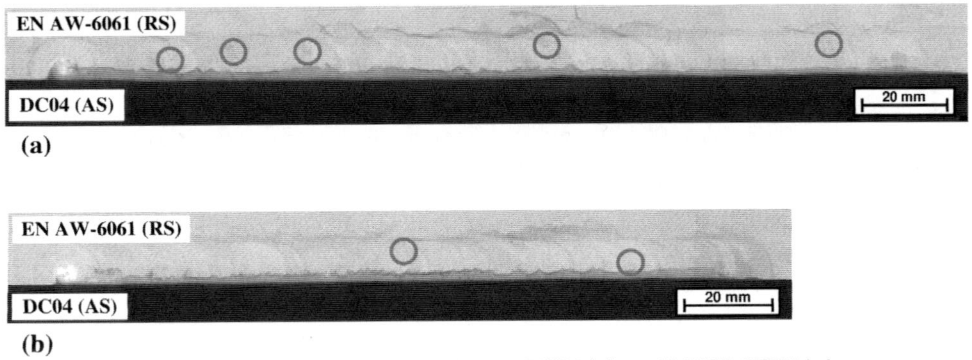

Figure 6. Computer laminography (a) FSW-joint, (b) USE-FSW-joint.

The CL-investigations confirm the flawlessness for both joints. Also no fundamental differences between FSW (Figure 6 (a)) and USE-FSW (Figure 6 (b)) can be found in the welds. But both joints contain some irregularities, probably larger steel particles and the USE-FSW joint shows a smaller amount of these irregularities. This can be a consequence of the power ultrasound and the related reduction of larger intermetallic phases in the welding area. These approaches will be checked in further investigations.

4. Summary

It could be shown, that it is possible, to realize Al/steel-joints by FSW as well as by USE-FSW. Therefore central parameters like axial force, tool rotation speed, travel speed, tilt angle and lateral offset were optimized.

For both FSW variants, the microstructure of the welding zone were investigated. For conventional FSW two types of particles, particles out of intermetallic phases and steel particles were detected in the nugget. Particles characterized by a fine stripped layer structure consisting of 83 % iron and 17 % aluminum and a two layer partial shell consisting of the two intermetallic phases $FeAl_3$ and Fe_2Al_5, whereas USE-FSW-joints only contained steel particles with an nearly closed $FeAl_3$ shell. This is a clear indicator that the additionally introduced power ultrasound reduces the amount of particles and intermetallic phases in the nugget for EN AW-6061/DC04-joints.

By computer laminography it could be shown that the realized joints for both variants of FSW are sound free. Small irregularities were detected for FSW- and USE-FSW-joints, whereas the amount of them were lower for ultrasound enhanced friction stir welding.

References

[1] Achar D. R. G., Ruge J. and Sundaresan S 1980 Verbindung von Aluminium mit Stahl besonders durch Schweißen *Aluminium* 147–9

[2] Radscheit R. R. 1996 Laserstrahlfügen von Aluminium mit Stahl *Dissertation* Universität Bremen

[3] Besharati Givi M K and Asadi P 2014 *Advances in friction- stir welding (Woodhead Publishing In Mechanical Engineering)* (Cambridge, England, Waltham, Massachusetts, Oxford, England: Woodhead Publishing)

[4] Watanabe T, Takayama H and Yanagisawa A 2006 Joining of aluminum alloy to steel by friction stir welding *Journal of Materials Processing Technology* **178** 342–9

[5] Uzun H, Dalle Donne C, Argagnotto A, Ghidini T and Gambaro C 2005 Friction stir welding of dissimilar Al 6013-T4 To X5CrNi18-10 stainless steel *Materials & Design* **26** 41–6

[6] Tanaka T, Morishige T and Hirata T 2009 Comprehensive analysis of joint strength for dissimilar friction stir welds of mild steel to aluminum alloys *Scripta Materialia* **61** 756–9

[7] Shen Z, Chen Y, Haghshenas M and Gerlich A P 2015 Role of welding parameters on interfacial bonding in dissimilar steel/aluminum friction stir welds *Engineering Science and Technology, an International Journal* **18** 270–7

[8] Liu X, Lan S and Ni J 2015 Electrically assisted friction stir welding for joining Al 6061 to TRIP 780 steel *Journal of Materials Processing Technology* **219** 112–23

[9] Ferrando W. A. Concept of Electrically Assisted Friction Stir Welding

[10] Merklein M and Giera A 2008 Laser assisted Friction Stir Welding of drawable steel-aluminium tailored hybrids *Int J Mater Form* **1** 1299–302

[11] Bang H, Bang H, Jeon G, Oh I and Ro C 2012 Gas tungsten arc welding assisted hybrid friction stir welding of dissimilar materials Al6061-T6 aluminum alloy and STS304 stainless steel *Materials & Design* **37** 48–55

[12] Klag O, Wagner G and Eifler D 2010 Mechanical Properties of Friction Stir Welded Mg/Mg- and Mg/Al-Joints *Adv. Eng. Mater.* **12** 1212–9

[13] Strass B, Wagner G and Eifler D 2014 Realization of Al/Mg-Hybrid-Joints by Ultrasound Supported Friction Stir Welding *MSF* **783-786** 1814–9

[14] Li X, Scherf A, Heilmaier M and Stein F 2016 The Al-Rich Part of the Fe-Al Phase Diagram *J. Phase Equilib. Diffus.* **37** 162–73

Acknowledgements

We would like to thank the "German Research Foundation (DFG)" for the support of the present work as a part of the Priority Program 1640 "Joining by Plastic Deformation".

IOP Conf. Series: Materials Science and Engineering **181** (2017) 012004 doi:10.1088/1757-899X/181/1/012004

Hybrid joints manufactured by ultrasound enhanced friction stir welding (USE-FSW) - corrosion properties

S Benfer[1,4], W Fürbeth[1], M Thomä[2], G Wagner[2], B Straß[3] and B Wolter[3]

[1]DECHEMA-Forschungsinstitut, Theodor-Heuss-Allee 25, 60486 Frankfurt am Main, Germany
[2]Institute of Materials Science and Engineering, Chemnitz University of Technology, 09125 Chemnitz, Germany
[3]Fraunhofer Institute for Nondestructive Testing IZFP, Campus E3 1, 66123 Saarbrücken, Germany
[4] Corresponding author: benfer@dechema.de

Abstract. To realize lightweight structures of material combinations like aluminum/magnesium and aluminum/steel an Ultrasound Enhanced Friction Stir Welding (USE-FSW) process was used. This process has a beneficial influence on the resulting microstructure (elimination of the brittle intermetallic phase Al_3Mg_2 as coherent layer) and the mechanical properties (increased tensile strength) of Al/Mg-joints and was now also applied for Al/steel-hybrid joints. Besides the mechanical properties the corrosion properties of the hybrid joints may play a significant role concerning the later use of the hybrid materials. Therefore, the corrosion properties of various hybrid joints have been investigated by different methods. With the Scanning Kelvin Probe (SKP) Volta potential differences between the base alloys and the welded area were investigated in air. The two-dimensional color-plots illustrate not only the Volta potential differences between the different phases but also their oxidation properties in air during the measurement time. Electrochemical measurements (open circuit potential and potentiodynamic polarization) have been carried out for the investigation of the corrosion properties of the FSW and USE-FSW hybrid joints in 0.5 molar NaCl solution. A three electrode setup within a mini-cell was used to enable measurements on different areas of the joints. This allows to observe the corrosion activity of the base alloys and the nugget phase separately. Differences between Al/steel-hybrid joints processed with and without ultrasound enhancement are discussed and compared with Al/Mg-hybrids.

1. Introduction

The rapid development of transport applications suggests adaptive and hybrid structures. Therefore, the employment of lightweight materials in steel structures is of great interest. In this regard a very advantageous material combination is aluminum/steel. The established method of fusion welding may not be used for aluminum/steel because of the formation of brittle intermetallic (IM) phases which decrease the joint strength [1]. Due to this, pressure welding methods which are characterized by welding temperatures below the melting points of the materials to join are an interesting alternative with the aim to avoid the development of brittle phases in the welding zone. A well established welding method in this group is friction stir welding (FSW) [2-4]. The research work on FSW to produce aluminum-to-steel joints is summarized in a review article by Hussein et al. [5]. The relatively low process temperature of the FSW results in a decreased amount of brittle intermetallic phases compared to fusion welded joints but cannot avoid the formation of intermetallics completely. To minimize this negative effect a hybrid welding method called "Ultrasound Enhanced Friction Stir Welding" (USE-FSW) was developed. In this process additional ultrasonic energy is transmitted in the welding parts during FSW with the aim to crack interlayers and to support the stirring in the joining area. This process was successfully applied on joints of aluminum and magnesium alloys [6, 7]. The combination of different metals in hybrid structures may cause corrosion problems in the welding area because of the formation of a galvanic couple which leads to severe corrosion of the less noble element. In the case of Al/Mg-hybrids an accelerated corrosion of Mg in chloride containing electrolyte was found [8, 9]. This paper will focus on the corrosion investigation of FSW and USE-FSW Al/steel joints. With the Scanning Kelvin Probe (SKP) Volta potential differences between the base alloys and the welded area were investigated in air. Electrochemical measurements (open circuit potential and potentiodynamic polarization) have been carried out for the investigation of the corrosion properties of the FSW and USE-FSW hybrid joints in 0.5 molar NaCl solution using a mini-cell. Differences between Al/steel-hybrid joints processed with and without ultrasound enhancement are discussed and compared with Al/Mg-hybrids.

2. Experimental

2.1 Materials

For the investigations the commercially available aluminum wrought alloy EN AW-6061 T6 (AlMg1SiCu) and the deep-drawing steel DC04 (1.0338) were used in a sheet geometry of 280 mm length, 100 mm width and 3 mm thickness. The chemical composition of the materials according to the manufacturer´s specifications is given in table 1.

Table 1. Chemical composition of the investigated materials.

Material	Elements (wt%)								
EN AW-6061 T6	Si	Fe	Cu	Mn	Mg	Cr	Zn	Ti	Al
	0.64	0.51	0.21	0.14	0.89	0.15	0.04	0.05	bal.
DC04	C	Si	Mn	P	S	Al	Ti	Nb	Fe
	0.041	0.015	0.3	0.01	0.0077	0.49	0.0007	0.0037	bal.

2.2 Joining procedure

For the process of friction stir welding a universal four axis machining center DMU80T from DMG MORI got extended with a pneumatic clamping for the metal sheets as well as four load cells from Kistler, so that a force-controlled process-sequence was possible. A butt-joint configuration was chosen for the joints, whereas steel was placed on the advancing side and aluminum on the retreating side. The welding direction was perpendicular to the rolling direction of EN AW-6061 and parallel to the rolling direction of DC04. For the hybrid-joining method of USE-FSW the machining center was additionally equipped with a Lab VIEW-controlled ultrasonic roll seam module of Schunk Sonosystems, which runs synchronously and parallel to the tool. The instrument and the used joining parameters are described in detail in the papers of Thomä et al. [10, 11].

2.3 Corrosion investigations

Prior to all measurements the surface of the samples was ground with wet SiC-paper (up to P1200), rinsed with ethanol and dried to achieve comparable conditions for all samples.

One method that gives space-resolved information about the corrosion properties of materials is the Scanning Kelvin Probe (SKP). The Kelvin Probe uses a contactless vibrating condensator technique which allows to measure the changes in the Volta potential or work function of electrons from metals, semi conductors or at phase boundaries of liquids. The method is described in detail in [12-14]. For our investigations of Volta potential differences between the Al alloy and steel and the formed phases after the FSW process a SKP-setup with height control (SKP_KM, Wicinski-Wicinski GbR) was used.

The measuring probe was a Cr-Ni-wire with a diameter of 100 µm. The Volta potential of the probe was not calibrated. Therefore the measured values are no absolute values and only Volta potential differences are discussed which also allow to draw first conclusions about the corrosion properties. Measurements were carried out in air with a relative humidity of about 33-35 %.

Electrochemical measurements have been carried out for the investigation of the corrosion properties of the FSW and USE-FSW Al/steel-joints in chloride containing electrolyte. A three electrode setup within a mini-cell (measurement area 0.071 cm^2) was used to enable measurements on different areas of the joints. The investigated samples were used as working electrode, a platinum electrode as counter electrode and a standard calomel electrode (+ 246 mV vs NHE) as reference electrode. The recorded values were converted to values against the normal hydrogen electrode (NHE).

The open circuit potential (OCP) measurements and the potentiodynamic polarization have been done with a Zahner IM 6 measurement unit. The open circuit potential was recorded for 60 min prior to the polarization experiment. The potentiodynamic curves were started -0.1 V relative to the open circuit potential and recorded up to +0.5 V relative to it with a scan rate of 1 mV/s. The electrochemical measurements were carried out in 0.5 molar NaCl solution.

3. Results and Discussion

3.1 SKP measurements

Figure 1 shows the measured Volta potential differences of the welding zones of an Al/steel FSW-joint (a) and an USE-FSW-joint (b) as two-dimensional colour-plots. The recording of the whole plot of one sample takes about 36 h. Therefore the plots illustrate not only the Volta potential differences between the different phases but also their oxidation properties in air during the measurement time. At the beginning of the measurements the Volta potential difference between the Al alloy and the steel is about 1.0 V. Both materials exhibit a shift of the Volta potential by about 0.2 to 0.3 V to more positive values within the measurement time due to surface oxidation processes, but the difference between the materials is nearly constant.

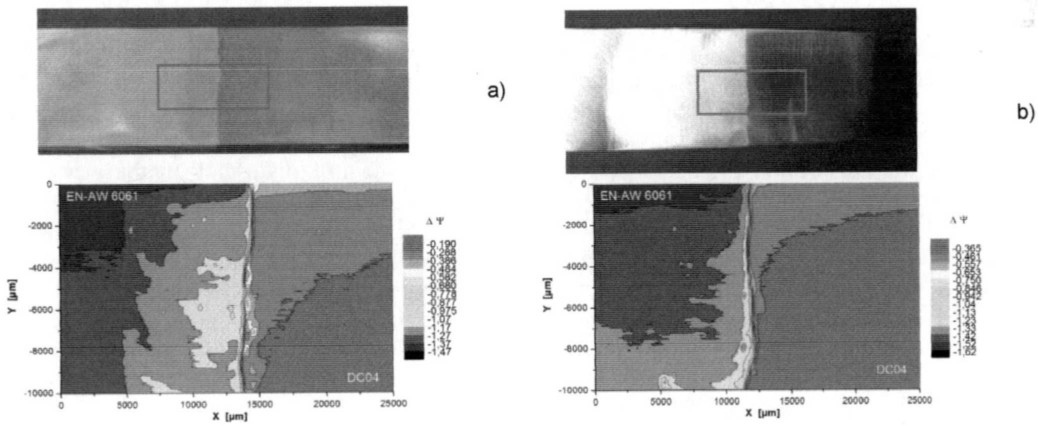

Figure 1. Volta-Potential maps of a FSW-joint (a) and an USE-FSW-joint (b) measured in air (relative humidity 33-35 %); time of each measurement: 36 h starting from the top of the plot.

The values of Volta potential differences of the main phases of FSW- and USE-FSW-joints are nearly the same. Differences between the two joints are obvious directly at the welding area. The FSW-joint shows a 1-2 mm broad seam with Volta potentials that are 0.1 to 0.3 V more negative than the steel. In addition there is a broader area in the nugget region (Al side) with a potential which is 0.1 to 0.2 V more positive than the Al alloy but considerably more negative than the steel. For the USE-FSW-joint the 1-2 mm seam was not observed. In the nugget area some kind of hook is visible which has a more positive Volta potential than the surrounding metal surface.

The Volta potential maps clearly show an influence of the ultrasound on the welding area of Al/steel hybrid joints but it is much less pronounced than in the case of the Al/Mg-hybrid joints [9]. This can be explained with the different joining parameters for the different joints. For the Al/Mg-hybrid joints the pin-offset was 1 mm to the Mg which results in a mechanical mixing of both materials during the process. In case of the Al/steel-hybrid joints, because of the high difference in the melting points of the materials, the pin only rotates within the Al with a distance of 0.3 mm to the faying surface of the steel. Therefore the influence of the ultrasound introduced via the steel sheet is less visible. Only the shoulder of the rotating tool has mechanical contact to the steel and may transport material from one side to the other.

The SKP measurements result in about 1 V more negative Volta potential values for the EN AW-6061 Al alloy compared to the DC04 steel. This is a similar difference to that measured between AZ80 Mg alloy and AC-48000 Al alloy where the Al alloy has shown the more positive values [9]. From that result one would expect that the EN AW-6061 Al alloy will show stronger corrosion in corrosive media than the steel particularly if both phases are in simultaneous contact to an electrolyte.

3.2 Open circuit potential (OCP)

The measurements in air are not directly transferable to that in a liquid electrolyte, as a comparison of the potential values for the steel and the Al alloy shows. During the SKP measurements in air the Volta potential of the steel was always about 1 V more positive than that of the Al alloy. The values change significantly in the sodium chloride electrolyte (figure 2).

The OCP of the EN AW-6061 alloy exhibits a shift from about -0.7 V to -0.47 V within the first 5 to 10 min and then stabilises around that value. The OCP of the steel declines continuously from about -0.23 V to -0.45 V within the first 10 min and then further to -0.52 V. After 15 to 20 min the OCP value of the DC04 steel is more negative than that of the EN AW-6061 Al alloy. This behaviour is different to that of the Al/Mg-hybrid joints where an OCP difference of 0.9 V was measured between the base materials in the sodium chloride electrolyte similar to the Volta potential difference in air (SKP measurement) [9].

The nugget area and the Al/steel transitional area show nearly constant OCP values immediately after the immersion in the electrolyte. The values are similar to that of the EN AW-6061 Al alloy. The OCP values of the base materials and the nugget phase within the hybrid joints are independent of the joining conditions (with or without ultrasound enhancement) as a comparison of figures 2a) and 2b) shows. Only the OCP value of the Al/steel transitional area seems to depend on the processing conditions, because the value for the FSW-joint shows a little shift to more negative values (-0.49 V after 60 min). But this may also be an effect of the area ratio Al/steel in the measured area, because of the uneven weld seam and positioning of the measurement cell. The OCP value of the DC04 measured near the weld shows within the first minutes the same decline as the base material but then reaches an OCP value between the DC04 and the EN AW-6061 after 60 min.

After the immersion time of 60 minutes the OPC difference between the materials is only about 50 mV or less. Therefore, the risk of an enhanced corrosion caused by a galvanic element is much less pronounced than in the case of the Al/Mg-hybrid joints.

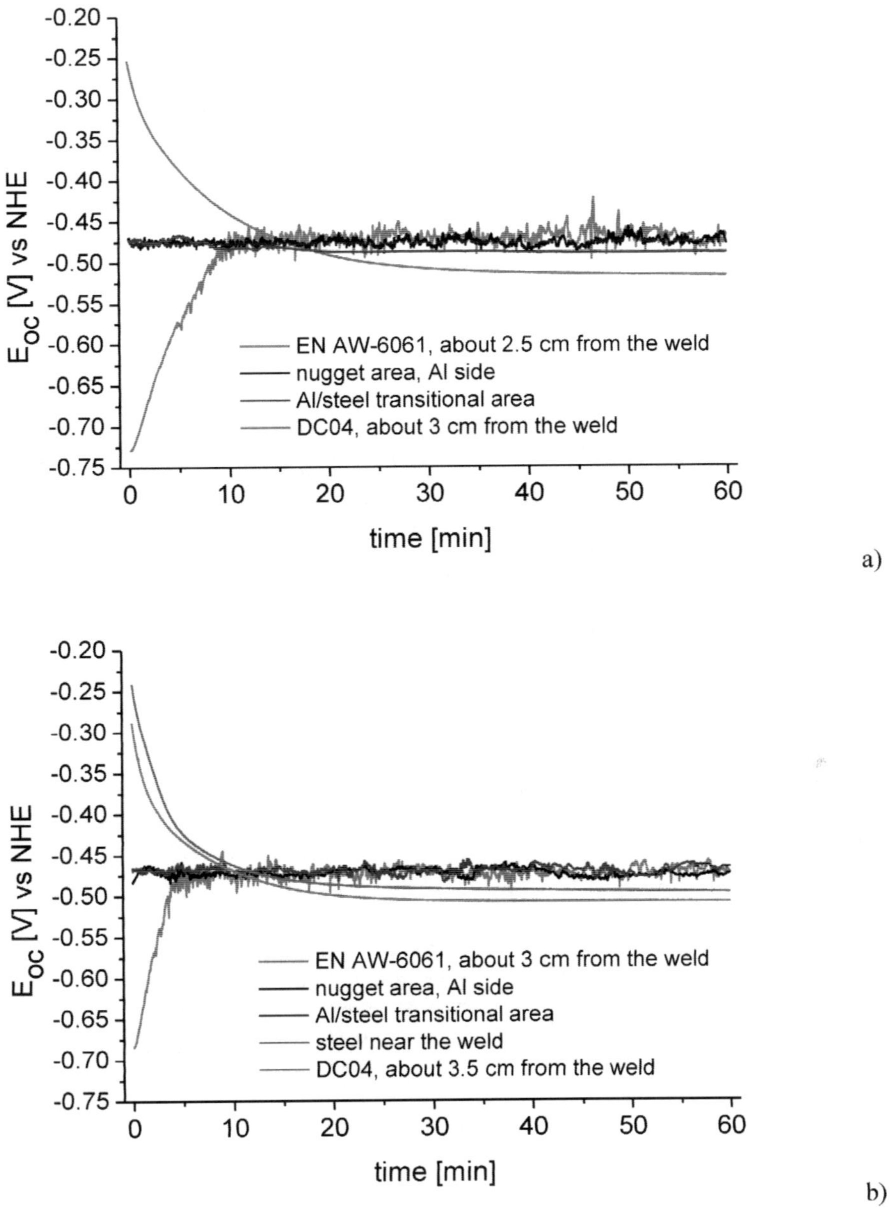

Figure 2. Open circuit potential measured on different positions of a FSW-joint (a) and an USE-FSW-joint (b) in 0.5 molar NaCl solution.

3.3 Potentiodynamic polarization

Figure 3 shows the results of the potentiodynamic polarization measurements on different positions of a FSW (figure 3a) and an USE-FSW (figure 3b) Al/steel-hybrid joint in 0.5 molar NaCl solution. The measurements were started 60 minutes after immersion at open circuit potential. The EN AW-6061 Al alloy exhibits a free corrosion potential (E_{oc}) around -0.46 V and a pitting potential around -0.42 V which is only slightly more positive than the free corrosion potential. The corrosion current density measured for the Al alloy at the E_{oc} is around 1-2 µA/cm^2 indicating a low corrosion rate. At the pitting potential the current density increases immediately up to values of several hundred µA/cm^2.

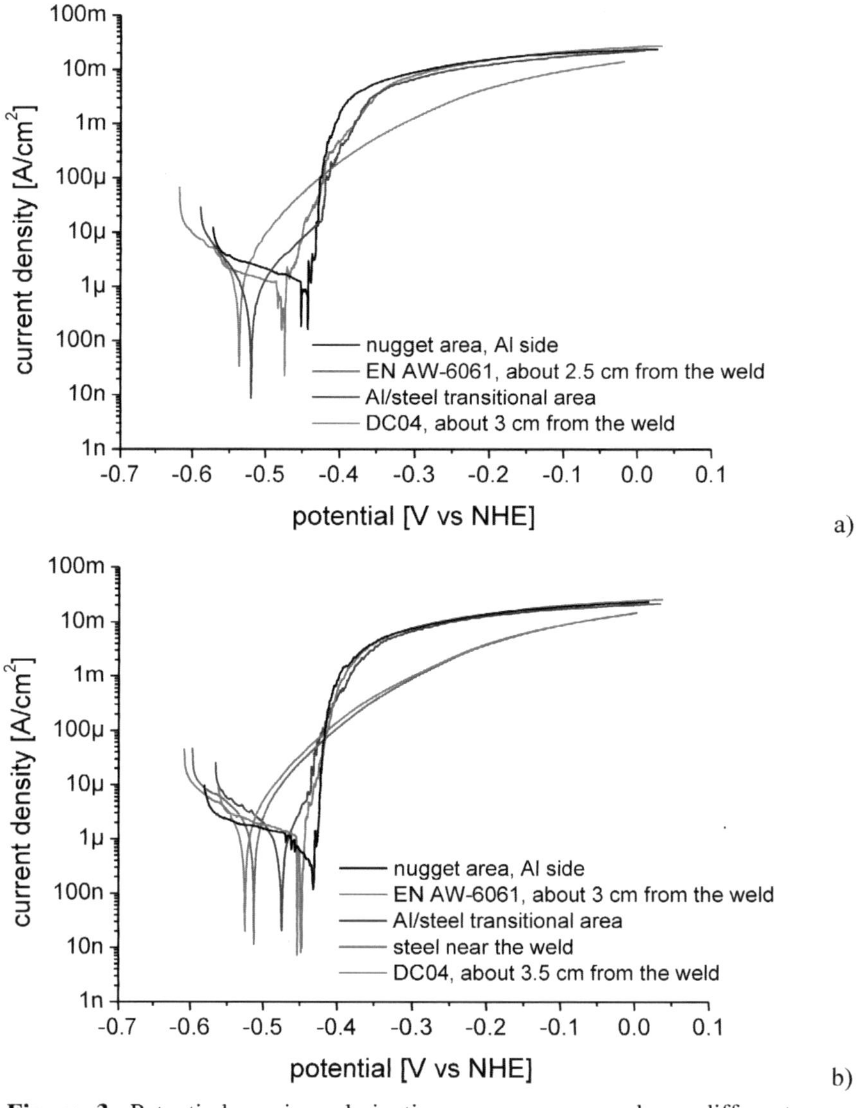

Figure 3. Potentiodynamic polarization curves measured on different positions of a FSW-joint (a) and an USE-FSW-joint (b) in 0.5 molar NaCl solution.

The free corrosion potential of the DC04 steel is around -0.54 to -0.53 V. The polarization curve of the steel shows a continuous increase in the anodic region. The corrosion current density at the E_{oc} is slightly higher than that of the EN AW-6061 Al alloy (2-3 µA/cm^2). The steel surface near the weld (figure 3b) only shows a small shift of the E_{OC} otherwise the polarization curve is nearly identical to that of the DC04 base material (greater distance from the weld).

The free corrosion potential of the nugget area shows during the potentiodynamic polarization a small shift to the anodic direction compared to the OCP measurement. The pitting potential of the nugget area is identical to that of the EN AW-6061 Al alloy. The corrosion current density of the nugget area of the FSW-hybrid joint at E_{OC} is somewhat higher than that of the USE-FSW-hybrid joint and higher than that of the EN AW-6061 Al alloy. This may be attributed to the higher amount of intermetallic phases within the measured area. These phases show a higher corrosion tendency (figure 4).

The E_{OC} value of the Al/steel transitional area of the FSW-joint is about 50 mV more negative than that of the USE-FSW-hybrid joint, which may be a result of a different Al to steel ratio or different phases within the measured area. However, both polarization curves exhibit in the anodic range first a continuous increase of the current density - comparable to the DC04 base material - and then a pitting potential and further progress comparable to that of the EN AW-6061 Al alloy. The corrosion current density at the E_{OC} is lower than that of the DC04 base material. This is a distinct difference to the Al/Mg-hybrid joints where an increase of the corrosion current density of several orders of magnitude was observed at the Mg/nugget transitional area [9].

3.4 Microscopic investigations

The measured areas were investigated by light microscopy after the polarization experiment. The pictures are presented in figure 4.

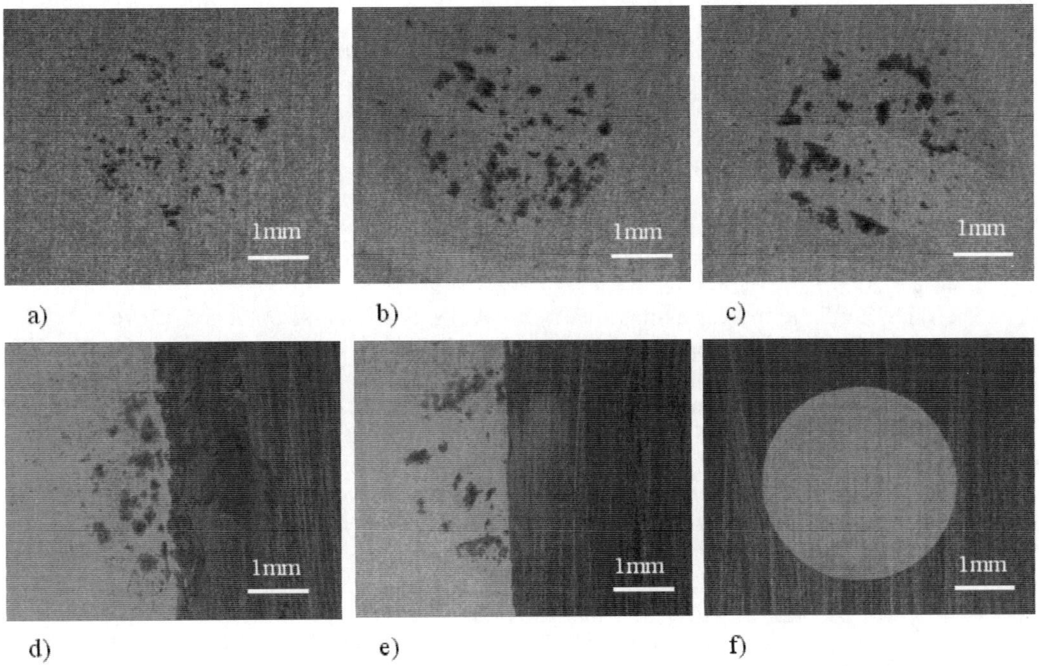

Figure 4. Microscopic pictures of the EN AW-6061 Al alloy (a), the nugget on the Al side of the FSW- joint (b) and the USE-FSW-joint (c), the Al/steel transitional area of the FSW- joint (d) and the USE-FSW-joint (e) and the DC04 steel (f) after the polarization measurements.

It is obvious that the EN AW-6061 Al alloy (4 a) shows typical signs of pitting corrosion whereas the DC04 steel (4 f) surface exhibits a whitish oxidation layer after the polarization experiment. In the welding area the corrosion properties change. The nugget area seems to consist of two phases which differ not only in colour but also in their corrosion properties (figures 4 b and c). The darker area shows a significantly stronger corrosion attack than the brighter one. There are not only small pits like in the case of the EN AW-6061 alloy but larger corroded areas. The corrosion properties at the Al/steel transitional area (figures 4 d and e) also differ significantly from that of the base materials. The steel in this area does not show the whitish oxidation layer. Especially in the case of the FSW-joint (4d) most parts of the steel area seem to be corroded. The first 1-2 mm at the boundary to the Al side show a different colour than the rest of the steel. This may be explained with the formation of intermetallic layers of Al and iron. This is in accordance with the SKP results which show a small seam at the weld area of the FSW-joint with a more negative Volta potential than the steel (figure 1 a). This more corroded area is not observed for the USE-FSW-joint (figure 4 e). The additional ultrasound seems to prevent the formation of this phase.

Detailed analyses of the corrosion products and the phases that show a preferred corrosion attack are still ongoing.

4. Summary

The innovative hybrid joining process of ultrasound enhanced friction stir welding was applied on EN AW-6061 and DC04 Al/steel-hybrid joints. The corrosion properties of the joints have been investigated in air (SKP) as well as in an aqueous electrolyte. It turned out that the two materials behave very different in the two media. The SKP measurements in air result in a difference of the Volta potentials of 1 V whereas the difference of the OCPs in 0.5 molar NaCl solution was only about 50 mV. Moreover the DC04 steel exhibits a more negative potential than the EN AW-6061 Al alloy after 60 min immersion in the electrolyte whereas it has shown the more positive Volta potential in air. As a result of the low potential difference in the NaCl solution no enhanced galvanic corrosion was observed for the EN AW-6061/DC04-hybrid joints as it has been the case for the Al/Mg-hybrid joints. Differences in the corrosion properties of the welded area (nugget, Al/steel transitional area) compared to the base materials may be attributed to the formation of intermetallic phases in this region. The amount of the intermetallic phases seemed to be reduced by the introduced ultrasound as a comparison of FSW- and USE-FSW-hybrids showed.

References

[1] Liu W, Ma J, Mazar Atabaki M and Kovacecic, R 2015 *Mater. Des.* **68** 146-57
[2] Khaled T 2005 *Fed. Aviat. Adm.* ANM-112N-05-06.
[3] Sheikhi S, Zettler R and dos Santos J. F. 2006 *Materialwiss. Werkstofftech.* **37(9)** 762
[4] Schofer E 1999 *Materialwiss. Werkstofftech.* **30** 693
[5] Hussein S A, Tahir A S M and Hadzley A B 2015 *Mater. Today Commun.* **5** 32-49
[6] Klag O, Wagner G and Eifler D 2010 *Adv. Eng. Mater.* **12** 1212–9
[7] Strass B, Wagner G and Eifler D 2014 *MSF* **783-786** 1814–9
[8] Straß B ,Wagner G, Conrad C, Wolter B, Benfer S and Fürbeth W 2014 *Adv. Mat. Res.* **966-967** 521-35
[9] Benfer S, Straß B, Wagner G and Fürbeth W 2016 *Surf. Interface Anal.* **48(8)** 843-52
[10] Thomä M, Wagner G, Straß B, Conrad C, Wolter B, Benfer S and Fürbeth W 2017 TMS 146th Annual Meeting & Exhibition *Proc. Symp. Friction Stir Welding and Processing IX* February 26 - March 2 2017 San Diego in print
[11] Thomä M, Wagner G, Straß B, Wolter B, Benfer S and Fürbeth W 2017 *IOP Conf. Ser.-Mat. Sci.* 19th Chemnitz Seminar on Materials Engineering March 16-17 2017 Chemnitz in print
[12] Stratmann M and Streckel H 1990 *Corr. Sci.* **30(6-7)** 681
[13] Stratmann M and Streckel H 1990 *Corr. Sci.* **30(6-7)** 697
[14] Wapner K, Schoenberger B, Stratmann M and Grundmeier G 2005 *J. Electrochem. Soc.* **152(3)** E114

Acknowledgements

The investigations were supported by the DFG Priority Program 1640: "Joining by plastic deformation" which is gratefully acknowledged.

IOP Conf. Series: Materials Science and Engineering **181** (2017) 012005 doi:10.1088/1757-899X/181/1/012005

Development of a Cu-Sn based brazing system with a low brazing and a high remelting temperature

M Schmieding, U Holländer and K Möhwald

Institut für Werktoffkunde, Leibniz Universität Hannover, An der Universität 2, 30823 Garbsen

Abstract. Objective of the project presented is the development of a joining process for hot working steel components at low brazing temperatures leading to a bond with a much higher remelting temperature. This basically is achieved by the use of a Cu-Sn melt spinning foil combined with a pure Cu foil. During brazing, the Sn content of the foil is decreased by diffusion of Sn into the additional Cu resulting in a homogenious joint with a increased remelting temperature of the filler metal. Within this project specimens were brazed and diffusion annealed in a vacuum furnace at 850 °C varying the processing times (0 – 10 h). The samples prepared were studied metallographically and diffusion profiles of Sn were recorded using EDX line scans. The results are discussed in view of further investigations and envisaged applications.

1. Introduction

Braze filler compound systems that contain of at least two metallic components, for example in form of multilayer coatings or foils are characteristic for transient liquid phase bonding (TLP-bonding). The thermal joining process takes place by interdiffusion of the components resulting in a homogeneous filler alloy with a higher remelting temperature compared to the melting temperature of the pure filler metal. This, of course, depends on dwell times and temperatures applied to the TLP-system chosen [1].

In the present study this fundamental mechanism is used to create durable joints at moderate joining temperatures. As shown in figure 1, two Cu75-Sn25 foils (wt.%) were applied with an additional copper foil to join hot working steel components.

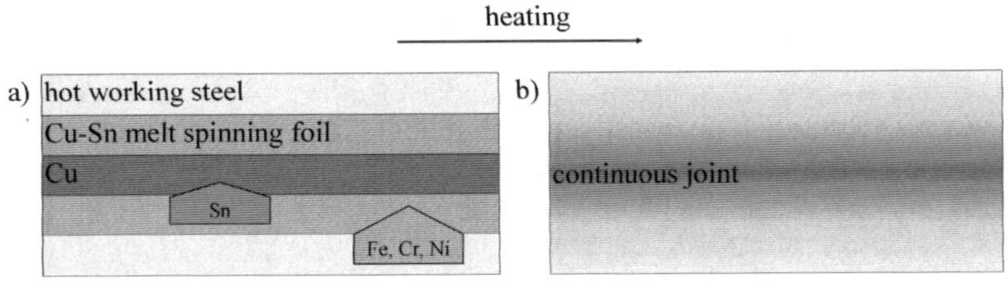

Figure 1. a) initial state of the brazing process; b) completed brazing process.

At first a conventional brazing process is performed, where the liquid Cu-Sn foil wets the steel and the copper surfaces. With increasing dwell time more and more copper is dissolved in the liquid braze metal, since an isothermal solidification of the braze metal takes place. Then a further diffusion heat treatment is performed in order to decrease the Sn content further by solid state diffusion. When reaching a homogenized joining zone with the remelting temperature envisaged, the joining process can be stopped.

Objective of the described work is the investigation and description of the diffusion process as a function of the brazing parameters in order to determine the optimal process conditions for this kind of joining process. Prior to the experimental details a short introduction to the Cu-Sn brazing system and the mathematical description of Cu-Sn interdiffusion is given below.

Cu-Sn system

The binary alloy containing 75 wt.% Cu and 25 wt. % Sn (85 at.% Cu, 15 at.% Sn) was chosen for this study in the form of a melt spinning foil with a thickness of 40 µm. It was supplied by the company Vacuumschmelze. As shown in figure 2, the brazing process was performed at 850 °C varying the processing times (0 – 10 h).

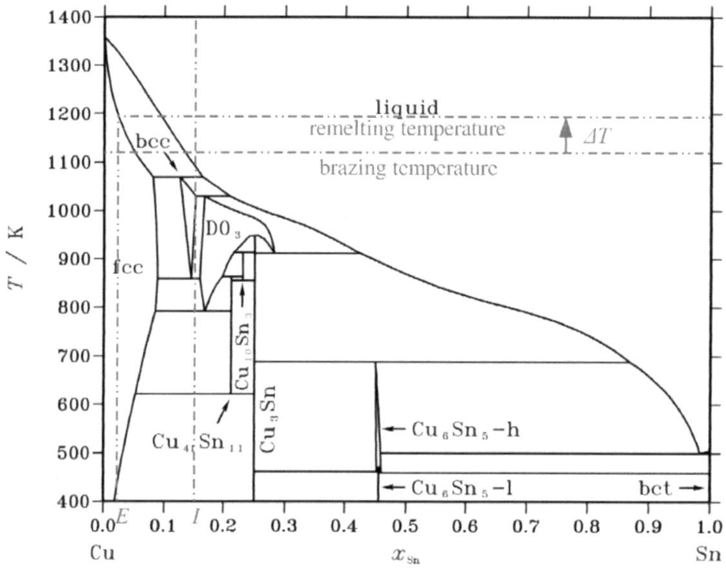

Figure 2. Cu-Sn phase diagram [2].

During the brazing process the initial composition *I* changes to a composition *E* with a decreased Sn content by the diffusion of Sn into the additional copper foil, which causes a shift ΔT in melting temperature.

Diffusion processes

For the present brazing system, the composition within the resulting joint is determined by the brazing time and temperature. Therefore a mathematical description of the diffusion processes is essential to understand the nature of the joining process and the resulting microstructure of the brazed seam and to forecast appropriate process conditions for envisaged applications.

The temporal evolution of a given concentration distribution c is described by the diffusion equation

$$\partial c(x,t)/\partial t = \partial/\partial x \cdot D(c, t) \cdot \partial c/\partial x(x,t) \tag{1}$$

with the diffusion coefficient $D(c,t)$.

Analytical solutions do exist in special cases, particularly for a constant diffusion coefficient [3]. But in the present case there is no analytical solution available since the interdiffusion of copper and tin depends highly on the actual tin concentration. Hoshino et al. [4] studied the diffusion of Sn in the Cu-Sn system intensively. Based on their data the concentration and temperature dependence of the diffusion coefficient of Sn in Cu can be expressed as

$$D_{Sn\ in\ Cu} = D_0 \cdot \exp(-E/RT) \cdot \exp(a \cdot c) \qquad (2)$$

with

diffusion coefficient $D_0 = 1.45 \cdot 10^{-5}$ m²/s,
activation energy $E = 179.8$ kJ/mol,
universal gas constant $R = 8.314$ J/mol K,
temperature $T = 973 - 1125$ K,
and concentration coefficient $a = 0.269$.

This data were used to solve the diffusion equation numerically for the boundary conditions given by the geometry and brazing conditions of the samples processed in this work.

2. Experimental details

Filler metal
The filler metal 75Cu-25Sn was supplied by the company Vacuumschmelze as a melt spinning foil with a thickness of 40 µm. It was used with a pure copper foil with a thickness of 500 µm as a filler system (figure 1).

Base metal
Hot working steel 1.2714 (A681 L6) was used as base metal for the joining process. Its composition is shown in table 1. This material is an air-hardening steel and it is typically solution annealed with the same temperature, which is used here for the brazing and diffusion treatment of the joint. So the joining process may be coupled with the hardening of the material in one furnace process.

Table 1: Composition of the base metal used.

1.2714	C	Cr	Mo	Ni	V
content in wt.%	0,55	1,10	0,50	1,70	0,10

Heat treatment
The samples studied were brazed in a vacuum furnace at 850 °C varying the processing times (0 – 10 h). Figure 3 shows the temperature during the heat treatment of the specimens as measured using companion specimens instrumented with thermocouples.

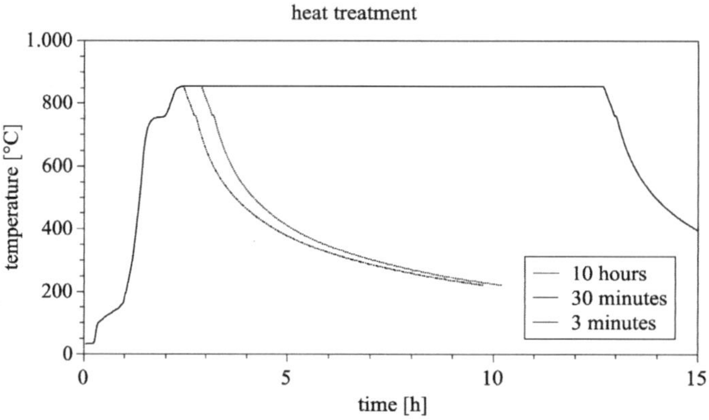

Figure 3. Temperature during heat treatment.

Interaction between melt spinning foil and additional copper

Joining using the filler system employed requires a fundamental understanding of the interaction between the Cu-Sn foil used and the additional copper upon heating, which can be studied best on samples with simple and defined joining geometries. Hence the diffusion of Sn in Cu was studied on brazing gap specimens shown in figure 4.

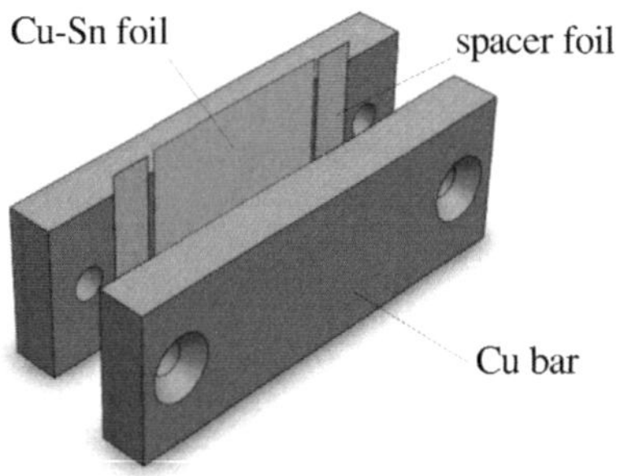

Figure 4. Cu gap specimen.

Two copper bars (45 mm x 15 mm x 5 mm) were held together by stainless steel screws. Stainless steel foils with a thickness of 40 µm were used as spacers and placed in between the copper as well as the Cu-Sn melt spinning foil. Once the brazing process was performed, the surface was polished to investigate the distribution of elements within the brazed joint by means of EDX line scans.

One set of specimens was brazed for 3 minutes at 850 °C to investigate the interaction between melt spinning foil and additional copper during the melting and isothermal solidification process. No diffusion annealing was performed.

In a further step the influence of annealing on the resulting Sn concentration was investigated. Therefore brazing processes were performed for 30 minutes and 10 hours.

Brazing hot working steel specimens

The behavior of the brazing system in the case of application is of special interest. Hence specimens corresponding to figure 5 were brazed and investigate similar to specimens mentioned in section 2.4.

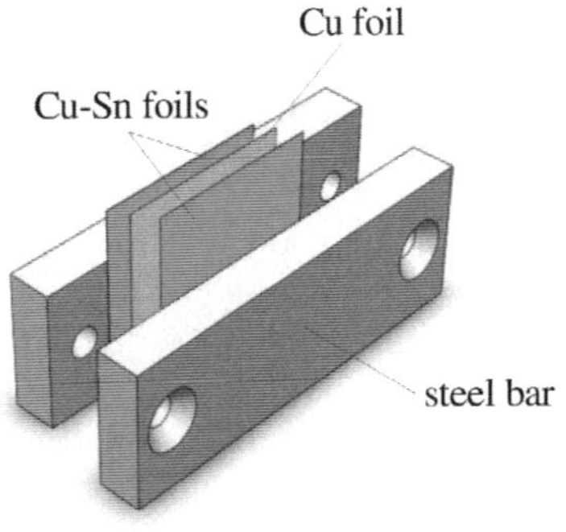

Figure 5. Hot working steel specimen.

Two hot working steel bars (45 mm x 15 mm x 5 mm) were held together by stainless steel screws. Two pieces of melt spinning foil (40 µm) combined with a copper foil (500 µm) were placed in between the hot working steel as a stack prior to brazing.
Metallographic cross sections were prepared from the brazed specimens in order to investigate the microstructure of the brazed seem as well as the distribution of elements within the brazed joint by means of EDX line scans.

3. Results

Interaction between melt spinning foil and additional copper
The left picture in figure 6 shows the microstructure of a Cu/Cu-Sn brazing joint specimen after brazing for 3 minutes at 850 °C. On cooling to room temperature Cu_3Sn phases formed in the centre of the brazed seam. These phases did not appear, if the joining process was performed for 30 minutes (middle image in fig. 6) due to the advanced dilution of Sn within the brazed seam.

After brazing at 850 °C for 10 hours the brazed seam was homogenized and some small pores are visible now in the centre (right image), which is the well known Kirkendall effect, resulting from solid-state interdiffusion of atoms of different size.

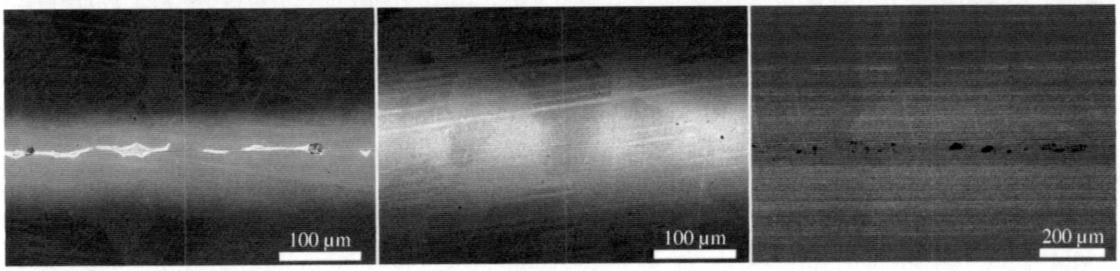

Figure 6. Microstructure of Cu specimens brazed at 850 °C using Cu-Sn melt spinning foil. After 3 minutes (left), 30 minutes (centre) and 10 hours (right).

A closer look at the evolution of the Sn diffusion profile throughout the joining zone is given by EDX line scans shown in figure 7. The rectangular plot illustrates the start condition corresponding to the melt spinning foil with a Sn content of 15 at.%. In the following measured diffusion profiles of Sn in Cu are compared to calculated profiles.

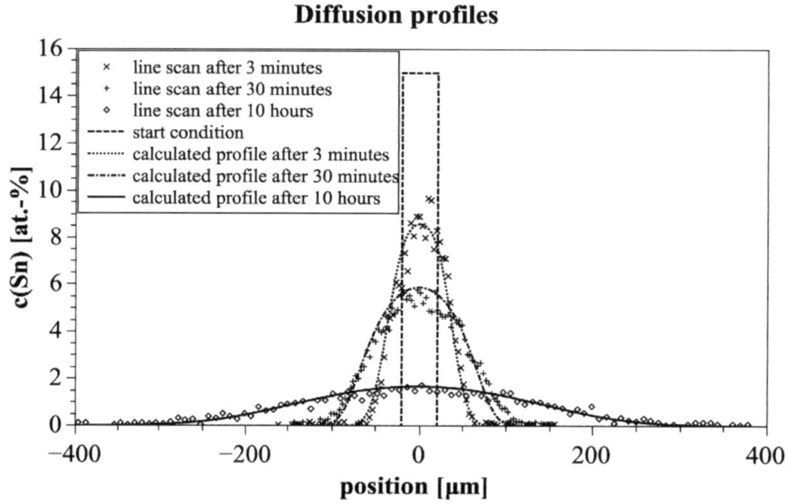

Figure 7. Diffusion profiles of Sn in Cu after several processing times.

In all cases there is a good match between the measured and calculated profiles, so the observed diffusion of Sn in Cu agrees well with the numerically solved model when using the diffusion coefficient given in equation (2).

Brazing hot working steel specimens

Figure 8. Microstructure of hot working steel specimens brazed at 850 °C using Cu-Sn melt spinning and copper foil. After 3 minutes (left), 30 minutes (centre) and 10 hours (right).

The left picture in figure 8 shows the microstructure of a hot working steel brazing joint specimen after brazing for 3 minutes at 850 °C. On cooling to room temperature Cu_3Sn precipitations formed at the interface to the steel. These precipitations did not appear that distinct, if the joining process was performed for 30 minutes (middle image in fig. 8) due to the forced dilution of Sn within the brazed seam. After brazing at 850 °C for 10 hours the brazed seam was homogenized (right image).

A closer look at the evolution of the Sn diffusion profile throughout the joining zone is given by EDX line scans shown in figure 9. The positions of the Cu-Sn and Cu foil prior to brazing are indicated by vertical lines. Depending on the annealing time the maximum Sn concentration at the steel interface decreased and therefore the remelting temperature increased. After 30 minutes of annealing the remelting temperature already is 930 °C according to the solidus temperature of a Cu alloy with max. 5 at.-% Sn content, see figure 2. An annealing for 10 hours yield to a further decrease of the Sn content to finally ca. 2 at.-% which corresponds to a solidus temperature of 980 °C.

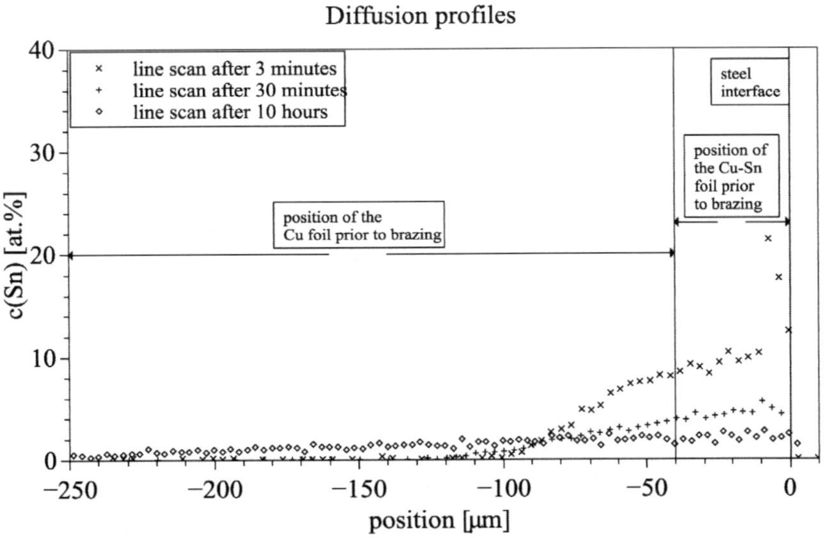

Figure 9. Diffusion profiles of Sn in hot working steel specimens after several processing times.

4. Conclusion and outlook

The Cu75-Sn25 melt spinning foil in combination with pure copper as additional solid filler metal is a promising candidate for joining hot working steel via TLP-bonding. Especially the high diffusion coefficient of Sn in Cu enables quite moderate brazing temperatures (850 °C) and annealing times (< 10 h) and results in homogenous, high melting copper based joints in a vacuum brazing process. Even after 30 minutes of brazing at 850 °C a remelting temperature of 930 °C was achieved and it can get even higher on further annealing.

In the ongoing studies the brazing time will be optimized with the aim of avoiding the formation of Cu_3Sn during the cooling process. So annealing times will be varied between 30 minutes and 10 hours until the minimum annealing time is found. Findings gained will be compared to numerical calculations.

In the final steps hot working steel specimens will be brazed and hardened in one furnace process and their technological properties like hardness, tensile strength and corrosion behaviour of the joint as well as the base material will be assessed.

Acknowledgment

The IGF Project IGF 18.706 N/DVS Number.: 07.081 of the research association "Forschungsvereinigung Schweißen und verwandte Verfahren des DVS, Aachener Straße 172, 40223 Düsseldorf" was, on the basis of a resolution of the German Bundestag, promoted by the German Ministry of Economic Affairs and Energy via the Federation of Industrial Research Associations (AiF) within the framework of the programme for the promotion of joint industrial research and development (IGF).

References

[1] Cook G O and Sorensen C D 2011 Overview of transient liquid phase and partial transient liquid phase bonding *J. Mater. Sci.* **46** pp 5305-32
[2] Landolt H and Börnstein R 2005 *Thermodynamic properties of inorganic materials compiled by SGTE* part 3 (Berlin: Springer) pp 54-57
[3] Crank J 1967 *The mathematics of diffusion* (Oxford: Clarendon Press) pp 137-59
[4] Hoshino K and Iijima Y (1980) Interdiffusion and Kirkendall Effect in Cu-Sn Alloys *Trans. J. I. Met.* **21** pp 674-82

IOP Conf. Series: Materials Science and Engineering **181** (2017) 012006 doi:10.1088/1757-899X/181/1/012006

Investigation of the effect of rapidly solidified braze ribbons on the microstructure of brazed joints

K Bobzin[1], M Öte[1], S Wiesner[1], P Rochala[1], J Mayer[2], A Aretz[2], R Iskandar[2] and A Schwedt[2]

[1]Surface Engineering Institute, RWTH Aachen University, Kackertstraße 15, 52056 Aachen, Germany
[2]Central Facility for Electron Microscopy, RWTH Aachen University, Ahornstraße 55, 52074 Aachen, Germany

Abstract. Shrinkage and warpage due to melting and solidification are crucial for the geometric precision of related components. In order to assure a high geometric precision, the formation of the microstructure in the joint during brazing must be taken into consideration. An extensive interaction can occur between liquid melt and base material, resulting in the formation of distinctive phases. This interaction depends on the parameters of the brazing process. However, the consequences of the interaction between phase formation and process parameters in terms of geometric precision cannot be estimated yet. Insufficient quality of the joint can be a result. In this study, investigations focus on the process of solidification in terms of time dependent diffusion behavior of elements. Therefore, microcrystalline and amorphous braze ribbons based on Ti are produced by rapid solidification and are used for joining. The microstructure of the braze ribbons as well as the melting behavior and phase formation during brazing are considered to be of particular importance for the mechanical properties of the brazed components.

1. Introduction

Brazing is defined as a joining process using a molten filler metal that has a liquidus temperature lower than the solidus temperature of the base material. During the brazing process, the liquid filler metal is drawn into the gap between the parts to be joined and solidifies during cooling. As a result, a firm bonding between the parts is established. Diffusion processes between filler metal and base material play an important role during the brazing process [1]. Numerous factors, such as the process parameters during brazing, the properties of substrate and brazing material, as well as the geometry of the components have an impact on the process of melting and solidification. The interaction of these factors can lead to the formation of an uncontrollable microstructure of the joint with various properties. It is hardly possible to affect or to control the above-mentioned factors during the brazing process. As a result, shrinkage, misalignment and warpage can occur resulting in rejected parts during quality control. For base materials with defined microstructure and tailored properties, there is also the risk of altering the material's microstructure during the brazing process due to the heat treatment in the heat affected zone. Thus, the base material is meant to be thermally stressed as few as possible during brazing. An interdisciplinary research group of RWTH Aachen University (SFB 1120) targets on producing brazed joints with high geometric precision by controlling and affecting the factors that have impact on the final joint properties.

During brazing, a molten liquid phase, characterized by intrinsic melt dynamics, is formed in the braze gap and solidifies subsequently. Boundary conditions of the braze process, such as irregular cooling due to inhomogeneous transport of energy, different thermal expansion coefficients of base material and braze material, non-uniform volume contraction or the formation of unwanted phases in the braze gap can affect the geometric precision of the brazed joint and can lead to joint flaws like pores or stress cracking. As these effects depend on multiple factors such as material, process parameters and component geometry, every single braze joint must be considered as an individual issue in terms of process design, when high precision in production shall be reached.

The complex metallurgical processes between filler metal and base material are yet to be fully understood and subject to intensive researches. For this reason, it is necessary to identify, analyze and understand precision-relevant factors in brazing in order to control them and manufacture high-precision components. Therefore, an increased repeatability of the process zone that is formed by braze material and heat influenced zone in the substrate is needed. Phase composition, thermal properties and remelting temperature of joints have to be investigated in order to understand the mechanisms that lead to geometric flaws during brazing and to prevent the substrate from altering its tailored properties during the brazing process.

In this work, Ti-based braze ribbons are produced by melt spinning (MS) and used for brazing. Melt spinning is a process to produce thin metallic ribbons from a molten liquid by rapid cooling, with cooling rates up to $\dot{T} = 10^6$ K/s. The braze ribbons can thus solidify in amorphous, semicrystalline or fully crystalline state. According to the objectives mentioned above, these braze ribbons are analysed in terms of their initial microstructure and material properties. Additional joining experiments with a hot-working steel enable the time- and temperature-resolved analysis of bonds, as well as the determination of the process parameters that affect the geometric precision. The aim of the investigations is thus to design brazing processes that allow to manufacture high-quality joints with high geometric precision.

2. Materials

For filler metal development, the production of corrosion-resistant brazed joints with high toughness was the primary target. Additionally, the filler metal should be processable via melt spinning, producing flexible ribbons. Furthermore, it should be possible to use it for joining the hot-working steel with the material number 1.2343. This steel is commonly used in tool construction. It is characterized by high heat resistance, high toughness and good thermal conductivity. Thus, this steel is very well suited for tools for injection moulding or permanent mould casting. The composition of the steel is shown in table 1.

Table 1. Metallurgical composition of the hot-working steel 1.2343.

1.2343	C	Si	Mn	Cr	Mo	V	Fe
[wt.-%]	0.38	1.00	0.40	5.30	1.20	0.40	91.32

The hot-working steel's austenitization temperature is $T_{\alpha/\gamma}$ = 1,030 °C. Since the phase transformation of this material has to be avoided during brazing, a potential filler metal has to have a melting temperature $T_m < T_{\alpha/\gamma}$. As starting point for the filler metal development, the alloy Ti70Ni30 with the composition of m_{Ti} = 70 wt.-% Ti and m_{Ni} = 30 wt.-% Ni was used. According to the corresponding binary phase diagram, the liquidus temperature amounts to T_{liq} = 942 °C. In this study, an alloy was sought with a lower liquidus temperature and more favorable properties. As a result, the alloy Ti67Cu15Ni15Si3 was developed.

3. Experiments

In this section, the melt spinning process used to produce the braze ribbons is presented as well as the high vacuum brazing process. Modern techniques for characterizing ribbons and joints on an atomic level are used, including scanning electron microscopy (SEM), transmission electron microscopy (TEM) and light microscopy. Additionally, differential scanning calorimetry (DSC) measurements for analyzing the thermal properties of the braze ribbons are conducted. Finally, the hardness testing method for investigating the mechanical properties of the joints is presented.

3.1 Melt Spinning Process

All the braze ribbons used in this work are produced by melt spinning. The Melt Spinner SC, manufactured by Edmund Bühler GmbH, Hechingen, Germany, was used for ribbon production. The basic set-up of such a system is shown in figure 1.

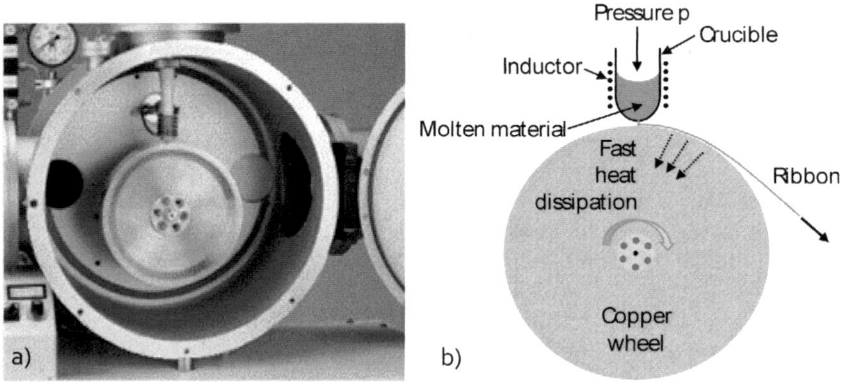

Figure 1. a) Recipient of the Melt Spinner SC with copper roll, induction coil and crucible. [2] b) Schematic of a melt spinning system.

For ribbon production, the components of the alloy are prealloyed in an arc furnace. Subsequently, the pre-alloyed material is melted inductively under argon atmosphere in the crucible. After the melt is homogenized, it is ejected from the crucible by protective gas pressure and hits a rotating, water-cooled copper roll. The melt cools down with a cooling rate up to $\dot{T} = 10^6$ K/s by rapid heat dissipation, due to the large contact area with the cooled roll. The rotational velocity of the roll correlates with the cooling rate of the melt [3, 4]. By controlling the rotational velocity, an amorphous, semi-crystalline or fully crystalline microstructure can be generated in the ribbon. In general, high cooling rates suppress the formation of a crystalline structure [5]. Besides the rotational velocity of the copper roll, the pressure during ejection also has a great impact on the formation of microstructure in the braze ribbon. After the ejection of the melt, the ribbon is carried along with the roll and a fully solidified ribbon is released from it due to the centrifugal force [3].

In this work, braze ribbons were produced by melting the alloy in a quartz crucible with a nozzle. The slit of the nozzle was s = 5 mm in width and d = 0.4 mm in thickness. The distance between copper wheel and slit nozzle was h = 0.5 mm and the rotational velocity of the roll was f = 25 Hz. For the alloy Ti70Ni30 the eject temperature was T_{eject} = 1,100 °C. For Ti67Cu15Ni15Si3 T_{eject} = 1,200 °C was chosen. For all experiments, the ejection pressure was p = 300 mbar.

3.2 Transmission Electron Microscopy (TEM)

Transmission electron microscopy, a special operating mode of electron microscopy, allows to investigate the lattice structure of a metallic sample. Microstructures down to the atomic level can be depicted and analyzed, as well as phase conditions, microstructural compositions, segregations, diffusion zones, grain boundaries and crystallographic defects such as dislocations or stacking faults.

For TEM analysis in this work, lamellae of the Ti70Ni30 and Ti67Cu15Ni15Si3 braze ribbons were prepared by using the focused ion beam (FIB) FIB Strata 205 by FEI, Hillsboro, USA. To allow electrons to fully permeate the samples, they are not allowed to have more than $d = 300$ nm in thickness. The investigations were carried out using STEM-HAADF mode (scanning transmission electron microscopy – high-angle annular dark field) of a Zeiss LIBRA 200 FE microscope by Carl Zeiss AG Microscopy, Oberkochen, Germany. Aim of the investigations was to detect the amorphous or crystalline structure of the braze ribbons.

3.3 Differential Scanning Calorimetry (DSC)
Differential scanning calorimetry is a thermal analysis method for measuring heat flows into a sample while heating or cooling the sample in a defined temperature program. With this method, investigations on phase transitions, chemical reactions or sintering reactions can be conducted. Furthermore, thermodynamic material properties, such as the specific heat capacity, can be determined. [6, 7] In the present work, the DSC system Setsys Evolution from Setaram Instrumentation AG, Caluire, France, was used to analyze the Ti70Ni30 and Ti67Cu15Ni15Si3 braze ribbons. Investigations targeted on the determination of temperatures for phase transitions, recrystallization and melting. Heating and cooling rates were chosen to be $\dot{T} = 10$ K/min. All investigations were conducted in argon atmosphere.

3.4 Process of brazing
For brazing experiments, the furnace MOV 553T from PVA TePla AG, Wettenberg, Germany, was used. Brazing experiments were conducted with the Ti70Ni30 and Ti67Cu15Ni15Si3 braze ribbons as well as the steel 1.2343. Brazed joints were produced, applying a static pressure of $p = 5.3$ N/mm^2 on the sandwich construction during brazing. All brazing experiments were conducted in high vacuum. In a first step, the samples were heated to $T = 860$ °C with a heating rate of $\dot{T} = 20$ K/min. This temperature was maintained for $t = 10$ min. Subsequently, further heating was carried out up to the brazing temperature of $T_{brazing} = 1{,}010$ °C with a heating rate of $\dot{T} = 10$ K/min. At this temperature, the samples were brazed for $t = 30$ min. Finally, the samples were cooled with a cooling rate of $\dot{T} = 32$ K/min until room temperature was reached.

3.5 Hardness testing
For micro-hardness testing, cross sections of the brazed joints, according to chapter 3.4, were prepared. Measurements were conducted, using the Fischerscope HM 2000 by Helmut Fischer GmbH, Sindelfingen, Germany. A maximum load of $F = 500$ mN was chosen. The maximum depth of indentation of the testing pyramid was $h = 5$ µm.

4. Results and discussion
In this section the results of the analysis of the braze ribbons and the brazed joints are presented and discussed. In addition, relations between braze ribbon properties, substrate and brazed joint are sought, discussed and evaluated with regard to precision-relevant aspects.

4.1 Melt Spinning Prozess
Starting from the alloy Ti70Ni30, Cu was added in order to suppress the formation of intermetallic phases during the brazing process. Furthermore, Si was added to the alloy for lowering its melting temperature and for increasing the alloy's diffusional activity [8]. As a result, the alloy Ti67Cu15Ni15Si3 was developed. Both Ti-based alloys could successfully be processed in the melt spinning process, obtaining braze ribbons with uniform thickness and width, figure 2.

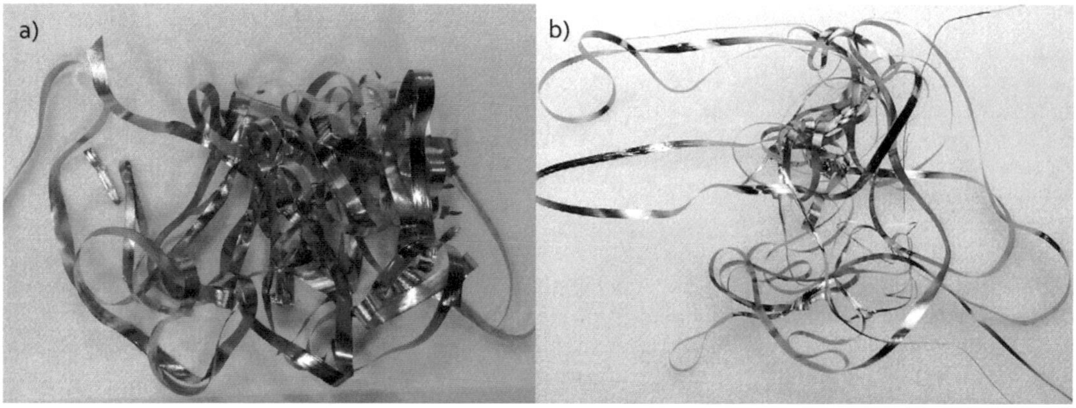

Figure 2. a) Ti70Ni30 and b) Ti67Cu15Ni15Si3 braze ribbons produced by melt spinning.

4.2 Transmission Electron Microscopy (TEM)
Figure 3 shows TEM images of the Ti70Ni30 and Ni67Cu15Ni15Si3 braze ribbons in their initial state after melt spinning. The corresponding selected area diffraction pattern (SADP) is depicted in the upper right corner of each figure.

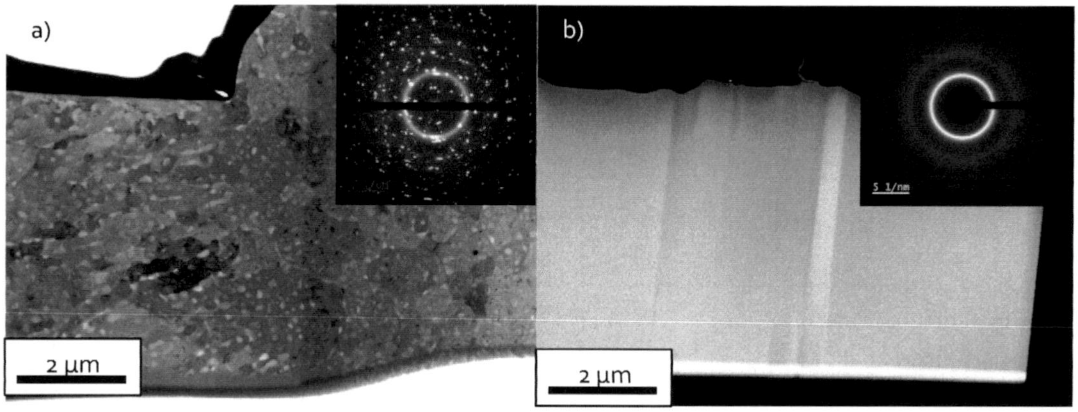

Figure 3. TEM images of the braze ribbons with SADP as insert in the right upper corner. a) Ti70Ni30 b) Ti67Cu15Ni15Si3.

Within the Ti70Ni30 braze ribbon, a grain structure can be found, figure 3a. Grain sizes in the sample are homogeneous and the distribution of grains is uniform. Referring to the corresponding SADP, the bright spots in the image represent a discrete diffraction condition of the sample's crystal structure. Thus, the crystalline structure of the material can be clearly detected. Amorphous areas could not be found. Figure 3b shows the TEM image of the Ti67Cu15Ni15Si3 braze ribbon as well as the corresponding SADP. In the image, the sample is depicted as a grey object without any structure, i.e. no grains can be found in this ribbon. In addition, the corresponding SADP shows no point reflections or clearly distinguishable diffraction rings as well, leading to the assumption that the ribbon has no lattice structure, meaning that the material is fully amorphous.

4.3 Differential Scanning Calorimetry (DSC)
Figure 4 shows the results of the DSC analysis of the Ti70Ni30 braze ribbon. In addition, the image shows the DSC curve of the Ti67Cu15Ni15Si3 ribbon in its initial state as well as after annealing at T = 860 °C for t = 10 min in vacuum.

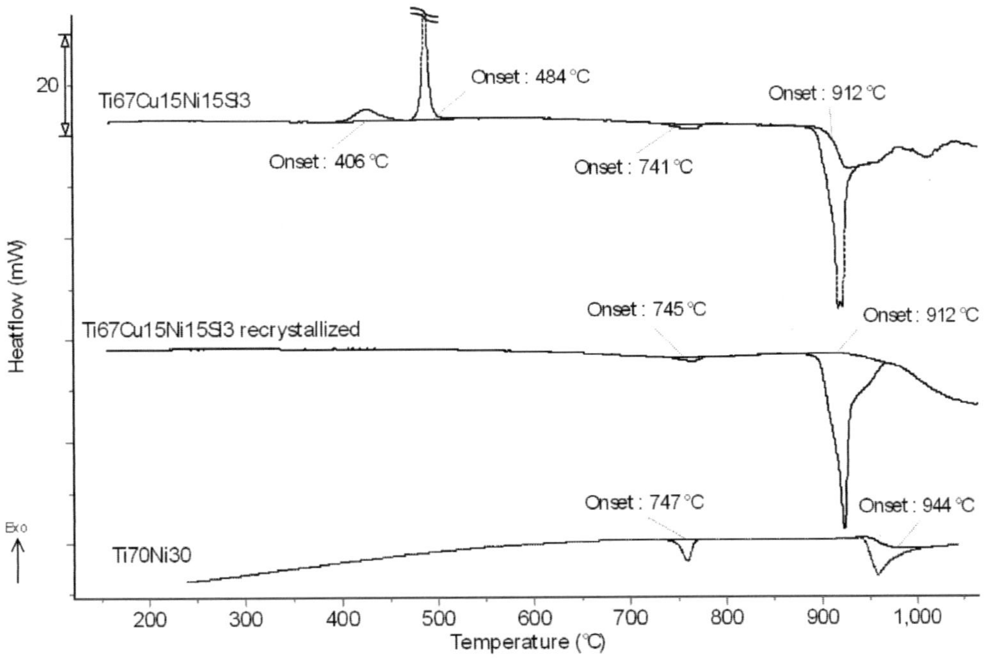

Figure 4.: DSC results of the Ti70Ni30 braze ribbon, the Ti67Cu15Ni15Si3 ribbon in its initial state as well as recrystallized after annealing in vacuum atmosphere at T = 860 °C for t = 10 min.

The Ti70Ni30 braze ribbon undergoes a phase transition at T = 747 °C, indicated by an endothermic reaction starting at this temperature as shown in figure 4. The binary Ti-Ni phase diagram shows that the initial α-Ti + Ti_2Ni microstructure turns into β-Ti + Ti_2Ni at T = 765 °C [9]. Referring to the DSC analysis, the melting temperature of the alloy is T_m = 944 °C, marked by an endothermic peak in figure 4. Thus, the alloy melts below the austenitization temperature $T_{\alpha/\gamma}$ = 1,030 °C of the steel 1.2343.

Figure 4 also shows the results of the DSC analysis of the Ti67Cu15Ni15Si3 ribbon in its initial state. While heating the amorphous ribbon up to T = 1,100 °C, the material shows multiple thermic effects. At T = 406 °C, a small exothermic peak can be observed, indicating the start of recrystallization of the ribbon. A further significant exothermic peak can be observed starting at T = 484 °C. The narrow, steep peak marking this reaction in figure 4 indicates a spontaneous, thermally induced "flipping" of the initial amorphous structure of the ribbon into a crystalline structure. Continuing the DSC analysis, a small endothermic peak can be observed at T = 741 °C. This reaction represents a phase transition of the alloy. Due to the complex composition of the ribbon, it is difficult to designate the specific phase reaction that happens at this temperature. The Ti70Ni30 alloy showed an endothermic phase transition at T = 747 °C, in which the mixed phase α-Ti/Ti_2Ni transduced to β-Ti + Ti_2Ni. A similar phase transition can be assumed for Ti67Cu15Ni15Si3, with Si presumably shifting the transition reaction to slightly lower temperatures. Furthermore, the influence of Cu and Si must be taken into account in terms of thermodynamics of this reaction as well as its influence on the probability of the formation of further phases. The melting temperature of the alloy Ti67Cu15Ni15Si3 is T_m = 912 °C, indicated by an endothermic peak of the corresponding graph in figure 4. In comparison to the melting temperature of the Ti70Ni30 braze ribbon at T_m = 944 °C, a reduction in the melting temperature of ΔT = 32 K could be achieved by the addition of Cu and Si to the alloy.

To investigate the influence of the amorphous structure on the phase transition temperature of the Ti67Cu15Ni15Si3 ribbon, a further experiment was conducted. A sample of the amorphous braze ribbon was annealed at T = 860 °C for t = 10 min. Referring to figure 4, the ribbon should fully recrystallize at T = 486 °C. The non-amorphous ribbon was then analyzed via DSC, see figure 4. During DSC analysis, no exothermic reactions, as observed for the amorphous ribbon of same composition, can be found. The experiment validates that the exothermic reactions of the non-annealed Ti67Ni15Ni15Si3 ribbon represents its temperature of recrystallization. Furthermore, the previously annealed ribbon shows a weak endothermic reaction at T = 745 °C, assumably indicating the phase transition of α-Ti to β-Ti, that could be observed for the amorphous braze ribbon as well at slightly lower temperatures. Above T = 484 °C, both ribbons are recrystallized and show the same phase transition temperature and thus have identical thermal properties. The DSC analysis of the previously annealed Ti67Cu15Ni15Si3 braze ribbon shows that it also has a melting temperature of T_m = 912 °C. Thus, the amorphous braze material recrystallizes at temperatures much lower than its melting temperature, causing the material to lose its special properties that come with the amorphous structure.

4.4 Process of brazing

Using the brazing process parameters from chapter 3.4, sound joints with the steel 1.2343 could be realized, using Ti70Ni30 and Ti67Cu15Ni15Si3 braze ribbons. Figure 5 shows scanning electron microscopy (SEM) images of cross sections of brazed joints that were each produced at $T_{brazing}$ = 1,010 °C.

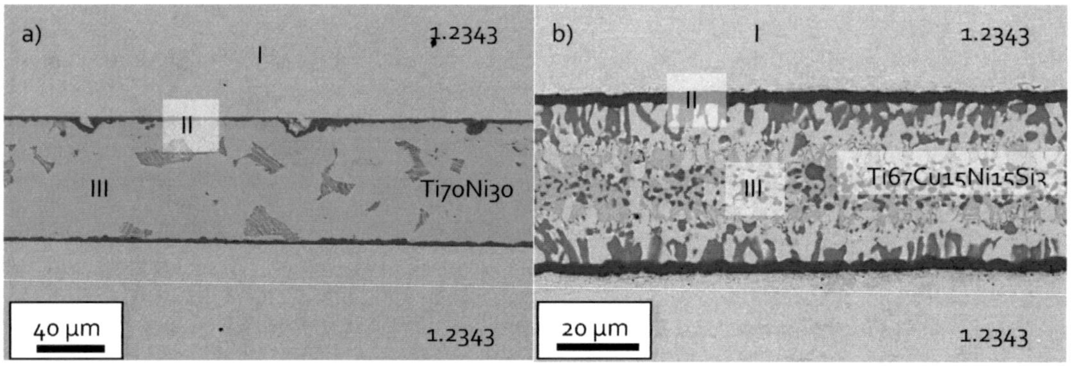

Figure 5. a) SEM images of the joint 1.2343 – Ti70Ni30 – 1.2343, $T_{brazing}$ = 1,010 °C b) SEM image of the joint 1.2343 – Ti67Cu15Ni15Si3 – 1.2343, $T_{brazing}$ = 1,010 °C.

Both joints in figure 5 are characterized by an individual microstructure. The joint that was produced using the Ti70Ni30 braze ribbon, shows a mostly homogeneous microstructure with two phases that can be well distinguished from each other. The interconnection between substrate and brazed seam is established by a Ti-rich reaction layer that is depicted as a dark grey phase between both components marked with "II" in both images. The brazed seam that was produced using the Ti67Cu15Ni15Si3 braze ribbon is characterized by a multiplicity of phases, each distinguished by a different grey tone in figure 5b. The interconnection between braze seam and substrate is also established by a Ti-rich reaction layer at the interface of the joint ("II"). The very complex microstructure of the seam correlates with the complex alloy composition of the braze ribbon, because both joints were produced, using the same process parameters. The effect of the amorphous structure of the Ti67Cu15Ni15Si3 ribbon on the phase formation can also assumed to be nonexistent, because the ribbon fully recrystallizes at temperatures near T = 484 °C, as discussed in section 4.3.

Since the formation of numerous different phases in the brazed seam is difficult to control, the braze alloy Ti67Cu15Ni15Si3 is considered to be critical in terms of precision-relevant aspects.

Additionally, it must be considered that in the brazed seam numerous intermetallics can be found with different thermal expansion coefficients. Thus, a strongly heterogeneous stress zone is created in the joint during cooling at the end of the brazing process, leading to intense residual stresses. With regard to the problems discussed above, the braze alloy Ti70Ni30 is to be preferred, since a much more homogeneous brazed seam is reached during the application of this braze material. A more detailed analysis of the microstructure of the bonds is subject of a paper currently in publication. It is also to be clarified, if during the application of the amorphous braze materials an increased diffusivity of atoms can be observed, whether from the braze material into the substrate or vice versa, compared to the application of non-amorphous braze materials.

4.5 Hardness testing

Brazed joints that were produced using the Ti70Ni30 and Ti76Cu15Ni15Si3 braze ribbons were tested in terms of micro-hardness. For each joint, hardness measurements were conducted at three characteristic positions: d = 50 μm from the interface in the substrate, at the interface and in the center of the brazed seam. These three positions are marked in Roman numerals in figure 5. The results of hardness testing are shown in figure 6.

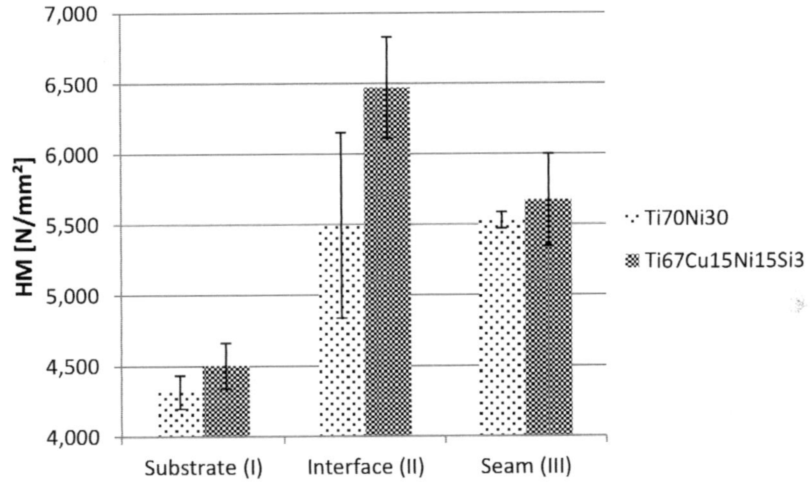

Figure 6. Hardness values of different areas of the joints, as shown in figure 5.

Via hardness testing, the brazed joints can be characterized in terms of their mechanical properties. For the substrate material, hardness values of $HM_I = 4{,}400$ N/mm² on average could be measured. The slight deviation of hardness values measured in this region could result from measuring errors or the marginally increased diffusion of atoms from the Ti67Cu15Ni15Si3 braze material into the substrate, compared to the Ti70Ni30 braze material. The hardness value of the substrate serves as a reference for the comparability of the hardness of the phases at the interface and in the brazed seam. The hardness value in the center of both brazed seams is also roughly the same, with approximately $HM_{III} = 5{,}600$ N/mm². It is noticeable that the hardness values at position III in the joint produced with the braze material Ti67Cu15Ni15Si3 fluctuate more strongly than in the joint produced with Ti70Ni30. For Ti67Cu15Ni15Si3, the standard deviation of the hardness values at position III is $s_{III} = 326$ N/mm², whereas for Ti70Ni30 the standard deviation at this point is $s_{III} = 58$ N/mm². The reason for this difference in standard deviation is the different microstructure of the two joints. A large number of phases is present in the joint produced with the ribbon Ti67Cu15Ni15Si3, see figure 5b. Each of these phases has a different hardness value. During the hardness testing in the center of the brazed seam, no single phase could be tested individually due to the size of the indentation area. As a

result, the hardness value measured for the seam represents an average value of all phases in this region. The joint produced with Ti70Ni30 has a much more homogeneous microstructure, see figure 6a, and thus much more consistent hardness values could be measured in this region.
At the interface of the bond, the relationship between hardness and microstructure is different. At the interface of the joint produced with Ti67Cu15Ni15Si3 hardness values of HM_{II} = 6,469N/mm² could be measured. The corresponding standard deviation is s_{II} = 359 N/mm². For the joint produced with Ti70Ni30 the hardness value is HM_{II} = 5,495 N/mm², having a standard deviation of s_{II} = 658 N/mm². The thickness of the reaction layer at both brazed joints could be a reason for the different hardness values measured at the interface. The reaction layer of the joint in figure 5a amounts to d = 2.5 µm in thickness. When applying the Ti67Cu15Ni15Si3 braze ribbon, a reaction layer with an average thickness of approximately d = 3.1 µm is established. For micro-hardness testing, the thicker reaction layer of the Ti67Cu15Ni15Si3 joint is easier to detect as the distortive influence of adjacent phases can be estimated to be lower. For this reason, lower hardness values could be measured for the Ti70Ni30 joint, while the corresponding standard deviation increases, as can be seen in figure 6. Figure 7 shows the indents from the micro-hardness testing at the interface of a sample brazed with the Ti67Cu15Ni15Si3 braze ribbon.

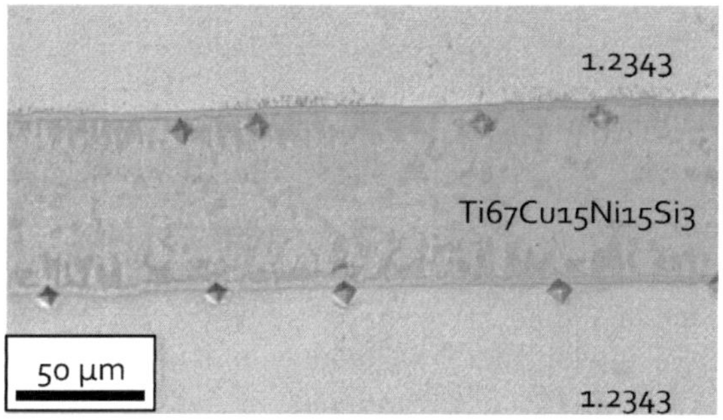

Figure 7. Light microscope image of the joint 1.2343 – Ti67Cu15Ni15Si3 –1.2343 with indentations from micro-hardness testing at the interface.

Although both investigated joints show high hardness values that are higher than the hardness of the substrate material, subsequent investigations of the indentations via light microscope revealed no evidence of brittle material failure. Therefore, further investigations on the mechanical properties of the compounds are necessary for a more precise characterization.

5. Conclusion

Ti-based braze alloys, having a melting temperature below T = 1,030 °C and thus below the austenitization temperature of the steel 1.2343, could be developed and successfully processed into braze ribbons via melt spinning. A fully amorphous ribbon could also be realized. Furthermore, the developed braze ribbons have been successfully used for brazing the steel 1.2343. Investigations on the thermal behavior of the amorphous braze ribbon showed that the amorphous structure disappears when reaching a material-dependent recrystallization temperature which is approximately ΔT = 430 K below the melting temperature T_m of the material. After exceeding the recrystallization temperature, the initially amorphous braze ribbon shares the same melting temperature as the crystalline ribbon of the same composition. First investigations on the mechanical properties of the ribbons showed that solid, non-brittle joints can be produced when applying the braze ribbons developed in this work.

Future studies will aim at a more precise characterization of the brazed joints in terms of mechanical properties, in order to identify and qualify further correlations between brazing process and brazed joint microstructure with regard to precision-relevant parameters. In addition, the influence of the amorphous structure of a braze material on the diffusion process during brazing shall be investigated.

References

[1] DIN 857-2:2007-03
[2] Edmund Bühler GmbH 2011 brochure Bühler Melt-Spinning system (SC-Version) (Hechingen)
[3] Sahm R; Egry I 1999 *Schmelze, Erstarrung, Grenzflächen. Eine Einführung in die Physik und Technologie flüssiger und fester Metalle* (Braunschweig: Vieweg)
[4] Uhrig E 2005 *Quasikristalle im System Zink-Magnesium-Seltene-Erden: Materialpräparation und Einkristallzüchtung* (Frankfurt a. M.: Universität Frankfurt a. M.)
[5] Kalinichenka S 2011 *Rascherstarrte nanokristalline Magnesiumlegierungen für die Wasserstoffspeicherung* (Dresden: TU Dresden)
[6] Gill P, Moghadam T, Ranjbar B 2010 Differential Scanning Calorimetry Techniques: Applications in Biology and Nanoscience *J. Biomol. Techn.* **21** 167-193
[7] Hoehne G W H, Hemminger W, Flammersheim H J 1996 *Differential Scanning Calorimetry: An Introduction for Practitioners* (Heidelberg: Springer-Verlag Berlin)
[8] Hoyer I 2005 *Beitrag zur Entwicklung von Hochtemperaturloten auf Eisenbasis* (Chemnitz: TU Chemnitz)
[9] ASM International 1992 *Handbook of Ternary Alloy Phase Diagrams* **3** (Ohio: Materials Park)

Acknowledgments

All presented investigations were conducted in the context in the Collaborative Research Centre SFB 1120 "Precision Melt Engineering" at RWTH Aachen University and funded by the German Research Foundation (DFG). For the sponsorship and the support we wish to express our sincere gratitude.

IOP Conf. Series: Materials Science and Engineering **181** (2017) 012007 doi:10.1088/1757-899X/181/1/012007

Feasibility study of fluxless brazing cemented carbides to steel

W Tillmann and N Sievers

Lehrstuhl für Werkstofftechnologie, TU-Dortmund, Germany

Abstract. One of the most important brazing processes is the joints between cemented carbides and steel for the tool industry such as in rotary drill hammers or saw blades. Even though this technique has already been used for several decades, defects in the joint can still occur and lead to quality loss. Mostly, the joining process is facilitated by induction heating and the use of a flux to enhance the wetting of the filler alloy on the surface of the steel and cemented carbide in an ambient atmosphere. However, although the use of flux enables successful joining, it also generates voids within the joint, which reduces the strength of the connection while the chemicals within the flux are toxic and polluting. In this feasibility study, a fluxless brazing process is used to examine the joint between cemented carbides and steel for the first time. For this, ultrasound is applied during induction heating to enable the wetting between the liquid filler metal and the surfaces of the cemented carbide and steel. The ultrasound generates cavitations within the liquid filler metal, which remove the oxides from the surface. Several filler metals such as a silver based alloy Ag449, pure Zn, and an AlSi-alloy were used to reduce the brazing temperature and to lower the thermal residual stresses within the joint. As a result, every filler metal successfully wetted both materials and led to a dense connection. The ultrasound has to be applied carefully to prevent a damage of the cemented carbide. In this regard, it was observed that single grains of the cemented carbide broke out and remained in the joint. This positive result of brazing cemented carbides to steel without a flux but using ultrasound, allows future studies to focus on the shear strength of these joints as well as the behavior of the thermally induced residual stresses.

1. Introduction

Ultrasonically-assisted brazing enables the production of dense joints in an ambient atmosphere without the use of a flux. For this, a sonotrode initiates vibrations in the brazing gap. These high frequency vibrations have two positive effects on the wetting behavior of the materials by generating cavitation in the liquid filler metal. This cavitation break and remove oxide layers and additionally reduce the surface tension of the liquid filler metal. This allows the liquid metal to wet the surfaces. According to Crawford [1], flux-free soldering with the aid of ultrasound was already applied in Germany in 1936, but most recently this approach became a niche application used only by a few companies [2, 3]. In the past decades, several investigations were conducted to analyze the mechanisms of a fluxless soldering and brazing process. Table 1 provides a summary of published studies regarding fluxless brazing with the assistance of ultrasound. It is obvious that most lightweight materials such as Al, Ti, Mg, and its alloys as well as materials, which also resisted oxidations in an ambient joining process such as SiC or WC were analyzed. The brazing temperature for the process depends on the used filler alloy and ranges between 380 °C and 750 °C.

Table 1. Summary of published investigations regarding fluxless brazing with the assistance of ultrasound.

No.	Material	Filler alloy	Temperature (°C)	Year	Source
1	1. Al, 2. Copper	1. Zn-Al, 2. Pb-Sn	482	1976	[1]
2	Al-Si-Mg	-	Melting Point	1994	[4]
3	Al-MMC	Zn with SiC	420	2011	[5]
4	Ti_6Al_4V and Al_4Cu1Mg	Zn-based	383-403	2011	[6]
5	Al	$Zn_{86}Al_{14}$	410-470	2012	[7]
6	WC-Co and BeCu	AgCuZnCd	640-750	2012	[8]
7	Ti_6Al_4V	$Al_{88}Si1_2$	600	2013	[9]
8	Al an Cu	$Zn_{97}Al_3$	420	2013	[10]
9	Al an Cu	$Zn_{97}Al_3$	400-480	2014	[11]
10	SiC and Ti_6Al_4V	AlSnZnMg	620	2013	[12]
11	Mg	Zn-based	350	2013	[13]
12	Al_2O_3	$AlCu_{4.5}Mg_{1.5}$	700	2014	[14]
13	Al and Ti	AlSiZnCuNi+Sn	517	2016	[15]
14	SiC	$Al_{88}Si_{12}$	620	2015	[16]
15	$TiAl_6V_4$	$Al_{97.3}Mg_{2.5}Cr_{0.3}$	660-680	2015	[17]

The publications mainly show the significant advantages of an ultrasonically assisted brazing process, since it produces an almost void-free joint even with materials that are difficult to wet, such as SiC, WC, or sapphire. But none of the performed studies deals with the joining process between cemented carbides and steel. A successful application of this ultrasonically assisted technology on brazing cemented carbides to steel would significantly enhance the quality of the joint as a dense structure without any voids would provide a much better mechanical fatigue behavior. Cemented carbides are further known for their complex wetting behavior, which could be improved by ultrasound. Additionally, the reduction of the use of toxic materials such as the flux would be an advantage for many producing companies.

One challenge when developing a steady bonding process is the oxidation behavior of the steel surface during heating. On the one hand, the cavitation will remove the oxides, but on the other hand too many oxides will decrease the joint quality. The presented papers do not provide any experience concerning this circumstance.

2. Experimental

The cemented carbide of the grade K10 (DIN ISO 513:2014-05) and the tool steel UNS T11302 were selected as base materials due to their wide distribution for industrial applications. The properties of the cemented carbide are listed in table 2. The cemented carbide consists of tungsten carbide with 5.6 m% of cobalt as binder, and has a hardness of at least 1730 HV30. The thermal expansion coefficient in a temperature range between 20 and 400 °C is $4.9*10^{-6}K^{-1}$ due to the low proportion of the binder metal. The composition of the tool steel is listed in table 3. This steel grade is an alloy with tungsten, molybdenum, chromium, and vanadium. These alloying elements function as carbide formers and reduce the grain size to enhance the wear resistance of this material. The thermal expansion coefficient of the steel is $12.5*10^{-6}K^{-1}$ for the same temperature range as measured for the cemented carbide. Due to the difference in the expansion coefficients, residual stresses will occur after the solidification of the liquid filler alloys and affect the strength of the formed joint, especially during the cooling stage.

Table 2. Properties of the cemented carbide K10 DIN ISO 513:2014-05 [18].

Grade	Carbide	Grain size (μm)	Binder	Proportion of the binder (m%)	Density (g/cm³)	Hardness (HV)
K10	WC	0.8-1.3	Co	5.6	14.95	1730

In the past, filler alloys, which contain a certain amount of cadmium, were used to reduce the brazing temperature and to ensure a strong joint. Since the restriction of using cadmium, higher temperatures are needed to produce sufficient joints with other filler metals, but the process became more complex by handling the residual stresses. Using the ultrasonically assisted brazing process, the lower melting point of other filler alloys could reduce the residual stresses and, additionally, might increase the strength of the joint by providing a larger bonding area due to the reduction of voids.

Table 3. Composition of the tool steel UNS T11302 in m% [19].

Steel	C	W	Mo	Cr	V	Mn	Si
T11302	0.9	6.4	5.0	4.3	1.9	≤0.4	≤0.45

For this investigation, three different filler alloys were selected to analyze their behavior in an ultrasonically assisted brazing process joining the cemented carbides to the tool steel. The chosen filler alloys are listed in table 4. A pure zinc foil was chosen as first filler metal due to its low melting temperature of 420 °C. The second filler was a eutectic aluminum and silicon alloy with a melting temperature of 577 °C. The last filler, a silver based alloy named Ag 449 according to DIN EN ISO 17672 has a melting range from 680 to 705 °C and is commonly used in industrial applications of these joints.

Table 4. Used filler alloys and melting range.

Filler alloy	Melting range (°C)	Application
Zn	420	250 μm - foil
$Al_{88}Si_{12}$	577	100 μm - foil
Ag 449	680-705	100 μm - foil

The experiments were carried out in two different setups to analyze the characteristics of each process. The cemented carbides had a size of 10x10x5 mm and the steel of 10x10x3 mm. The specimens were polished and stacked together. The foils of the filler alloys were cut to fit the brazing gap.

The different brazing setups are shown in figure 1. An induction coil provided the heat for the process. In the first setup A the coil was placed below the sample holder and the heat was transferred through the holder to the specimen. The specimen was only fixed with the holder without any additional force to prevent a movement during the applied ultrasound. This setup was successfully tested in different studies for ultrasonically assisted brazing processes [7, 11, 14] and the reason, why it was used for the first brazing experiments of this investigation. A thermocouple measured the temperature close to the brazing gap. The sonotrode was placed on the specimen when the specimen reached the melting temperature of the filler alloy and the ultrasound was activated for 20 s. The heating device was switched off 5 s before the ultrasound, so that the vibrations were still active during the cooling process. The sonotrode was set to a frequency of 20 kHz and an amplitude of 6 μm. In the second setup B, the induction coil was placed directly around the specimen to ensure a faster heating. Screws were used to fix the specimen. The sonotrode was placed on the upper plate to induce the ultrasonic vibrations. The steel plate was placed on the top to induce the ultrasound into the joint and to protect the brittle cemented carbides from vibrations.

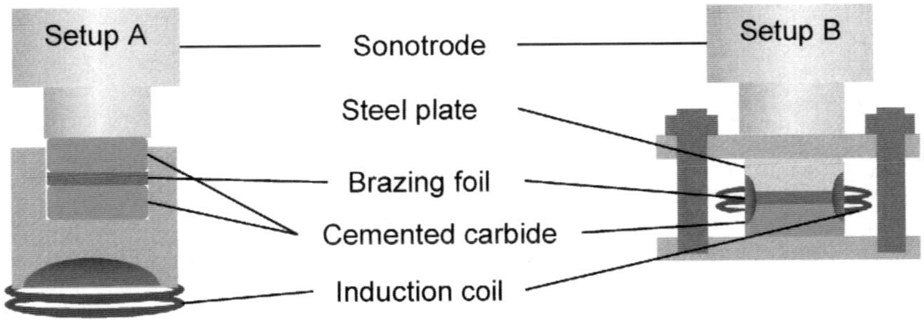

Figure 1. Experimental setups of the ultrasonically assisted brazing process.

3. Results and discussion

In figure 2, the temperature measurement of the thermocouple is visualized for both brazing setups. The heat needs at least 420 s to heat the specimen to 700 °C in setup A as the specimen holder has to be heated up as well. Due to the long heating time, many oxides were formed on the surface of the steel and exhibit a proper wetting of the filler metal, even with the assistance of the ultrasound. In this regard, the heating speed was significantly increased by the brazing setup B so that the temperature of 700 °C was reached in 7 s.

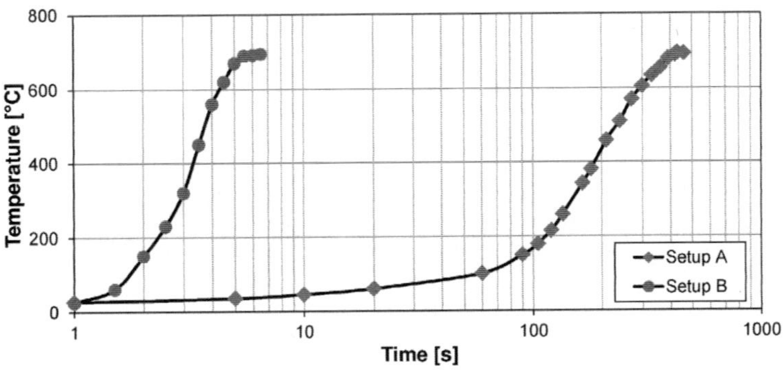

Figure 2. Trend of the heating stage of both brazing setups.

The first setup A was used to form a brazed joint between WC-Co and WC-Co with the selected filler alloys to study the effect of the ultrasound on this material, which is usually difficult to wet with liquid fillers. Anyway, it was not possible to form a dense joint between steel and a cemented carbide, because the heating was to slow and too many oxides were formed on the surface of the steel. Figure 3 shows the joint that was produced with a zinc filler metal and a process temperature of 450 °C. The ultrasonic vibration was activated until the zinc solidified. The cross-section shows the dense structure of the joint. Both surfaces of the cemented carbides were well linked to the solidified filler metal. The width of the joint decreased to 145 µm. Within the joint, a lamellar structure is noticeable. This structure resembles a freeze-frame of the vibrations within the joint, respectively, the interferences of the ultrasonic waves in the liquid metal. Here, particles of the cemented carbide and of ZnO have accumulated on the dead centers of the waves to form this unique structure.

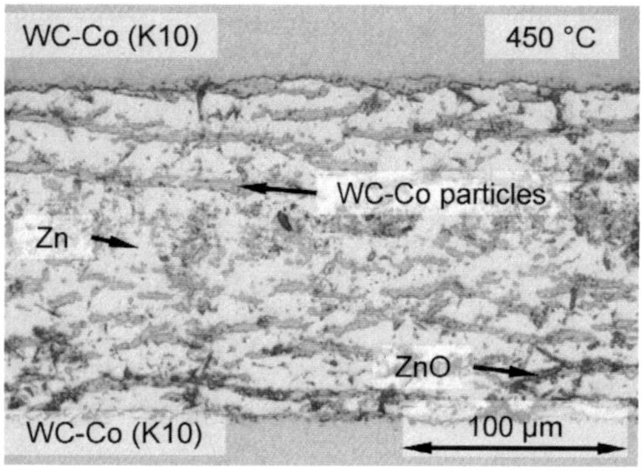

Figure 3. Cross-section of a brazed joint between two cemented carbides, consisting of WC-Co, which was formed with a Zn filler metal, a process temperature of 450 °C, and a noticeable lamellar structure.

By switching to setup B, it was also possible to braze the cemented carbide to steel by using different filler metals. In this regard, figure 4 shows the cross-section of the brazed joint, which was produced by using the Zn filler metal at a process temperature of 450 °C. This process formed a dense structure without any voids. Both the steel and the cemented carbide were connected to the filler metal. At the interface of the cemented carbide, single broken WC grains can be identified within the brazements. The joint gap has a width of 250 µm. The shadows at both interfaces are related to the preparation as the filler metal is much softer than the cemented carbide and the steel, so that a hollow was formed in the joint.

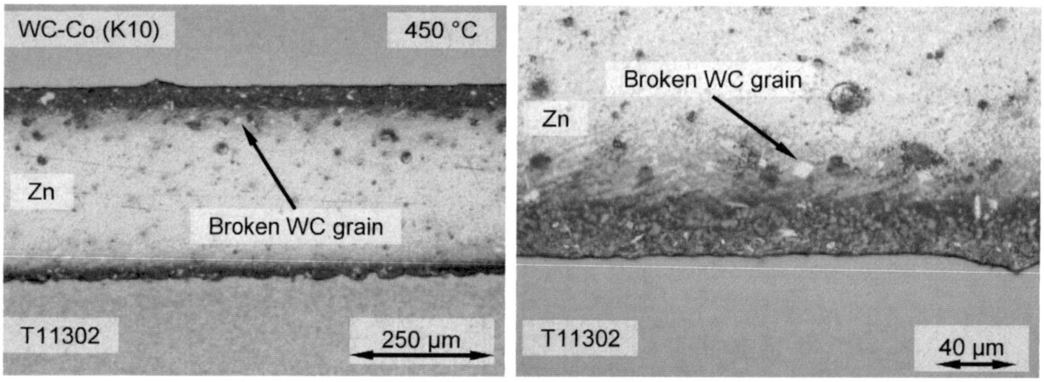

Figure 4. Cross-section of a brazed joint between a cemented carbide and T11302 steel, which was formed with a Zn filler metal and a process temperature of 450 °C.

The cross-section of the joint with the $Al_{88}Si_{12}$ filler alloy and an increased process temperature of 600 °C in figure 5 reveals a dense almost eutectic structure without any voids as well. Some solid solutions of Al and Si, which have a different composition than the eutectic of Al and Si, can be recognized in the filler. Actually, no dense intermetallic layer of iron aluminide was formed at the T11302 interface. This was probably because the brazing process was fast enough or the cavitation helped to prevent a forming of these brittle phases, which decreased the strength. In this brazement, the brazing gap reached a width of 54 µm. Additionally, some WC-Co particles can be found in the joint, which certainly broke from the surface of the cemented carbide due to the ultrasonic vibrations.

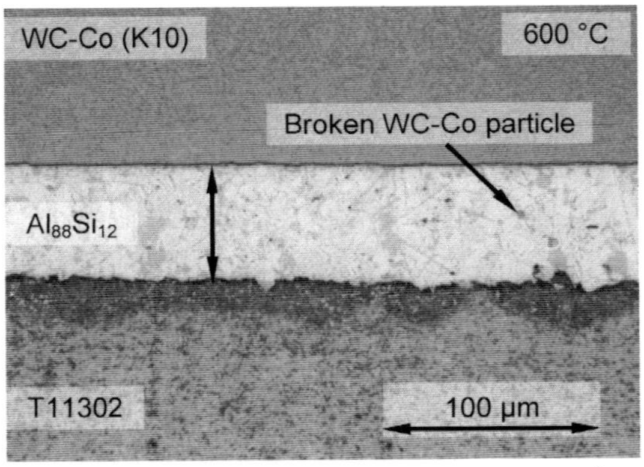

Figure 5. Cross-section of the brazed joint between a cemented carbide, consisting of WC-Co and T11302 steel, which was formed with $Al_{88}Si_{12}$ filler alloy and a process temperature of 600 °C.

Ag 449 formed a tight joint at a brazing temperature of 850 °C, figure 6. In this case, the brazing gap reached a width of 33 µm. It is obvious that no single WC grains but large WC-Co particles were broken out of the surface of the cemented carbide and stuck in the joint. In fact, the used cemented carbide is very sensitive to ultrasonic vibrations and cracks are easily induced during the process due to the brittleness of the cemented carbide.

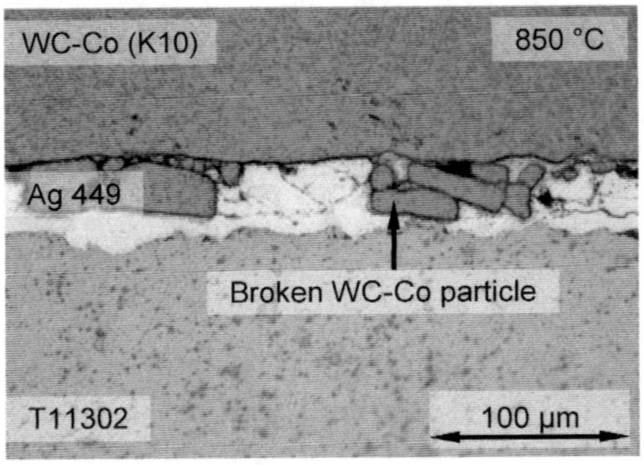

Figure 6. Cross-section of the brazed joint between a cemented carbide, consisting of WC-Co and T11302 steel, which was formed with an Ag 449 filler alloy and a process temperature of 850 °C.

4. Conclusion and Outlook

This feasibility study analyzed the ultrasonically assisted brazing process to join cemented carbides to steel. Starting from a brazing setup that was used in former studies for ultrasonically assisted brazing processes, the heating was too slow to ensure a fast brazing process and the oxidation of the steel surface prevented a tight joint due to the long heating time. The oxidization does not affect the cemented carbide during the heat treatment and the cavitation of the induced ultrasonic vibrations enabled a sufficient wetting of the liquid filler metal instead of using a hazardous flux. By improving the brazing setup to accelerate the heating, the oxidation of the steel was minimized, and a successful brazing to cemented carbide was realized with different filler alloys. The different filler enable an improvement of the residual stresses within the joint due to the lower brazing temperatures compared to commonly used silver based alloys. In addition, a rupture of some WC grains or bigger cemented carbide particles was observed within the joint.

In this regard, it is necessary to study if the vibrations affect the strength of the cemented carbides by micro cracking or if the particles are just released from the surface of the cemented carbides. In this case, the vibration parameters have to be improved to enable a good wetting of the filler alloys on both surfaces and to minimize the weakening of the cemented carbide. In addition, shear tests will be performed to measure the strength of the joint and to investigate if an improved performance can be obtained utilizing this new process handling when compared to the use of fluxes.

Acknowledgments
Parts of the investigation were conducted at the State Key Laboratory of Advanced Welding and Joining of the Harbin Institute of Technology. In this regard, we thank Professor Peng He and Associate Professor Zhiwu Xu for their support.

References

[1] Crawford A E 1955 *Ultrasonic Engineering* Academic Press (New York)
[2] Antonevich J N 1976 Fundamentals of Ultrasonic Soldering *J. Weld. Res. Suppl.* pp 200-207
[3] Rass I 2016 Ultrasonic soldering of metals, light metals, glasses and ceramics, *proceedings of the 11th International Conference Brazing, High Temp. Brazing and Diff. Welding*
[4] Eskin G I 1995 *Cavitation mechanism of ultrasonic melt degassing* Ultrasonics Sonochemistry
[5] Yan J, Xu Z, Shi L, Ma X and Yang S 2011 Ultrasonic assisted fabrication of particle reinforced bonds joining aluminum metal matrix composites *Materials and Design*
[6] Ma Z, Zhao W, Yan J and Li D 2011 Interfacial reaction of intermetallic compounds of ultrasonic-assisted brazed joints between dissimilar alloys of TiA6AlA4V and AlA4CuA1Mg *Ultrasonics Sonochemistry*
[7] Xiao Y, Ji H, Li M, Kim J and Kim H 2013 Microstructure and joint properties of ultrasonically brazed Al alloy joints using a Zn–Al hypereutectic filler metal *Materials and Design*
[8] Ji H, Li M, Lu Y and Wang C 2012 Mechanical properties and microstructures of hybrid ultrasonic resistance brazing of WC-Co/BeCu *Journal of Materials Processing Technology*
[9] Chen X, Yan J, Gao F, Wei J, Xu Z and Fan G 2013, Interaction behaviors at the interface between liquid Al–Si and solid Ti–6Al–4V in ultrasonic-assisted brazing in air *Ultrasonics Sonochemistry*
[10] Xiao Y, Ji H, Li M and Kim J 2013 Ultrasound-assisted brazing of Cu/Al dissimilar metals using a Zn–3Al filler metal *Materials and Design*
[11] Xiao Y, Ji H, Li M and Kim J 2014 Ultrasound-induced equiaxial flower-like CuZn5/Al composite microstructure formation in Al/Zn–Al/Cu joint *Mater. Sci. Eng A*
[12] Chen X, Yan J, Ren S, Wei J and Wang Q 2013 Ultrasonic-assisted brazing of SiC ceramic to Ti–6Al–4V alloy using a novel AlSnSiZnMg filler metal *Materials Letters*
[13] Wielage B and Weis S 2013 Corrosion behaviour of ultrasound-assisted resistance soldered magnesium joints *proceedings of the 11th International Conference Brazing, High Temp. Brazing and Diff. Welding*
[14] Cui W, Li S, Yan J, He J and Liu Y 2015 Ultrasonic-assisted brazing of sapphire with high strength Al–4.5Cu–1.5Mg alloy *Ceramic International*
[15] Chen X, Xie R, Lai Z, Liu L, Zou G and Yan J 2016 Ultrasonic-assisted brazing of Al–Ti dissimilar alloy by a filler metal with a large semi-solid temperature range *Materials and Design*
[16] Chen X, Xie R, Lai Z, Liu L, Yan J and Zou G 2016 Interfacial Structure and Formation Mechanism of Ultrasonic-assisted Brazed Joint of SiC Ceramics with Al—12Si Filler Metals in Air *Journal of Materials Science & Technology*
[17] Tillmann W, Zimpel M, Dias N F L, Pfeiffer J, Wojarski L and Xu Z 2015 Mechanical and microstructural analysis of ultrasonically assisted induction-brazed TiAl6V4 joints *Welding in the World*
[18] N.N. 2016 *datasheet of the cemented carbide K10* Hartmetall Gesellschaft
[19] N.N. 2016 *datasheet of the steel UNS T11302* Premium Stahl

IOP Conf. Series: Materials Science and Engineering **181** (2017) 012008 doi:10.1088/1757-899X/181/1/012008

Combustion synthesis of reactive nickel-aluminum particles as an innovative approach for thermal joining applications

S Schreiber[1,2], G D Theodossiadis[1] and M F Zaeh[1]

[1]Institute for Machine Tools and Industrial Management (*iwb*), Technical University of Munich, Boltzmannstrasse 15, 85748 Garching, Germany
[2]Author to whom any correspondence should be addressed

Abstract. Reactive systems, which are widely used in combustion synthesis, represent a promising solution for challenging joining tasks. They are able to undergo a self-sustaining, highly exothermic reaction when exposed to an external energy source. Reactive foils are the only systems that are currently commercially available. However, their industrial use is limited due to the brittle nature of the material and the restriction to planar geometries. Reactive particles represent a more flexible format, but are currently not commercially available. Therefore, a two-step electroless plating process has been developed to synthesize nickel-aluminum core-shell structures. These structures function as microreactors, which provide the energy for the thermal joining process. Ignition tests with electromagnetic waves were performed in order to investigate the overall reactivity. Energy input and particle size significantly influence the activation and the reaction behavior of the core-shell structures. Furthermore, a general approach to use reactive particles as a heat source in joining applications is proposed.

1. Introduction

Combining materials with dissimilar thermo-physical properties, like fiber-reinforced plastics, light metals or ceramics, within only one component is an effective means for functional and lightweight design. Both are becoming increasingly important in the automotive and aerospace industry as the reduction of mass, fuel consumption, and emissions have become core demands [1]. The manufacturing of hybrid joints is a major challenge in production engineering and requires efficient, adaptable, and innovative joining techniques. Reactive systems, which can function as a heat source in the thermal joining of distinct materials, represent a promising solution to meet these demands.

1.1 Characteristics of reactive systems

Reactive systems gained increased attention in the past few years by both, academia and industries [2; 3]. The unique ability to undergo a self-sustaining, highly exothermic reaction is successfully used in the field of combustion synthesis for producing advanced materials [4]. The exothermic reaction can be initiated by a locally confined energy input, e.g. an electrical impulse, mechanical impulse, or laser radiation [5; 6]. Moreover, the activation energy can also be provided by a volumetric energy input, e.g. uniform heating in an oven [7] or electromagnetic waves [8]. These two basic reaction modes are denoted as self-propagating high temperature synthesis (SHS) and thermal explosion (TE) or volume combustion synthesis (VCS) [9]. Generally, a reactive system is composed of at least two reactants in a defined stoichiometric ratio which determines the amount of the released energy [10]. Besides a great variety of diverse reactants, the combination of nickel and aluminum is of growing interest due to the beneficial characteristics of the reaction product [11; 12]. Nickel aluminides can be used as highly resistant coatings [13] or as lightweight material, since for example nickel aluminide (NiAl) features a low density combined with high strength as well as good oxidation and corrosion resistance [14]. These promising material properties gave rise to various sophisticated investigations in order to understand the mechanisms and kinetics of the combustion synthesis [15; 16]. The heat of reaction for the formation of NiAl, released within only a few milliseconds, results in a maximum temperature of 1810 °C [16] and motivates the use of the reactive system Ni + Al in thermal joining processes.

1.2 Joining with reactive systems

The research focus for joining with reactive systems is currently on reactive foils, also known as multilayer systems [17; 18]. Reactive foils consist of alternating layers in the range of nanometers of at least two metallic components. The thicknesses of the single layers and the overall thickness of the reactive foil significantly influence the reaction behavior [19]. Depending on the manufacturing process, which can be either cold rolling [20] or magnetron sputtering [21], the number of single layers ranges between a few up to thousand layers. The ignition of the reactive multilayer system via an electrical or mechanical impulse (SHS mode) leads to the formation of a reaction wave expanding circularly from the point of initiation with a maximum speed exceeding 10 m/s [22]. Consequently, maximum reaction temperatures are reached within and only for a few milliseconds and thus allow joining of thermosensitive materials.

In the context of thermal joining, the reactive multilayer system is placed between the two joining partners, an adequate joining pressure is applied and an external energy input is used for ignition. The released energy serves as a heat source, which melts the surface layers of low-melting materials. For high-melting materials, the joining partners are coated with an additional solder layer in order to form an adhesive bond [23]. Joining distinct materials, like different metals [24], metal and ceramic [25], or bonding on wafers [26] has been studied by many researchers. Reactive multilayer systems, consisting of nickel and aluminum, are commercially available under the trade name NanoFoil®. However, there are several disadvantageous characteristics of the multilayer systems for joining processes, such as the brittle nature of the material, the limitation of applications to planar geometries and the lack of automation of the joining process [27].

A promising approach to combine the beneficial reaction properties and overcome the restrictions of multilayer systems are reactive systems in a more flexible format, namely reactive particles. Heterogeneous, stoichiometric mixtures of distinct powders are commonly used for the combustion synthesis of advanced materials [3]. Yet, there are additional forms of reactive particles, which are under scientific investigation. The basic structure is similar to the multilayer system described above and consists of two metallic reactants which are able to undergo an exothermic reaction. The main difference is that reactive particles are reactive themselves as they combine both reactants within one particle. The morphology can be described as core-shell structure, as shown in figure 1a), or as lamellar composite, as shown in figure 1b). The former are synthesized via wet chemical processes [15; 28] whereas the latter are synthesized via ball milling [29] or by shredding cold rolled multilayer systems [30].

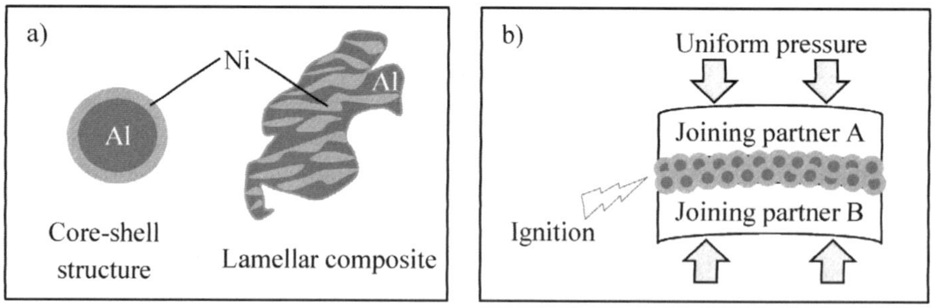

Figure 1. Schematic representation of a) reactive particles and b) thermal joining.

The exothermic reaction can be initiated similarly to multilayer systems. In contrast to multilayer systems or pressed pellets, loose particles are not in direct contact with each other. Therefore, the second ignition mode (TE mode) plays an important role in activating the reactive system. In order to guarantee the ignition, particles can be uniformly heated in an oven [31] or exposed to electromagnetic waves [32]. Electromagnetic waves are well absorbed by metals [33], such as nickel and aluminum. It has been shown that microwave radiation allows selective heating of reactive powder mixtures and has a positive effect on the reaction characteristics [8; 32].

The joining process for reactive particles is the same as with reactive multilayer systems. The ignition of the reactive particles between the desired joining partners results in an energy release which can be used to join via welding, bonding, or adhesion. Despite numerous studies on the reaction behavior of compressed powder mixtures [14; 31] and of lamellar composites [34], comparatively little attention is being paid to reactive core-shell structures and to joining via reactive particles.

Therefore, this research focusses on reactive core-shell structures and on joining via reactive particles in general. A two-step wet chemical process to synthesize reactive core-shell structures is studied. Furthermore, a general approach to combine the beneficial characteristics of microwave ignition in the thermal joining process with reactive particles is presented.

2. Experimental procedure

The synthesis of reactive nickel-aluminum core-shell structures requires the following chemicals: nickel(II) chloride (Sigma-Aldrich, 98 %), ammonium chloride (Sigma-Aldrich, 99.998 %), sodium citrate dihydrate (Sigma-Aldrich, ≥ 99 %), sodium fluoride (Fluka, ≥ 99 %), nickel(II) acetate tetrahydrate (Sigma-Aldrich, 98 %), glycolic acid (Sigma-Aldrich, 99 %), ethylenediaminetetraacetic acid tetrasodium salt hydrate (Sigma-Aldrich, ≥ 99.0 %), hydrazine hydrate solution (Fluka, 78-82 %) and sodium hydroxide (Honeywell, ≥ 98.0 %). All chemicals were used as received. Elemental aluminum powder (d_{50} = 23.5 µm, 99.7 % purity) was purchased from New Materials Development.

2.1 Wet chemical synthesis

Reactive nickel-aluminum core-shell structures were synthesized in a two-step electroless plating process. First, the oxide layer of the elemental aluminum powder was removed and substituted with a thin nickel layer. Second, the initial thin nickel layer was coated with additional nickel until a stoichiometric ratio of 1:1 was reached and reactive particles, which can be considered as microreactors, were obtained.

As mentioned before, the first step is a pre-treatment of the elemental aluminum powder, which is covered with an oxide layer as it is not handled under an inert gas atmosphere. Moreover, the pre-treatment serves to remove possible contaminations, activate the surface for subsequent nickel coating and prevent further oxidation [35]. The composition of the pre-treatment solution is shown in table 1.

Table 1. Composition of the pre-treatment solution for elemental aluminum powder [35].

Chemical	Formula	Pre-treatment solution
Nickel(II) chloride	$NiCl_2$	30 g/l
Ammonium chloride	NH_4Cl	7 g/l
Sodium citrate dihydrate	$C_6H_5Na_3O_7 \cdot 2H_2O$	20 g/l
Sodium fluoride	NaF	0.5 g/l

The purpose of the second step is to deposit more nickel onto the pre-coated particles. The advantage of electroless plating processes is that coatings can be deposited precisely and uniformly. Besides, the composition of the deposited material plays a significant role as it is supposed that impurities affect the reaction characteristics of the core-shell structures and hence influence the thermal joining process. Therefore, the selection of the reducing agent is of major importance. It has been shown that phosphorous (e.g. sodium hypophosphite) or boric (e.g. sodium borohydride or dimethylamine borane) reducing agents lead to significant co-deposition of phosphor and boron [36]. In this study, hydrazine is used as reducing agent since it allows nickel depositions with high purity. The composition of the plating solution is shown in table 2 and based on the findings of DINI & CORONADO [37].

Table 2. Composition of the plating solution for pre-treated aluminum powder [37].

Chemical	Formula	Plating solution
Nickel(II) acetate tetrahydrate	$C_2H_6NiO_4 \cdot 4H_2O$	40 g/l
Glycolic acid	$C_2H_4O_3$	60 g/l
EDTA tetrasodium salt	$C_{10}H_{12}N_2Na_4O_8 \cdot 4H_2O$	25 g/l
Hydrazine hydrate solution	$N_2H_4 \cdot H_2O$	100 ml/l

2.2 Ignition tests

To test the reactivity of the synthesized core-shell structures, ignition tests with electromagnetic waves were carried out. Based on the internationally defined ISM-frequencies (industrial, scientific, and medical applications), microwave heating was applied at a frequency of 2.45 GHz.

3. Results and discussion

In this section, the results of the wet chemical plating process are shown and discussed. In addition to that, a general procedure for joining with reactive particles in combination with microwaves is presented.

3.1 Synthesis and ignition of reactive core-shell-structures

The starting point of the two-step plating process is elemental aluminum powder with a mean particle diameter of 23.5 µm. In contrast to the material data sheet, particles do not have a spherical but an irregular shape as demonstrated in figure 2. Furthermore, particles significantly vary in size and feature small adhesions of elemental aluminum on the surface.

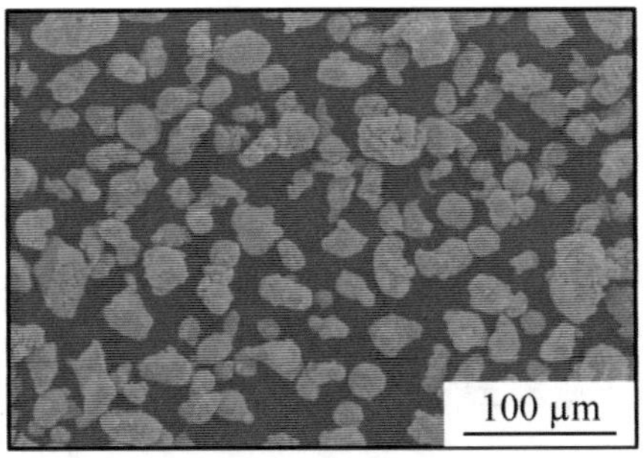

Figure 2. Scanning electron micrograph of elemental aluminum powder.

In order to obtain reactive core-shell structures, the pre-treated particles were plated with nickel until a stoichiometric ratio of nickel and aluminum of 1:1 was reached. The scanning electron micrograph in figure 3a) presents the nickel plated particles. Due to the broad particle size distribution of the elemental aluminum powder, the synthesized core-shell structures vary in size as well. Furthermore, the formation of agglomerates during the plating process is visible. The metallographic section in figure 3b) demonstrates that the deposited nickel layers are closed and uniform. The composition of the nickel layers was examined with the help of energy dispersive X-ray spectroscopy and does not feature any significant amounts of impurities due to the efficient pre-treatment and the beneficial characteristics of the chosen reducing agent. However, the broad particle size distribution results in core-shell structures with individual compositions even though the overall stoichiometric ratio of nickel and aluminum of the synthesized batch was adjusted via the process parameters to 1:1.

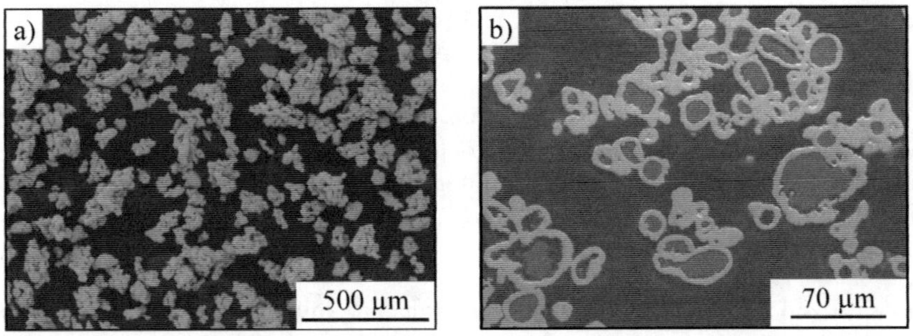

Figure 3. Scanning electron micrographs of a) reactive core-shell microparticles and b) a metallographic section.

The reactivity of the synthesized core-shell structures was tested with microwaves at a frequency of 2.45 GHz and an output power of 800 W. The reaction during microwave heating takes place after five to ten seconds and can be described as a short lighting of the particles. Yet, not all particles were ignitable which is probably due to the limitation of the output power of the experimental setup, as smaller particles were ignitable. Moreover, pressed pellets were also not ignitable even though the loose powder itself reacted in an exothermic way. Further studies to investigate the reaction characteristics of reactive core-shell structures have to be conducted. For this reason, a general approach for thermal joining with reactive particles, which are ignited with the help of microwave energy, is presented.

3.2 Thermal joining with reactive particles

As reactive particles can be used as a heat source in thermal joining of similar and dissimilar materials, further investigations on the ignition of core-shell structures via microwave heating are required. In order to have precise control over the energy input, a test rig as shown in figure 4 is set up.

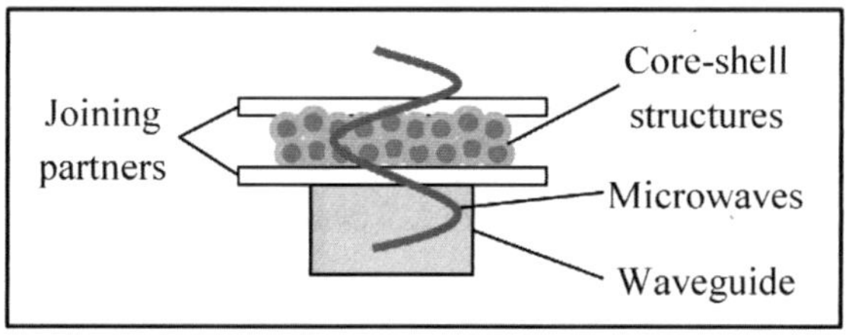

Figure 4. Schematic representation of the thermal joining process with reactive particles via microwaves (not drawn to scale).

Microwaves are guided directly into the joining zone and allow efficient heating of loose powder or pressed pellets. The electromagnetic waves are adjusted in such a way that exposure times can be reduced. Interesting challenges, that have to be addressed, are process monitoring during microwave heating, the cause-effect relationship between core-shell structures and microwaves and the application of reactive particles to freeform surfaces. Generally, joining with reactive core-shell structures can be realized via welding, soldering, or adhesion.

4. Conclusions

Thermal joining with reactive systems is a promising approach in the field of production engineering and lightweight design. As reactive particles represent a flexible format that can be adapted to the specific joining task, a two-step electroless plating process to synthesize reactive core-shell structures has been developed. Reactive nickel-aluminum core-shell structures with high purity have been obtained. The reactivity of the system was proven with the help of ignition via microwave energy. Further studies regarding the synthesis of the reactive core-shell structures are required in order to obtain defined particles. Furthermore, a general approach to ignite reactive particles via microwave energy, which allows for an efficient transfer of energy into the joining zone, has been shown.

References

[1] Koffler C 2014 Life cycle assessment of automotive lightweighting through polymers under US boundary conditions *Int. J. Life Cycle Asess.* **19** 538–545

[2] Merzhanov A G 1995 History and recent developments in SHS *Ceramics International* **21** 371–379

[3] Morsi K 2012 The diversity of combustion synthesis processing *J. Mat. Sci.* **47** 68–92

[4] Merzhanov A G 2004 The chemistry of self-propagating high-temperature synthesis *Journal of Materials Chemistry* **14** 1779

[5] Moore J J and Feng H J 1995 Combustion synthesis of advanced materials *Prog. Mat. Sci.* **39** 243–273

[6] Fritz G M, Spey S J, Grapes M D and Weihs T P 2013 Thresholds for igniting exothermic reactions in Al/Ni multilayers using pulses of electrical, mechanical, and thermal energy *J. Appl. Phys.* **113** 14901

[7] Thiers L, Mukasyan A S and Varma A 2002 Thermal explosion in Ni-Al system *Combust. Flame* **131** 198–209

[8] Rosa R, Veronesi P and Leonelli C 2013 A review on combustion synthesis intensification by means of microwave energy *Chem. Eng. Process.* **71** 2–18

[9] Varma A, Rogachev A S, Mukasyan A S and Hwang S 1998 Combustion Synthesis of Advanced Materials: Principles and Applications *Adv. Chem. Eng.* **24** 79–226
[10] Ansara I, Dupin N, Lukas H L and Sundman B 1997 Thermodynamic assessment of the Al+Ni system *Journal of Alloys and Compounds* **247** 20–30
[11] Naiborodenko Y, Lavrenchuk G and Filatov V 1982 Self-propagating high-temperature synthesis of aluminides *Sov. Powder Metall. Met. Ceram.* **21** 909–912
[12] Philpot K A, Munir Z A and Holt J B 1987 An investigation of the synthesis of nickel aluminides through gasless combustion *J. Mat. Sci.* **22** 159–169
[13] Rasmussen A J, Agüero A, Gutierrez M and José Landeira Østergård M 2008 Microstructures of thin and thick slurry aluminide coatings on Inconel 690 *Surface and Coatings Technology* **202** 1479–1485
[14] Pascal C, Marin-Ayral R and Tédenac J 2002 Joining of nickel monoaluminide to a superalloy substrate by high pressure self-propagating high-temperature synthesis *Journal of Alloys and Compounds* **337** 221–225
[15] Andrzejak T, Shafirovich E and Varma A 2007 Ignition mechanism of nickel-coated aluminum particles *Combust. Flame* **150** 60–70
[16] Zhu P, Li J and Liu C 2002 Reaction mechanism of combustion synthesis of NiAl *Mater. Sci. Eng.* A **329-331** 57–68
[17] Politano O, Baras F, Mukasyan A S, Vadchenko S G and Rogachev A S 2013 Microstructure development during NiAl intermetallic synthesis in reactive Ni–Al nanolayers *Surface and Coatings Technology* **215** 485–492
[18] Gunduz I E, Fadenberger K, Kokonou M, Rebholz C and Doumanidis C C 2008 Investigations on the self propagating reactions of nickel and aluminum multilayered foils *Appl. Phys. Lett.* **93** 134101
[19] Knepper R, Snyder M R, Fritz G, Fisher K, Knio O M and Weihs, T P 2009 Effect of varying bilayer spacing distribution on reaction heat and velocity in reactive Al/Ni multilayers *J. Appl. Phys.* **105** 83504
[20] Battezzati L, Pappalepore P, Durbiano F and Gallino I 1999 Solid state reactions in Al/Ni alternate foils induced by cold rolling and annealing *Acta Mater.* **47** 1901–1914
[21] Barbee T W and Weihs T P 1994 *Method for fabricating an ignitable heterogeneous stratified metal structure* US Patent 5,547,715
[22] Alawieh L, Knio O M and Weihs T P 2011 Effect of thermal properties on self-propagating fronts in reactive nanolaminates *J. Appl. Phys.* **110** 13509
[23] Duckham A, Levin J and Weihs T P 2006 Soldering and brazing metals to ceramics at room temperature using a novel nanotechnology *Advances in Science and Technology* **45** 1578–1587
[24] Wang J, Besnoin E, Knio O M and Weihs T P 2004 Investigating the effect of applied pressure on reactive multilayer foil joining *Acta Mater.* **52** 5265–5274
[25] Grieseler R, Welker T, Mueller J and Schaaf P 2012 Bonding of low temperature co-fired ceramics to copper and to ceramic blocks by reactive aluminum/nickel multilayers *Phys. Status Solidi* **209** 512–518
[26] Braeuer J, Besser J, Wiemer M and Gessner T 2012 A novel technique for MEMS packaging *Sens. Actuators A-Phys.* **188** 212–219
[27] Boettge B, Braeuer J, Wiemer M, Petzold M, Bagdahn J and Gessner T 2010 Fabrication and characterization of reactive nanoscale multilayer systems for low-temperature bonding in microsystem technology *J. Micromech. Microeng.* **20** 64018
[28] Mukasyan A S, White J, Kovalev D Y, Kochetov N A, Ponomarev V I and Son S F 2010 Dynamics of phase transformation during thermal explosion in the Al–Ni system *Physica B* **405** 778–784
[29] Takacs L 2002 Self-sustaining reactions induced by ball milling *Prog. Mater. Sci.* **47** 355–414

[30] Stover A K, Krywopusk N M, Gibbins J D and Weihs T P 2014 Mechanical fabrication of reactive metal laminate powders *J. Mater. Sci.* **49** 5821–5830
[31] Wu W W, Gubarevich A V, Wada H and Odawara O 2011 Volume combustion synthesis of NiAl as applied to ceramics joining *International Journal of Self-Propagating High-Temperature Synthesis* **20** 94–99
[32] Rosa R, Colombini E, Veronesi P, Poli G and Leonelli C 2012 Microwave ignited combustion synthesis as a joining technique for dissimilar materials *J. Mater. Eng. Perform.* **21** 725–732
[33] Buchelnikov V D, Louzguine-Luzgin D V, Xie G, Li S, Yoshikawa N, Sato M, Anzulevich A P, Bychkov I V and Inoue A 2008 Heating of metallic powders by microwaves *J. Appl. Phys.* **104** 113505
[34] Kawanaka H, Taguchi M, Hirose S and Ogura S 2016 *Reactive powder, bonding material using reactive powder, bonded body bonded with bonding material and method for producing bonded body* US Patent 20160121395A1
[35] Ramaseshan R, Seshadri S and Nair N 2001 Electroless nickel-phosphorus coating on Ti and Al elemental powders *Scr. Mater.* **45** 183–189
[36] Mallory G O and Hajdu J B 1990 *Electroless plating* (Orlando, Fla) The Society p 4
[37] Dini J W and Coronado P R 1967 Thick nickel deposits of high purity by electroless methods *Plating* **54** 385–390

Interfacial microstructure and mechanical properties of brazed aluminum / stainless steel - joints

V Fedorov, M Elßner, T Uhlig and G Wagner

Institute of Materials Science and Engineering, Chemnitz University of Technology, 09125 Chemnitz, Germany

Abstract. Due to the demand of mass and cost reduction, joints based on dissimilar metals become more and more interesting. Especially there is a high interest for joints between stainless steel and aluminum, often necessary for example for automotive heat exchangers. Brazing offers the possibilities to manufacture several joints in one step at, in comparison to fusion welding, lower temperatures. In the recent work, aluminum / stainless steel - joints are produced by induction brazing using an AlSi10 filler and a non-corrosive flux. The mechanical properties are determined by tensile shear tests as well as fatigue tests at ambient and elevated temperatures. The microstructure of the brazed joints and the fracture surfaces of the tested samples are investigated by SEM.

1. Introduction

Due to the good mechanical properties and the high corrosion resistance of the stainless steel in the combination with the low specific weight and the excellent corrosion resistance of aluminum alloys, aluminum/stainless steel - joints are of special interest for the automotive industry. Hence, the significant difference of the melting ranges of aluminum and stainless steel leads to the demand of an exact temperature control during the brazing process. The main problem in joining these materials is the formation of brittle Fe-Al intermetallic compounds (IMC) at the interface [1]. This is highly dependent on central parameters, like temperature and duration, of the used process. To realize a high-quality joint between stainless steel and aluminum, several joining processes have been investigated [2]. In the last few years, the welding techniques like friction stir welding [3-5], arc weld-brazing [6-8] and laser weld-brazing [9-11] offer a great potential for aluminum alloy / steel joining. However, these processes are often limited to special part geometries and designs of the welded joints. In comparison to welding, brazing offers the possibilities to manufacture several high-quality joints with a complex geometry in one step at low temperatures. Especially, induction brazing allows a short brazing time and a local heat input. Consequently, good mechanical properties are achievable. The knowledge of the static properties and the fatigue behavior at application temperatures is necessary to determine the potential lifetimes of the dissimilar joints [12]. In this work, the mechanical properties of the aluminum / stainless steel brazed joints are characterized by tensile shear tests as well as fatigue tests at ambient and elevated temperatures.

2. Experimental procedures

Aluminum alloy (AA 3003) sheets with dimensions of 5 mm × 20 mm, 1.5 mm thick, and austenitic stainless steel (AISI 304) sheets with 40 mm × 20 mm and a thickness of 1.5 mm are used as base materials. The AlSi10 filler (AA 4045) is applied as a paste. The thickness of the produced brazed

IOP Conf. Series: Materials Science and Engineering **181** (2017) 012009 doi:10.1088/1757-899X/181/1/012009

joint is adjusted at 100 µm. Single lap shear samples with an overlap length of 5 mm are used for mechanical testing. The investigated sample geometry is presented in figure 1.

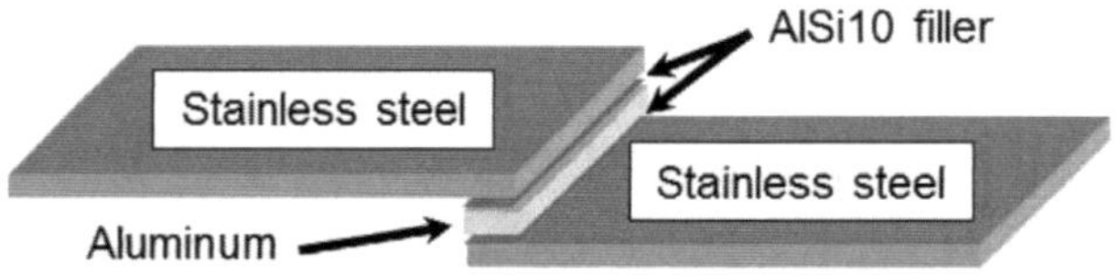

Figure 1. Investigated sample geometry.

The aluminum / stainless steel - joints are produced by induction brazing using a non-corrosive flux in an argon atmosphere at 600 °C. The temperature is measured by a two-channel pyrometer. The brazing process including cooling time takes about 2 min. The mechanical properties are determined by monotonic tensile tests as well as fatigue tests at 20, 100 and 200 °C. The tensile shear tests are conducted in a material testing machine Zwick Allround-Line 20 kN. The fatigue tests are carried out on a RUMUL resonance pulsator under load controlled condition with a load ratio of R = 0.1 at a frequency of 55 Hz. The microstructure of the brazed joints and the fracture surface of the tested samples are investigated by SEM.

3. Results and discussion

3.1 Microstructure

Figure 2 shows the microstructure of a produced aluminum / stainless steel - joint. According to previous work [13], the resulting braze metal consists of a primary Al solid solution and an Al-Si eutectic. Due to diffusion and reaction of Fe, Al and Si, an Al-Fe-Si layer is formed at the interface to the stainless steel. The EDX analyses indicate that the composition of the Al-Fe-Si layer matches the Al_7Fe_2Si phase, figure 3. Literature data about the Al-Fe-Si system also confirm this assumption [14, 15]. During the short-time induction brazing, no rapid growth of the IMC layer occurs. It can be seen that the Al-Fe-Si layer is about 1 µm thick. Additionally, Al-Fe-Si precipitates (light gray particles) in the braze metal are identified, figure 3. They are formed due to the diffusion of Fe into the braze metal. The EDX analyses show that the precipitates are most probably Al_7Fe_2Si phases.

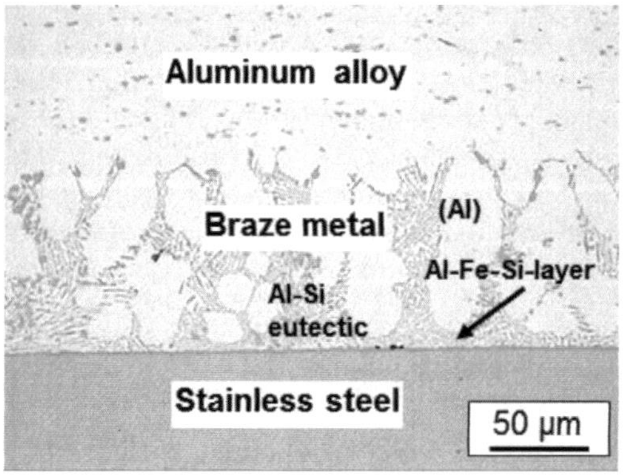

Figure 2. Microstructure of the aluminum / stainless steel brazed joint (OM).

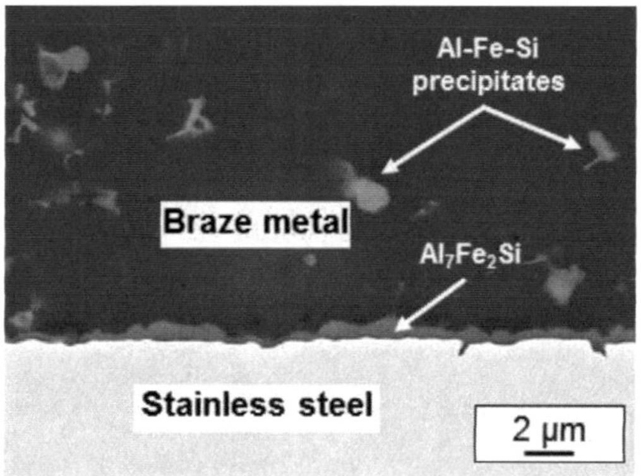

Figure 3. Interface to stainless steel (SEM, BSD).

3.2 Tensile shear test

To evaluate the mechanical properties of the joints, tensile shear tests are carried out at ambient (20 °C) and elevated temperatures (100 and 200 °C). Three single lap shear samples per testing temperature are produced with an overlap of 5 mm. In figure 4 the results of the tensile shear tests of the brazed samples are presented. At ambient temperature, a tensile shear strength of 53 MPa is determined. The values are significantly higher than the results of Roulin et al. (21 MPa) [16]. Roulin et al. produced aluminum / stainless steel joints in a furnace at a temperature of 600 °C with a holding time of 10 min. The low strength values can be explained by the longer holding time compared to induction brazing. The interfacial zone of the brazed joints has a complex structure, it consists of two intermetallic layers: $FeSiAl_5$ and $FeAl_3$. The formation of the second, more brittle intermetallic layer is a main cause for the mechanical degradation of the joints. The solution of this problem is a short holding time to avoid the formation of the intermetallic $FeAl_3$ layer. With an increase of testing temperature, a strength decrease is registered. The joining strength is 52 MPa at 100 °C and 50 MPa at 200 °C. The decrease of the strength can be explained by dislocation movements like slip of screw dislocations and climb of edge dislocations. The diffusion, activated by the temperature, increases the effect of dislocation movement. This leads to a higher brittleness of the brazed joint.

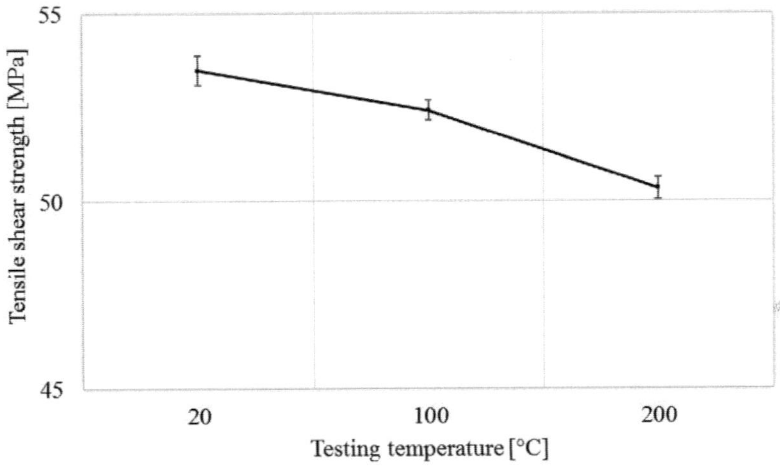

Figure 4. Tensile shear strength of brazed joints as a function of the testing temperature.

The fracture surfaces of the tensile shear tested samples are investigated using the top view and the cross sections. It can be seen that the brazed samples fail in the joint interface, especially in the Al_7Fe_2Si layer. The rough surface of the IMC indicates a fracture within the layer, figure 5. There are some residues of braze metal adhering at the IMC, figure 6. Due to the heat input during the brazing process, thermal induced stresses occur at the interface because of the high differences in hardness of the Al_7Fe_2Si layer (1020 HV0.0025), the braze metal (105 HV0.0025) and the stainless steel (245 HV0.0025). Consequently, this causes the crack initiation at the interface of the aluminum / stainless steel brazed joints.

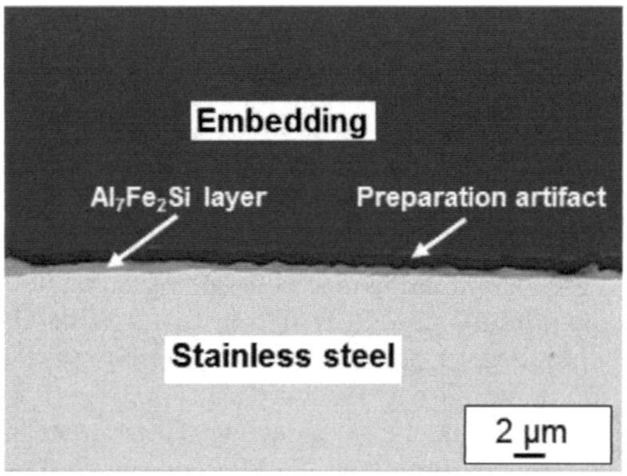

Figure 5. Cross section of the fracture surface of the tensile shear tested samples.

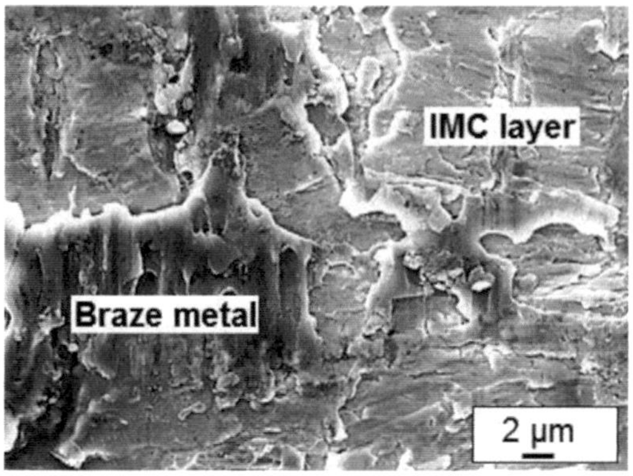

Figure 6. Fracture surface of the tensile shear tested samples.

3.3 Fatigue tests

The fatigue tests are carried out up to a fatigue endurance limit of 10^7 cycles. Based on the measured tensile shear strength of the brazed joints four stress levels (90, 70, 50, 30 %) are set. Three single lap shear samples per stress level are tested at ambient (20 °C) and elevated temperatures (100 and 200 °C). The S-N curves given in figure 7 represent the results of the fatigue tests of the brazed joints. The curves show three distinct fatigue ranges: low cycle fatigue (LCF), high cycle fatigue (HCF) and long life fatigue (LLF). It can be seen that the aluminum / stainless steel brazed joints, tested at ambient temperature, reach a fatigue life of 10^7 cycles at a stress amplitude of 7 MPa, whereas for elevated temperatures the fatigue endurance life of 10^7 cycles is not reached. Therefore, tests at a stress level of 20 % equivalent to 5 MPa are successful carried out. In addition to the operation of the basic fatigue mechanism, creep has an influence on the damage mechanism of the brazed joints. Due to the cyclic loading and the thermal activation, a slip of the grain boundaries and changes in the dislocation movement take place. In these places, an agglomeration of microstructural defects and pores occurs, which favors the progressive crack growth.

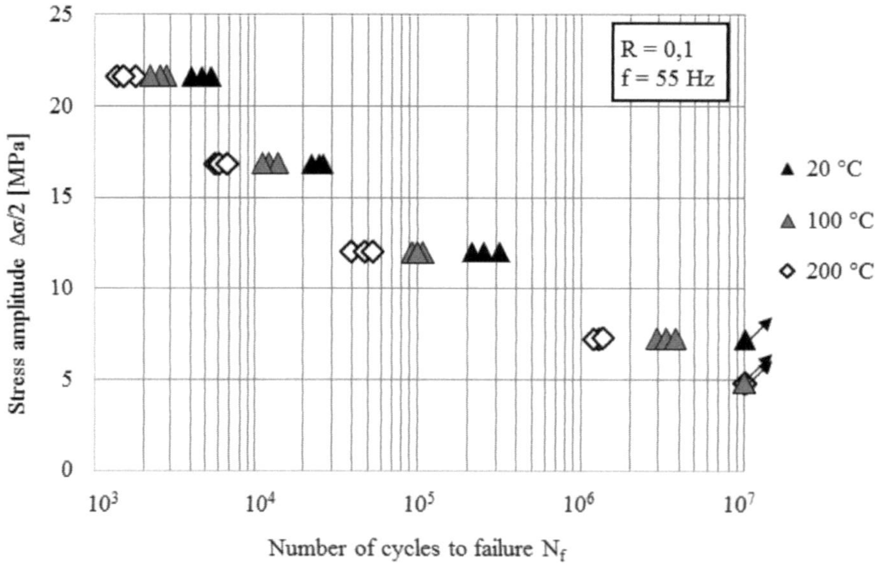

Figure 7. S-N data of the aluminum / stainless steel brazed joints.

In figures 8 and 9 a cross section and a top view of the fracture surfaces of the tested samples are presented. It is can be supposed that the cracks start in ductile braze metal and run through. After fatigue tests, the entire Al_7Fe_2Si layer as well as residues of the braze metal adhere at the stainless steel, figure 8. The amount of residues of braze metal at the IMC surface is higher than after monotonic tensile testing, figure 9. This indicates that the fracture occurs in the joint interface near the Al_7Fe_2Si layer, where the thermal induced stresses are concentrated. Consequently, the Al-Fe-Si intermetallic layer has an influence on the fatigue behavior of the aluminum / stainless steel brazed joints at high stress amplitudes as well as at a low stress amplitudes.

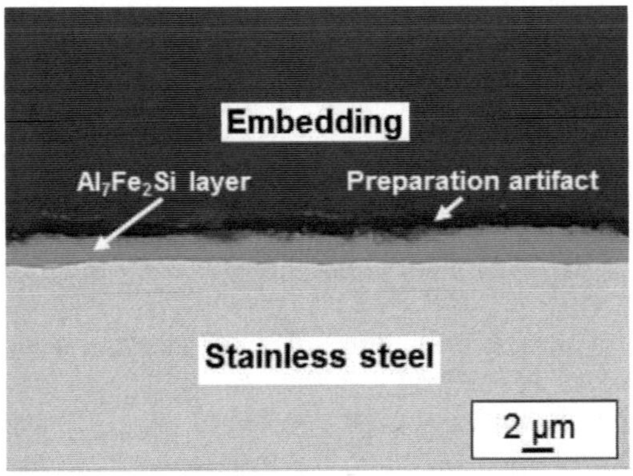

Figure 8. Cross section of the fracture surface of the fatigue tested samples.

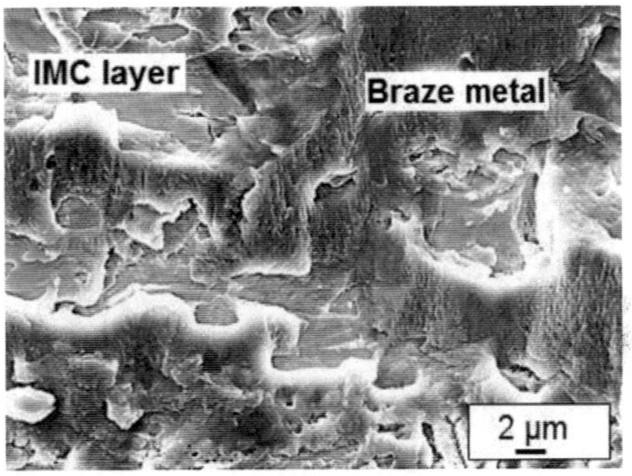

Figure 9. Fracture surface of the fatigue tested samples, S =°12°MPa, N =°2×10^5.

It can be summarized that the Al_7Fe_2Si layer influences the tensile shear strength and the fatigue behavior of the aluminum / stainless steel brazed joints predominantly. The degree of this influence must be determined by investigation of the initiation and the propagation of cracks depending on the number of cycles.

4. Conclusions

Aluminum / stainless steel - joints are produced by induction brazing using an AlSi10 filler. The brazed joints possess a tensile shear strength of 53 MPa at ambient temperature. With an increase of the testing temperature, a strength decrease is registered. The joining strength is 52 MPa at 100 °C and 50 MPa at 200 °C. Moreover, fatigue tests are carried out up to a fatigue endurance limit of 10^7 cycles. The aluminum / stainless steel brazed joints, tested at ambient temperature, reach a fatigue life of 10^7 cycles at a stress amplitude of 7 MPa. It is figured out that the fatigue life decreases with an increase of testing temperature. The aluminum / stainless steel – joints, tested at elevated temperatures, reach a fatigue life of 10^7 cycles at a stress amplitude of 5 MPa. The results of these investigations show that the Al_7Fe_2Si layer influences the tensile shear strength and the fatigue behavior of the aluminum / stainless steel brazed joints predominantly. In further investigations, the initiation and the propagation of cracks depending on the number of cycles will be observed.

Acknowledgments
The authors thank the Federal Ministry for Economic Affairs and Energy for the financial support of the presented work and the colleagues of MAHLE Behr GmbH & Co. KG for the cooperation.

References

[1] Agudo L 2007 Intermetallic FexAly-phases in a steel/Al-älloy fusion weld *Materials Science* **42** 4205–4214

[2] Martinsen K, Hu S J and Carlson B E 2015 Joining of dissimilar materials *CIRP Annals - Manufacturing Technology* **64** 679–699

[3] Uzun H and Donne C D 2005 Friction stir welding of dissimilar Al 6013-T4 to X5CrNi18-10 stainless steel *Materials and Design* **26** 41–46

[4] Watanabe T and Takayama H 2006 Joining of aluminum alloy to steel by friction stir welding *Journal of Materials Processing Technology* **178** 342–349

[5] Lee W B and Schmuecker M 2006 Interfacial reaction in steel–aluminum joints made by friction stir welding *Scripta Materialia* **55** 355–358

[6] Sierra G and Peyre P 2008 Galvanised steel to aluminium joining by laser and GTAW processes *Materials Characterization* **59** 1705–1715

[7] Dong H and Hu W 2013 Detachment of interfacial layers during arc-brazing of aluminum alloy to carbon steel with filler wire *Transactions of Nonferrous Metals. Society of China* **23** 1583-1588

[8] Yagati K and Bathe R 2014 Fluxless arc weld-brazing of aluminium alloy to steel *Materials Processing Technology* **23** 2949–2959

[9] Mathieu A and Shabadi R 2007 Dissimilar material joining using laser (aluminum to steel using zinc-based filler wire) *Optics & Laser Technology* **39** 652–661

[10] Lü J and Yang W 2014 Microstructure and mechanical properties of galvanized steel/AA6061 joints by laser fusion brazing welding *Acta Metallurgica Sinica* **27** 670-676

[11] Liu J and Jiang S 2015 Laser fusion–brazing of aluminum alloy to galvanized steel with pure Al filler powder *Optics & Laser Technology* **66** 1–8

[12] Kahl S, Ekström H E and Mendoza J 2014 Tensile, fatigue, and creep properties of aluminum heat exchanger tube alloys for temperatures from 293 K to 573 K (20° C to 300° C) *Metal lurgical and Materials Transactions* **A 45** 663-681

[13] Fedorov V, Weis S and Wagner G 2016 Mechanical and microstructural behavior of brazed aluminum/stainless steel mixed joints *IOP Conference Series: Materials Science and Engineering* **118**

[14] Gupta S P 2003 Intermetallic compound formation in Fe–Al–Si ternary system: Part I *Materials Characterization* **49** 269–291

[15] Du Y and Schuster J C 2008 A thermodynamic description of the Al-Fe-Si system over the whole composition and temperature ranges via a hybrid approach of CALPHAD and key experiments *Intermetallics* **16** 554–70

[16] Roulin M, Luster J W, Karadeniz G and Mortensen A 1999 Strength and structure of furnace-brazed joints between aluminum and stainless steel *Welding Journal* 151–55

Development of process data capturing, analysis and controlling for thermal spray techniques - SprayTracker

C Kelber[1], S Marke[2], U Trommler[3], C Rupprecht[4] and S Weis[1]

[1]Professur Füge- und Beschichtungstechnik, Institut für Produktionstechnik, Westsächsische Hochschule Zwickau, 08012 Zwickau, Germany
[2]IfU Diagnostic Systems GmbH, 09244 Lichtenau, Germany
[3]CBS Information Technologies AG, 09117 Chemnitz, Germany
[4]Fachgebiet Beschichtungstechnik, Institut für Werkzeugmaschinen und Fabrikbetrieb IWF, Technische Universität Berlin, 10587 Berlin, Germany

Abstract. Thermal spraying processes are becoming increasingly important in high-technology areas, such as automotive engineering and medical technology. The method offers the advantage of a local layer application with different materials and high deposition rates. Challenges in the application of thermal spraying result from the complex interaction of different influencing variables, which can be attributed to the properties of different materials, operating equipment supply, electrical parameters, flow mechanics, plasma physics and automation. In addition, spraying systems are subject to constant wear. Due to the process specification and the high demands on the produced coatings, innovative quality assurance tools are necessary. A central aspect, which has not yet been considered, is the data management in relation to the present measured variables, in particular the spraying system, the handling system, working safety devices and additional measuring sensors. Both the recording of all process-characterizing variables, their linking and evaluation as well as the use of the data for the active process control presuppose a novel, innovative control system (hardware and software) that was to be developed within the scope of the research project. In addition, new measurement methods and sensors are to be developed and qualified in order to improve the process reliability of thermal spraying.

1. Introduction

The basic idea of the SprayTracker® system is the provision of a manufacturer independent diagnostic, evaluation and control system, which increases process reliability during thermal spraying and opens up new possibilities for quality assurance. In order to work efficiently, an implementation of this system in the respective spraying system concept is necessary. For this purpose, a data communication of the SprayTracker System and the spraying system is realized via PROFIBUS (Process Field Bus). This communication module is switched between the PLC control and the control panel (stand-alone computer with touch screen) of the spraying system. As a result, all spray parameters and spraying data recorded by the spraying system can be read, recorded and archived as well as actively controlled by SprayTracker. Until now, the process is only monitored by optical methods or sound emissions but not controlled actively based on the recorded real-time data. The connection with the plant parameters must be established in order to ensure the active process control, and thus quality assurance [1, 2].

IOP Conf. Series: Materials Science and Engineering **181** (2017) 012010 doi:10.1088/1757-899X/181/1/012010

The synchronous measurement of data from various source of the entire coating process is very important. Different types of parameters can be distinguished:

- System parameters
- Process parameters
- Coating parameters
- Compliance with work safety regulations

The SprayTracker serves as the central detection unit. All sensors are connected to the central detection unit via different interfaces. A software module accepts the manufacturing facility parameters completely automatically, regardless of the type and number. Additionally it provides a time stamp after a prior synchronization with the spraying system.

In parallel, other sensors can also be connected to this software module (e.g. optical sensor and acoustic data acquisition). These two measuring systems operate discontinuously. The NIR (Near-Infra Red) sensor provides particle velocities and particle temperatures. These have the greatest significance in the spray distance. A particle dosimeter can be also connected wirelessly.

To supply the system with data, numerous sensors are required, which are wired or linked wirelessly with the SprayTracker. The following table shows the possible diagnostic procedures, which can be combined with the SprayTracker.

Table 1. Selection of typical diagnostic procedures for thermal spraying, which can be combined with the SprayTracker system [3].

	Measuring method	Measured values
Passive procedures	CCD camera	v_{Pa}, T_{Pa}, particle distribution
	IR detection / IR camera	T_{flame}, T_{part}, temperature distribution
	Pyrometry of individual particles (eg DPV 2000, NIR sensor)	v_{Pa}, T_{Pa}, particle number density, particle size
	Particle Flux Imaging (PFI)	Open jet geometry or fluctuations
	Emission spectroscopy	Gas temperatures above 9000 K, statements on particle evaporation
Active procedures	CCD camera + laser lighting (PIV-Particle Image Velocimetry)	v_{Pa}, particle distribution (density)
	Laser Doppler Anemometry (LDA)	v_{Pa}, particle concentration density or distribution
	Phase Doppler Anemometry (PDA)	v_{Pa}, particle concentration density or distribution, particle size
	Particle Shape-Imaging (PSI)	v_{Pa}, particle concentration density or distribution, particle size and shape

Most of the methods mentioned in Table 1 have in common, that a spatially resolved and therefore restricted measurement takes place. In this respect, the combination of location-resolved optical measurement methods with integrating measurement methods, such as acoustic analysis for example, is recommended, which maps the overall process using further specific parameters. These acoustic analyzes are suitable for quality monitoring of spraying processes. The frequency spectrum allows a statement about the spray parameters as well as the detection of burst signals, which can indicate crack propagation in the layer [1].

The individual diagnostic methods have already been adequately researched [1, 2, 4-6]. The target was the combination of the methods and the mapping of the measured values on a common time beam. This allows a real-time analysis of the on-line diagnostics.

2. Experimental

Set-up of the SprayTracker system

To be able to use the SprayTracker as a stand-alone device, the concept illustrated in Figure 1 was expanded and necessary system components were procured. A passively cooled computer system serves as a central controller. Solid-state and conventional 2.5 "hard disks" were integrated into the electromagnetic compatible (EMC) housing. An additional dust and acoustic protection has been designed.

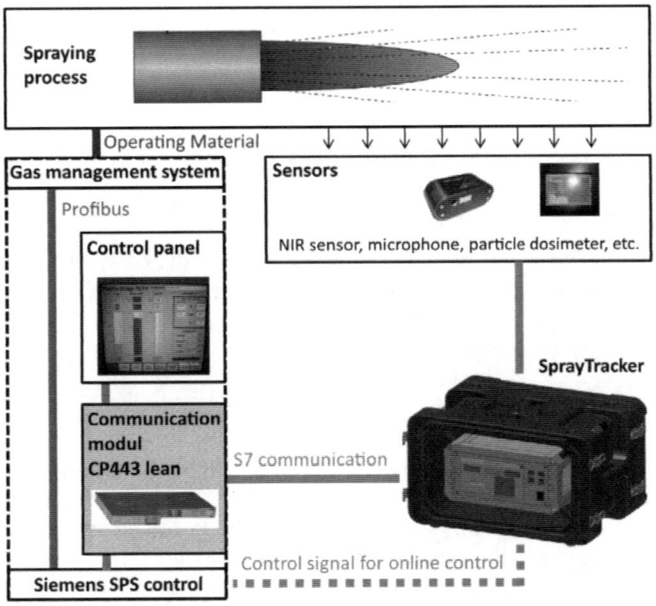

Figure 1. Scheme for the integration of SprayTracker into the spraying systems.

A touch screen is used for data control and manually input. The entire controller is fitted with a shock-resistant, screen-damped, air- and moisture-tight polymer-metal composite housing (Figure 2), whereby an additionally external monitor can be connected. Signal LEDs (light-emitting diode) indicate the operating state of the system. Figure 3 shows the user interface displayed on an external monitor while the system parameters are visible in the background.

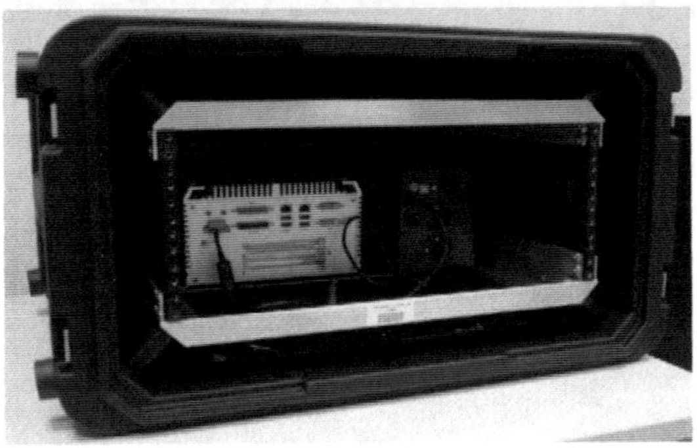

Figure 2. Housing with computer and uninterruptible power supply.

Figure 3. User interface of the SprayTracker next to the plant control.

There are three operating modes: user (pure data visualization), spraying specialist (with additional functions, such as limit setting for process control) and system administrator (manufacturer support, for example, for system expansion). No further user surfaces are accessible to the user in order to avoid unauthorized system malfunctions. There is another operating level for the expert and the supervisor.

Storage concept and data management system
For storing an SQL (Structured Query Language) database is used, which is located on an SQL server. This server is installed in the housing and can record data completely independently of the coating system. The software module communicates with all sensors and the process control. All data records have time information, which is synchronized with the process control. Different data sets (including different data formats) can be sorted and managed chronologically.The saving concept is designed to fulfil the following functions:

1. Complete data record during the coating process and archiving of data.
2. Alarm triggering in case of failure.
3. Visualization of current data as well as archive data.

In order to obtain useful information for process control or for the observation of the degree of the electrodes wear, the acoustic data must be matched with other system parameters. In addition, noise must be actively gated out by matching with background measurements. The SprayTracker system meets these requirements by using special quasi-digital and digital filter techniques (bandpass and notch filters).

3. Results

Numerous spray tests were carried out to illustrate the need of process monitoring and active process control. In plasma spraying tests, new and worn electrodes were used to compare measurement data of different wear conditions of the spraying system. Figure 4 shows the different wear conditions of the electrodes and their arrangement in the system. On the right hand side, the positioning of the spray gun in front of the sample clamping is observable.

Figure 4. New and worn electrodes, electrode setup in the plasma spray gun and spray gun setup.

By comparing the surface of the resulting cotings in Figure 5, hardly any differences in coating quality are visable, whereas different production conditions (new and worn electrodes) were used. This fact often leads to problems in quality assurance, because inferior coatings are interpreted as supposedly good coatings.

Figure 5. Steel samples with Al_2O_3 / TiO_2 coating, produced using new (left) and worn electrode (right) of a plasma spraying system.

Upon closer investigations on the micro structure of the coatings the effect of the new and worn electrodes on the layer quality is clearly visible, Figures 6 and 7. The coating deposition and the homogeneity of the layers decrease with increasing wear of the electrodes due to reduced deposition efficiency. This results in a reduced coating quality, appearing by clusters of pores, increased roughness and a significantly decreased hardness.

Figure 6. Al_2O_3 / TiO_2 layer sprayed with new electrodes with parameter set H8 Ar34 l/min (Optimized industrial parameter).

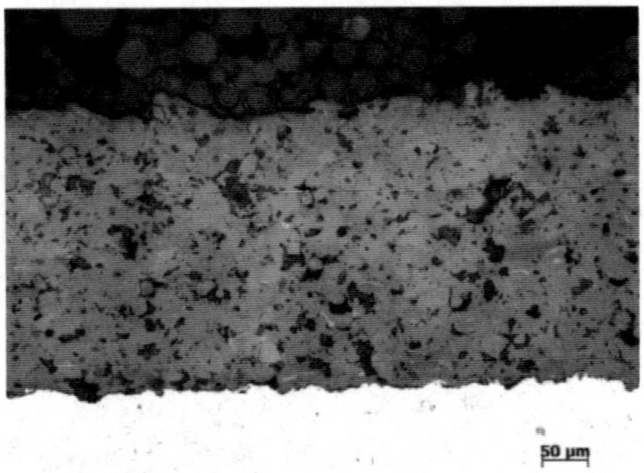

Figure 7. Al_2O_3 / TiO_2 layer sprayed with old electrodes with parameter set H8 Ar34 l/min.

Further tests show that small changes in the gas flow rate cause significant changes in the coating properties. With the reduction of the hydrogen flow rate, the layers become generally thinner, with the reduction of the argon flow rate, the layers become slightly thicker. This can be explained by the lower kinetic energy of the plasma jet, which is associated to an increasing particle presence time in the spray jet. This leads to a higher particle temperature and thus to an increased deposition rate. However, the layers become either more porous or the self-tension in the layer increases, which also promotes the formation of microcracks. To prove the fact, that the layer quality variations correlate with the process parameter fluctuations and the diagnostically ascertainable characteristic values, results of the NIR sensor and acoustic measurements during hydrogen flow rate variation are presented in Figure 8 and 9.

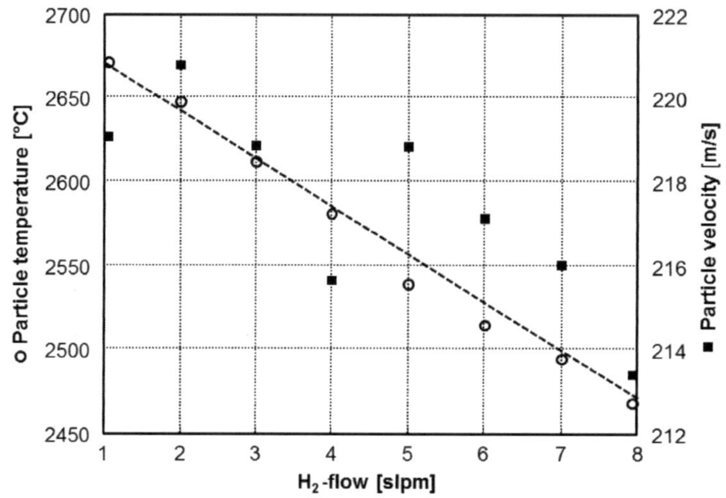

Figure 8. Particle temperture and particle velocity vs. hydrogen flow rate.

The variation of manufacturing facility parameters led to distinct differences in spray jet temperature, particle velocity and coating application (Figure 8). Both parameters were measured by the NIR sensor. This proved that the online process monitoring is basically suitable as a quality assurance tool. The user is thereby abled to recognize a changing coating quality in an early stage of the production process.

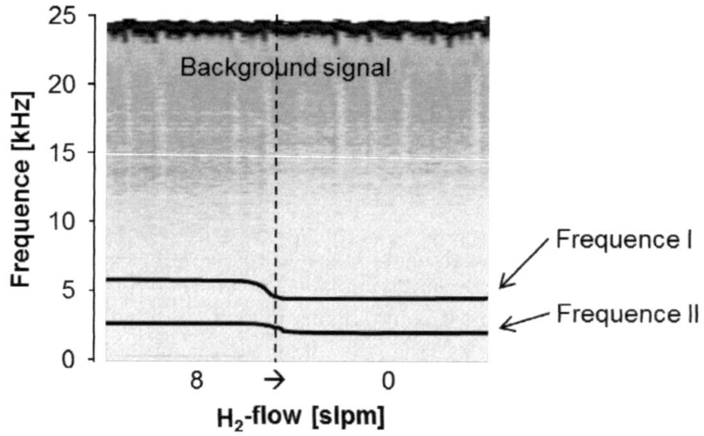

Figure 9. Shift of main frequency caused hydrogen flow rate reduction from 8 l/m to 0 l/min.

The process fluctuations were also visible in the acoustic spectrum. The characteristic main frequencies of the process and their shift to lower values with decreasing hydrogen flow rate are observable in Figure 9. Thus, the observation of these frequencies is suitable for monitoring the process gas flow.

4. Summary

The SprayTracker system, developed within the framework of a research project in cooperation with IfU, CBS and Westsächsische Hochschule Zwickau, is a completely new and innovative system that meets the need for new holistic process analysis and control systems by carrying out numerous measurement and control tasks simultaneously.

- Data recording of the real-time process parameters of the spraying system as well as comparison of the input variables to defined set points.
- Recording the position data and the kinetic parameters of the handling system (including the robot position and movement, the position and feed of the component manipulators, the linear feedforward and the rotational speed of the shaft rotary table).
- Workpiece monitoring (e.g. component temperature, deformation, layer thickness).
- Analysis of inflight particle parameters (particle velocity, temperature and particle distribution) as well as the spray jet profile (shape, divergence and angular deviation).
- Connection possibilities for user-specific sensors, particle dosimeters, gas detectors, etc..
- Visualization of monitored characteristics in customizable user profiles.
- Different operating modes or user rights to avoid incorrect operation.
- Complex database system with intelligent data filtering, redundant data storage and data transmission.
- User interface as an aid for the plant operator with limit visualization and alarm functions.
- Possibility of an autonomous plant control concept (initially with operating instructions, also as a closed-loop system in perspective).

Due to the above-mentioned application possibilities, the system is a sensible extension for existing plants as well as an attractive option for new plants. Through the SprayTracker system, spraying system manufacturers are able to offer their customers an innovative quality assurance tool that fulfills all diagnostic basic functions and can be customized.

Acknowledgments
The work carried out in this paper was supported by the Zentrale Innovationsprogramm Mittelstand (ZIM). We would like to thank all employees of the project and their associated companies.

References

[1] Wang G, Tillmann W and Zielke R 2011 Schallemissionsanalyse beim thermischen Spritzen *DGZfP-Jahrestagung (Bremen)*
[2] Zielke R, Tillmann W, Wang G and Abdulgader M 2012 Schallemissionsanalyse zur Prozessüberwachung beim thermischen Spritzen *DACH-Jahrestagung (Graz)*
[3] Rupprecht C 2013 Neue Methoden und Anwendungen der Thermischen Spritzens *Habilitation*
[4] Mauer G, Vaßen R, Stöver D, Kirner S, Marqués J-L, Zimmermann S, Forster G and Schein J 2010 Improving Powder Injection in Plasma Spraying by Optical Diagnostics of the Plasma and Particle Characterization *Journ. of Therm. Spray Technol.* **20**
[5] Zimmermann S 2007 Particle Shape Imaging (PSI) – eine innovative Methode der Partikeldiagnostik bei thermischen Beschichtungsverfahren *Dissertation*
[6] Wank A, Wielage B, Grund T, Rupprecht C, Zimmermann S and Landes K 2005 Characterization of the particle jet of different HVOF guns by means of LDA *ITSC (Basel)*

IOP Conf. Series: Materials Science and Engineering **181** (2017) 012011 doi:10.1088/1757-899X/181/1/012011

Robot based deposition of WC-Co HVOF coatings on HSS cutting tools as a substitution for solid cemented carbide cutting tools

W Tillmann[1], C Schaak[1], D Biermann[2], R Aßmuth[2] and S Goeke[2]

[1]TU Dortmund, Institute of Materials Engineering, Germany
[2]TU Dortmund, Institute of Machining Technology, Germany

E-mail: wolfgang.tillmann@udo.edu

Abstract. Cemented carbide (hard metal) cutting tools are the first choice to machine hard materials or to conduct high performance cutting processes. Main advantages of cemented carbide cutting tools are their high wear resistance (hardness) and good high temperature strength. In contrast, cemented carbide cutting tools are characterized by a low toughness and generate higher production costs, especially due to limited resources. Usually, cemented carbide cutting tools are produced by means of powder metallurgical processes. Compared to conventional manufacturing routes, these processes are more expensive and only a limited number of geometries can be realized. Furthermore, post-processing and preparing the cutting edges in order to achieve high performance tools is often required.

In the present paper, an alternative method to substitute solid cemented carbide cutting tools is presented. Cutting tools made of conventional high speed steels (HSS) were coated with thick WC-Co (88/12) layers by means of thermal spraying (HVOF). The challenge is to obtain a dense, homogenous, and near-net-shape coating on the flanks and the cutting edge. For this purpose, different coating strategies were realized using an industrial robot. The coating properties were subsequently investigated. After this initial step, the surfaces of the cutting tools were ground and selected cutting edges were prepared by means of wet abrasive jet machining to achieve a smooth and round micro shape. Machining tests were conducted with these coated, ground and prepared cutting tools. The occurring wear phenomena were analyzed and compared to conventional HSS cutting tools. Overall, the results of the experiments proved that the coating withstands mechanical stresses during machining. In the conducted experiments, the coated cutting tools showed less wear than conventional HSS cutting tools. With respect to the initial wear resistance, additional benefits can be obtained by preparing the cutting edge by means of wet abrasive jet machining.

1. Introduction

Cemented carbide, also called hard metal, is a metal matrix composite (MMC) based on a mixture of a hard material, commonly tungsten carbide (WC), and a metal e.g. cobalt (Co). In this case, cobalt is the metallic binder which imbeds WC particles. Cemented carbide components are produced by means of powder metallurgical processes. These hard cutting materials can be classified in conventional hard metal (HW), fine grain hard metal (HF), and coated hard metals (HC) [1]. Cemented carbide tools show a superior high temperature strength and low wear. In general, there are two different design types of these tools. The first tool type is fully made of cemented carbide (solid cemented carbide tools). The second type of tools consists of a steel tool body (e.g. high speed steel) that is improved by bonding cemented carbide onto the cutting edges [2]. The morphology of the cemented carbide component strongly influences the properties of the final cutting tool. Especially, fine carbide and grain sizes (< 1 μm) improve the wear resistance and the toughness of the cemented carbide. The low toughness is one big disadvantage of conventional cemented carbide components [2,3]. The second disadvantage are the higher production costs of solid cemented carbide tools compared to HSS tools. The reason for this are the higher raw material costs as well as the expensive processes of fine powder production and powder conditioning. Furthermore, the post-processing and preparation of the cutting edges is more extensive, because of the high hardness and strength of the material. One alternative to reduce the need of cemented carbide and to improve the toughness of the tool is the usage of cemented carbide inserts on steel tool bodies. Furthermore, the effort of machining or grinding the hard material is reduced. These cemented carbide inserts are bonded to the surface of the tool by means of brazing or diffusion bonding [4-6]. The main disadvantages of this multi-stage method are the complex processing (positioning, loading, etc.) and the thermal load during processing.

Based on the concept of WC-Co cemented carbide tools, a partial reinforcement with a cemented carbide could also be realized by means of thermal spraying. WC-Co and WC-CoCr deposited by means of thermal spraying are well established coating systems [7-9]. These coatings can be deposited by atmospheric plasma spraying (APS) or high velocity oxygen fuel (HVOF) spraying. Overall, HVOF shows the lowest porosity and the lowest wear rates resulting in the highest coating quality [10].

The benefits of tools coated by means of HVOF with WC-Co are a tough substrate combined with an extremely hard and wear resistant surface layer. The coating process is easy to operate and suitable for mass production. Furthermore, a worn out tool can be reconditioned by depositing a new WC-Co layer on it. Yet, near net shape thermally sprayed coatings are still a challenge to solve [11,12]. Especially, complex parts with radii, gaps, and edges are hard to coat with a homogenous coating thickness. Modern robot systems can help to solve these challenges. However, the spray strategy has to be optimized for this purpose and the robot kinetics have to be taken into account. The challenge of the planed application is to get a homogenous, thick, full covering sprayed WC-Co layer on a small HSS cutting tool. Considering the use of cutting tools, especially the cutting edge has to be coated very well. Otherwise, a fatal delamination would occur in this area because of the high cutting forces which act on the coating. Additionally, in order to reduce the post-processing effort and time, the coating should be near-net-shaped. A post-processing cannot be avoided, because of the high roughness of the as-sprayed coating. The cutting edges and flanks have to be ground and prepared prior to the usage. The post-processing generates the final shape of the cutting edge and thus of the tool.

In the present paper, different coating process strategies were tested to coat a HSS cutting tool. The results were analyzed and discussed. Furthermore, two different post-processing methods were investigated. To complete the experiments, the coated and post-processed cutting tools were used to cut a steel sample of AISI 1060 in an experimental rig.

2. Experimental Details

2.1 Materials and Coating Process

A tungsten carbide cobalt base powder was used (Type: Woka 3102. Supplier: Oerlikon Metco) as feedstock material. The powder is composed of 88WC-12Co (12% cobalt powder). The powder is agglomerated and sintered and the tungsten carbides have an average grain size of 2,5 μm. The range of the agglomerate size ranged between 15 μm and 45 μm (-45 +15).

For the coating experiments, a Woka Jet 400 system (supplier: Sulzer Metco) were used with a Sulzer Metco MultiCoat controller system and a Twin-120-H powder feeder. The spraying parameters are summarized in table 1. The spray gun was mounted on an ABB IRB 4600 60/2.05. The software ABB RobotStudio 5.13 was used for path programming.

Table 1. HVOF spraying parameters.

Powder type	Woka 3102 / Oerlikon
Powder size [µm]	15/45
Spray gun	WOKA Jet 400
Kerosine [l/h]	23
O2 [NLPM]	876
Carrier gas [NLPM]	4,6
Disc speed setting [%]	10
Feed rate [g/min]	-
Stirrer speed [%]	50
Set of distance [mm]	300
Surface Speed [m/min]	300

HS6-5-2C (1.3343), ASP2023 (1.3395) and X38CrMoV5-3 (1.2367) were used as substrate material (table 2). These steels are typical tool steels. A sketch of the used tool shape and their dimensions is shown in figure 1. For the planned application of WC-Co HVOF coatings as cutting tool coatings, it is important to get a covering coating at the cutting edge and the flanks. For this purpose, different coating strategies were investigated. For spray strategy I, the spray gun is leant down 30.5° and the spray gun path follows the cutting edge. This ensures that the spray jet directly targets at the cutting edge. The aim of this strategy is to obtain a good coating at the cutting edge. For spray strategy II, both flanks including the cutting edge were coated subsequently. The third spray strategy combines strategy I and strategy II. Figure 1 shows spray strategy III. The spray path covers the flanks and the cutting edge. In this case, the spray gun follows the contour of the tool's body with a distance of 300 mm. The run up distance was kept long, thus guaranteeing a constant speed. Additionally, a long turning distance was used between the overruns for an extended cooling time. For strategy II and III the spray angle was set to 90 °. Prior to the coating process, the samples were grit blasted with corundum (EKF 100, 4 bar). The full experimental plan is summarized in table 2.

Table 2. Experimental plan.

Sample number	Material	Spray strategy	Stand of distance	Number of overruns
P2	HS6-5-2C (1.3343)	Type I	300 mm	30
P4	HS6-5-2C (1.3343)	Type II	300 mm	20
P6	HS6-5-2C (1.3343)	Type III	300 mm	10
P7	ASP2023 (1.3344)	Type III	300 mm	1
P8	ASP2023 (1.3344)	Type III	300 mm	2
P9	ASP2023 (1.3344)	Type III	300 mm	4
P10	X38CrMoV5-3 (1.2367)	Type III	300 mm	1
P11	X38CrMoV5-3 (1.2367)	Type III	300 mm	2
P12	X38CrMoV5-3 (1.2367)	Type III	300 mm	4

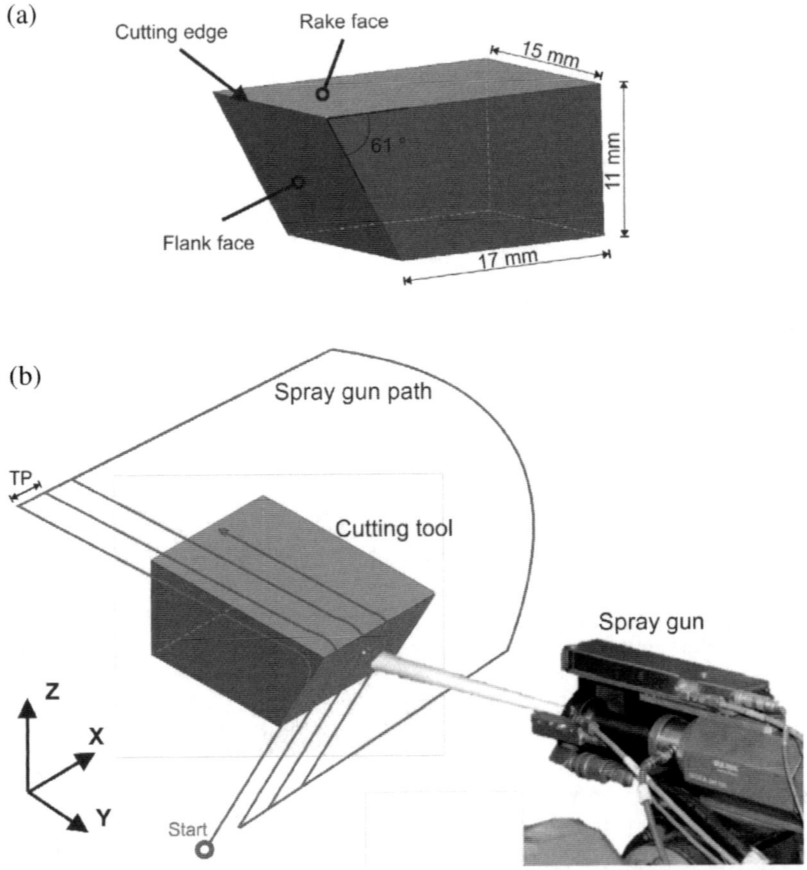

Figure 1. a) Sketch of the used cutting tool (substrate) with its dimensions and shape. b) Sketch of the spray strategy III. The red lines indicate the pathways of the spray gun.

2.2 Post-Processing and Analysis Methods

After the coating processes, several samples were used to investigate the coating morphology and to measure the coating thickness. For this purpose, metallurgical cross-sections were produced and analyzed with a light microscope (LIMI) and scanning electron microscope (SEM). A few samples were post-treated after the coating process. The cutting edge of these samples were ground or ground and prepared. Grinding the coated samples results in sharp cutting edges. However, to remove minor defects along the cutting edge and to achieve a favorable cutting edge micro-shape in terms of a rounding, some edges were prepared by means of wet abrasive jet machining. The preparatory work was conducted with corundum ZWSK F220 in water, pressurized by air at $p_{jet} = 2.5 \ldots 6$ bar. The results of the post-treatment were investigated by means of SEM. Finally, all samples with different conditions (without post-treatment, ground, ground + wet abrasive jet machined) were tested in a cutting test (see figure 2). The fundamental chip formation machine at the Institute of Machining Technology allows detailed investigations of cutting processes with cutting speeds up to $v_c = 180$ m/min. In this regard, bars of AISI 1060 with a length of $l = 400$ mm were machined. The uncut chip thickness was set at $h = 0.05$ mm and the cutting speed was $v_c = 30$ m/min. Five cuts were made for each tool. The cutting process was observed with a high speed camera and the cutting forces were recorded. The cutting test parameters were kept constant for all tested tools.

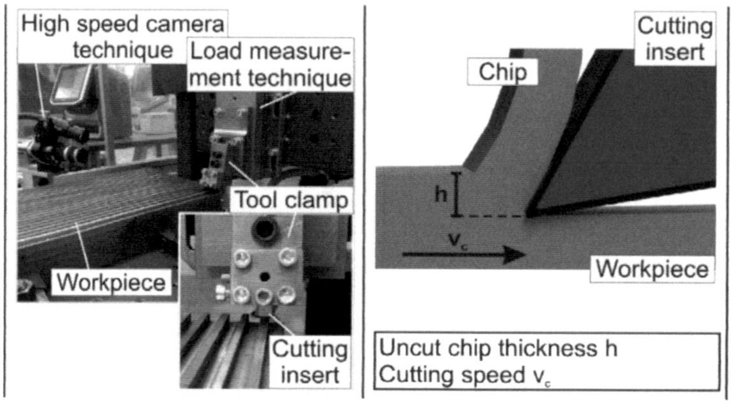

Figure 2. Images of the cutting test setup.

3. Results and Discussion

3.1 Coating

Representative light microscope images of cross-sections of the coated tools are visible in figure 3. The different spray strategies (spray gun paths) lead to completely different results. For strategy I, the coating is almost non-existent (see figure 3a). Even with many overruns, only small amounts of WC-Co were deposited at the flank faces. This result is insufficient and this strategy is not expedient. The reason for these bad results is the high kinetic gas flow during HVOF spraying. The cutting edge deflects the gas flow, thus the particles do not adhere on the surface. This effect is comparable to spraying with very low spray angles. The deposition rate correlates directly with the spray angle [13]. Spray strategy II leads to significantly better results (see figure 3b). The flank faces are well-coated but the cutting edge is still not fully covered with WC-Co, even though the spray path was set to coat the cutting edge. Both coating strategies showed a deformation of the cutting edge (see figure 3a and 3b). In the initial state, all cutting edges had the same shape. This deformation could be caused by the kinetic and thermal energy during the HVOF coating process.

The best result as shown in figure 3c were obtained with spray strategy III. The flanks and the cutting edge are fully covered with WC-Co, no delamination is visible, and the interface is free of gaps. It is visible that the coating thickness is smaller in the area of the cutting edge. This effect might be related to the changing spray angle caused by the rotation of the robot around the cutting edge. The spray gun follows the contour of the tool but the spray angle might not be ideal (90°) due to the kinetics of the robot. Overall, this spray strategy (see also figure 1) is more complex than strategy I and II. Furthermore, the industrial robot needs a larger movement area. However, overall the results are much better.

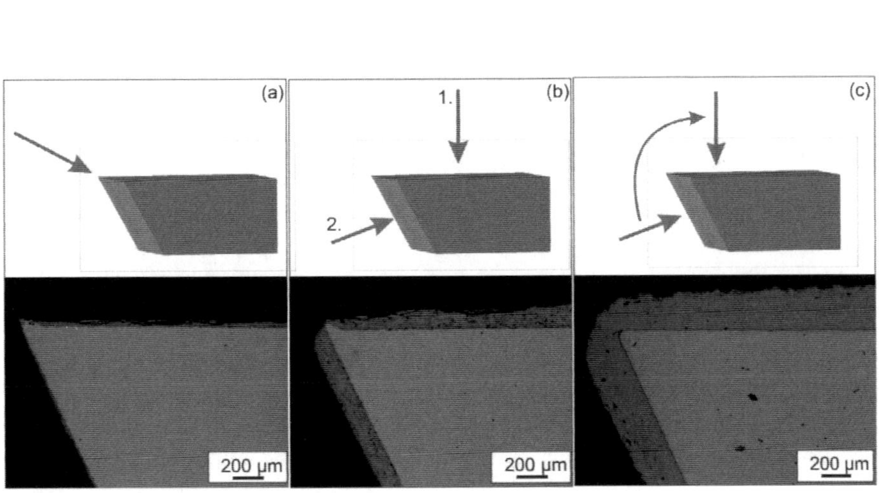

Figure 3. Cross-section image (light microscope) of the WC-Co coated cutting tools. Spray strategy: a) type I (sample P2, 30 overruns); b) type II (sample P4, 20 overruns); c) type III (sample P6, 10 overruns).

The third spray strategy was chosen as the best strategy of the three tested strategies. Additional coating experiments were conducted with strategy III and with different numbers of overruns (see table 2). To obtain quantifiable values, the coating thickness was measured at different positions (No. 1 to 11) on the flank faces and the cutting edge (position No. 7). The distance between the different measuring positions was 1000 µm. The results of the coating thickness measurements are given in the diagram in figure 4.

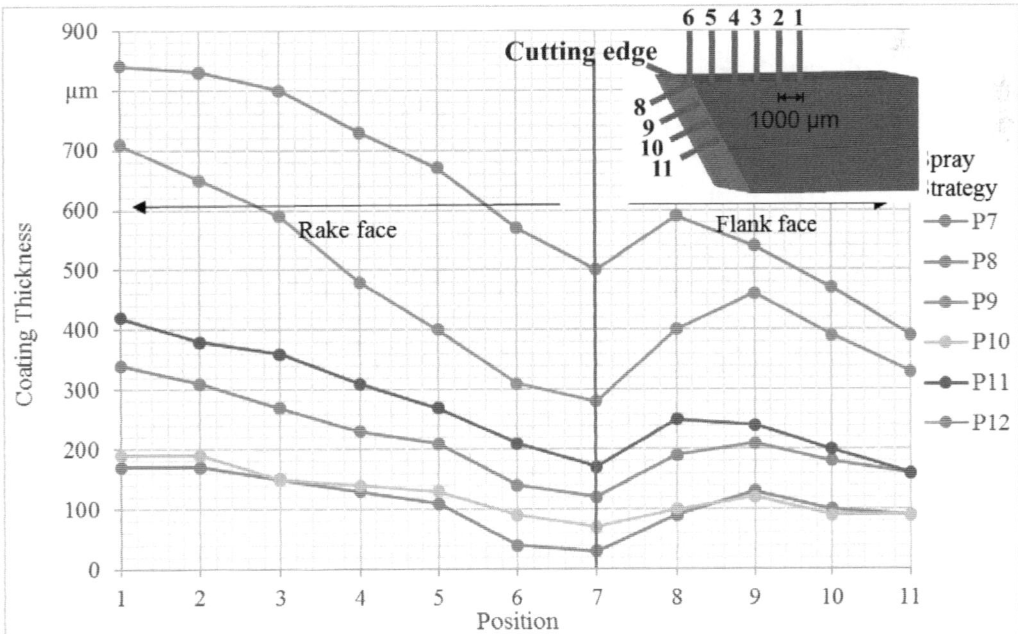

Figure 4. Results of the coating thickness measurements at different positions for the experiments P7-P12.

It can be seen that the lowest coating thickness is always found in the area of the cutting edge (figure 4). This effect is more visible for a higher number of overruns. Furthermore, the coating thickness is more inhomogeneous with more overruns. In the diagram, the left part next to the cutting edge represents the rake face (see figure 1) of the cutting tool. The rake face shows an increasing and large coating thickness (up to 850 µm). The flank face, the right part in the diagram, shows a decrease of the coating thickness for areas not close to the cutting edge. It is suspected that these effects are related to the suboptimal spray angle when the robot rotates the spray gun around the small cutting edge. Only at the back of the flank (position 1 – 3), the spray angle seems to be 90°, thus resulting in a high deposition rate. To obtain an adequate coating thickness (coating thickness t_{coat} > 100 µm) at the cutting edge, at least two overruns with spray strategy III should be realized. Overall, the results of the experiments reveal that the cutting edge can be coated with a covering and dense WC-Co layer if the spray strategy is adapted to the geometry. Of course, there are further different spray strategies which may provide even better results. For this purpose, the kinematics of the robot should be taken into account.

3.2 Post-Processing and Cutting Test

After the coating process, the produced WC-Co layer were post-treated by means of grinding or grinding and wet abrasive jet machining. Both methods were described in chapter 2.1. Representative SEM images of the coated and post-treated tools are shown in figure 4. After grinding, the cutting edge is very sharp (figure 4a/c). The bright area is the WC-Co coating, see figure 4. The gray area is the HSS tool body. The coating does not show any cracks after processing. With the grinding process, a constant coating thickness can be adjusted. After grinding and wet abrasive jet machining, the cutting edge is round and smooth and the surface tends to be slightly rougher (figure 4b/d).

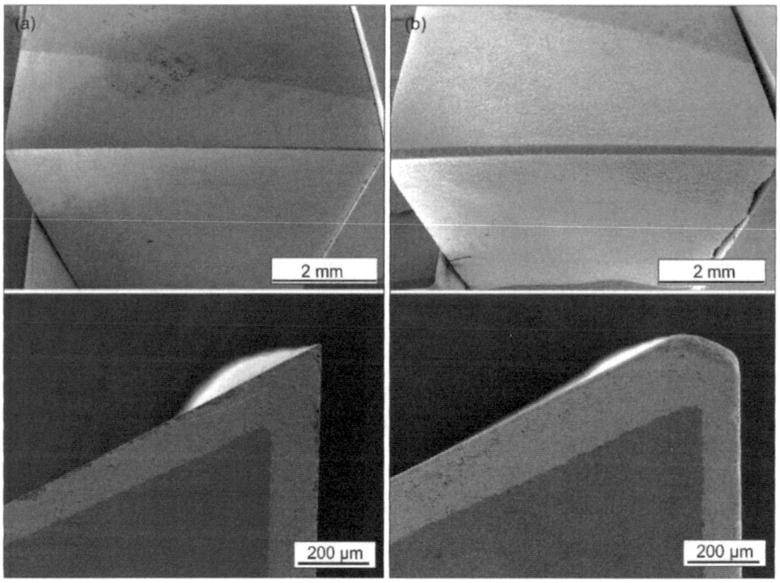

Figure 4. SEM image of the cutting edge and cross-section of WC-Co coated cutting tools after post-treatment by grinding a) and grinding + grit blasting b).

Figure 5 shows the results of the cutting tests of the WC-Co coated and uncoated HSS tools. The uncoated HSS tools shows large deformations and outbursts at the cutting edge as well at the flank and rake face (figure 5b). Energy-dispersive X-ray spectroscopy (EDS) did not show adhesions of the workpiece at the tool surface. The measured loads change intensely due to the breakouts during cutting (figure 5a). Peaks up to 1800 N are visible. The ground and ground + prepared samples show significant lower and almost constant forces (figure 5c and 5e). For the WC-Co coated and ground tools, the cutting force is approx. 350 N at the beginning. At the end of the process time, the loads change slightly and increase up to approx. 430 N. This effect might have occurred due to breakouts at the cutting edge that are visible in figure 5d. Compared to the uncoated HSS tool, the breakouts are smaller and only located at the cutting edge.

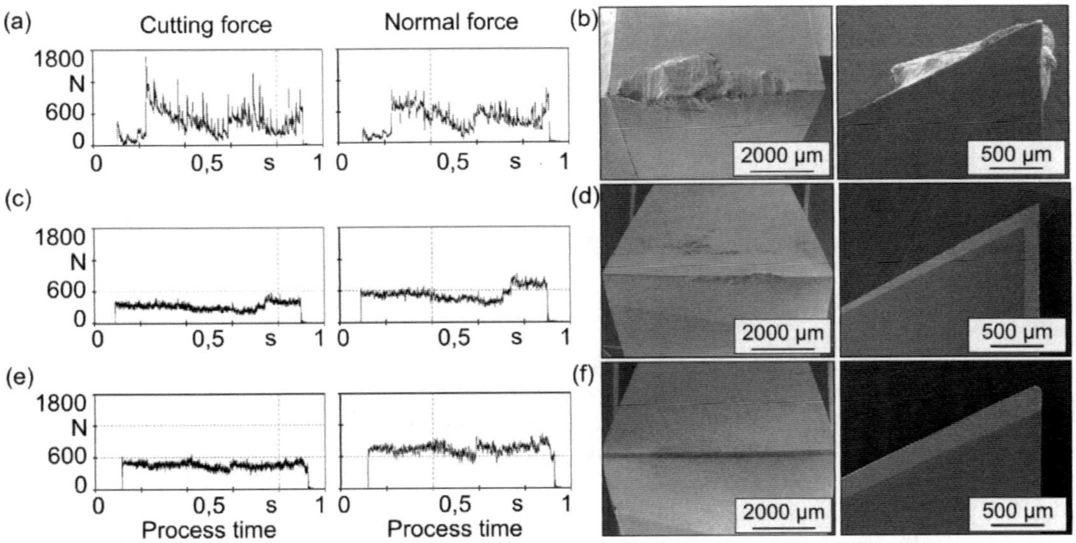

Figure 5. SEM image of the cutting edge of WC-Co coated cutting tools after the machining test. a) + b) HSS tool without coating c) + d) HSS tool with ground WC-Co coating e) + f) HSS tool with ground and prepared WC-Co coating.

The WC-Co coated, ground and prepared HSS tools show minor wear. In figure 5e, the coating on the rake face was nearly removed due to the grinding process. Because of the inhomogeneous coating thickness after spraying (figure 4), the rake face showed only a very small WC-Co coating after grinding. Otherwise, a homogenous cutting tool surface was not realizable.

The measured forces are slightly higher than for the WC-Co coated tools, which were only ground. This effect depends on the shape of the cutting edge. The ground cutting edge is sharper and has a lower contact area during cutting. Hence, this cutting edge provides a higher contact pressure and reduced loads are measured when cutting the steel sample. The prepared cutting edge is round and thus it has a higher contact area during cutting. This results in higher forces up to 600 N (figure 5e) that are necessary for cutting the steel bar. However, the round cutting edge seems to be more stable, probably due to a better force transmission into the coating and from the coating to the substrate (HSS tool body). Breakouts and deformations of the WC-Co coating are not visible.

4. Summary and Conclusion

HSS cutting tools were coated with WC-Co (88/12) by means of HVOF. Different spray strategies were tested to obtain a close and homogenous coating on the cutting edge, the rake and flank face. The best results were achieved when the spray gun path follows the contour of the tool as accurately as possible. This was achieved by a large rotation of the spray gun around the cutting edge by the robot system. Although the movement speed of the robot system was kept constant, the coating thickness on the flanks was not homogenous. This effect was related to the kinematics of the robot, the resulting suboptimal spray angle, and the emerging flow turbulences. To achieve more constant coatings (near-net-shape coatings), these effects have to be taken into account for future investigations.

The coated HSS tools were post-treated by means of grinding or grinding and wet abrasive jet machining. Grinding leads to very sharp cutting edges whereas grinding combined with subsequent preparations leads to a round cutting edge. Conducted cutting tests with the produced HSS tools show that ground and prepared tools perform most sufficiently. The round shape of the WC-Co cutting edge seems to lead to an increased stability. Reasons for this finding could be a higher contact area during cutting and hence, a lower area pressure. Furthermore, a better load transmission related to the round shape could be another reason for this effect. Overall, the behavior under load and the detailed failure mechanisms need to be investigated in comprehensive future research.

Literature

[1] DIN ISO 513:2012 Classification and application of hard cutting materials for metal removal with defined cutting edges – Designation of the main groups and groups of application.

[2] Paucksch E 2008 *Zerspantechnik. Prozesse, Werkzeuge, Technologien; mit 45 Tabellen* (Wiesbaden: Vieweg+Teubner Verlag)

[3] Denkena B and Tönshoff H K 2011 *Spanen. Grundlagen* (Berlin, Heidelberg: Springer-Verlag)

[4] Stahlhut C 2011 *Laserstrahllöten von Stahl und Hartmetall für zerspanende Werkzeuge mit definierter Schneide* (Dissertation Leibniz Universität Hannover)

[5] Chiu L H, Wang H F, Huang C. P, Hsu C T and Chen T C 2008 Effect of Brazing Temperature on the Microstructure and Property of Vacuum Brazed WC-Co and Carbon Steel Joint *Adv Mat Res* **47-50** p 682–85

[6] Cottenden A M and Almond E A 1981 Hardmetal interlayered butt-joints made by diffusion bonding and pressure bonding *Met Technol* **8 1** p 221–33

[7] Selvadurai U, Hollingsworth P, Baumann I, Hussong B, Tillmann W, Rausch S and Biermann D 2015 Influence of the Handling Parameters on Residual Stresses of HVOF-sprayed WC-12Co Coatings *Surf. Coat. Technol.* **268** p 30-35

[8] Reignier C, Sturgeon A, Lee D and de Wet D 2002 HVOF sprayed WC-Co-Cr as a generic coating type for replacement of hard chrome plating (*Proc. ITSC 2002*)

[9] Tillmann W, Vogli E, Baumann I, Matthaeus G and Ostrowski T 2008 Influence of the HVOF Gas Composition on the Thermal Spraying of WC-Co Submicron Powders (−8 + 1 μm) to Produce Superfine Structured Cermet Coatings *J Therm Spray Tech* ***17* 5** pp 924–932

[10] Geng Z, Hou S, Shi G, Duan D and Li S 2016 Tribological behaviour at various temperatures of WC-Co coatings prepared using different thermal spraying techniques *Tribol Int* **104** pp 36-44

[11] DVS-Bericht 245 2007 *Thermisches Spritzen. Potentiale, Entwicklungen, Märkte* (Düsseldorf, DVS-Media)

[12] Candel A and Gadow R 2006 Optimized Multiaxis Robot Kinematic for HVOF Spray Coatings on Complex Shaped Substrates *Surf. Coat. Technol.* **201 5** p 2065-2071

[13] Tillmann W, Baumann I, Hollingsworth P and Laemmerhirt I-A 2013 Influence of the Spray Angle on the Properties of HVOF Sprayed WC–Co Coatings Using (−10 + 2 μm) Fine Powders *J Therm Spray Tech* **22 2** p 272–79

Manufacture of iron-based, amorphous coatings with high fracture toughness

K Bobzin, M Öte and T Königstein

Surface Engineering Institute of RWTH Aachen University, Kackertstr. 15, 52072 Aachen, Germany

Abstract. Amorphous iron-based material have excellent corrosion behaviour, show good tribological performances and exhibit interesting thermophysical properties. The deposition as a coating system by thermal spraying technology is an innovative approach to manufacture these materials. In this study, the mechanical properties of three iron-based amorphous coatings with different chromium content x_{Cr} = 0, 5 and 15 at.% are presented deposited by means of High Velocity Oxygen Fuel Spraying. For the determination of the amorphous content the linear relationship between crystallization energy and amount of amorphous structures is used. Comparing the crystallization energies of amorphous ribbons manufactured by melt spinning to those of feedstock materials and free standing coatings, assumptions regarding the amorphous contents are drawn. The results show that the amorphous content in the feedstock material is influenced by the amount of chromium content. Furthermore, the amorphous content of all coatings do not exceed those of the feedstock materials. Powder x_{Cr} = 15 at.% and the corresponding coating exhibit smallest amount of amorphous structure, presumably due to a not fully melted state of the impacting particles. The values of fracture toughness of the coatings are determined by means of indentation and subsequent measurement of the crack lengths. Furthermore, values of indentation modulus and hardness are measured and compared to each other. While length of indentation cracks decreases with increasing chromium content, an increase in indention modulus and hardness is observed. In comparison to ceramic reference YSZ and the steel reference 1.4404, all amorphous coatings show promising properties such as low indentation crack lengths and high hardness.

1. Introduction

In order to produce amorphous materials, extremely high cooling rates in the order of $\dot{T} = 10^6\text{-}10^7$ K/s during the solidification are necessary. High cooling rates suppress the formation of crystalline structures and enable the formation of amorphous metastable phases. By means of thermal spraying, the above mentioned cooling rates can be reached and coatings with fully or partially amorphous microstructures can be manufactured. The chemical composition of the alloy is another factor which is decisive for the formation of amorphous structures. Alloys with a chemical composition consisting of FeBSiNb have been reported to be especially suitable for the formation of amorphous phases [1]. Due to the lack of crystalline structures, amorphous materials offer a good protection against selective corrosion, especially due to the absence of grain boundaries in the microstructure [2, 3, 4, 5]. Furthermore, amorphous materials exhibit high hardness and are therefore frequently discussed regarding their deployment in tribologically loaded components [4]. A major disadvantage of amorphous materials, on the other hand, is their low fracture toughness. Use of materials with high fracture toughness might be necessary for numerous applications, i.e. for systems which operate under cyclic thermomechanical loading. Apart from corrosion resistance and tribological behavior of the applied coatings, thermophysical properties, i.e. thermal conductivity, have been examined in this study. The measured thermal conductivities amount, in some instances, are lower than $\kappa = 3{,}0$ W/(mK) [6, 7]. In comparison, the thermal conductivities of plasma sprayed Yttria stabilized Zirconia (YSZ) can be lower than $\kappa = 1{,}0$ W/mK [8]. Therefore, it can be stated that metallic glasses might also come into use in order to protect thermally loaded components, depending on the maximum temperatures, which the components are exposed to. The operating temperatures should be well below the melting temperature of the coatings and even less than their crystallization temperature, so that no crystallization occurs, which would change the thermal conductivity of the coating.

Within the scope of this study, FeCrBSiN coatings with varying Cr and Fe contents have been developed by altering the chemical composition of amorphous material presented in [1]. In this regard, the chromium content (x_{Cr}) of the coatings has been increased stepwise between $x_{Cr} = 5$ at. % and 15 at.%, while iron content has been decreased correspondingly. The aim was, to improve the oxidation resistance of the coatings by increasing the Cr content. The proportions of the amorphous phase in the developed coatings have been assumed by comparing the crystallization energy of examined specimens with that of a ribbon manufactured by means of melt spinning. The procedure, as introduced in [2, 5], will be explained in chapter 2 more in detail. Furthermore, the fracture toughness of the coatings has been analyzed. In order to evaluate the fracture toughness (K_{Ic}) of materials, a wide number of different approaches exist in the literature [9]. In this study the model according to Anstis et al. has been employed, see equation (1) [10]. In order to calculate the K_{Ic} factor the values for hardness (H), elastic modulus (E), indentation force (F) and crack length need to be determined. The experiments used in order to determine these values will be introduced in chapter 2. Subsequently the results will be discussed and conclusions will be given.

$$K_{Ic} = 0.016 \cdot \left(\frac{H}{E}\right)^{0.5} \cdot \frac{F}{c^{1.5}} \tag{1}$$

2. Test Methods

Three different $Fe_{72-x}Cr_xB_{20}Si_4Nb_4$ ($x_{Cr} = 0, 5, 15$) feedstock materials featuring different chromium contents were produced by means of gas atomization. The powders were sieved with a targeted powder fraction of -45 to +10 µm. It was confirmed by the measurements conducted using a specialized optical particle analyze system; Morphologi G2 by Malvern Instruments Ltd (Worcestershire, England), that the resulted powder fraction corresponds to the targeted one. Furthermore, single powder particles made of different feedstock materials were analyzed metallographically in cross-sections using scanning electron microscope (SEM) LEO 1530 VP Gemini by the company Carl Zeiss AG (Oberkochen, Germany). Subsequently, the inner crystalline structures of the powder particles were analyzed by means of X-Ray Diffractometry (XRD) using Seifert XRD 3003 from GE Sensing & Inspection Technologies GmbH (Hürth, Germany). For the XRD measurements a CuKα target was selected. The angle of incidence $\omega = 10°$, step with $2\Theta = 0.05°$, holding time $t_h = 10$ s and measurement interval between $2\Theta = 20°$ to $80°$ remained constant for all measurements. Moreover, all powders were analyzed by means of dynamic Differential Scanning Calorimetry (DSC) using SETSYS Evolution System of SETARAM Instrumentation (Caluire, France) in order to obtain melting temperatures. For the DSC measurements a sample weight of $m_s = 20$ mg and heating rates of $\dot{T} = 10$ K/min were selected. Afterwards, the crystallization energies of the powders were determined by calculating the baseline integral value. A description of how to determine the energy values from DSC curves can be found in [11, 12]. In order to assess the proportions of amorphous structures within the feedstock material, said value is compared to the crystallization energy of a reference material manufactured by melt spinning process using the Melt Spinner SC by Edmund Bühler GmbH (Hechingen, Germany). Same approach can be found in [1]. The melt spinning process is designed such way that extremely high cooling rates can be realized. During the melt spinning process the material is liquidized by means of induction. Subsequently, by increasing argon pressure the liquid material is pressed through a ceramic nozzle onto a rotating copper roller. Due to high thermal conductivity of copper high cooling rates can be realized allowing the manufacture of amorphous ribbons. For the manufacture of the reference specimens a slit-nozzle with the geometry of width w = 5 mm and broadness b = 0.7 mm was utilized. The distance between the nozzle and copper roller was s = 0.4 mm. The rotating frequency of the copper roller was set to f = 25 Hz. Within this study, it must be noted that this evaluation method regarding the amorphous content is based on the assumptions that the reference specimen exhibits a fully amorphous state and that the crystallization energy and amount of amorphous structure have a linear relationship. The calculation of the amorphous content of the powder is done by following equation (2). In this formula $x_{am,P}$ represents the amorphous content of the powder, Δh_P the crystallization energy of the powder and Δh_R the crystallization energy of the ribbon.

$$x_{am.P} = \frac{\Delta h_P}{\Delta h_R} \tag{2}$$

All amorphous iron-based feedstock materials were sprayed by High Velocity Oxygen Fuel spraying (HVOF) using K2 system of the company GTV Verschleißschutz GmbH (Luckenbach, Germany). Spraying distance s = 300 mm, Oxygen flow rate $\dot{V}_O$ = 800 SLPM and the traverse speed of the robot v_R = 1,000 mm/s were kept constant throughout the experiments. The nozzle has a length of l = 145 mm and a diameter of d = 12 mm. Flat discs made from AlSi10 were used as substrate. Prior to the coating process, the substrate's surface was roughened by means of grit blasting with corundum F24 exhibiting particle size of x_p = 600 - 850 µm. By means of an SEM, the coatings were examined with regard to their microstructures in metallographic cross-sections. All coatings were analyzed with regard to their crystalline composition via XRD. Free standing coatings were manufactured for DSC measurements and to determine the crystallization energy (Δh_C). The amorphous content of the coatings are calculated, by comparing the crystallization energy of the coatings to the ribbons.

For the determination of fracture toughness, crack network was initiated via indentation in a cross-section using a Vickers geometry. The tests were conducted according to DIN EN ISO 6507-1. For the generation of the networks of cracks, a test load of m = 3 kg (F = 29.43 N) was chosen. After the maximum load is reached, the load was kept constant for t = 10 s. Subsequently, the lengths of the horizontally running cracks were measured under a light microscope. Moreover, the indentation modulus and the universal hardness of the coatings were determined via Fischerscope of the type HM2000 by the company Fischer (Sindelfingen-Maichingen, Germany). All measurements were conducted with n = 10 repetitions for each material, thus ensuring statistical accuracy. For the purpose of comparison, an established ceramic thermal barrier coating system was used. It consists of a MCrAlY bond coat and a top layer made of YSZ. Apart from the ceramic material, a stainless steel coating made from 1.4404 was chosen for comparison as well.

3. Research Results

The SEM powder characterization has shown that all particles exhibit a spherical shape. The powder particles consisting of x_{Cr} = 15 at. % show a clear microstructure indicating the presence of crystalline structures. The presence of crystalline structure in this case has also been verified via XRD showing a high intensity peak at 2Θ = 45°, see figure 1a. This indicates a ferritic structure. Measurements via DSC have shown that the melting temperature of the feedstock material increases with increasing chromium content. The differences between the powder containing no chromium and that containing x_{Cr} = 15 at. % chromium, amounts to about ΔT = 70 °C. The calculated amorphous contents using equation (1) for different powders are presented in figure 1b. These results underline the XRD pattern, showing smallest amorphous content for powder x_{Cr} = 15 at. %. Feedstock material x_{Cr} = 5 % exhibits calculated values of almost x_{am} = 120 % indicating very high amount of amorphous structures. Here it must be noted that values above x_{am} = 100 % does not have a physical meaning but shows that the powder exhibits a higher amorphous content than that of the manufactured ribbons. This can be attributed to high cooling rates during powder manufacturing process by gas atomization process.

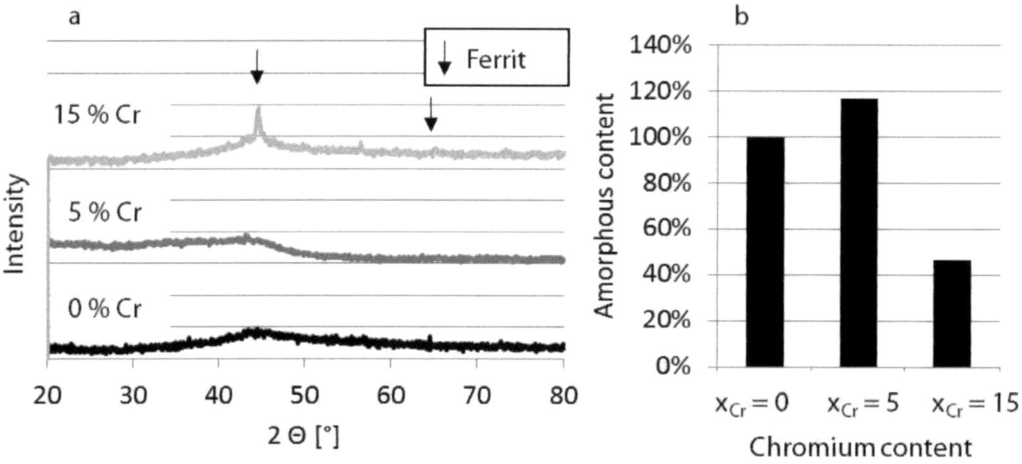

Figure 1. (a) XRD measurements and (b) calculated amorphous content of feedstock materials with different chromium contents.

In figure 2, cross-sectional views of different coatings applied in the context of this study have been presented. All coatings exhibit a dense microstructure as shown in figure 2. It can be observed that with increasing chromium content, the appearance of the voids and pores in the coating microstructure changes. Moreover, it must be noted that the coating thickness is significantly lower in case of the coating with highest chromium content, although all the process settings were the same in all experiments. Furthermore, for x_{Cr} = 15 at. % a high amount of spherical particles can be observed within the coating microstructure. It is likely, that the high melting temperature of this feedstock material, which is determined by means of DSC measurements to be almost ΔT = 70 °C higher than that of the other feedstock materials, restrict the melting behavior of some powder particles during their flight. Some of the arriving particles, which do not exhibit convenient melting degree, are not incorporated into the coating build-up and bounce back from the coating surface. These are exhausted from the coating chamber as overspray. Therefore, the reduction in the coating thickness in case of the feedstock material with a chromium content of x_{Cr} = 15 at. % can be attributed to the loss of some unmolten particles in overspray. Moreover, compression of the underlying splats due to the impact of arriving high-velocity unmolten particles might also be contributing to the reduction in coating thickness. Furthermore, the size, orientation and positions of the pores in the coating with x_{Cr} = 15 at. % differs compared to those with x_{Cr} = 0 at. % and 5 at. %. For the coating with x_{Cr} = 15 at. %, the majority of the pores range in size of a few micrometers. These cannot only be found between the splats, but also inside the unmolten particles. The low deformability of the feedstock materials together with the high impact velocity of the particles might be responsible for the formation of crack like structures within the deposited unmolten particles. These have in general a perpendicular orientation to the substrat's surface.

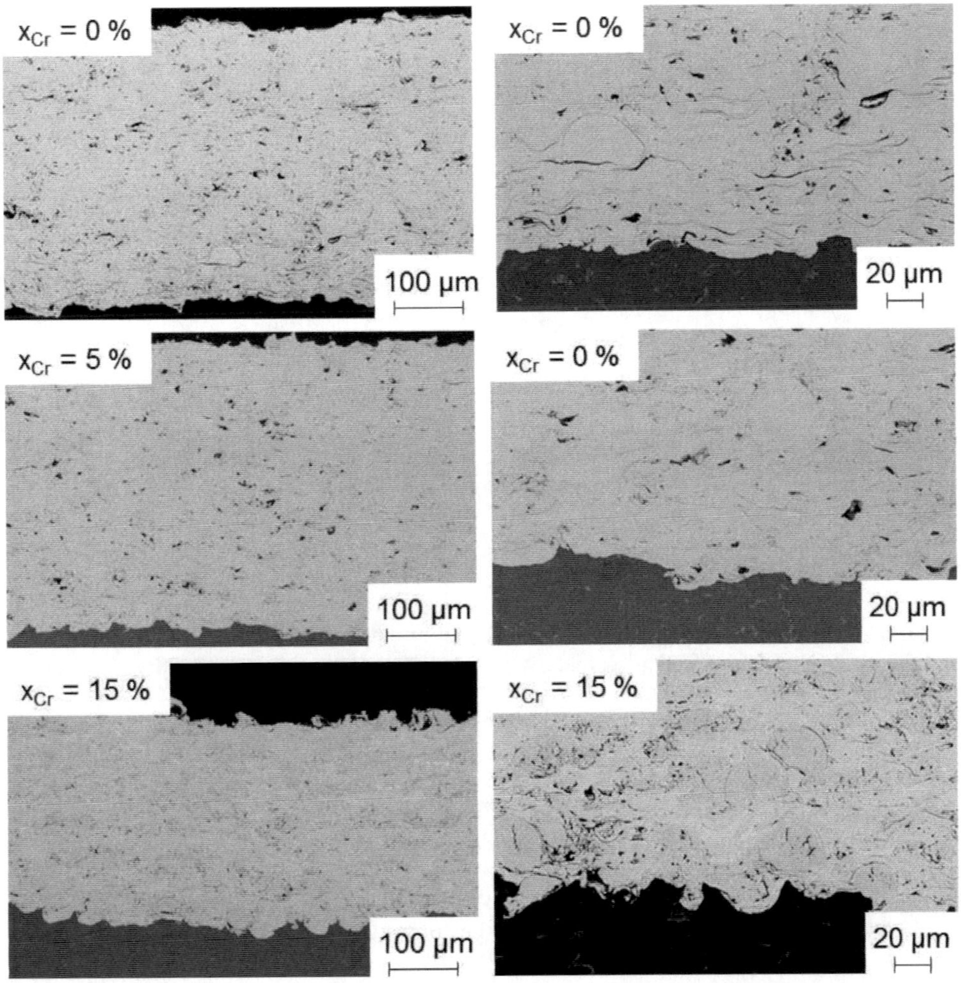

Figure 2. Cross-sections of coatings captured by means of SEM: left magnification 400x, right magnification 1,000x.

XRD measurements conducted on the deposited coatings do not reveal any phase transitions, as shown in figure 3. Coating deposited with x_{Cr} = 15 at. % remains crystalline or at least partly crystalline. Coatings deposited with x_{Cr} = 0 and 5 at. %, show a largely amorphous/nanocrystalline structure. This can be concluded from the characteristic halo pattern and a sharper peak at $2\Theta = 45°$ compared to the feedstock material. Additionally, DSC measurements were executed for quantification of the amorphous content. Comparing the results between the feedstock materials and coatings it can be observed, that the amorphous content of all coatings is significantly lower and amounts less than 50 % of the feedstock material. It is assumed that during coating process crystallization of the feedstock material occurs and nanocrystalline structures are formed. For the coating with x_{Cr} = 15 at. %, lowest amorphous content was observed, which is line with the XRD measurements.

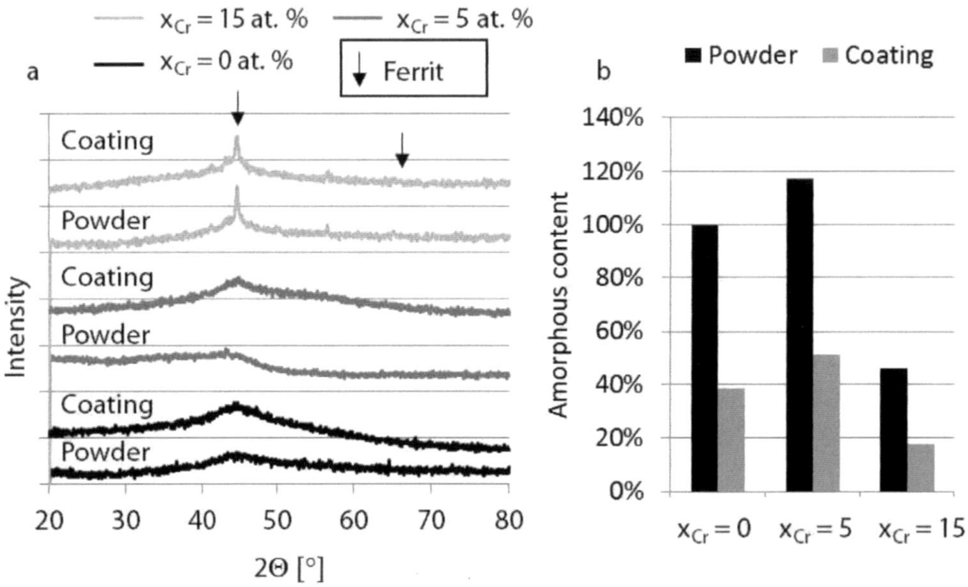

Figure 3. (a) XRD measurements and (b) calculated amorphous content of free standing coatings in comparison to feedstock materials with different chromium contents.

All coatings indicate the presence of partially amorphous structures. In the following for the sake of simplification these will be referred as amorphous coatings although these are not fully amorphous. For comparing the fracture toughness of the amorphous coatings, two references, a plasma sprayed YSZ coating and a plasma sprayed stainless steel 1.4404 coating, were manufactured. For both feedstock materials APS is the most frequently used spraying technology. From polished cross-section in figure 4 it can be seen that the coating made of 1.4404 exhibits areas of dark contrast. These areas most likely correspond to oxide inclusions. Part of these oxides are formed on the particle surface during particle flight in free jet. After impact with the substrate these oxides are incorporated into the coating, forming brittle phases. In comparison, the YSZ coating exhibits a porous microstructure. Neither lamellar coating build-up can be identified nor different phases can be distinguished by observing the cross-section. For the iron-based amorphous coating no oxide inclusions can be observed. The metallographic cross-section show highly deformed splats and a lamellar coating build up which is characteristic for metallic coatings deposited by HVOF.

By indentation into the polished cross-sections crack networks were initiated. Figure 4 shows both reference materials and exemplarily the coating with x_{Cr} = 5 at.% prior and subsequent to the indentation. Since for the reference material 1.4404, the determination of the crack length is difficult to follow in the illustration, it is marked with bars. As the oxides arrangement is largely interlamellarly, crack formation occurs at these brittle phases. For the iron-based amorphous coatings material separation occurs between single splats. As expected, largest cracks were observed in the coating made of YSZ.

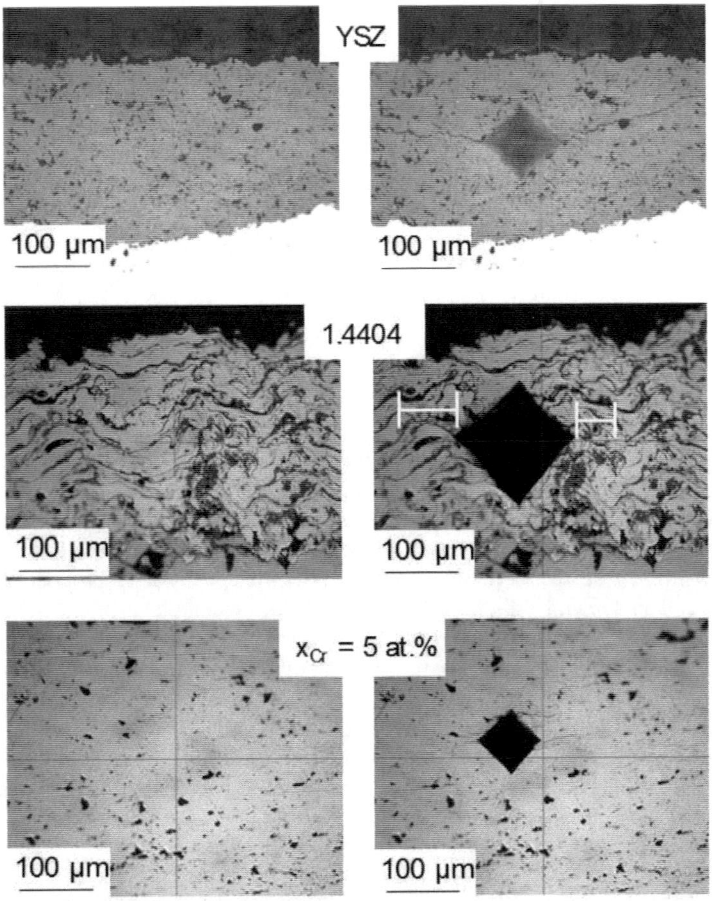

Figure 4. Examples of cross-sections prior and subsequent to indentation test.

For the calculation of the fracture toughness according to Anstis et al., among other values indentation modulus and hardness are required, which were determined by indentation into polished cross sections. In table 1 all results are summarized including standard deviations. It can be observed that for the amorphous iron-based coatings, indentation modulus and hardness increase with rising chromium content while the crack length decreases. The references 1.4404 and YSZ exhibit lower indentation modulus and hardness compared to amorphous coatings, while crack length is smallest for 1.4404 and largest for YSZ. Due to the fact that 1.4404 is rather ductile and YSZ rather brittle, these results were expected. For the determination of fracture toughness according to equation (1), the average values were used. Comparing the results from different iron-based amorphous coatings, it can be stated that the fracture toughness increases significantly with increasing chromium content in the coating. Due to the fact that the first term in equation (1), the ratio of hardness to the indentation modulus, does not show any significant changes between the examined samples, crack length must be the dominating factor affecting fracture toughnesses of the iron-based amorphous coatings. As expected YSZ coating exhibits the smallest fracture toughness with a value of $K_{Ic} = 0.5$ MPa·m$^{0.5}$. Similar results with regard to fracture toughness of YSZ coatings ranging between $K_{Ic} = 1\text{-}2$ MPa·m$^{0.5}$ were reported in the literature [13]. Comparing the fracture toughness of the iron-based amorphous coatings to the YSZ coating it can be seen that the iron-based coating exhibits 2 to 10 times higher fracture toughness values.

Table 1. Average values and standard deviations of indentation modulus (E), hardness (H), crack length (c) and average values of fracture toughness (K_{Ic}).

	E (GPa)	H (GPa)	c (μm)	$(E/H)^{0.5}$	K_{Ic} (MPa·$m^{0.5}$)
x_{Cr} = 0 at.%	173.0 ± 6.4	9.2 ± 0.9	111.5 ± 42.3	4.3	1.7
x_{Cr} = 5 at.%	178.7 ± 9.5	10.1 ± 1.2	84.3 ± 27.2	4.2	2.6
x_{Cr} = 15 at.%	191.8 ± 13.2	10.2 ± 1.4	56.4 ± 14.0	4.3	4.9
YSZ	129.9 ± 13.3	8.4 ± 1.1	238.5 ± 68.4	3.9	0.5
1.4404	110.7 ± 11.0	3.1 ± 0.7	34.8 ± 22.4	5.9	14.1

4. Summary and outlook

Amorphous iron based materials are highly innovative materials and can be applied by means of Thermal Spray technology in form of coatings. The application areas have yet to be explored further, especially with regard to their deployment in tribological systems. Within this study, coatings manufactured by HVOF technology exhibit a partially amorphous structure. Comparing the amorphous content of the feedstock material and that of deposited coatings, a reduction in amorphous content is observed which can be explained by crystallization process during coating deposition. Nevertheless, a high amount of amorphous phase in the case of coating with a chromium content of x_{Cr} = 5 % could be realized. All amorphous coatings show high fracture toughnesses compared to a reference YSZ coating. These findings underline the potential of amorphous coatings with regard to application in tribological systems which will be part of future research.

Further studies will focus on how different thermal spray technologies influence the formation of amorphous structures. Special attention will be paid on air plasma spraying. Due to the higher processing temperatures, it is assumed that particles made of the feedstock material with x_{Cr} = 15 at. % will be molten entirely, thus enabling the formation of highly amorphous structures as well. Furthermore, the possibility to manufacture a wire with the same chemical composition will be discussed, which could then be used in Plasma Transferred Wire Arc (PTWA) technology for the deposition of coatings on inner geometries.

Acknowledgement

The authors gratefully acknowledge the financial support of the German Research Foundation (DFG) within the Project 'Basic research on the applicability of Fe-based TS-coating with the purpose of thermal insulation' (BO 1979/34-1).

References

[1] Amiya K, Urata A, Nishiyama N and Inoue A 2004 *Materials Transactions* **45(4)** 1214–18
[2] Zhou Z, Wang L, Wang F C, Zhang H F, Liu Y B and Xu S H 2009 *Surface & Coating Technology* **204** 563–70
[3] Liu G, An Y, Guo Z, Chen J, Hou G and Chen J 2012 *Applied Surface Science* **258** 5380–86
[4] Liu X Q, Zheng Y G, Chang X C, Hou W L, Qang J Q, Tang Z and Burgess A 2009 *Alloy and Compounds* **484** 300–7
[6] Shin D I, Gitzhofer F and Moreau C 2005 *Journal of Thermal Spray Technology* **16** 118–27
[5] Liu G, An Y, Guo Z, Chen J, Hou G and Chen J 2012 *Applied Surface Science* **258** 5380–86
[7] Shin D I, Gitzhofer F and Moreau C 2007 *Journal of Material Science* **42** 5915–23
[9] Sergejev F and Antonov M 2006 Comparative study on indentation fracture toughness measurements of cemented carbides *Proc. Estonian Acad. Sci. Eng.* **12(4)** 388–98
[8] Dutton R, Wheeler R, Ravichandran K S and An K 2000 *Journal of Thermal Spray Technology* **9** 204–9
[10] Anstis G R, Chantikul P, Lawn B R and Marshall D B 1981 *Journal of American Ceramic Society* **64(9)** 533–8

[11] Desiraju G R, Vittal J J and Ramanan A 2011 *Crystal Engineering* World Scientific Publishing Co. Pte. Ltd. Singapore
[12] Gabbott P 2008 *A Practical Introduction to Scanning Differential Calorimetry* Blackwell Publishing Ltd. United Kingdom
[13] Vassen R, Cao X, Tietz F, Basu D and Stöver D 2000 *Journal of American Ceramic Society* **83(8)** 2023–28

Tailoring the heat transfer on the injection moulding cavity by plasma sprayed ceramic coatings

K Bobzin[1], Ch Hopmann[2], M Öte[1], M A Knoch[1], I Alkhasli[1,*], H Dornebusch[2] and M Schmitz[2]

[1]Surface Engineering Institute at RWTH Aachen, Kackerstr. 15, 52072 Aachen, Germany
[2]Institute of Plastics Processing at RWTH Aachen, Seffenter Weg 201, 52074 Aachen, Germany
[*]alkhasli@iot.rwth-aachen.de

Abstract. Inhomogeneous material shrinkage in injection moulding can cause warpage in thermoplastic components. To minimise the deformations of the injection moulding parts, the heat transfer during the cooling phase can be adjusted according to the local cooling demand on the surface of the mould cavity by means of plasma sprayed coatings with locally variable thermal resistance over the surface of the mould. Thermal resistance is a function of thermal conductivity and thickness of the coatings, where thermal conductivity of thermal barrier coatings can be adjusted by altering the chemical composition and the microstructure, which is depending on the thickness. This work evaluates the application of plasma sprayed coatings with variable thickness as thermal barrier coatings in the mould cavity. The thermal resistance of the coating and thereby the heat transfer from the melt into the mould will be influenced locally by varying the coating thickness over the cavity area according to the local cooling demand. Using the laser flash method, the thermal conduction of coatings with different thicknesses will be determined. On the basis of the experimentally determined thermal conduction, the effect of the coatings on the temperature field of the mould cavity will be numerically calculated and the required thickness distribution of the coating for an optimal temperature gradient will be determined.

1. Introduction

Injection moulding is a manufacturing process which allows the production of plastic parts with complex geometries and high reproducibility. During the injection moulding process, plastics granulates are plasticised under high temperatures and shearing stresses. The melt is afterwards injected under high pressure into the mould. As the molten plastic flows from the cavity gate deeper into the mould, it cools down as a result of the heat transfer from the hot molten plastic to the comparably cool cavity walls. Due to the difference in flow length and geometric shapes, this typically results in a non-uniform temperature distribution within the mould cavity. The plastic part is created by the solidification of the molten plastic inside the mould. Due to the inhomogeneous material shrinkage, caused by the non-uniform temperature and pressure distribution within the mould cavity, warpage is induced [1]. Therefore, the avoidance of different shrinkage potentials inside a moulded part is necessary to ensure a dimensionally accurate part. One way to achieve net shape production of the plastic parts is to compensate the warpage by altering the mould cavity geometry. Warpage can also be reduced by maintaining a more appropriate temperature distribution in the mould cavity as the cavity is filled with molten plastic and cooled down. Figure 1 illustrates the warpage behaviour for varying temperatures of the cavity while keeping a constant core temperature. Due to the asymmetric cooling, different shrinkages occur locally resulting in warpage. In figure 1 it can be observed, that even low differences in temperature can result in respectively high warpages.

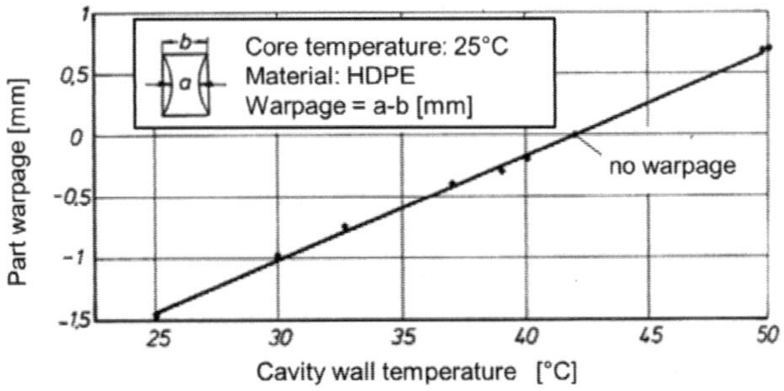

Figure 5: Warpage at constant core temperature and variable cavity wall temperature [2].

To ensure uniform cooling of the part, great effort is spent on the design of cooling channels. Highly complex and conformal cooling channels are implemented inside the injection mould to enable an adequate temperature distribution [3]. New approaches are the inverse thermal mould design, where the location of the cooling channels in the mould is based on the local cooling demand of the plastic part and the use of a self-optimising temperature control to adjust the solidification actively within the process runtime [4]. For both of these techniques, a manipulation of the local specific volume is performed, since the shrinkage potential of the plastics material is directly correlated to its density. This manipulation is achieved by controlling the local temperature by a static tempering layout or an active tempering technique.

This work aims at creating a ceramic coating on the mould cavity to serve as a thermal insulation layer to be used numerous times in production. This paper presents a novel approach for achieving an adjusted temperature distribution on the surface of the mould cavity by employing an air plasma sprayed coating with its thickness adapted to the cooling demand of the given injection moulding process.

Air plasma spraying (APS) is a technological variation of the thermal spraying process which is used to coat various surfaces with a great variety of materials. Electrical energy is converted into heat energy within a plasma generator with the help of an electrical arc, resulting in a highly energetic plasma flame at the nozzle, which can reach temperatures up to 20,000 K. Powder particles are injected into the plasma flame, accelerated and heated up and then impinged on the surface of the part creating a coating. This results in a complex coating microstructure, which does not necessarily exhibit constant porosity throughout the coating thickness. Therefore, the thermal conductivity of the coatings can show slight differences depending on their thicknesses [8, 9].

High temperatures in the APS allow the processing of ceramic powders and manufacturing thermal barrier coatings (TBC). Due to their low thermal conductivity and high melting point, ceramic TBCs are often used for thermal insulation of parts that are subjected to high-temperature environments. In this study, a ceramic coating with varying thickness will be developed to be used in injection moulding. To influence the cooling rate of an injection moulded part, a mould insert can be locally coated to achieve a targeted heat transfer. By adapting the thickness distribution of the coating according to the cooling demand of the plastic part, a more uniform temperature distribution inside the cavity can be achieved. This can enable the reduction of undesirable warping of the plastic components.

The thickness distribution of the coating necessary to achieve a more uniform temperature distribution on the surface of the mould cavity is first determined by simulating the injection moulding process and obtaining the temperature distribution at the mould cavity. Figure 6 illustrates the temperature distribution in the mould cavity which refers to a simulation from a previous work [9], which was further developed in this work. It can be seen that the temperature of the mould cavity decreases along the direction from the gate opening towards the end of the cavity. In order to compensate the heat loss towards the end of the cavity, a linear coating thickness distribution is applied, which has its minimum thickness at the gate opening and maximum at the cavity end.

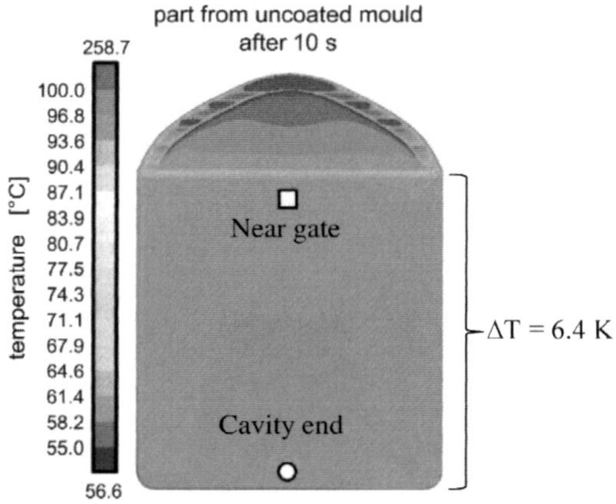

Figure 6: Temperature distribution on the cavity mould after 10 seconds of injection [10].

It should be noted that the thermal conductivity of the coating is expected to be slightly different at different coating thicknesses. This is due to the fact that over the course of the spraying process the substrate is heated, thus influencing the splats formed at the particle impact, which in turn can affect the porosity of the coating. In order to identify the significance of this difference, the thermal conductivity of the coating will be measured for 5 different coating thicknesses. Based on the experimentally determined thermal conductivity of the coating, the injection moulding process will be simulated with the determined thermal insulation of the coating in order to analyse the effect of the coating on the temperature distribution on the surface of the mould cavity. The following section will address the manufacturing process of the TBC with varying thicknesses and the determination of its thickness dependent thermal conductivity. The next section will present the simulation of the injection moulding process and illustrate the effect of the varying thickness TBC on the process. The paper will be concluded with a summary and an outlook to further applications of the varying thickness TBCs in injection moulding process.

2. TBC with variable thickness

Yttria stabilised zirconia (YSZ) powder with particle size of -90+16 µm is used for the coating of an iron-based substrate with the dimensions of 95x20x3 mm with the plasma generator TriplexPro^{TM}-210 (Oerlikon Metco Europe GmbH, Hattersheim, Germany) using the parameters given in Table 2. The varying thickness distribution was realised by adapting the passage numbers to the locally required thickness. The thickness distribution of the coating with varying thicknesses was determined with a 3D laser scanning confocal microscope Keyence VK-X210 (Illinois, USA) and is shown in Figure 7. The thickness distribution at the thinnest part of the coating is not displayed due to the limited measurement range of the confocal microscope. The varying thickness distribution of the coating was assumed to be linear for simplicity. It is possible to achieve a smoother thickness variation by polishing the coating surface, which is a standard procedure depending on the application. The manufactured coating with varying thickness will serve as a basis for the simulation of the injection moulding with coated mould cavity in the following section.

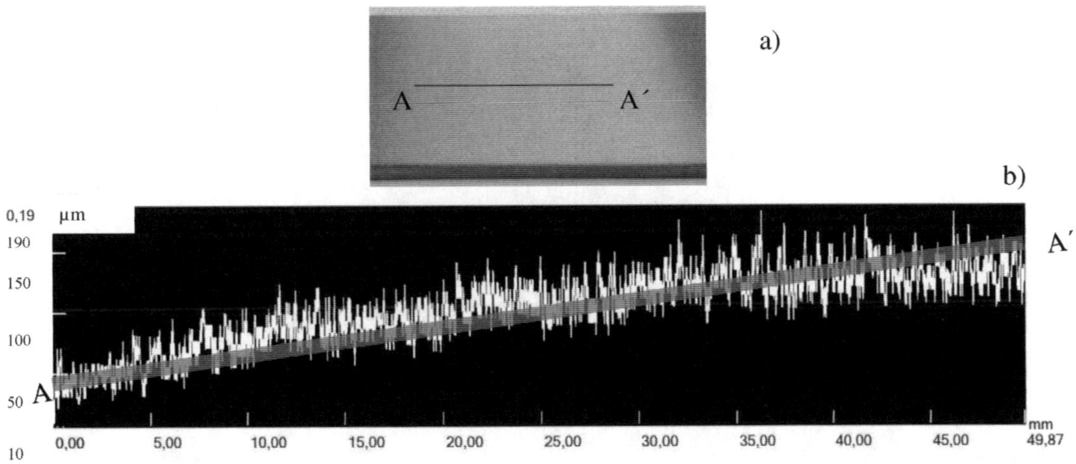

Figure 7: YSZ coating with varying thickness and its thickness distribution measurement the line AA´.

In order to take into account the deviations in the thermal conductivity of coatings with different thicknesses, coatings with uniform thicknesses of 24, 48, 72, 120 and 170 µm were sprayed with the same process parameters as the coating with the varying thickness. An example of a cross-section image of the coatings is illustrated in Figure 8. The coating illustrates a fairly homogeneous thickness distribution over its length and a homogenous porosity distribution over its thickness.

Table 2: Spraying Parameters.

Parameter	Value
Current	450 A
Argon	60 SLPM
Nitrogen	8 SLPM
Spraying distance	120 mm
Nozzle	9 mm

The thermal diffusivity of each coating is determined according to the laser flash method with a Flashline™ 4010 (Anter Corporation, PA, USA) at 80°C which is a typical temperature during the cooling phase in the injection moulding. Each of the 5 coatings with different thicknesses and an uncoated substrate were prepared as specimens with a dimension of 10 mm x 10 mm x 3.1 mm. The top and the bottom faces of the specimens were sprayed with a thin layer of graphite suspension in order to improve the absorbance of the laser. The laser flash method consists of shooting a laser beam at the top surface of the probe and measuring the temperature on the bottom surface. Thermal diffusivity of the probe is then calculated by means of the time to the half maximum of the temperature of the bottom face and the thickness of the probe. Thermal conductivity was calculated using the measured thermal diffusivity, the density and the specific heat capacity of the material [11]. The effect of the substrate was taken into account with the help of the measurement of the uncoated substrate. Thermal conductivities of the coatings with different thicknesses did not show a significant correlation with their thicknesses because the scattering of the measurements was rather high in comparison to the slight variation in thermal conductivity values that was expected for the coatings with different thicknesses. The average value of the measured thermal conductivities for all measurements was determined to be $\lambda_{coating}$ = 1.1 W/mK. This value corresponds well with porous YSZ coatings sprayed by air plasma spraying by Schlichting et al. [12] and is used in the simulation of the injection moulding process, which is described in the following section.

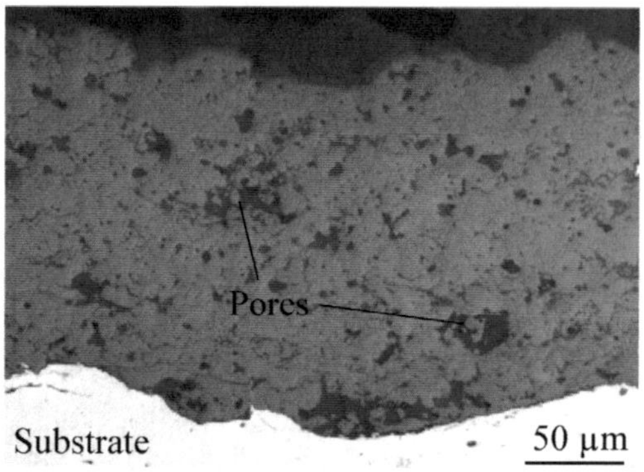

Figure 8: Cross-section picture of the coating with a thickness of 120 µm.

3. Effect of variable thickness coating on the cavity temperature

The manufactured coating with varying thickness was taken as the basis for the simulation of the injection moulding process with a coated mould cavity. In the simulation, the thickness of the coating is zero at near the gate opening and increases linearly in 10 steps up to 170 µm at the cavity end. Its thermal conductivity was set to $\lambda_{coating}$ = 1.1 W/mK according to the measurement with the laser flash method.

To analyse the effects of the coatings on the processing of injection moulded parts, simulations have been performed using injection moulding simulation software SIGMASOFT by SIGMA Engineering GmbH, Aachen. Modelling extremely thin elements with finite element and finite volume methods leads to a high element count as well as numerical precision issues. This is due to the mapping of the elements to a standardised element, which is used for the discretisation of the governing differential equations. To avoid both issues, the impact of the resulting thermal conductivities is modelled by a representative heat transfer coefficient (HTC). At first, a representative HTC was chosen to model a macroscopic contact without the coating. The initial heat transfer coefficient was chosen in a range recommended by the simulation software. For the first trials, a heat transfer coefficient of $\alpha = 1{,}000\ W/m^2K$ has been used for the whole contact area between part and mould. The heat transfer occurring as a result of the employed HTC value is translated into a reference thickness for a hypothetical insulation layer made of air. Figure 9 shows the geometric approach, which has been used to describe the heat transfer between part and mould. The representative HTC between the part and the mould steel should include the effect of the air layer and the coating. This is achieved by coupling in series the thermal resistances of the air layer and the coating, and then translating the combined thermal resistance into locally varying representative HTC values.

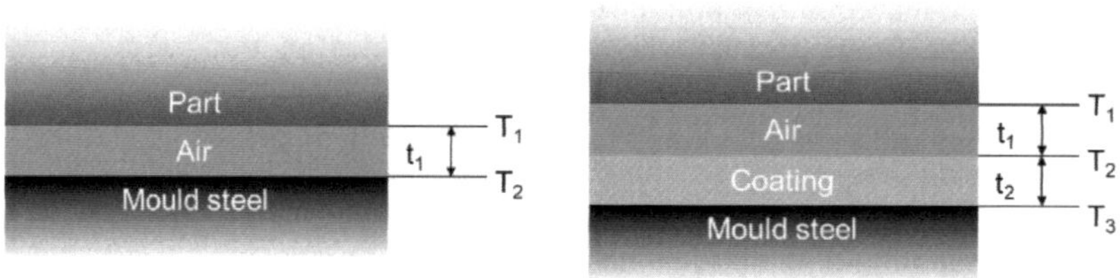

Figure 9: Geometrical simplification for the determination of heat transfer coefficients.

In a first step, based on the prescribed heat transfer coefficient of 1,000 W/m^2K and equations (1) and (2), resulting from the Fourier's law and the definition of the heat transfer coefficient, a representative air layer thickness of 0.0262 mm is determined.

$$\dot{q} = \alpha \cdot (T_1 - T_2) = \frac{\lambda_{\text{air}}}{t_1} \cdot (T_1 - T_2) \quad (1)$$

$$t_1 = \frac{\lambda_{\text{air}}}{\alpha} \quad (2)$$

where $\dot{q}$ denotes the heat flow, λ thermal conductivity, t_{air} the thickness of the air layer and $T_1 - T_2$ the temperature difference. In a second step, using Equation (3) and (4), the heat transfer is defined with the help of the total thermal resistance R.

$$\mathrm{R} = \sum \frac{t}{\lambda} = \frac{t_1}{\lambda_{\text{air}}} + \frac{t_2}{\lambda_{\text{coating}}} \quad (3)$$

$$\alpha_{total} = \frac{1}{R} \quad (4)$$

To include the heat transfer coefficients into the injection moulding simulation the mould surface has been separated into 10 discrete regions. Each region with different theoretical coating thicknesses is assigned with corresponding heat transfer coefficient. The theoretical coating thicknesses have been selected to be linearly increasing along the direction from the gate opening up to the end of the cavity. The corresponding position, coating thicknesses and calculated HTCs are given in Table 3.

Table 3: Discretised coating thickness and heat transfer coefficient.

Distance from P near the gate opening in Figure 10 [mm]	Coating thickness [µm]	Heat transfer coefficient [W/m^2K]
8.5	8.5	992.33
25.5	25.5	977.34
42.5	42.5	962.80
59.5	59.5	948.68
76.5	76.5	934.98
93.5	93.5	921.66
110.5	110.5	908.72
127.5	127.5	896.13
144.5	144.5	883.89
161.5	161.5	871.98

For the injection moulding simulation, a plate shaped sample geometry has been used, which is covered by two mould inserts. The plastic material used for the simulation is polypropylene "Sabic PP505P". For the discretisation of the coating layer, one of the mould inserts is split into 10 elements. Afterwards, for each part of the insert, the previously defined heat transfer coefficient is prescribed for the contact between the part and mould insert. Additionally, to cover the whole part, a solid mould is added. The geometric setup of the injection moulding simulation is illustrated in Figure 10.

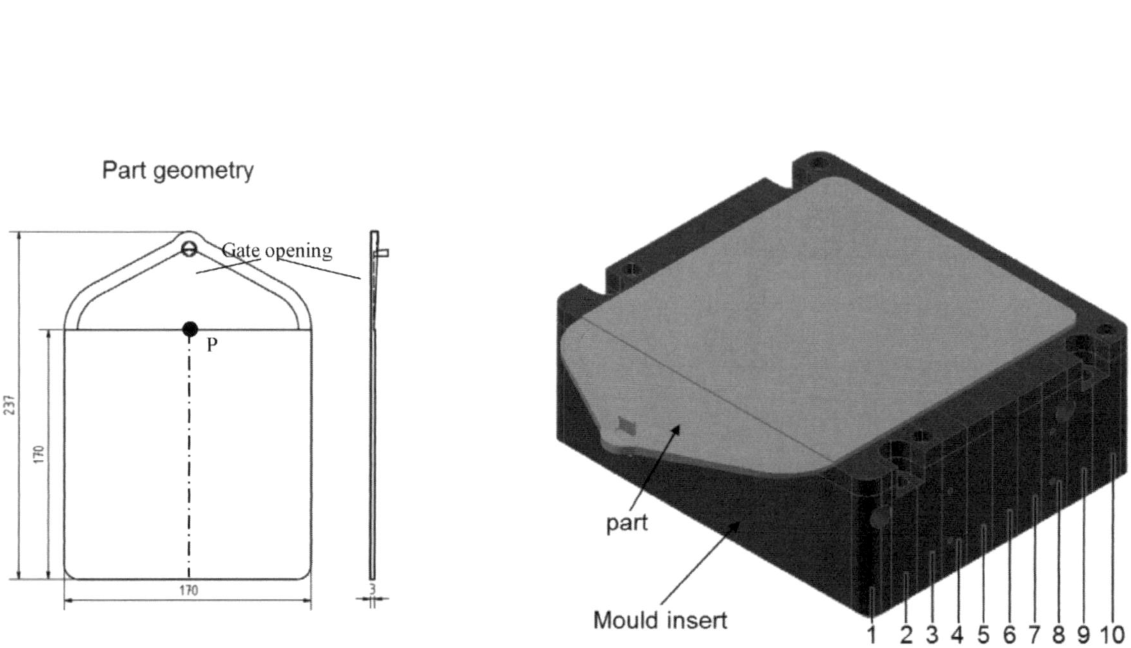

Figure 10: Geometric setup for the injection moulding simulation.

4. Results of the injection moulding simulation

The simulation shows a positive impact on the temperature distribution of the part. In the area where the coating was applied, a more homogeneous temperature distribution is achieved. Figure 11 illustrates the temperature distribution on the part surface, where the coating was applied to the mould insert. The simulation results for the modified mould exhibit a more homogeneous temperature profile. Implementation of linearly rising heat transfer coefficient allows a targeted manipulation of the heat transfer resulting locally at the part surface. As illustrated in figure 1, reducing the temperature difference, which is occurring locally by 2.5 K can lead to significant reductions of warpage, especially considering manufacturing of high precision parts.

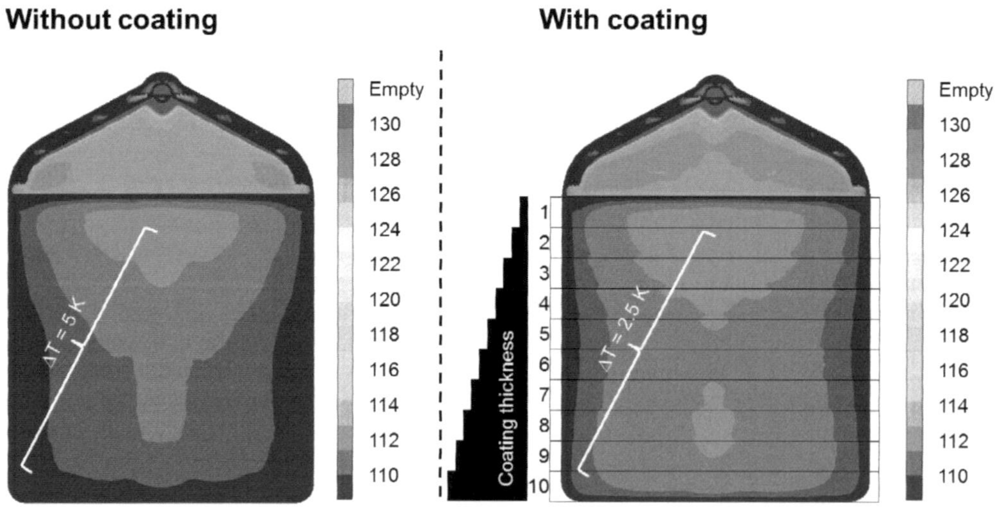

Figure 11: Simulated temperature distribution of the part after 10 s cooling.

5. Summary and outlook

A novel concept for varying thickness TBC for the application in injection moulding was presented in this work. Production of such a coating was presented and the thickness dependent thermal conductivity values for the presented coating were determined by the laser flash method. The contribution of the varying thickness TBC on the heat transfer coefficient in injection moulding process was numerically determined with the help of the experimentally determined thermal conductivity value. Employing the varying thickness TBC influenced the injection moulding process positively by helping to achieve a more uniform temperature distribution on the surface of the mould cavity, thus promising a reduction in warpage of plastic parts. In the future, validation experiments will be performed for these insulation coatings within the injection moulding process. The focus of these investigations will be the analysis of the resulting temperature distribution at the part thickness as well as the durability of these surface coatings with respect to the high temperature and pressure loads occurring in this process. Furthermore, coatings with a more complex thickness distribution will be developed, that will be able to reflect the cooling demand of more complex mould geometries. Due to the local application possibility, only critical positions have to be treated using air plasma spraying. Therefore, when only addressing cold temperature regions, the overall cycle time is not influenced, such that an efficient process with the same cycle time is still given.

Acknowledgments

All presented investigations were conducted in the context of the Collaborative Research Centre SFB1120 "Precision Melt Engineering" at RWTH Aachen University and funded by the German Research Foundation (DFG). For the sponsorship and the support we wish to express our sincere gratitude.

References

[1] Menges G, Haberstroh E, Michaeli W and Schmachtenber E 2002 *Werkstoffkunde der Kunststoffe* (München, Wien: Carl Hanser Verlag)

[2] Schürmann E 1979 *Abschätzmethoden für die Auslegung von Spritzgießwerkzeugen* PhD Thesis, RWTH Aachen

[3] Menges G, Michaeli W and Mohrern P 2007 *Spritzgießwerkzeuge. Auslegung, Bau, Anwendung* (München, Wien: Carl Hanser Verlag)

[4] Hopmann C, Filz P, Nikoleizig P and Schmitz M 2016 Precision melt engineering - Controlling the solidification process in injection moulding *28. Int. Koll. Kunststofftechnik* Aachen 306-350

[5] Bobzin K, Öte M, Knoch M A, Liao X, Hopmann C and Ochotta P 2016 Investigation of influencing factors on the transfer of wire arc sprayed Zn coatings for the metallization of plastic parts *Int. Thermal Spray Conf.* Shanghai 24-29

[6] Bobzin K, Öte M, Linke T F, Liao X, Hopmann C and Ochotta P 2015 Metallisierung von Kunststoffoberflächen – ein neues Verfahren: „In-Mould Metal Spraying (IMMS)“ *Jahrbuch Oberflächentechnik* Duisburg 67-73

[7] Bobzin K, Öte M, Linke T F, Liao X, Hopmann C and Ochotta P 2015 Anwendung des thermischen Spritzens für die Herstellung von Metall-Kunststoff-Bauteilen *Therm. Spr. Bull.* 28-31

[8] Bobzin K 2013 *Oberflächentechnik für den Maschinenbau* (Weinheim: Wiley-VCH Verlag,)

[9] Tucker R C Jr 2013 *ASM Handbook: Volume 5A* (Ohio: ASM International)

[10] Bobzin K, Öte M, Linke T F, Alkhasli I, Hopmann C, Nikoleizig P and Schmitz M 2015 Development of simulative approaches for precisely designing the properties of plasma sprayed coatings for application in injection moulding *Proc. 3rd ECCOMAS Young Investigators Conf.* Aachen 6-10

[11] Tojo T, Atake T, Mori T and Yamamura H 1999 Heat capacity and thermodynamic functions of zirconia and yttria-stabilized zirconia *J. Chem. Therm.* **31** 831-845

[12] Schlichting KW, Padture NP and Klemens PG 2001 Thermal conductivity of dense and porous yttria-stabilized zirconia *J. Mat. Sci.* 3003-3010

IOP Conf. Series: Materials Science and Engineering **181** (2017) 012014 doi:10.1088/1757-899X/181/1/012014

Application of TiC reinforced Fe-based coatings by means of High Velocity Air Fuel Spraying

K Bobzin, M Öte, M A Knoch, X Liao and J Sommer

Institut für Oberflächentechnik, RWTH Aachen University, Germany

E-Mail: sommer@iot.rwth-aachen.de

Abstract. In the field of hydraulic applications, different development trends can cause problems for coatings currently used as wear and corrosion protection for piston rods. Aqueous hydraulic fluids and rising raw material prices necessitate the search for alternatives to conventional coatings like galvanic hard chrome or High Velocity Oxygen Fuel (HVOF)-sprayed WC/Co coatings.

In a previous study, Fe/TiC coatings sprayed by a HVOF-process, were identified to be promising coating systems for wear and corrosion protection in hydraulic systems. In this feasibility study, the novel High Velocity Air Fuel (HVAF)-process, a modification of the HVOF-process, is investigated using the same feedstock material, which means the powder is not optimized for the HVAF-process. The asserted benefits of the HVAF-process are higher particle velocities and lower process temperatures, which can result in a lower porosity and oxidation of the coating. Further benefits of the HVAF process are claimed to be lower process costs and higher deposition rates. In this study, the focus is set on to the applicability of Fe/TiC coatings by HVAF in general. The Fe/TiC HVAF coating could be produced, successfully. The HVAF- and HVOF-coatings, produced with the same powder, were investigated using micro-hardness, porosity, wear and corrosion tests. A similar wear coefficient and micro-hardness for both processes could be achieved. Furthermore the propane/hydrogen proportion of the HVAF process and its influence on the coating thickness and the porosity was investigated.

1. Introduction

An important point for components in hydraulic applications is their lifetime. In the hydraulic steelwork, a lifetime of 30-50 years is a common demand [1]. To fulfil this requirement, the used piston rod has to be protected against wear and corrosion. Typical coatings for highly loaded hydraulic components are WC/CoCr and Cr_3C_2/NiCr sprayed by HVOF. These coatings exhibit a good corrosion and wear protection. Raw material prices have, however, risen over the last years. Therefore, researchers and the industry are investigating possible alternative cost efficient systems with comparable resistance to mechanical and corrosive attack. One alternative is the reduction of the carbide content of the coating. Another alternative is to investigate cost-efficient and ecological [2] Fe-based matrix materials with alternative hard materials, like TiC. Moreover, it can be noted that the additives in lubricants are often optimized for Fe-based materials, in order to increase the corrosion protection of the technical components [3].

In a previous study of the authors, TiC reinforced Fe-based (Fe/TiC) coatings with a relatively low carbide contend (34 wt.-%) has been developed considering the above-mentioned issues. Fe/TiC coatings were applied on the carbon steel C45 using two different HVOF systems; one operated with liquid fuel and one operated with gaseous fuel. Electroplated hard chrome and WC/CoCr HVAF coatings were used as references. The corrosion properties of the coatings system were investigated using polarization tests in artificial sea water (ASW) and in two typical flame retardant hydraulic lubricants. The wear properties were tested using a linearly reciprocating tribometer. In both the corrosion and wear tests, the Fe/TiC coatings showed similar or even better behaviour than the references. Of particular note is that the TiC in the Fe/TiC coating was partially converted into TiO_2. It was furthermore observed that Cr reacted with C and formed Cr_3C_2. A partial oxidation of matrix material was observed as well. The results show that HVOF sprayed Fe/TiC coatings can be used as an alternative coating system for protecting highly loaded piston rods in marine environments. [4]

The HVAF process is a modification of the HVOF process [5]. Compared to a HVOF process, the HVAF process has some different characteristics, e.g. a lower process temperature. The lower process temperature is achieved by using air instead of oxygen as oxidant [5]. The lower combustion temperature can result in coatings with a lower content of oxides. A number of studies confirm this hypothesis [5, 6, 7, 8, 9, 10]. The variety of different torch designs must, however, not be neglected in this consideration. Thus, no general statement concerning the oxygen content of HVOF and HVAF coating can be made [7]. In [11], HVAF WC/Co(Cr) coatings showed less oxidized phases and the amount of dissolved cemented carbides was lower, in comparison to HVOF coatings. This effect is ascribed to the lower combustion temperature. The gas flow rates in HVAF are higher than the one in HVOF. This can, in turn, enable a higher powder mass flow rate and consequently higher deposition rates [12]. The coating quality of coatings produced using parameters optimized for high deposition rates have yet to be investigated.

As stated in [4], using the HVOF process for applying Fe/TiC coatings, the following reactions took place. The TiC particles of the powder were partially converted into TiO_2 phases in the coating and the Cr parts of the powder reacted with C and formed Cr_3C_2 in the coating. With the assumption that, these reactions can be influenced by the combustion temperature, the idea of this study was to use the HVAF process to deposit Fe/TiC coatings. In this study, the focus lies on the feasibility of producing such HVAF Fe/TiC coatings. Here it must be noted that the powder fraction was not optimized for the HVAF process. To the knowledge of the authors, the applicability of Fe/TiC coatings by means of HVAF process and the corrosion and wear protection of these coatings applied with this technology have not been investigated in the literature, yet.

2. Experimental setup and description

The behaviour of coatings applied by means of HVOF and HVAF under corrosive and tribological loading was investigated using electro-chemical corrosion tests and a PoD tribometer. The coatings were investigated metallographically in the as-sprayed state and after the corrosion tests. Furthermore, the coatings' microhardness was measured. Before the microhardness measurements, the electro-chemical corrosion and the PoD tests, the coated samples were ground and polished to a surface parameter $Rz \approx 3$ µm.

1.0503 (C45) is a typical piston rod material and is, therefore, used as substrate material. The samples have the dimensions 50 x 30 x 8 mm^3. An iron-based matrix (51.5 wt.-% Fe, 27 wt.-% Cr, 18 wt.-% Ni and 3 wt.-% Mo) and 34 wt.-% of TiC was used as feedstock material. Quite high Cr and Ni contents were chosen to ensure a good corrosion resistance of the coating. The powder was agglomerated and sintered with a grain size fraction of -45 +10 µm (Oerlikon Metco WOKA GmbH, Barchfeld, Germany). The agglomerates have a spherical shape. The TiC particles have an average size of approximately 2 µm and are evenly distributed in the powder. For both the HVOF and the HVAF process, the same feedstock material was used.

The HVAF coatings were produced using an AK-07 system (Kermetico Inc., Benicia, USA). The first attempts did, however, clog the nozzle. To avoid the clogging, the propane flow rate was set at minimum and the air flow rate at maximum. This parameter set can be designated "cold" and, therefore, the particles exhibit relatively low temperatures resulting in reduction in nozzle clogging. This parameter set is called "HVAF 1". The SEM-image of parameter set "HVAF 1" shows a high porosity, which may be unfavourable for the use as a corrosion protection coating. To produce denser coatings, the temperatures of the sprayed particles need to be increased, which would influence flattening behaviour of the feedstock material. Two strategies for changing the flattening behaviour were used. The first strategy was to increase the propane gas flow rate and the second was to add hydrogen into the combustion chamber, additionally. If the propane flow rate is increased, the combustion temperature increases. If a hydrogen flow rate is added into the combustion chamber, the heat transfer between the spray particles and the gas-mixture is improved [13]. For both mechanisms, stable process parameters could be identified. In the following, one sample each for both mechanisms will be presented in addition to sample HVAF 1. The samples are named HVAF 2 and HVAF 3. The selection was based on the achieved porosity and the wear and corrosion behaviour. For the sake of completeness, all process parameters are given in table 1.

Table 1. HVAF-Spray parameters for the AK-07.

Constant Parameter	Value
Nitrogen flow rate	19 SLPM
Powder feed rate	40 g/min
Air pressure	7.3 bar

Sample Name	Propane Pressure (bar)	Hydrogen (SLPM)	Stand-off distance (mm)
HVAF 1 (Ref.)	**5.5**	**0**	**175**
HVAF 2 (P1.1)	**5.8**	**0**	**175**
(P1.2)	5.8	0	210
(P2)	6.1	0	175
(H1)	5.5	2.5	175
(H2)	5.5	5	175
(H3)	5.5	10	175
HVAF 3 (H4)	**5.5**	**15**	**175**
(H5)	5.5	20	175

In order to assess the microstructure, cross sections of the HVAF coatings were prepared. The cross section were analysed using scanning electron microscopy (SEM; LEO 1530 VP Gemini, Carl Zeiss AG Microscopy, Oberkochen, Germany) and energy dispersive X-ray spectroscopy (EDS; Quantax 200 from Bruker Nano GmbH, Berlin, Germany). The porosity of the coating was determined with the evaluation software Image Analysis and a light microscope Zeiss Axiophot (both Carl Zeiss AH, Oberkochen, Germany).

The microhardness of the coatings was measured with the measurement system; Micromet 1 (Buehler, Illinois, USA). For this purpose, the coating was loaded with a weight of $m = 300$ g for $t = 15$ sec. Every coating of each parameter set was measured 10 times in order to generate statistically relevant data.

As stated in the introduction, in a previous study by the authors, Fe/TiC HVOF coatings were investigated [4]. In this study, the best corrosion and wear properties could be achieved using the parameter set shown in table 2 using a K2 HVOF torch (GTV Verschleißschutz GmbH, Luckenbach, Germany).

Table 2. Spray parameters for the K2 [4].

Parameter	Value
O_2 flow rate	870 SLPM
Kerosene L/h	20 L/h
Oxygen/Fuel ratio λ	1.4
Stand-off distance	300 mm
Transversal Speed	1,000 mm/s
Powder feed rate	29 g/min

For the investigation of the wear properties, a Pin-on-Disc (PoD) tribometer (CSM Instruments SA, Freiburg, Germany) was used. An Al_2O_3 ball with a diameter $ø = 6$ mm was used as counter body.

Al_2O_3 was chosen to characterise the wear properties of the coatings. The use of Al_2O_3 counter body material ensures the comparability as Al_2O_3 is widely used for wear tests. For more application oriented tests, the counter body material should be changed, e.g. to the seal material. The further PoD parameters are shown in table 3. The 15 x 15 mm² samples were tested using a hydrofluorocarbon (HFC) lubricant as well as without a lubricant. The experiments with lubricant were done to investigate the influence of a typical hydraulic fluid used in low flammability applications. The HFC lubricant was provided by Fuchs Hydraulik GmbH & Co. KG (Kirn, Germany), the exact composition is confidential. Subsequent the wear test, the wear coefficient K for the samples was determined using equation (1) [2]. Therefore, the volumetric wear V and the measured wear track diameter d_t was determined using confocal laser scanning microscopy (VKX 200, Keyence Corporation, Osaka, Japan).

Table 3. PoD test parameters.

Parameter	Value
Wear track diameter d	5 mm
Wear track length s	500 m
Normal load w	10 N
Rotation speed	10 cm/s
Test temperature	RT

$$K\left[\frac{mm^3}{Nm}\right] = V\cdot\left[w(s\cdot\frac{s}{d_t})\right]^{-1} \quad (1)$$

To analyse the corrosion resistance of the samples, polarization tests were performed in ASW at room temperature in accordance to DIN 50905-4. ASW is an electrolyte, which simulates the real environment in marine applications while maintaining the comparability. Therefore, ASW is important for testing coatings used in coastal areas and offshore constructions. A Calomel-(Hg_2Cl_2) electrode was used as reference electrode and the potential scan rate was set at 0.5 mV/s. The rest potentials and the corrosion current densities were investigated using the Tafel slope analysis.

3. Results and discussion

Figure 1 shows a selection of SEM images of the HVAF coatings. All coatings HVAF 1, HVAF 2 and HVAF 3 exhibit a good bonding to the substrate. The coating thickness d_{ct} of the samples was measured at d_{ct} = 232 – 296 µm. Coating thicknesses differ significantly from each other, although all samples were sprayed with the same number of passes and the same robot parameters. The sample HVAF 1 exhibits the biggest area with local inhomogeneities, like porosity and imperfections. The magnification of HVAF 1 shows a high porosity area near the substrate. To minimize the porosity, the parameters were varied as described in chapter "2. Experimental setup and description". The sample HVAF 2 was sprayed with a higher propane flow rate to increase the temperature in the combustion chamber. In general, this coating shows significantly less local inhomogeneities than the sample HVAF 1. The measured porosity of this sample was 1.9 %. In the high magnification view of sample HVAF 2, some small pores can be seen, especially in the top layers. These small pores can be assumed to be the result of a reduced compression of the coating applied during the last passes. For the following corrosion and wear tests, the samples were polished. Thus, the small pores in the top layer will not influence the corrosion tests. Such a finishing is common practice throughout the industry and therefore, the increased inhomogeneity displayed in the magnification is assumed to be irrelevant. Generally speaking, this sample shows a much denser coating than the sample HVAF 1. For sample HVAF 3, the hydrogen flow rate was increased in order to increase the heat transfer between the particles and the gas flow. The cross section of this coating shows a dense coating. In high magnification view, one unmolten particle can be seen. The porosity of this coating was about 2.6 %.

Areas with local inhomogeneities were not detected for sample HVAF 3. In conclusion, the local inhomogeneities of the coating could be decreased by increasing the propane gas flow rate and adding a hydrogen gas flow.

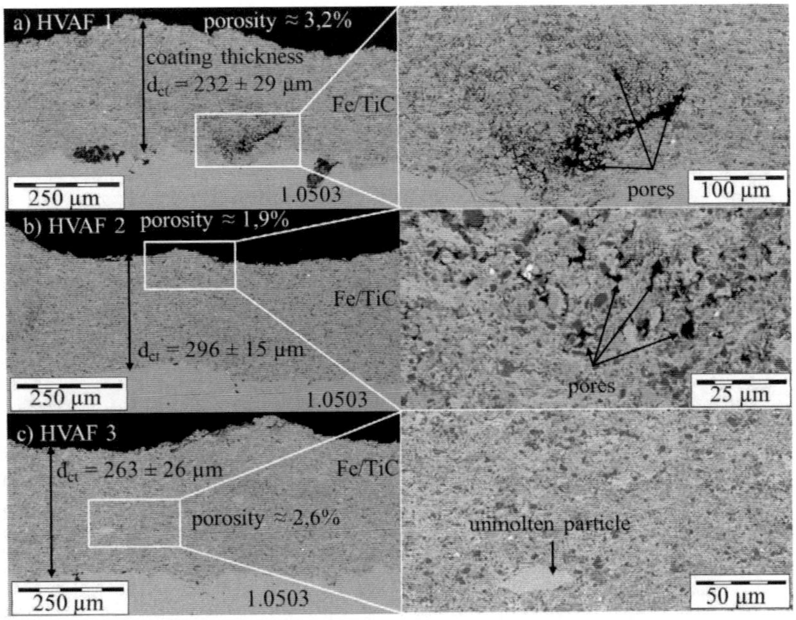

Figure 1. SEM images of the samples HVAF 1 (a), HVAF 2 (b) and HVAF 3 (c).

In figure 2, the SEM images as well as the results of the EDS analysis of the coatings HVAF 1 and HVAF 2 are shown. In the coatings, Cr carbides, Ti carbides, mixed oxide phases and a solid solution were detected. The solid solution consists of Cr, Fe, Ni and Mo. The mixed oxide phases have a high content of Ti. An explanation for the oxidized matrix parts with a high Ti content might be decarburisation and the high kinetic energy of the TiC particles. If the TiC particles hit the surface, they can break brittlely, react with the air and phases with a high TiO_2 content can form. In the sample HVAF 1, an additional Mo phase could be detected.

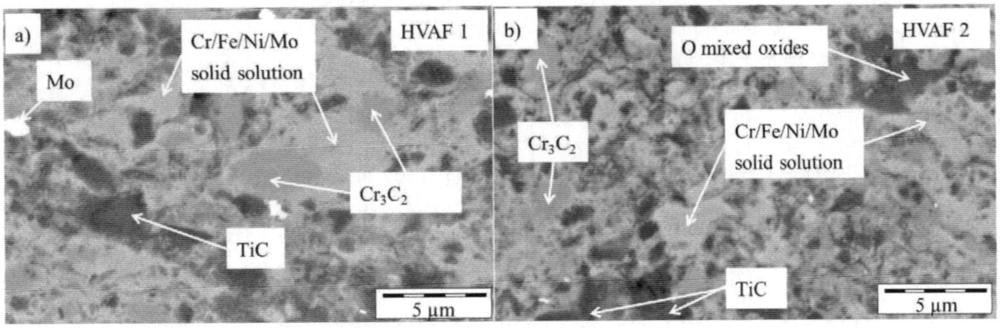

Figure 2. SEM/EDS analysis of the sample HVAF 1 (a) and HVAF 2 (b).

In figure 3 the microhardnesses of the considered samples are shown. Ten measurements were conducted for every sample. All HVAF coatings have a much higher hardness than the substrate. The sample HVAF 1 shows the lowest microhardness of 630 HV0.3. The microhardness of sample HVAF 3 has a value of 672 HV0.3. The samples HVAF 2 and HVOF show a similar microhardness of 694 HV0.3 and 730 HV0.3. The lower microhardness of the samples HVAF 1 and HVAF 3 is assumed to be a result of the higher porosity of these coatings compared to that of the other coatings HVAF 2 and HVOF. It can be noted that a lower microhardness was measured for samples, which have a higher porosity. The sample HVOF has a porosity about 1 %. All HVOF and HVAF samples have a high standard deviation with regard to their microhardness, which is typical for thermally sprayed coatings. This means the microhardness of the coatings varies, depending on the position of the indentation.

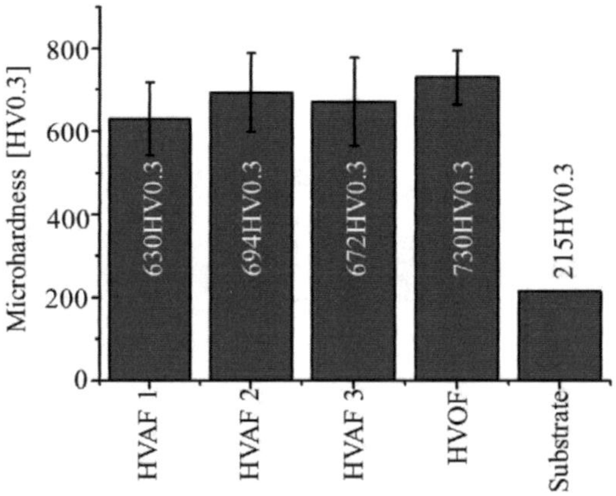

Figure 3. Microhardness of the samples HVAF 1, HVAF 2, HVAF 3, HVOF and the substrate.

In figure 4 the polarization curves of the coatings and the uncoated substrate material in ASW are shown. The rest potential (U_R) and the corrosion current density (i_c) of the samples are shown in table 4. In ASW, the substrate has the lowest corrosion resistance. It has the lowest rest potential with a value of U_R = -457 mV_H and the highest corrosion current density of i_c = 0.0096 mA/cm². The curves of the HVOF and the HVAF Fe/TiC coatings are comparable. Among these, the rest potential of the coating HVAF 3 is the highest, meaning it is the noblest coating, whereas the HVOF coating is the least noble one. The HVOF coating has the lowest corrosion current density, whereas the coatings HVAF 2 and HVAF 3 show a slightly higher corrosion current density. The sample HVAF 1 shows the highest corrosion current density of all coated samples. An interesting aspect can be found for sample HVAF 2, which has the lowest current density until the breakthrough at about U = 100 mV_H. Within the potential range of U = 400 – 550 mV_H, HVAF 1 and HVAF 2 have almost identical current densities as the substrate. For higher overpotentials, a slight drop can be seen for both samples. Over a large area, the corrosion rates of the coatings HVAF 2 and HVAF 1 is significantly smaller than the rate of the substrate. The samples HVAF 3 and HVOF did not exceed the current density of the substrate. It should be noted that this behaviour can only be detected after the breakthrough. Since there are many unknown factors, which may influence the current density in this area, e.g. the temperature, electrolyte aging, etc. closer attention should be paid to the area before the breakthrough potential. It should be noted that the HVOF sample exhibits lower current densities after the breakthrough and thus, the overall corrosion rate after the polarization test is assumed to be lower for the HVOF sample. As a result of the corrosion tests, it can be stated that the coating HVAF 2 shows the lowest corrosion rates of all samples until the breakthrough. The results of the polarization curves can be summarized as followed. The HVAF coatings have the highest rest potential and, until the breakthrough, these coatings show similar or even better corrosion behaviour than the HVOF coating.

Sample HVAF 2 has the lowest porosity and overall the lowest current density of all HVAF coatings. Despite the high overpotential of $\Delta U > 1.25$ V, all coatings show a lower corrosion rate than the substrate. This clearly indicates that all coatings protect the substrate during the entire corrosion test.

Table 4. Current density curves data.

Sample	Rest potential U_R [mV]	Corrosion current density i_c [mA/cm²]
Substrate	-457	0.0096
HVAF 1	-316	0.0050
HVAF 2	-300	0.0022
HVAF 3	-286	0.0035
HVOF	-360	0.0014

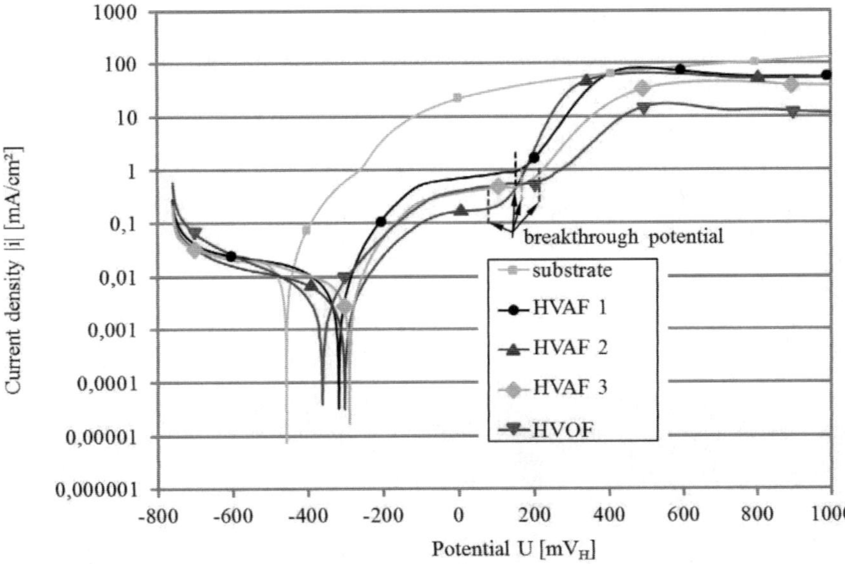

Figure 4. Corrosion current potential curves of the coatings in ASW at RT.

In figure 5, cross sections of the samples HVOF, HVAF 1, HVAF2 and HVAF 3 after the corrosion tests are shown. The cross sections confirm the results of the polarization test. For all samples, it can be seen that there was no corrosion of the substrate. The oxidation of the matrix is more distinct for the HVAF than the one of the HVOF sample. Consequently, the HVOF coating has a bigger remaining thickness of uncorroded coating material than the HVAF coatings. This corrosion behaviour can be traced to the lower overall current density of the HVOF sample.

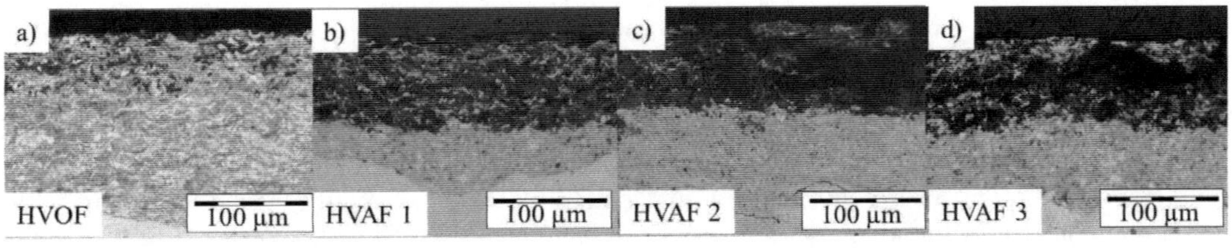

Figure 5. Cross-section polishes of the samples HVOF (a), HVAF 1 (b), HVAF 2 (c) and HVAF 3 (d). In figure 6, the results of the POD tests are shown. In figure 6 a), the wear tracks for the samples HVAF 2 and HVAF 3 are displayed. The images of the unlubricated wear tracks show non-homogeneous wear tracks with similar characteristics for all HVOF and HVAF coatings. All samples have abrasive wear behaviour and splat breakages. The samples HVOF and HVAF 3 exhibit qualitatively less outbreaks than the sample HVAF 2. A high porosity of the coating promotes breakages of the coating, which correlates with the images shown in Figure 6. The porosity offers a target for the counter body and a chunk of the coating can break out. For the lubricated tests, wear tracks with homogeneous shape and no significant outbreaks were detected.

In figure 6 b), the wear coefficient K for the unlubricated and the lubricated wear tracks is shown. All coatings show a lower wear rate than the substrate. This applies to the unlubricated and lubricated experiments. For the unlubricated coatings, the sample HVAF 1 shows the highest wear coefficient of $K = 2.7 \cdot 10^{-5}$ mm³/Nm. With the parameter set HVAF 3, a similar wear coefficient $K = 1.4 \cdot 10^{-5}$ as for the HVOF sample was achieved. In general, the lubricated wear coefficients K are much lower than the unlubricated ones. The highest wear coefficient K was again measured for the substrate. The sample HVAF 1 achieved a wear coefficient of $K = 2.1 \cdot 10^{-6}$ mm³/Nm. The lowest wear coefficient K was achieved for the sample HVOF 3 with a value of $K = 5.2 \cdot 10^{-7}$ mm³/Nm. For the HVOF sample, the lubricated wear could not be determined due to an instrumental failure. The images of the wear tracks and the detected wear coefficient K correlate. The sample with the lowest wear has the lowest amount of outbreaks.

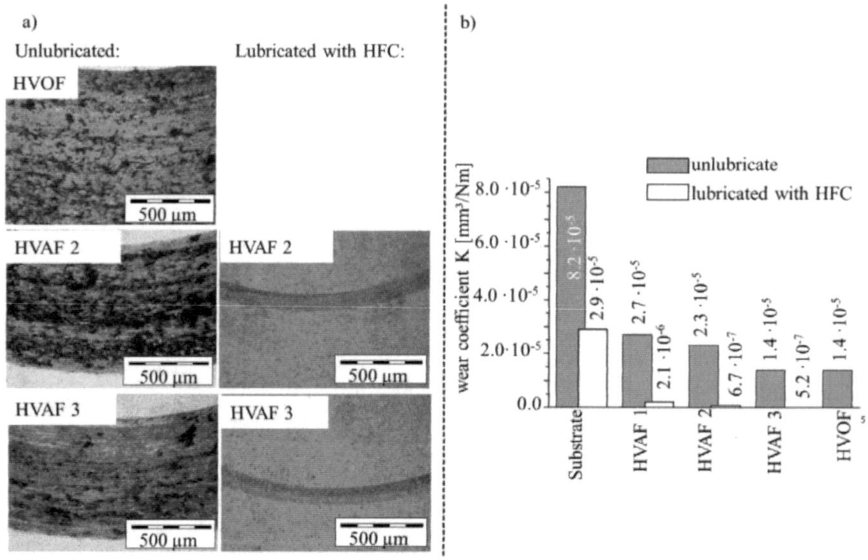

Figure 6. Wear tracks of sample HVOF, HVAF 2 and HVAF 3, (a) Wear of friction without lubrication and with lubrication (b).

4. Conclusion and outlook

The aim of this study was to proof the feasibility of applying Fe/TiC coatings using the HVAF process. Fe/TiC coatings were successfully applied onto a 1.0503 substrate. The first set of parameter shows larger areas with inhomogeneities. After a parameter variation, these inhomogeneities could be reduced. In the EDS analysis, mixed oxides could be detected in the HVAF coatings. In a further study, the oxide percentage of the HVAF and HVOF coatings can be compared as this might explain the differences in the corrosion and wear behaviour. All HVAF coatings achieve a better wear behaviour than the substrate in lubricated and unlubricated tests. The corrosion behaviour of all coatings is better than that of the substrate. Furthermore, similar trends can be observed for moderate overpotentials for all coatings. Nonetheless, the differences in the polarization curves clearly show that the process parameters are an important factor. For low overpotentials, the coating HVAF 2 shows the best corrosion behaviour of all coating systems until the breakthrough. This coating does also exhibit the lowest porosity of all HVAF coatings indicating the porosity influences the corrosion behaviour and that lower porosity is favourable. The broad range of the feedstock material's grain size is assumed to be one reason for the uncharacteristically high porosity of all HVAF coatings. For a further coating optimization, it could therefore be useful to use powders with a narrower grain size than the one used in this study. Another aspect which may consider is the optimization of the HVAF torch. To avoid nozzle clogging, a smaller combustion chamber and smaller nozzles can be used.

Acknowledgments
This work is based on the DFG project BO 1979/45-1. The authors gratefully acknowledge the financial support of the German Research Foundation (DFG).

References

[1] Burhart H F 2011 Kolbenbeschichtung, Verfahren, Qualitätsbeurteilung und Trends *IFAS-Kolloquium, Aachen Germany*
[2] Milanti A, Koivuluoto H, Vuoristo P and Bolelli G2014 Microstructural characteristics and tribological behavior of HVOF-Sprayed novel Fe-based alloy coatings *Coatings* **4**
[3] Findeisen D and Feldhuser S2015 *Ölhydraulik* (Berlin: Springer)
[4] Bobzin K, Öte M, Linke T F and Malik, K A2016 Wear and corrosion resistance of Fe-based coatings reinforced by TiC particles for application in hydraulic systems *Journal of Thermal Spray Technology* **25** pp. 365-74
[5] Guo R Q, Zhang C, Chen Q and Yang Y 2011 Study of structure and corrosion resistance of Fe-based amorphous coatings prepared by HVAF and HVOF *Corrosion Science* **53** 2351-56
[6] Wang Q, Zhang S, Cheng Y and Xiang J 2013 Wear and corrosion performance of WC-10Co4Cr coatings deposited by different HVOF and HVAF spraying processes *Surface & Coating Technology* **218** pp. 127-36
[7] Bolelli G, Berger L M, Boerner T and Koivuluoto H 2015 Tribology of HVOF- and HVAF-sprayed WC-10Co4Cr hardmetal coatings: a comparative assessment *Surface & Coating Technology* **265** 125-44
[8] Hulka I, Serban V A, Niemi K and Vuoristo P 2012 Comparison of structure and wear properties of fine-structured WC-CoCr coatings deposited by HVOF and HVAF spraying Processes *Solid State Phenomena* **188** 422-27
[9] Bolelli G, Hulka I, Koivuluoto H and Lusvarghi L 2014 Properties of WC-FeCrAl coatings manufactured by different high velocity thermal spray processes *Surface & Coating Technology* **247** 74-89
[10] Berger L M, Puschmann R, Spatzier J and Matthews S 2013 Potential von HVAF-Spritzprozessen *Thermal Spray Bulletin* 17-20
[11] Hashmi S 2014 Comprehensive materials processing: thermal spray coating process *Newnes* 230-266
[12] Salman A, Gabbitas B, Cao P and Zhang D 2009 Tribological properties of Ti(AlO/Al_2O_3) composite coating by thermal spraying *International Journal of Modern Physics* **23** 1407-12
[13] Verstak A and Baranovski V 2006 AC-HVAF sprayed tungsten carbide: properties and application *Proc. ITSC Seattle, USA*

IOP Conf. Series: Materials Science and Engineering **181** (2017) 012015 doi:10.1088/1757-899X/181/1/012015

Processing of AlCoCrFeNiTi high entropy alloy by atmospheric plasma spraying

M Löbel[1], T Lindner[1], C Kohrt[2] and T Lampke[1]

[1]Institute of Materials Science and Engineering, Chemnitz University of Technology, 09125 Chemnitz, Germany
[2]Rhein-Ruhr Beschichtungs-Service GmbH, 47495 Rheinberg, Germany

Abstract. High Entropy Alloys (HEA) are gaining increasing interest due to their unique combination of properties. Especially the combination of high mechanical strength and hardness with distinct ductility makes them attractive for numerous applications. One interesting alloy system that exhibits excellent properties in bulk state is AlCoCrFeNiTi. A high strength, wear resistance and high-temperature resistance are the necessary requirements for the application in surface engineering. The suitability of blended, mechanically ball milled and inert gas atomized feedstock powders for the development of atmospheric plasma sprayed (APS) coatings is investigated in this study. The ball milled and inert gas atomized powders were characterized regarding their particle morphology, phase composition, chemical composition and powder size distribution. The microstructure and phase composition of the thermal spray coatings produced with different feedstock materials was investigated and compared with the feedstock material. Furthermore, the Vickers hardness (HV) was measured and the wear behavior under different tribological conditions was tested in ball-on-disk, oscillating wear and scratch tests. The results show that all produced feedstock materials and coatings exhibit a multiphase composition. The coatings produced with inert gas atomized feedstock material provide the best wear resistance and the highest degree of homogeneity.

1. Introduction

The term of HEA was introduced by Yeh et al. [1]. By definition, those new alloy systems are composed of at least five elements with a share of 5-35 at.% each. Usually the formation of brittle, intermetallic and complex phases is expected. Nevertheless, investigations have shown the formation of simple solid solutions with face centered cubic (FCC) and body centered cubic (BCC) structures due to the stabilization by high entropy effect. No intermetallic or complex phases were formed [1, 2]. At the same time, independent investigations were carried out by Cantor et al. [3], who proved that a single phase FCC structure forms in the equimolar alloy FeCrMnNiCo. Especially the usage of transition metals in the alloy system prevents the formation of intermetallic phases or complex compounds [1, 2]. However, structural investigations on most HEAs revealed a multiphase structure consisting of several phases [4]. Therefore the term compositionally complex alloys (CCA) was introduced for alloys, which are not composed of one single simple solid solution [5]. CCAs with a high share of Al and Ti are especially convenient to form lightweight HEAs due to their low density. An alloy system, which has already been investigated in bulk state, is AlCoCrFeNiTi. First investigations of Zhou et al. [6] revealed a dendritic structure consisting of two BCC solid solutions. Furthermore, high fracture strength of 2.58 GPa and plastic strain of 8.8 % were achieved in compression test. These properties were attributed to solid solution strengthening by Ti atoms [6].

Further investigations concentrated on the structure and influence of alloying elements, especially Ti and Al. Al showed strong influence on phase composition. For alloys with high Al content, mainly strong and brittle BCC phases were formed. Increasing the amount of Ti leads to strengthening by solid solution hardening and if added in high amounts to the formation of intermetallic phases [7-10]. A comparison of wear performance between the CCA system AlCoCrFeNiTi and conventional steels (SUJ2, SKH51) showed promising properties [10, 11]. So far, mainly bulk properties of HEAs and CCAs have been investigated. Only few investigations have been carried out to apply HEAs and CCAs by thermal spray technology. First investigations mainly focused on microstructural characterization and phase composition of coatings [12, 13]. However, some research has been carried out to investigate the wear resistance of thermal spray HEA coatings, revealing good wear resistance in comparison with conventional steel (SUJ2, SKD61) [14]. The focus of this study is the development of an AlCoCrFeNiTi coating for wear protection applications. Surface engineering by thermal spray processing enables a resource efficient use of material by functional division into surface and bulk parts. This is necessary for possible applications due to the application of expensive alloying elements in HEA/CCA alloy systems.

2. Experiments

2.1 Preparation and characterization of feedstock material

In this study different feedstock materials of the CCA system AlCoCrFeNiTi with equimolar composition are investigated. The first approach to supply feedstock for thermal spray processes was the production of a powder blend. Pure elemental powders were utilized, with the exception of Fe, which was added as pre-alloy FeCr13. The different powders were mixed in a tumbling mixer WAB Turbula. This powder blend was directly used as feedstock material for APS. Furthermore, the powder blend was used as a raw material for the production of alloyed powder with a homogeneous composition by high-energy ball milling (HEM) in a Zoz Simoloyer CM08 mill. Milling parameters are listed in table 1.

Table 1. Milling parameters for the production of alloyed powder by HEM utilizing a powder blend.

Ground material / powder mass	0.8 kg
Grinding medium	Steel balls (100Cr6 / Ø 5 mm)
Atmosphere	air
Ball to powder ratio (BPR)	10:1
Process control agent	Stearic Acid (SA)
Rotor speed	400-700 rpm cyclic)
Total milling time	14 h

Another attempt to produce homogeneous thermal spray powder was the application of inert gas atomization. In a first step, a homogeneous alloy was produced by melting elements under vacuum. Subsequently the alloy was atomized with Argon as inert process gas to avoid reactions. After atomization the powder was separated into several particle size fractions. The fraction with D_{90}< 100 µm was used for thermal spray.

All feedstock materials were investigated regarding their particle size distribution by laser diffraction analysis in a Cilas 930 device. For the milled and atomized powder, metallographic cross sections were prepared and investigated in the scanning electron microscope (SEM) Leo 1455VP to examine the structure, homogeneity and particle morphology. A backscattered electron detector (BSD) has been utilized to show material contrast. Furthermore, the chemical composition was investigated by energy dispersive X-Ray spectroscopy (EDX) with EDAX Genesis. Phase analysis was carried out by X-Ray diffraction (XRD) with a D8 Discover/Bruker AXS diffractometer using Co-K_{α}-radiation.

2.2 Coating process and characterization

Prior to coating, steel flat substrates (S235) with a diameter of 40 mm were prepared by corundum blasting using corundum with grain size 425 – 600 µm, a pressure of 3.5 bar, a distance of 100 mm and angle of 60°. After ultrasonic cleaning in an ethanol bath, the substrates were coated with the APS system GTV-F6 to produce coatings with a thickness of approximately 200 µm. The parameters were adjusted according to optimize coating porosity and hardness. The resulting parameters for the different feedstock materials and are shown in table 2.

Table 2. APS parameters for applied feedstock materials.

	Powder blend	**HEM powder**	**Atomized powder**
Ar	38 l/min	38 l/min	54 l/min
H_2	14 l/min	12.5 l/min	9 l/min
I	530 A	600 A	530 A
Powder feed rate	50 g/min	25 g/min	50 g/min
No. of layers	9	12	12
Spray distance	120 mm	120 mm	120 mm
Relative traverse speed	100 m/min	70 m/min	100 m/min
Spray path offset	4 mm	4 mm	4 mm

Cross sections of all coatings were prepared and examined by SEM to investigate the microstructure and phase composition. The chemical composition was investigated by EDX. In a next step, the Vickers hardness (HV) was measured in the cross sections with a Wilson Tukon 1102 device. Furthermore, the wear behavior under different tribological conditions was tested by ball-on-disk wear tests with a Tetra Basalt Tester, oscillating wear tests with a Wazau SVT 40 device and scratch tests with a CSM Revetest-RST device. The wear test parameters are shown in table 3.

Table 3. Wear test parameters for APS coatings.

Ball-on-disk test		**Oscillating wear test**		**Scratch test**	
force	20 N	force	26 N	mode	progressive
radius	5 mm	frequency	40 Hz	force	1-200 N
speed	96 rpm	time	900 s	speed	2.5 mm/min
cycles	15916	amplitude	0.5 mm	length	5 mm
Ø Al_2O_3	6 mm	Ø Al_2O_3	10 mm	tip	Truncated diamond cone, r: 200 µm

Wear marks resulting from scratch and oscillating wear tests were investigated with a laser scanning microscope (LSM) Keyence VK-X200. Ball-on-disk wear tracks were analyzed with the stylus instrument Hommelwerke LV-50E. Furthermore, wear marks made by scratch tests were investigated with the digital microscope Keyence VHX-500.

3. Results and Discussions

3.1 Feedstock characterization

Metallographic cross sections of the milled and atomized powders were investigated and are shown in figure 1.

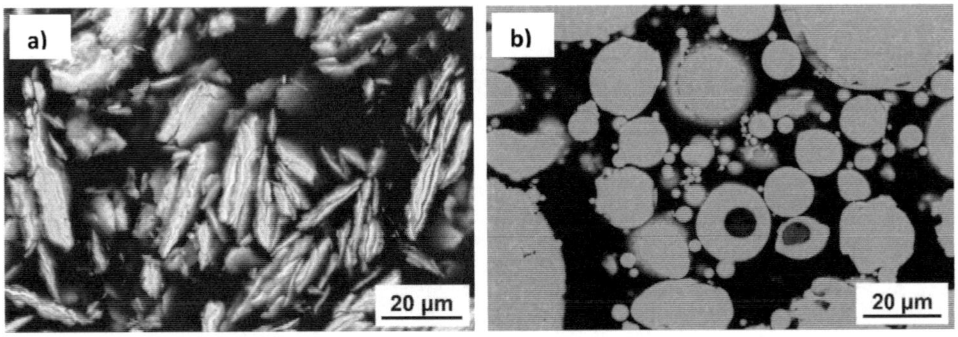

Figure 1. SEM (BSD) images of: a) HEM powder and b) atomized powder.

The HEM powder consists of plate-like particles with a heterogeneous structure. Within particles, lamella can be distinguished. The contrast in BSD images indicates that the particles do not have a chemically homogeneous composition.

In contrast, the cross section of atomized powder shows spherical particles and a more homogeneous microstructure. Several phases can be distinguished within particles. This microstructure is in accordance with earlier work on bulk samples [6, 15]. In a next step, the overall chemical composition has been analyzed by EDX. The results, in comparison with the nominal composition, are shown in table 4.

Table 4. Chemical composition of HEM and atomized powder measured by EDX.

	Al (at.%)	Co (at.%)	Cr (at.%)	Fe (at.%)	Ni (at.%)	Ti (at.%)
nominal	*16.7*	*16.7*	*16.7*	*16.7*	*16.7*	*16.7*
HEM powder	24.3	19.8	15.7	13.7	13.8	12.7
atomized powder	18.3	16.0	16.9	16.7	16.1	16.0

The chemical composition of the HEM powder deviates from the nominal composition. Especially the content of Al and Ti shows strong deviation. The chemical analysis might be influenced by the inhomogeneous state of powder. In contrast, the composition of the atomized powder is in good accordance with nominal composition. Only the Al content shows a deviation which exceeds 1 at.%. The particle size distribution of all applied feedstock materials was investigated by laser diffraction analysis. The results are displayed in figure 2.

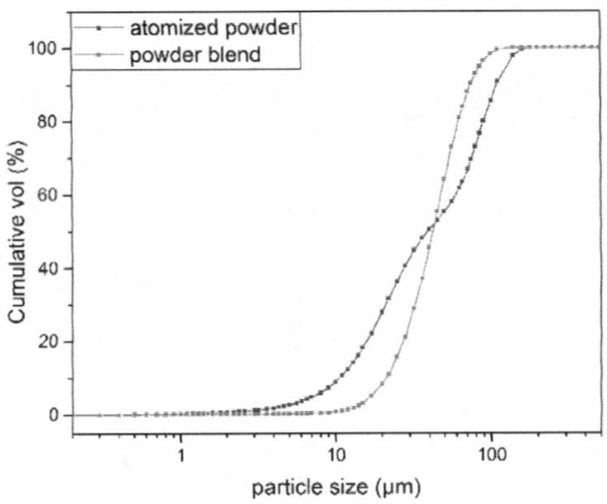

Figure 2. Cumulative particle size distribution of powder blend and atomized powder feedstock.

Due to the irregular particle shape of HEM powder, an analysis by laser diffraction was not possible. The evaluation of particle size analysis reveals a d_{50} value of 41.3 µm and a d_{90} value of 75.0 µm for the powder blend. In comparison, the atomized powder shows a coarser size distribution with a d_{50} value of 42.9 µm and d_{90} value of 109.9 µm. Phase analysis was carried out by XRD. Resulting diffractograms are shown in figure 3.

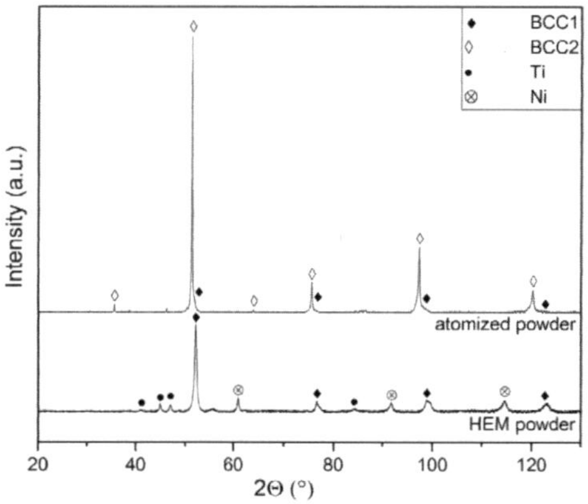

Figure 3. XRD patterns of HEM and atomized powder.

The XRD pattern of the HEM powder exhibits minor peaks of elemental Ti and Ni. Consequently, no homogeneous and fully alloyed powder was obtained. These peaks indicate the existence of unmilled Ni and Ti particles. However, a BCC structure (PDF 00-006-0694 / Im-3m (229)) is formed, which is denoted as BCC1.

The diffractogram of the atomized powder displays peaks, which can be ascribed to another BCC phase (PDF 03-065-4198 / Pm-3m (221)) labeled as BCC2. The high intensity diffraction peaks exhibit a shoulder towards higher diffraction angles, indicating the overlapping of diffraction peaks ascribed to the BCC1 phase. Furthermore, powder of the same lot with a coarser powder size distribution has been investigated. Diffraction patterns of this powder clearly reveal peaks, which can be ascribed to the BCC1 phase. Microstructural investigations also show a multiphase structure indicating the existence of the BCC1 phase.

3.2 Coating characterization

The first step of coating characterization was the preparation of metallographic cross sections and investigation in SEM. Metallographic cross sections of coatings produced with powder blend, HEM powder and atomized powder feedstock are displayed in figure 4.

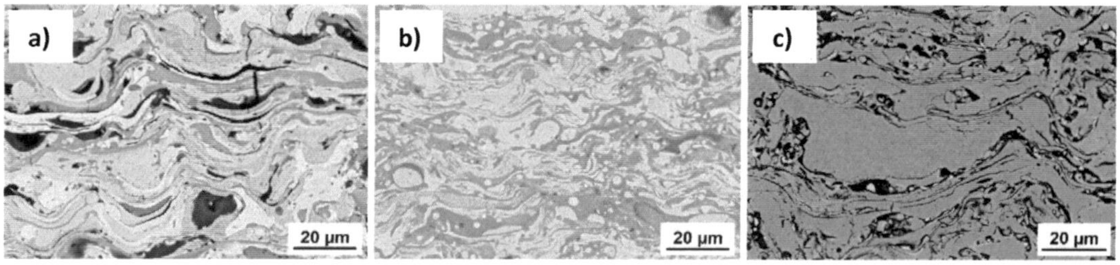

Figure 4. SEM (BSD) images of coating cross sections deposited with APS of: a) powder blend, b) HEM powder and c) atomized powder.

The coating type produced with powder blend feedstock exhibits a lamellar and heterogeneous microstructure. Distinct contrast between single lamella can be observed, which indicates strong deviation of local chemical composition. Furthermore, dark oxide lamella exist, which are a result of high temperatures during the APS process and contact of molten powder particles with ambient air.

For the coating deposited with HEM powder, a similar microstructure is revealed. However, this coating type shows much smaller phases and no distinct black oxide lamella. Some round particles can be seen in the coating, which are a result of incomplete particle melting and mixing during coating process.

In the last coating type, deposited with atomized powder, no distinct contrast can be observed between single lamella, indicating a more homogeneous microstructure. However, there is visible contrast within single lamella, which suggests that several phases exist. Furthermore, oxide lamella and pores exist. The overall composition of all coatings has been analyzed by EDX and is displayed in table 5.

Table 5. Chemical composition of coatings produced with different feedstock material measured by EDX.

feedstock	Al (at.%)	Co (at.%)	Cr (at.%)	Fe (at.%)	Ni (at.%)	Ti (at.%)
nominal	*16.7*	*16.7*	*16.7*	*16.7*	*16.7*	*16.7*
powder blend	28.5	14.5	15.2	14.7	12.0	15.0
HEM powder	16.1	16.9	11.2	16.1	16.9	22.8
atomized powder	18.7	16.3	15.2	16.3	16.5	17.0

For the coating type produced with powder blend feedstock, a distinct deviation between measured and nominal composition occurs. Especially the Al content differs from the nominal composition. The difference might be caused by vaporization during the coating process as well as the inhomogeneous state of the coating. Coatings produced with HEM powder show a distinctly changed chemical composition especially for Cr and Ti content. In contrast the coating type deposited with atomized powder is in better accordance with the nominal composition. Only the Cr and Ti contents are changed markedly in comparison with the feedstock material (table 4). The smaller deviation in comparison with the two other feedstock materials is probably an effect of the more homogeneous powder microstructure. The powder is already in an alloyed state, which avoids vaporization of single elements. Phase analysis by XRD has been carried out for all coatings. Resulting diffractograms are depicted in figure 5.

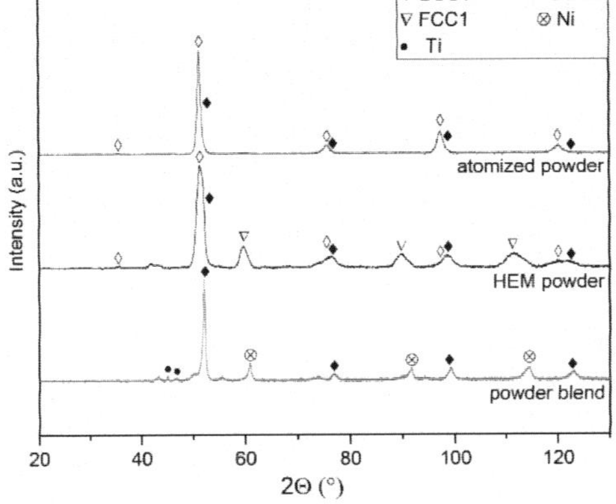

Figure 5. XRD patterns of coatings produced with powder blend, HEM powder and atomized powder feedstock.

For the coating type produced by APS of powder blend feedstock, diffraction peaks, which can be referred to several phases appear. A BCC structure (PDF 00-006-0694 / Im-3m (229)), which is denoted as BCC1, is formed. Furthermore, diffraction peaks of pure Ni and Ti appear, showing that no complete alloy formation has occurred during the thermal spray process. The short processing time does not allow for these particles to fully melt and mix with other elements.

The coating type deposited by applying HEM powder exhibits major peaks which can be ascribed to another BCC phase (PDF 03-065-4198 / Pm-3m (221)), labeled as BCC2. Furthermore, diffraction peaks of the BCC1 phase appear. Unlike the diffraction pattern of the HEM powder feedstock (figure 3) no diffraction peaks of pure Ni and Ti emerge, showing that a more homogeneous state is reached after the thermal spray process. Further diffraction peaks can be ascribed to a FCC phase (PDF 01-071-7594 / Fm-3m (225)) named FCC1. For the coatings produced with powder blend and HEM powder feedstock, several small diffraction peaks appear which cannot be ascribed. These diffraction peaks are possibly a result of oxide formation during the thermal spray process.

In comparison with atomized powder feedstock, the diffractogram of the coating type deposited with this powder displays the same major peaks. The peaks can be attributed to the BCC2 phase. According to the diffraction pattern of atomized powder, diffraction peaks exhibit a shoulder towards higher diffraction angles, indicating the overlapping of diffraction peaks ascribed to the BCC1 phase. The investigation of metallographic cross sections (figure 4) reveals a multiphase structure, which also suggests the existence of the BCC1 phase. Hardness HV0.1 was measured in cross sections of all coatings. The results are displayed in figure 6.

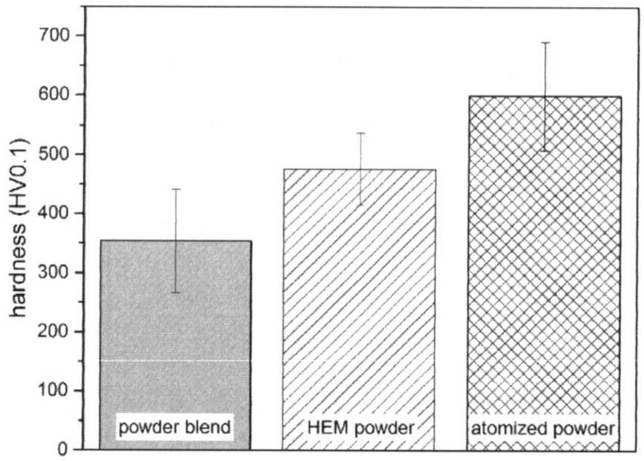

Figure 6. Hardness of APS-coatings, produced with different feedstock material.

For the coating type deposited with powder blend feedstock, an average hardness of 354 HV0.1 was measured. The coating type produced with HEM powder has an increased average hardness of 476 HV0.1, which is probably caused by the finer microstructure and the absence of pure Ni and Ti (XRD / figure 5). Both coatings show a strong standard deviation of over 10 %, which is caused by the inhomogeneous microstructure of the coatings. For the coating type deposited with atomized powder, an average hardness of 599 HV0.1 was measured. The results of wear tests under different tribological conditions are displayed in figure 7.

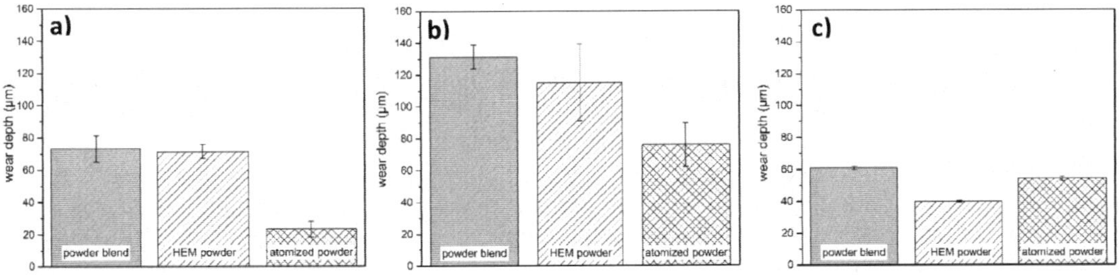

Figure 7. Measured resulting wear depth of coatings in: a) ball-on-disk wear, b) oscillating wear and c) scratch test.

The results of the ball-on-disk wear test only show minor difference in wear depth of coatings deposited with powder blend and HEM powder feedstock, indicating a similar wear resistance. In contrast, the coating type produced with atomized powder reveals a distinctly reduced wear depth and thus improved wear resistance.

The evaluation of the oscillating wear test shows that the highest wear depth and lowest wear resistance was achieved for the coating type deposited with powder blend feedstock. For the coating type produced with HEM powder, the wear depth is slightly reduced. However, this coating type exhibits the highest standard deviation, which elucidates the inhomogeneous state of this coating type. For the coating type deposited with atomized powder the wear depth is clearly reduced, which is caused by the fully alloyed state and a more homogeneous microstructure. Furthermore, this coating type exhibits the highest hardness, which is beneficial at the present wear mechanism.

The progressive mode scratch test has been analyzed by determining the maximum wear depth of the scratches. The highest wear depth was measured for the coating type deposited with powder blend feedstock. In comparison, a slightly reduced wear depth and increased wear resistance was determined for the HEM powder coating. No further decrease of wear depth could be achieved for the coating type deposited with atomized powder. The resulting wear resistance in the scratch test is in between the coatings produced with powder blend and HEM powder. Due to the deviation of wear behavior in the scratch test compared to ball-on-disk and oscillating wear tests, all scratches were investigated by optical microscopy. Figure 8 shows the scratch area where the maximum load was applied in the progressive mode scratch test.

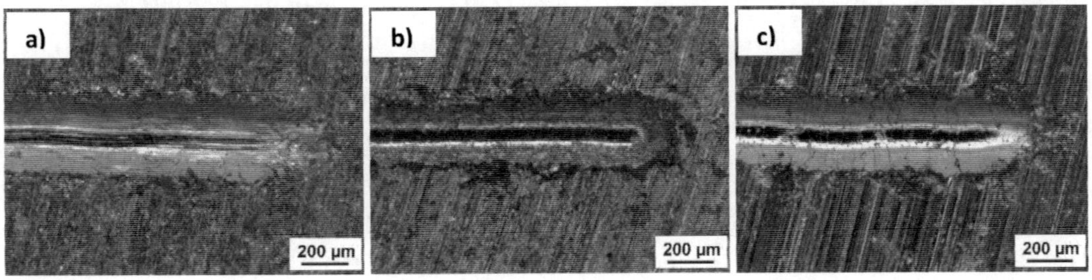

Figure 8. Surface of coatings produced with a) powder blend, b) HEM powder and c) atomized powder feedstock after progressive mode scratch test.

The scratches in the coatings produced with powder blend and HEM powder feedstock show a relatively smooth surface with only small spalling of material at the edge of the scratch. In contrast, the coating type produced with atomized powder shows distinct cracks on the bottom of the scratches. Furthermore, a spalling of material at the edge is visible. The increase of measured maximum wear depth and the appearance of cracks in the scratch test under abrasive wear stress indicates the brittle behavior of this coating type.

4. Conclusion

The applicability of the equimolar CCA AlCoCrFeNiTi for surface engineering has been analyzed in this study. Three different feedstock materials (powder blend, HEM and atomized powder) were investigated regarding their suitability to be deposited by APS. The investigation of different feedstock materials reveals that inert gas atomization is most suitable to produce homogeneous and spherical alloyed powder. Two BCC solid solutions are formed for this production method. According to earlier investigations on bulk material, a multiphase structure appears. Furthermore, APS coatings were produced for all feedstock materials. For the coatings produced with powder blend and HEM powder feedstock, a heterogeneous microstructure is formed. By applying atomized powder, a more homogeneous microstructure occurs. Within the coatings, several phases can be distinguished. For the coatings produced by applying atomized powder feedstock the same BCC solid solutions like in the feedstock material are formed. However, oxide lamella and pores exist, which might be reduced by further optimization of coating parameters. Furthermore, the application of other thermal spray technologies with lower heat input like high-velocity-oxygen-fuel spraying has to be investigated. The investigation of wear behavior revealed best wear resistance in ball-on-disk and oscillating wear test for the coatings deposited with atomized powder due to the more homogeneous microstructure. In the scratch test, a slightly lower wear resistance compared to the HEM powder coatings is observed. Furthermore, cracks appear for the coatings produced with atomized powder feedstock, indicating brittle behavior.

Acknowledgements

The authors gratefully acknowledge the Arbeitsgemeinschaft Industrieller Forschungsvereinigungen "Otto von Guericke" e.V. (AiF) for support of this work (AiF-No. KF2152612CK4) with funds from the German Federal Ministry for Economic Affairs and Energy. The authors thank Marc Pügner for conducting the XRD measurements and assistance during their evaluation and Ilka Günther for conducting the particle size analysis.

References

[1] Yeh J-W, Chen S-K, Gan J-Y, Lin S-J, Chin T-T, Shun T-T, Tsau C-H and Chang S-Y 2004 *Metall. Mater. Trans. A* **35** 2533-2536
[2] Yeh J-W, Chen Y-L, Lin S-J and Chen S-K 2007 *Mater. Sci. Forum* **560** 1-9
[3] Cantor B, Chang I T H, Knight P and Vincent A J B 2004 *Mater. Sci. Eng. A* **375-377** 213-218
[4] Murty B S, Yeh J-W and Ranganathan S 2014 *High-Entropy Alloys* (Elsevier Inc.)
[5] Miracle D B, Miller J D, Senkov O N, Woodward C, Uchic M D and Tiley J 2014 *Entropy* **16** 494-525
[6] Zhou Y J, Zhang Y, Wang Y L and Chen G L 2007 *Appl. Phys. Lett.* **90** 181904
[7] Wang Y, Ma S, Chen X, Shi J, Zhang Y and Qiao J 2013 *Acta Metall. Sin (Eng. Lett)* **26/3** 277-284
[8] Yu Y, Wang J, Li J, Kou H and Liu W 2015 *Mater. Lett.* **138** 78-80
[9] Zhang K and Fu Z 2014 *Intermetallics* **22** 24-32
[10] Chuang M-H, Tsai M-H, Wang W-R, Lin S-J and Yeh J-W 2011 *Acta Mater.* **59** 6308-6317
[11] Yu Y, Liu W M, Zhang T B, Li J S, Wang J, Kou H C and Li J 2014 *Metall. Mater. Trans. A* **45** 201-207
[12] Ang A S M, Berndt C C, Sesso M L, Anupam A, Praveen S, Kottada R S and Murty B S 2014 *Metall. Mater. Trans. A* **46** 791-800
[13] Wang L M, Chen C C, Yeh J-W and Ke S T 2011 *Mater Chem Phys* **126** 880-885
[14] Huang P-K, Yeh J-W, Shun T-T and Chen S-K 2004 *Adv Eng Mater* **6** 74-78
[15] Soare V, Mitrica D, Constantin I, Badilita V, Stoiciu F, Popescu A-M J and Carcea I 2015 *J. Mater. Sci. Technol.* **31** 1194-1200

Self-sharpening-effect of nickel-diamond coatings sprayed by HVOF

W Tillmann, A Brinkhoff, C Schaak and J Zajaczkowski

Institute of Materials Engineering, TU Dortmund, Dortmund, Germany

Abstract. The durability of stone working and drilling tools is an increasingly significant requirement in industrial applications. These tools are mainly produced by brazing diamond metal matrix composites inserts to the tool body. These inserts are produced by sintering diamonds and metal powder (e.g. nickel). If the wear is too high, the diamonds will break out of the metal matrix and other diamonds will be uncovered. This effect is called self-sharpening. But diamonds are difficult to handle because of their thermal sensitivity. Due to their high thermal influence, manufacturing costs, and complicate route of manufacturing (first sintering, then brazing), there is a great need for alternative production methods for such tools. One alternative to produce wear-resistant and self-sharpening coatings are thermal spray processes as examined in this paper. An advantage of thermal spray processes is their smaller thermal influence on the diamond, due to the short dwelling time in the flame. To reduce the thermal influence during spraying, nickel coated diamonds were used in the HVOF-process (high velocity oxygen fuel process). The wear resistance was subsequently investigated by means of a standardized ball-on-disc test. Furthermore, a SEM (scanning electron microscope) was used to gain information about the wear-mechanism and the self-sharpening effect of the coating.

1. Introduction

Stone working and stone drilling tools are extremely highly charged during use. Hence, only the hardest materials can be used for this field of application. Generally, diamond impregnated inserts or diamond reinforced tools are used for this purpose as diamonds are the hardest known material. For industrial applications, synthetic diamonds are used due to the costs of such tools. Synthetic diamonds can be produced with different processes such as the high temperature high pressure synthesis (HPHT), which is the most widely used process for this task. HPHT uses high temperatures and high pressures to produce diamonds from a carbon-rich melt. Metals such as Fe, Co, or Ni can be used for this process [1]. Furthermore, thin diamond coatings can be applied by means of CVD (chemical vapour deposition) [2], whereas detonation and ultrasonic methods [3] are more alternative techniques.

To obtain a tool or tool insert that can be used for stone processing, non-coated diamonds are not suitable, due to their high brittleness and the fact that there are no contouring possibilities. Normally, the diamonds are mechanically embedded into a metal matrix. These diamond metal matrix composites (MMC) are used for tools or tool inserts. The production of diamond MMC tools is mainly divided in two process routes. The first route is the production by means of powder metallurgy during which the diamonds are sintered with a metal powder. This method is mainly used for solid diamond MMCs or diamond inserts. The processing of synthetic diamonds is a challenge, because of the thermal sensibility of diamonds [4]. At high temperatures, the diamonds can react with their metal matrix, which leads to graphitization effects that damage the surface of the diamonds [5].

IOP Conf. Series: Materials Science and Engineering **181** (2017) 012016 doi:10.1088/1757-899X/181/1/012016

Long sintering times combined with long holding times in the presence of metallic matrix lead to uncontrolled solution reactions, dissociation, or degradation of the diamonds, which subsequently affects the life span of the tools. Because of these effects, fast sintering methods become more interesting [6].

The second main processing route of diamond MMC is based on brazing methods. Brazing pastes, impregnated with diamonds, can be applied on large components to form a closed, diamond impregnated coating. Furthermore, pick-and-place techniques allow to braze diamonds onto defined positions. For this purpose, active elements have to be used to achieve a wettable surface [7]. The yielded temperature has to be controlled as well to reduce thermally induced damages of the diamonds during brazing.

The deposition of diamond reinforced coatings by means of different thermal spraying processes has already been investigated in numerous studies [8-10]. Especially, thermal spray processes that use moderate to high kinetic energies and low to moderate thermal energies were used. The aim was to minimize the thermal influence during the coating process. Two processes are particularly suitable: The detonation spraying process (CD-gun) provides for large diamonds within a softer matrix such as aluminium or bronze [8]. The high velocity oxygen fuel (HVOF) process proves to be the suitable process for fine diamonds with a thin metallic coating (e.g. nickel) [11]. Additionally, the tribological wear behavior was investigated. Diamond reinforced coatings, deposited by means of flame spraying and detonation spraying process, were applied to improve the wear behavior [8] [12].

This paper scrutinizes the wear mechanisms of diamond reinforced coatings, produced by means of HVOF. Especially the self-sharpening effect of the coating that occurs when worn out diamonds break out and new sharp diamonds are exposed or which result from a partial breaking or cracking of the diamond [13], is investigated.

2. Experimental

2.1 Thermal spray processes

For this study, round specimens (diameter 40 mm, thickness 8 mm) made from 1.2249 (45CrSiV6) serve as substrates. The preparation was conducted with corundum (type EKF 100) with a grain size of 100 – 150 µm. The blasting parameters are a distance of 100 mm with an angle of 45° and a pressure of 4 bar.

The spray equipment used for the experiments was a high velocity oxygen fuel (HVOF) system type CJS K5.2 by Thermico GmbH & Co KG (Dortmund, Germany). This system uses the energy of a continuous combustion of an oxygen-fuel mixture to melt the powder feedstock material. The powder is injected behind the combustion chamber by a powder feeder (Thermico GmbH & CO. KG, type CPF 2). Table 1 summarizes the spray parameters. The parameters are composed of optimized parameters for nickel coatings, generated during former investigations. The parameters for the application of diamond reinforced coatings were tested in former research projects [14].

Table 1. Process Parameters used for the application of diamond reinforced coatings.

Parameters	Value
Oxygen	800 l/min
Hydrogen	60 l/min
Kerosene	14 l/h
Feeder gas (Nitrogen)	10 l/min
Feed rate	50 g/min
Gun movement	17.500 mm/min
Stand-off distance	250 mm
Distance between each path	5 mm

The spray gun was handled by a 6 axis robot type IRB 4600 2,05 m (ABB Automation GmbH, Zürich, Switzerland).

Figure 1. SEM image of an embedded diamond in a nickel matrix [14].

For this study, monocrystalline nickel coated synthetic diamonds type CM-M Ni 60 supplied by Ceratonia GmbH & Co. KG (Eltmann, Germany) were used to produce the diamond reinforced coatings. Figure 1 shows a SEM image of Ni-coated diamonds that serve as feedstock. For this study, blocky-shaped diamonds were used. The diamonds exhibit an average grain size of 10 µm (8 µm to 12 µm). The feedstock is composed of 60 wt. % of Ni.

2.2 Wear resistance tests

To obtain information about the wear-mechanism of the produced diamond reinforced coatings, a standardized ball-on-disc test was used (ASTM G99 [14]). A tribometer (Pin-Disc Machine) by CSEM Industries (Neuchâtel, Switzerland) was used to conduct tests with a constant radius of 15 mm. The parameters of the ball-on-disc test are summarized in table 2.

Table 2. Parameters of the ball-on-disc tests.

Parameters	Value
Rotation	120.000 U
Counter body	Al2O3
Load	10 N

To gather information about the wear coefficient, the volume of the wear track was measured. The wear coefficient ($\kappa_{wear\ track}$) results from the volume of the wear track ($V_{wear\ track}$) divided by the load (F) used for this test, multiplied by the total distance of the rotation (s). The volume of the wear track was measured with a 3D-profilometer type InfiniteFocus (Alicona Imaging GmbH, Graz, Austria). The InfiniteFocus is able to visualize the surface of the specimen in a 3D model.

$$\kappa_{wear\ track} = \frac{V_{wear\ track}[mm^3]}{F\ [N] * s\ [m]}$$

The volume of the wear track ($V_{wear\ track}$) results in an area of wear, multiplied by the circumference ($u_{wear\ track}$).

$$V_{wear\ track} = A_{wear} * u_{wear\ track}$$

Furthermore, the ball (counterpart) was investigated and a 3D image of the ball was created to visualize the wear of the ball. This image was used to determine the volume of the worn out parts of the ball ($V_{wear\ ball}$), which results in:

$$V_{wear\ ball} = \frac{\pi * {h_{wear\ ball}}^2}{3} * (3 * r_{ball} - h_{wear\ ball})$$

The height of the worn out part of the counterpart $h_{wear\ ball}$ is determined by

$$h_{ball} = r_{ball} - (r_{ball}^2 - r_{wear\ ball}^2)^{\frac{1}{2}}$$

The wear coefficient of the counterpart ($\kappa_{wear\ ball}$) is calculated similarly to the wear coefficient of the track, by the volume of the wear of the ball ($V_{wear\ ball}$) divided by the load (F) used for the test, multiplied by the total distance (s).

$$\kappa_{wear\ ball} = \frac{V_{wear\ couterpart}\ [mm^3]}{F\ [N] * s\ [m]}$$

2.3 Surface analysis

One of the main objectives of this study is the analysis of the wear mechanism of the diamond reinforced coating, with a special focus on the self-sharpening effect of the coating. Therefore, the wear track was investigated in the SEM (scanning electron microscope) after the wear tests. For this study, a SEM type JSM 70 1F (Jeol GmbH, Freising, Germany) was used. A 3D model was created to receive a more precise image of the surface of the specimen. This model was created by taking three images with a slightly different angle (0° and +- 5°) that were subsequently interpolated to a 3D model with the help of the software InfiniteFocus.

3. Results and discussion

3.1 Wear resistance tests

Figure 2 shows a cross-section image of the diamond reinforced coating. It shows the substrate, made of 1.2249, the coating, and the embedding resin in the upper section of the image. The diamonds are spread homogeneously in the coating. The image also shows an average coating thickness of 230 µm. The interface exhibits no slots between the coating and the substrate, which indicates a good bounding of the coating to the substrate.

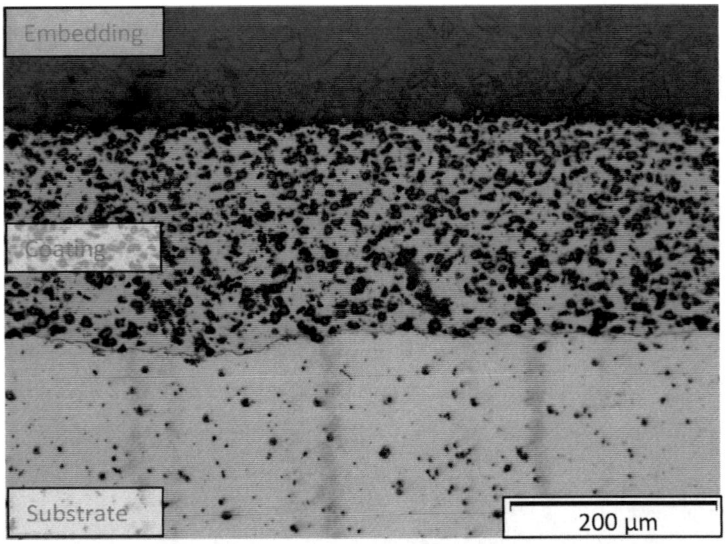

Figure 2. Cross-section image of the diamond reinforced coating.

Ball-on-disc wear tests were applied to the coating of the substrate. Table 3 summarizes the comparison of the wear coefficients of two different coating systems compared to the substrate without any coating. For the comparison a WCCo 88/12 coating, deposited by means of HVOF was used. As WCCo is a standard coating used for wear protection [8] it was used for the comparison with the diamond reinforced coating.

Table 3. Wear coefficients of the specimens.

Specimen	Wear coefficient $\kappa_{wear\ track}$	Wear coefficient counterpart $\kappa_{wear\ ball}$
Substrate 1.2249	$3.56*10^{-3}$ mm³/(N*m)	Not measurable
Diamond reinforced coating	$1.668*10^{-5}$ mm³/(N*m)	$1.809*10^{-6}$ mm³/(N*m)
WCCo coating	$0.1614*10^{-5}$ mm³/(N*m)	$0.01861*10^{-6}$ mm³/(N*m)

The results are averages from three different measurements. Table 3 shows that the diamond reinforced coatings increase the wear resistance when compared to the substrate and the wear coefficient gets close to regions of WCCo 88/12 coatings. The wear coefficient of WCCo coatings is the lowest in this comparison. This could be explained by the smaller carbides of the WCCo coatings, which are finely spread in the coating. The matrix material cobalt, is harder when compared to nickel, the matrix material of the diamond reinforced coatings. Whereas the matrix material of the diamond reinforced coating will wear out faster than the matrix material of the WCCo coating.

Table 3 also shows the wear coefficients of the counter body. The ball, used for the wear test of the substrate, indicates no wear. The comparison of the coefficients of the two different coatings reveals some differences. The wear coefficient of the diamond reinforced coating is higher than the coefficient of the WCCo coating. Due to the lower hardness of the counterpart made of Al2O3 (about 1800HV [16]), the counterpart wears out fist. The harder diamonds cause a high mass loss on the counterpart by cutting the Al2O3 from the ball. This can be seen as a first evidence for an outstanding abrasive behavior of the coating.

3.2 SEM analysis

The SEM was used to investigate the wear mechanism of the diamond reinforced coating. It was the aim to investigate if a self-sharpening effect of the coating does exist.

Figure 3. SEM image of the diamond reinforced coating.

Figure 3 shows the worn surface of the diamond reinforced coating after the wear test. Several diamonds stick out the surface. These diamonds were located in the track of the ball-on-disc test. Besides two almost uncovered diamonds, a third diamond is visible. This diamond is almost completely covered by a nickel coating. This observation can be seen as another evidence for the self-sharpening effect of the coating. A 3D model was generated to gain more information about the topography of the surface and the diamonds in the coating.

Figure 4. 3D-model of the diamond coating.

Figure 4 shows a SEM 3D model of the wear track with some exposed diamonds on the surface. The diamonds stick out the surface and cause, despite the high hardness of the Al2O3, a high abrasive wear on the surface of the ball. The ball further causes wear on the surface of the coating. It can be seen that the matrix metal is worn out and diamonds can be found in different depths. If the wear is high, the nickel matrix will first dig out. The initial state of this mechanism is visible in Figure 4. Between the diamonds, trenches are formed. Consequently, some diamonds will drop out or break out of the surface. This effect exposes new diamonds with a new sharp contour. The new, sharp diamonds verify the initial hypothesis of the self-sharpening effect of diamond reinforced coatings.

4. Conclusion

HVOF-sprayed, diamond reinforced coatings were tested with a ball-on-disc device according to ASTM G99. Subsequently, the volume loss was measured and the wear coefficient was calculated. The results show that the wear coefficient of the coating is extremely low in comparison to that of the uncoated specimen. However, the Al2O3 counterpart shows a higher wear coefficient compared to that of other coatings. It was shown that the matrix metal is removed and that the diamonds stick out of the surface. Consequently, this effect will lead to a drop out of diamonds and new diamonds are exposed. This effect can be labelled a self-sharpening effect of the diamond metal matrix coating. To obtain more information about the wear mechanism and the self-sharpening effect, the experiments should be further expanded. Therefore, experiments with a tribometer in a great chamber SEM are planned. This method allows an in-situ observation of the occurring mechanisms.

Acknowledgments

We thank the Deutsche Forschungsgemeinschaft (DFG) for their financial support (TI 343/94-1). In addition, we thank our technical staff for preparing our specimens and our students for assisting in these investigations.

References

[1] Bundy F P, Hall H T, Strong H M and Wentorf R H 1955 Man-made Diamonds *Nature* **176** pp 51-55

[2] Balmer R S et al. 2009 Chemical vapour deposition synthetic diamond: materials technology and applications *Journal of Physics: Condensed Matter* **21** pp 23

[3] Xu K, Xue Q 2004 A new method for deaggregation of nanodiamond from explosive detonation: graphitization oxidation method *Physics of the Solid State* **46** pp 649-650

[4] Evans T 1979 Changes produced by high temperature treatment of diamond *The properties of Diamond, Academic press* pp 403-424

[5] Tillmann W, Ferreira M, Steffen A, Rüster K, Möller J, Bieder S, Paulus M and Tolan M 2013 Carbon reactivity of binder metals in diamond-metal composites – Characterisation by scanning electron microscopy and X-ray diffraction *Diamond & Related Materials* **38** pp 118 - 123

[6] Tillmann W, Kronholz C, Ferreira M, Knote A, Theisen W, Schütte P and Schmidt J 2011 Diamond-metal matrix interaction in tools, fabricated by conventional and current induced sintering *International Journal of Powder Metallurgy* **47** pp 29 – 36

[7] Tillmann W, Osmanda A M, Yurchenko S and Theisen W 2005 Löten von Diamant – Grenzflächenreaktionen und Benetzung *Mat. Wiss. u. Werkstofftech.* **36** pp 370-376

[8] Nebel J 2012 Detonationsspritzen von Diamant-Bronze-Verbundschichten für die Gesteinbearbeitung *Dissertation* (Lehrstuhl für Werkstofftechnologie/ tu Dortmund)

[9] Steffens H D, Babiak Z and Wewel M 1991 Lichtbogenspritzen mit Fülldrähten. Möglichkeiten und Grenzen - Arc spraying using cored filler wires. Possibilities and limitations. *Proc. Conf. Thermische Spritzkonferenz, Internationale DVS Tagung* (Stuttgart, Germany) **136** pp 9295

[10] Verstak A, Ilyuschenko A, Vityaz P, Lugscheider P and Rass E 1993 Underwater plasma spraying of carbon contained hard coatings *Proc. Conf. Thermische Spritzkonferenz Internationale DVS Tagung* (Aachen, Germany) **152** pp 342-345

[11] Tillmann W, Vogli E, Rechlin R, Bach F-W, Möhwald K, Babiak Z and Rothardt Th 2004 Manufacturing diamond impregnated tools for stone machining through thermal spraying *Proc. Conf. International thermal spray conference* (Japan) pp 18 – 22

[12] Richardson A F, Neville A and Wilson J I B 2003 Developing diamond MMCs to improve durability in aggressive abrasive conditions *Wear* **255** pp 593-605
[13] Zhou Y, Atwood M, Golini D, Smith M and Funkenbusch P D 1998 Wear and self-sharpening of vitrified bond diamond wheels during sapphire grinding *Wear* **219** pp 42-45
[14] Tillmann W, Schaak C and Nebel J 2013 Production of versatile abrasive and wear resistant Ni-diamond composite layers by HVOF spraying *Proc. 2nd International Conference on Stone and Concrete Machining* (Dortmund, Germany) pp 59-65
[15] ASTM G99-05 2016 Standard test method for wear testing with a pin-on-disk apparatus
[16] Sommer K, Heinz R and Schöfer J 2010 *Verschleiß metallischer Werkstoffe* (Wiesbaden, Germany. Vieweg+Teubner) pp 365-422

IOP Conf. Series: Materials Science and Engineering **181** (2017) 012017 doi:10.1088/1757-899X/181/1/012017

Production integrated nondestructive testing of composite materials and material compounds – an overview

B Straß, C Conrad and B Wolter
Fraunhofer Institute for Nondestructive Testing IZFP, Campus E3.1, 66123 Saarbrücken, Germany

E-Mail: benjamin.strass@izfp.fraunhofer.de, bernd.wolter@izfp.fraunhofer.de

Abstract. Composite materials and material compounds are of increasing importance, because of the steadily rising relevance of resource saving lightweight constructions. Quality assurance with appropriate Nondestructive Testing (NDT) methods is a key aspect for reliable and efficient production. Quality changes have to be detected already in the manufacturing flow in order to take adequate corrective actions. For materials and compounds the classical NDT methods for defectoscopy, like X-ray and Ultrasound (US) are still predominant. Nevertheless, meanwhile fast, contactless NDT methods, like air-borne ultrasound, dynamic thermography and special Eddy-Current techniques are available in order to detect cracks, voids, pores and delaminations but also for characterizing fiber content, distribution and alignment. In Metal-Matrix Composites US back-scattering can be used for this purpose. US run-time measurements allow the detection of thermal stresses at the metal-matrix interface. Another important area is the necessity for NDT in joining. To achieve an optimum material utilization and product safety as well as the best possible production efficiency, there is a need for NDT methods for in-line inspection of the joint quality while joining or immediately afterwards. For this purpose EMAT (Electromagnetic Acoustic Transducer) technique or Acoustic Emission testing can be used.

1. Introduction

As the demand for environmental-friendly production and application of products is consequently increasing, composite materials like Carbon Fiber Reinforced Polymers (CFRP) or Metal Matrix Composites (MMC) are of gaining relevance for achieving efficient lightweight designs due to their high specific strengths. But as long as damage mechanisms and their progress of such complex materials are still under investigation, reliable NDT methods to detect a minimum damage size to still ensure safe conditions have to be identified and established in production as well as during the application. On the other hand the necessity of realizing lightweight construction concepts places high demands on joining technology as appropriate methods for example to weld dissimilar materials have to be established or even developed. NDT techniques offer the great potential not only to prove the joint quality after the welding process (post-process) but also to realize process monitoring concepts, which enable to identify necessary adjustments of different process parameters while welding (in-process) in order to achieve high quality and less waste. X-ray and ultrasound (US) techniques are still state of the art for the inspection of composite materials and material compounds. But both are rather difficult to integrate into a production line, as X-ray methods require a special shielding and conventional ultrasound methods need to work in contact and with coupling media (e.g. by water immersion). For this reason in most cases the nondestructive testing takes place in the post-process, where no more corrective actions can be done in case of defects or irregularities. However, X-ray and ultrasound are suitable and essential as reference methods for any other NDT method due to the achievable high resolution. But other NDT techniques like for example EMAT, eddy current and acoustic emission testing have great potential to speed up the inspection process by an automated implementation into production processes.

2. Post-process NDT methods for composite materials and material compounds

With respect to the achievable resolution and detection sensitivity conventional US techniques are often not able to detect sufficient small irregularities, as it is necessary in regard to efficient and safe lightweight design constructions. To overcome this, special high-frequency ultrasound techniques (HF-US) have been developed, which are working in the frequency range from 10 MHz to 200 MHz. This increase in ultrasound frequency and therefore in resolution and sensitivity can be achieved by performing the measurements in a water-filled immersion tank due to the difference in ultrasound velocity in comparison to air [1]. As a result, irregularities with a size up to 0.2 mm can be detected, what makes this method a fast alternative to destructive, metallographic investigations. But nevertheless HF-US is not applicable as inline NDT-method for most applications due to corrosive effects, which cause that the parts cannot be used or forwarded to the next production step anymore.

Figure 1 shows the results of HF-US measurements on magnetic pulse welded aluminum/steel-joints. On the left side the C-scan of a faultless weld seam is demonstrated, whereas the C-scan on the right exhibits significantly higher intensities of the reflected amplitudes in the area of the weld seam indicating the presence of welding defects. In this case so-called kissing bonds were present at the interface between aluminum and steel, which are generally very difficult to detect. HF-US proved itself to be capable to identify such bonding defects in welded joints.

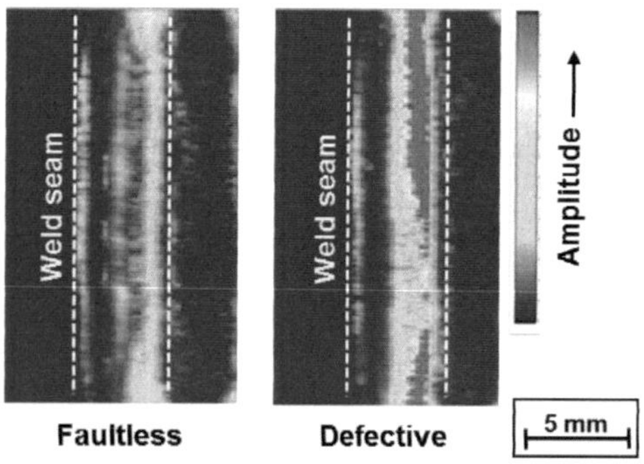

Figure 1. Nondestructive detection of kissing bonds in magnetic pulse welded aluminum/steel-joints using high-frequency ultrasound.

X-ray measurements are imaging methods using X-rays and mathematical algorithms to generate images or sequences of the investigated specimen [2]. While in X-ray radiography only 2D-images can be achieved, X-ray computed tomography (CT) can provide 3D-models of the samples. For this purpose numerous radiographs of the specimen are taken while it is rotating between the X-ray source and the detector. Afterwards the model of the sample gets calculated and visualized [3]. In this manner high resolutions can be achieved, which generally predestine X-ray measurements for the nondestructive identification of nearly every irregularity in materials and welding areas. But depending on the investigated material and its thickness such measurements are often very time-consuming and due to the required rotation of the parts in the course of the beam often not applicable with regard to the geometry of the parts. Therefore X-ray radiography and CT can often only be realized in the post-process as reference methods. An alternative to CT is the computed laminography (CL) which enables the inspection of bigger parts with high resolution. In contrast to CT the X-ray source and the detector are moved in CL-measurement which reduces the physical, geometric restrictions [3, 4]. Considering this, in addition with the adoption of enhanced reconstruction methods, the testing time can be considerably reduced and this method can generally be applied even for inline testing in the production line for several applications [4].

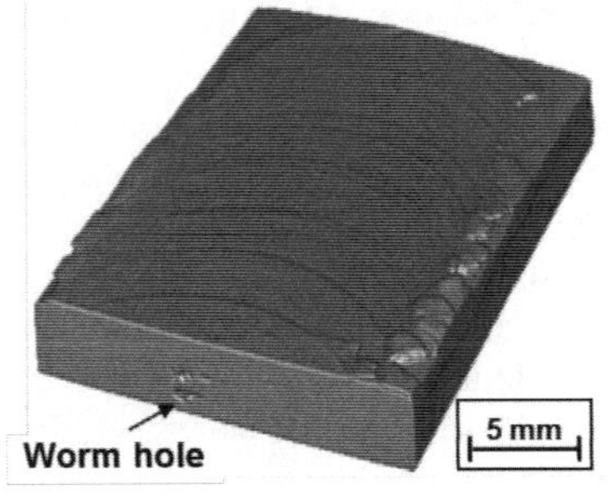

Figure 2. Detection of a worm hole in a friction stir welded Al-joint by 3D-CT.

As figure 2 demonstrates it is possible to detect very small-sized defects inside the volume of a specimen with X-ray techniques, like in this example a so called worm hole inside a friction stir welded Al/Al-joint by 3D-CT. However not only to detect existing defects but also to characterize different material or joint properties X-ray methods are suitable. For example the orientation of the fibers in composite materials or, like shown in figure 3, the degree of material mixing in welded joints can be visualized. In this case figure 3 compares the mixing levels of Al/Mg hybrid joints which are produced by conventional friction stir welding (a) and ultrasonic enhanced friction stir welding (b) using computed laminography with faded out aluminum contents. It can be clearly realized, that the superimposition of the friction stir welding process by ultrasonic energy leads to an intensified material mixing in the joining area [5].

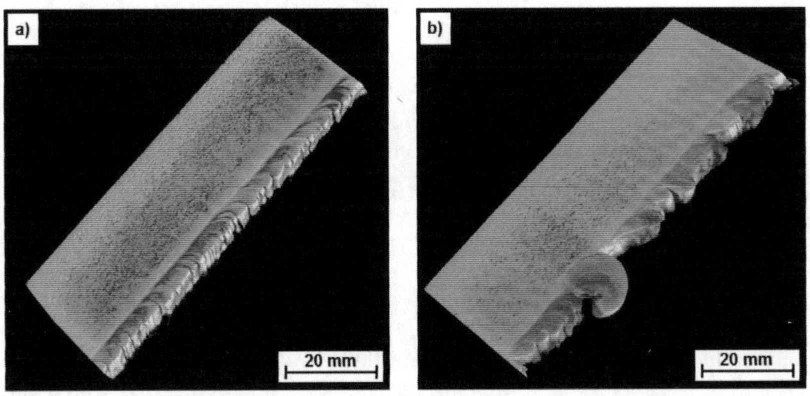

Figure 3. Visualization of the material mixing in the stirred area of a friction stir welded (a) and an ultrasonic enhanced friction stir welded (b) Al/Mg-joint by computed laminography (Al faded out) [5].

3. Production integrated NDT of composite materials

Composite materials like metal or polymer matrix composites, aim to combine the advantageous properties of different materials. Relating to lightweight construction this means to achieve the highest possible strength with the lowest possible density. In order to exploit the lightweight potential of such complex materials in an optimal way, appropriate NDT methods to monitor the material condition have to be developed and used during the application and already in the material or component production. Characteristic defects that may occur in composite materials are for example delaminations, debonds, porosity or fiber breakage. Following several methods will be presented which are suitable to realize a nondestructive testing of such reinforced materials.

3.1 Sampling Phased Array (SPA)
Sampling Phased Array (SPA) is an ultrasonic NDT method which operates in transmitter-receiver mode and uses the signals of different transducers in the array to characterize the material. A signal that is transmitted by one of these transducers of the array is received by all the other transducers. Due to this a great advantage of SPA is the ability to investigate comparatively big volume sections in very short time. In addition SPA enables to focus on different investigation depths simultaneously which is not possible in conventional phased array measurements. By combining this technique with the Synthetic Aperture Focusing Technique (SAFT) it is possible to generate a real time imaging of the investigated material area. Using the Reverse Phase Matching it is even possible to analyze complex and anisotropic materials, like for example CFRP or austenitic steel welds by determining the velocity distribution of the wave propagation using the elastic constants of the materials. Figure 4 illustrates the detection of various artificial defects of different size and at different depths on a CFRP specimen with a thickness of 14 mm [6].

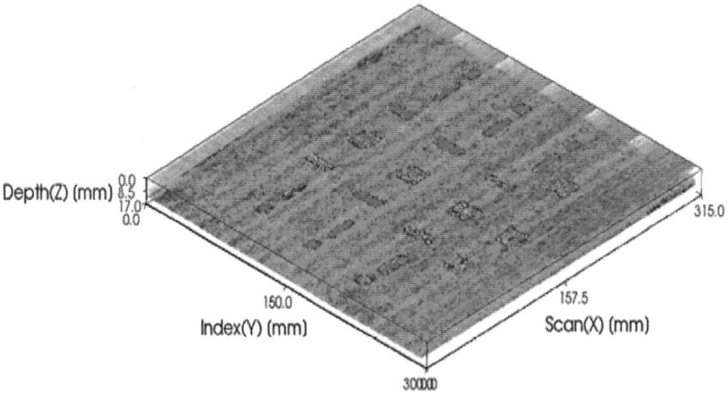

Figure 4. Detection of artificial defects in CFRP by Sampling Phased Array method with Reverse Phase Matching and 3D imaging [6].

3.2 Non-linear Ultrasound
Non-linear Ultrasound describes a method that uses the interaction between ultrasonic waves propagating in the sample and the mechanical properties of materials in defect-near areas to characterize the sample condition. Existing micro cracks, for example in CFRP components, may be closed if no stress or compressive stress is applied and therefore these defects are very difficult to detect in such an early stage of damage. By transmitting an ultrasonic wave of adequate intensity to the sample, tensile components are able to achieve temporary crack opening, which can be detected due to its effect on the amplitude of the received ultrasonic signal [7].

3.3 Eddy current
Eddy currents can be induced in conductive materials by placing a coil on or near the surface which is excited with an alternating current and generates an alternating magnetic field around the coil. These induced eddy currents generate an opposed magnetic field to the one resulting from the excitation coil. Any defect or irregularity in the investigated area of a sample leads to a disturbance in the flow of eddy currents and the resulting magnetic field, which can be measured by the impedance signal of the receiving coil [3]. In [3] and [8] the ability but also the restrictions of eddy current testing to detect defects in CFRP were investigated. In both cases defects which directly affect the conductive fibers, for example by fiber cracking, could be identified [3, 8]. In addition delaminations at the interface between fibers and matrix material could be proven. But nevertheless this technique is not able to detect inhomogenities which only affect the nonconductive components in composite materials, like

i.e. pores in the polymer matrix of CFRP [8]. This method is a very fast, cheap and robust NDT technique, but limited in detection depth as the penetration depth of the magnetic field depends on the operating frequency of the eddy current system [3].

3.4 Lock in thermography

As a result of a temporally or locally periodic modulation of the thermal excitation, i.e. sinusoidal, a so called "thermal wave" can be induced in a material or a component. This wave propagates through the material and gets reflected at the boundaries of defects. The result is an interference of the reflected proportion of the wave with the modulated excitation wave which leads to a measurable temperature variation at the surface and can be detected by a thermographic camera. By applying a Fourier transformation the amplitude and the phase information of the measured signals can be evaluated which enables an imaging of defects inside the volume of the investigated samples [3, 9]. As described in [3] this method is very suitable for application on complex structures but limited concerning the detection depth. In this case artificial defects with a dimension of 4 x 4 x 0.5 mm^3 could be proven up to 8 mm in depth. Figure 5 demonstrates exemplarily the ability of lock in thermography to detect artificial defects of different sizes in a CFRP plate.

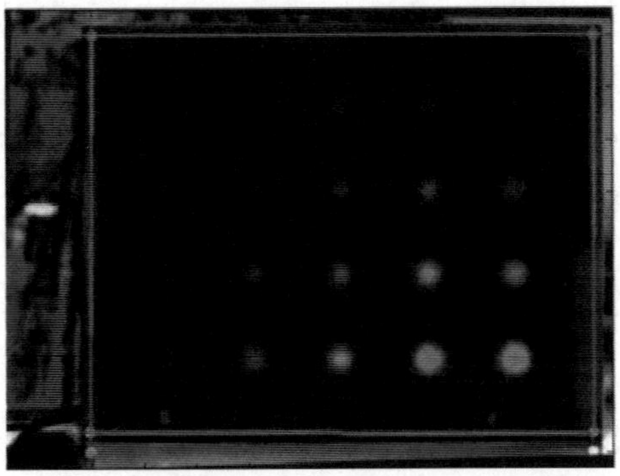

Figure 5. Detection of artificial defects in CFRP by lock in thermography [8].

4. Production integrated NDT of material compounds

To fulfill the requirements of modern lightweight construction, innovative joining technologies have to be developed or established that enable the realization of multi material design and avoid additional joining elements or overlapping material sections. But also conventional and widespread joining techniques have to be optimized regarding their efficiency in order to achieve an optimal resource usage. An important tool to optimize and to monitor or even control joining processes is the application of different NDT techniques, which will be presented exemplarily for different welding methods in the following.

4.1 Electromagnetic Acoustic Transducers (EMAT)

Nondestructive testing with Electromagnetic Acoustic Transducers (EMAT) describes a special method of ultrasonic testing. In contrast to conventional ultrasonic testing, EMAT enables contactless testing without any couplant as the ultrasonic waves are generated directly in the material, which has to be investigated, by electromagnetic induction. This is realized by a RF coil that is located below a permanent magnet and induces eddy currents in the material (see figure 6a). Due to this operating principle the EMAT technique is suitable for inline- and in-process inspection, but can only be used to investigate electrically conductive materials [10]. The principle of defect detection is similar to conventional ultrasonic testing. Figure 6 illustrates the investigation of weld seams by EMAT. Whereas a faultless weld seam exhibits no additional ultrasonic reflection between the echo of ultrasonic wave generation and the edge of the investigated sample (figure 6b), any defect in the weld seam causes reflections of the ultrasonic waves and weakens the proportion of ultrasonic energy which is transmitted through the welding area (figure 6c). In the case of a complete material separation, like shown in figure 6d, the ultrasonic waves are reflected at the interface of the defect completely and no edge echo of the sample can be attained.

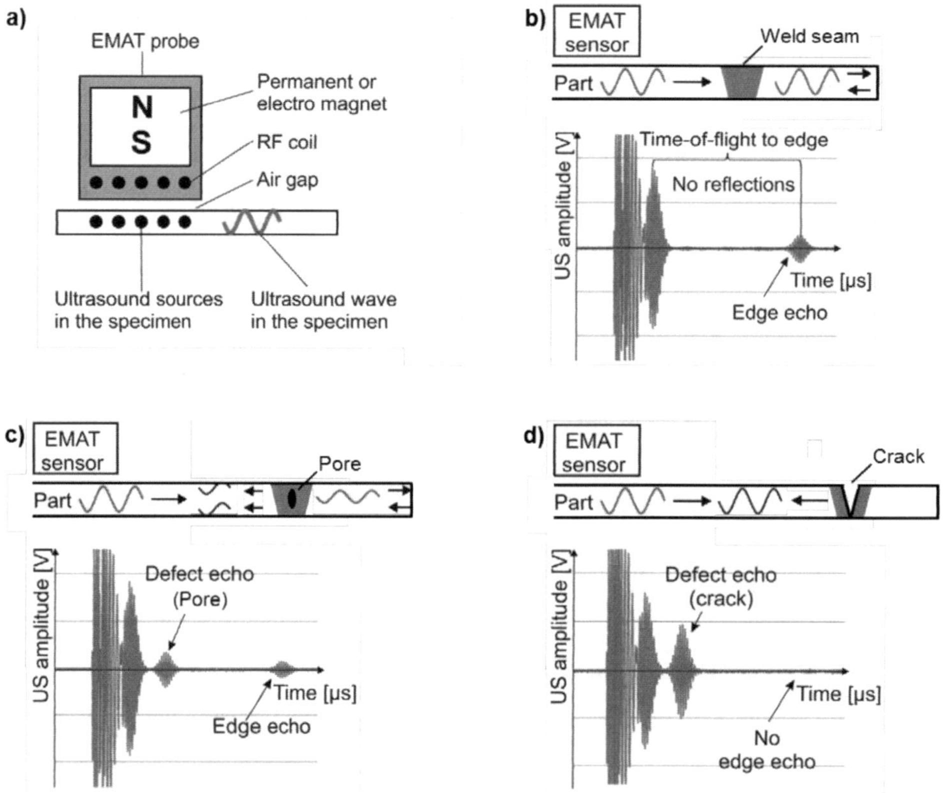

Figure 6. Schematic illustration of the detection of defects in weld seams by EMAT; a) physical principle, b) signals of a faultless weld seam, c) signals of a defective weld seam, d) signals of a weld seam with complete material separation.

4.2 Conventional ultrasonic testing

An example of a process monitoring concept in resistance spot welding is shown in figure 7. In contrast to EMAT, conventional ultrasound has to be transmitted to the investigated parts by a contacting transmitter. This is realized by implementing the US transmitter into one of the welding guns, which has to get in contact with one of the joining partners anyway. The other welding gun is equipped with an appropriate receiver to detect the ultrasonic waves, which pass the welding area during the process. In this case the ultrasonic testing is performed in transmission mode. The detection of defects and therefore the monitoring of the spot weld quality take place in a similar way than described for the EMAT´s as any irregularity in the welding process or the welding spot leads to measurable changes in the signal that is detected by the US receiver.

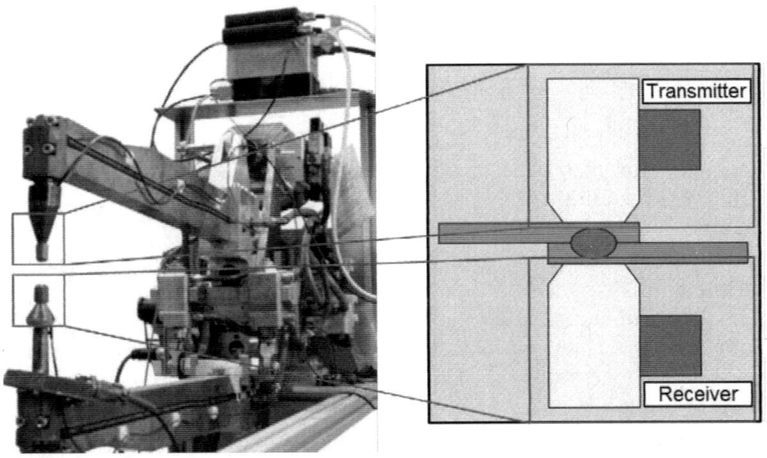

Figure 7. Ultrasonic Weld Monitor (USWM) in resistance spot welding.

4.3 Evaluation of process-intrinsic characteristics

A possibility to obtain direct feedback on possible process and thus quality variations of the corresponding component in the sense of an inline in-process procedure is the determination of process intrinsic characteristics. For example, the monitoring of the laser welding process on steels is carried out in [11, 12] by recording and anlalyzing the acoustic emissions of the process. This can be achieved by measuring the structure-borne acoustic emissions, e.g. by applying a corresponding sensor to the sheets to be joined or by measuring air-borne acoustic emissions. Particularly advantageous in this context is the detection of the signals in the high-frequency range from several hundred to approximately 1000 kHz, since this can eliminate disturbing environmental influences. Figure 8 illustrates this method using the example of the air-borne acoustic emissions of a laser weld with artificial as well as natural welding defects. It can be seen that the amplitude of the measured signal reacts significantly to deviations resulting from process variations or the presence of a defect and exceeds the limits of a previously determined tolerance band [11, 12].

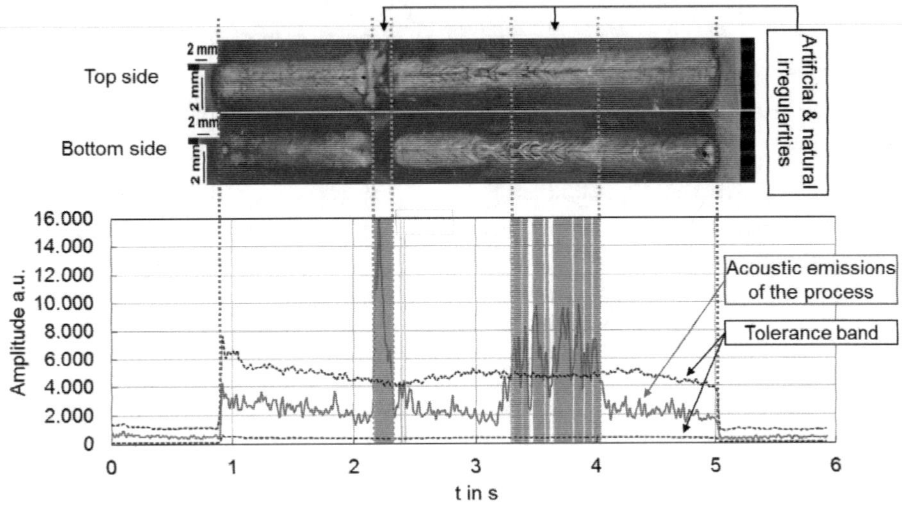

Figure 8. Process monitoring in laser welding by air-borne acoustic emissions [12].

Another example of the use of process-intrinsic characteristics for the purpose of process monitoring is described in [13] on the basis of the Friction Stir Welding process (FSW) of aluminum alloys. As FSW is a pressure welding method, a welding tool has direct contact to the joining partners, resulting in process forces in all three spatial directions. The axial welding force of the tool in the z-direction is a main process parameter of Friction Stir Welding, which is also used as control variable to realize a force-controlled process. However, the forces in the plane of the joining partners (x - and y-direction) can be recorded and evaluated in the sense of a real time process monitoring, as represented in figure 9. So, process fluctuations, as well as the occurrence of defects in the weld seam are expressed in deviations of these forces as illustrated by the black circled area. A subsequent analysis of the data by means of a short-time Fourier transform (STFT) facilitates the identification of relevant critical areas through the representation of temporal changes of the frequency spectrum which are highlighted by red circles in figure 9. The knowledge obtained in this way can also be used as the basis for the realization of appropriate process control algorithms. In addition to the schematic representation of this process monitoring approach, called "MonStir®", figure 9 shows the proof of the identified defect in the weld by means of ultrasonic testing and a microscopic view of a metallographic cross section [13].

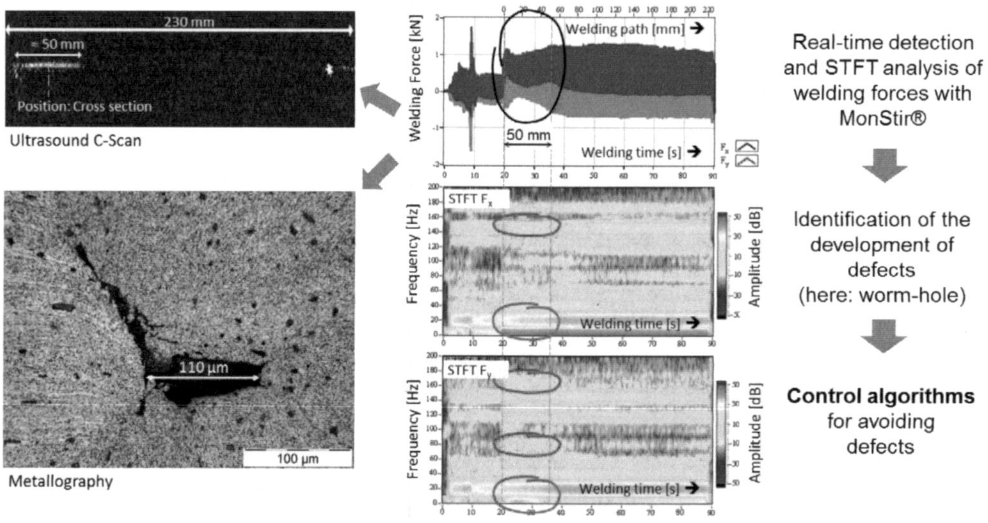

Figure 9. Schematic illustration of the MonStir® concept for process monitoring and control of the Friction Stir Welding process [13].

4.4 Measurement of mechanical and microstructural properties by micro-magnetic methods

In addition to the possibilities presented for the detection of imperfections, qualitative and quantitative determinations of mechanical and microstructural properties of materials can also be made using NDT methods. To achieve this, the 3MA technique (Micromagnetic Multiparameter Microstructure and stress Analysis) makes use of the correlation between the micro-magnetic properties of ferromagnetic materials and their mechanical or microstructural properties. On the basis of destructively determined reference values, a polynomial is calculated by means of a regression analysis which describes this relationship mathematically and thus enables the nondestructive determination of mechanical and microstructural properties such as, for example, hardness, tensile strength or diffusion layer thicknesses. In order to make the determination as accurate as possible, 41 measurement variables are measured by the combination of the methods: incremental permeability, harmonic analysis, Barkhausen noise and multi-frequency eddy current. Figure 10 shows on the left the comparison between destructive and nondestructive determination of the hardness over a laser welding seam. In the right-hand part of the figure, the residual stresses measured longitudinally and transversely to the seam are shown in a comparative manner over the weld seam. In both cases, there is a very good correspondence between destructive and nondestructive measurement values.

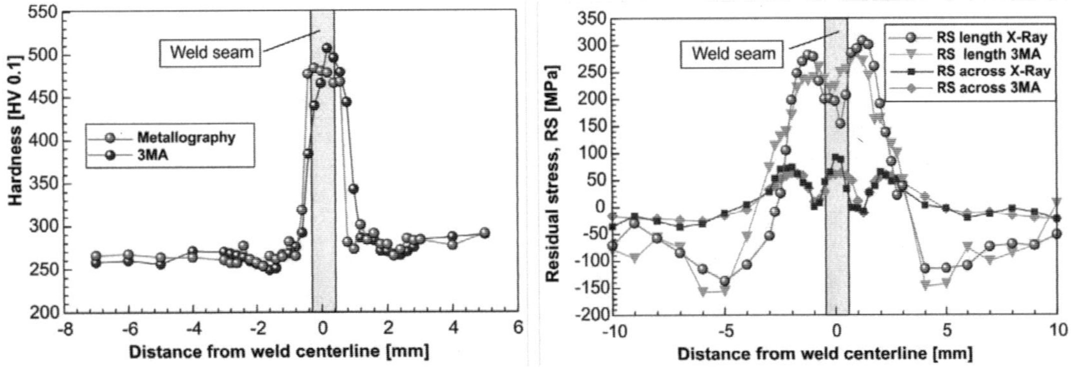

Figure 10. Nondestructive quantitative evaluation of hardness and residual stresses across a laser welded seam using 3MA technique.

5. Conclusions

As clarified by this article, the NDT has a high relevance to ensure adequate material quality as well as joint quality in the sense of lightweight construction. Due to the large number of available methods, as well as their constant further development and also new developments in the field of NDT, the challenges of the quality assurance of composite materials and material compounds can be overcome and thus a substantial contribution to a resource-saving product design can be provided.

Acknowledgments

The authors would like to thank all the colleagues at Fraunhofer IZFP who contributed to the outcome of this article.

References

[1] Netzelmann U 1993 Nondestructive testing of ceramic-metal joints by high-frequency ultrasound techniques *Joining ceramics, glass and metal* (Frankfurt/Main: Deutsche Glastechnische Gesellschaft) pp 168–175

[2] Verein Deutscher Ingenieure (VDI) 2009 Computertomografie in der dimensionellen Messtechnik - Grundlagen und Definitionen *VDI/VDE 2630 Blatt 1.1.* (Düsseldorf: VDI-Verlag)

[3] Beine C, Boller C, Netzelmann U, Porsch F, Venkat R, Schulze M, Bulavinov A and Heuer H 2010 NDT for CFRP aeronautical components a comparative study *2nd International Symposium on NDT in Aerospace (Hamburg)*

[4] Maisl M, Schorr C, Porsch F and Haßler U 2010 Computerlaminographie, Grundlagen und technische Umsetzung *Industrielle Computertomografie: Zerstörungsfreie Bauteilprüfung, 3D-Materialcharakterisierung und Geometriebestimmung* (Aachen: Shaker Verlag) pp 261–266

[5] Straß B 2015 Ultraschallunterstütztes Rührreibschweißen von Al/Mg-Verbunden: Prozess-analyse, mikrostruktureller, mechanischer und korrosiver Eigenschaftsvergleich *Werkstoffkundliche Berichte* vol 36 (Kaiserslautern: University of Kaiserslautern)

[6] Ramanan S V, Bulavinov A, Pudovikov S, Boller C and Wenzel T 2010 Quantitative non-destructive evaluation of CFRP components by sampling phased array *2nd International Symposium on NDT in Aerospace (Hamburg)*

[7] Helfen T B 2014 Nichtlinearer Ultraschall zur Charakterisierung von Ermüdungsschäden während der Hochfrequenz-Beanspruchung von C-Faser-Kunststoffverbunden *PhD Thesis* Universität des Saarlandes

[8] Müller W 1993 Erhöhung der Aussagefähigkeit von zfP-Prüfergebnissen an CFK-Platten durch Einsatz und Kombination unterschiedlicher Sensortechnik *Seminar über die Qualitätssicherung von Faserverbundwerkstoffen (Darmstadt)*

[9] Zöcke C, Stössel R, Langmeier A and Arnold W 2007 Automatische Erkennung und Charakterisierung von Defekten in CFK-Werkstücken unter Verwendung thermischer Aufnahmen *DGZfP-Berichtsbände* (Berlin: Deutsche Gesellschaft für zerstörungsfreie Prüfung (DGZfP))

[10] Salzburger H-J, Niese F and Dobmann G 2012 EMAT pipe inspection with guided waves *Welding in the world* **56** 35–43

[11] Bastuck M, Herrmann H-G, Wolter B, Böttger D and Zinn P-C 2016 AkuProLas: Acoustic Inline Process Monitoring for Laser Welding Applications *World Conference on Nondestructive Testing (19): WCNDT 2016* (Berlin: Deutsche Gesellschaft für zerstörungsfreie Prüfung (DGZfP))

[12] Bastuck M 2016 In-Situ-Überwachung von Laserschweißprozessen mittels höherfrequenten Schallemissionen (Saarbrücken: Universität des Saarlandes)

[13] Jene T 2008 Entwicklung eines Verfahrens zur prozessintegrierten Prüfung von Rührreibschweißverbindungen des Leichtbaus sowie Charakterisierung des Ermüdungsverhaltens der Fügungen *Werkstoffkundliche Berichte* vol 21 (Kaiserslautern: University of Kaiserslautern)

IOP Conf. Series: Materials Science and Engineering **181** (2017) 012018 doi:10.1088/1757-899X/181/1/012018

Requirements and testing methods for surfaces of metallic bipolar plates for low-temperature PEM fuel cells

P Jendras, K Lötsch and T von Unwerth

Department of Advanced Powertrains, Institute for Automotive Research, Technical University of Chemnitz, 09126 Chemnitz, Germany

Abstract. To reduce emissions and to substitute combustion engines automotive manufacturers, legislature and first users aspire hydrogen fuel cell vehicles. Up to now the focus of research was set on ensuring functionality and increasing durability of fuel cell components. Therefore, expensive materials were used. Contemporary research and development try to substitute these substances by more cost-effective material combinations. The bipolar plate is a key component with the greatest influence on volume and mass of a fuel cell stack and they have to meet complex requirements. They support bending sensitive components of stack, spread reactants over active cell area and form the electrical contact to another cell. Furthermore, bipolar plates dissipate heat of reaction and separate one cell gastight from the other. Consequently, they need a low interfacial contact resistance (ICR) to the gas diffusion layer, high flexural strength, good thermal conductivity and a high durability. To reduce costs stainless steel is a favoured material for bipolar plates in automotive applications. Steel is characterized by good electrical and thermal conductivity but the acid environment requires a high chemical durability against corrosion as well. On the one hand formation of a passivating oxide layer increasing ICR should be inhibited. On the other hand pitting corrosion leading to increased permeation rate may not occur. Therefore, a suitable substrate lamination combination is wanted. In this study material testing methods for bipolar plates are considered.

1. Introduction

Low temperature polymer electrolyte fuel cells (PEM FC) transform hydrogen into electrical current, thermal heat and water vapour as waste product, by an oxidation reaction with oxygen in the air. A single fuel cell consists of a proton conducting membrane with an electrode on each side (MEA), a porous gas diffusion layer (GDL) and adjoining current collectors. By stacking multiple cells the current collectors of two neighbouring cells act as bipolar plates. They fulfil the following main functions in the fuel cell stack [1]:

- Conduct current between neighbouring cells,
- Provide channels for reactant gases and for removing reaction products,
- Facilitate water through the cell,
- Thermal management through the cell and
- Isolation of the individual cells.

Bipolar plates can be graphite plates, polymer composite plates with carbon black particles or metallic sheets. Because of their high strength, metallic bipolar plates can be made of very thin sheets with a thickness of 0.1 mm or less. This enables a high power density, which is necessary for automotive fuel cell systems. Furthermore, sheet metals can be stamped, cut and welded in very short cycle times,

IOP Conf. Series: Materials Science and Engineering **181** (2017) 012018 doi:10.1088/1757-899X/181/1/012018

which is suitable for the needed mass production in the automotive industry. On the downside, base metals form a passive layer which increases the interfacial contact resistance to the GDL and corrode in acidic environments. Therefore, they have to be made of noble metal sheets or coated with a protective and conductive layer. Due to a lack of mass quality insurance methods and a need of a high operation safety in automotive applications, actual fuel cell stacks contain metallic bipolar plates made of expensive materials [2].

The objective of this paper is to describe the examination of the high requirements and their testing methods with the goal of reducing material costs and production costs by a specification of local functions and requirements and by avoiding too high technical efforts. Furthermore, the authors give a general overview of available testing methods and define local and measurable requirements.

2. Requirements

The general requirements for fuel cells and bipolar plates are described by the *United States Department of Energy (DOE).* The main requirements of the surface of the active area of metallic bipolar plates are low interfacial contact resistance (ICR) between the GDL and the bipolar plate and a good corrosion resistance against the acidic environment [3]. As shown in figure 1, the active area contains the flow field in the center of the bipolar plate, where the chemical reactions take place. The remaining parts of the bipolar plate are known as passive area. The common ratio between active and passive area ranges between 1/1 and 3/1 [2].

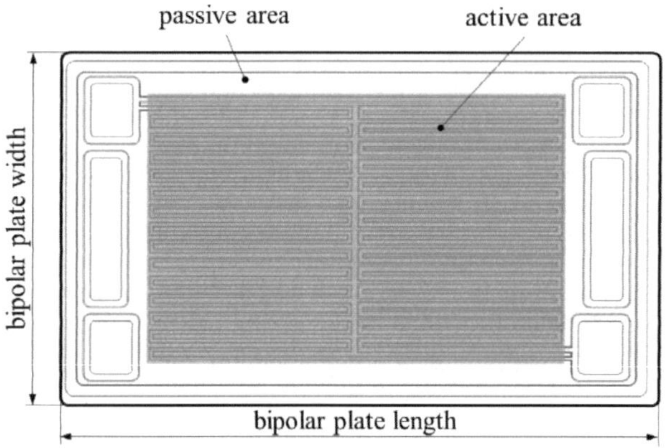

Figure 1. Active (grey) and passive area of a sample bipolar plate.

An additional requirement is the surface roughness of metallic bipolar plates that has an effect on the achievable ICR and the corrosion properties. The size ratio between a carbon fiber and a polished or a rough surface is illustrated in figure 2. The diameter of the carbon fiber in the GDL is slightly larger than the roughness peaks. Different roughnesses result in different amounts of contact points between GDL and bipolar plate, as well as a different pressure at the contact points since pressure equals force divided by area. Tawfik, Hung and Mahajan have described these effects accurately and depicted a range of average roughness [4]. Table 1 summarizes the exact values of ICR, corrosion resistance and surface roughness.

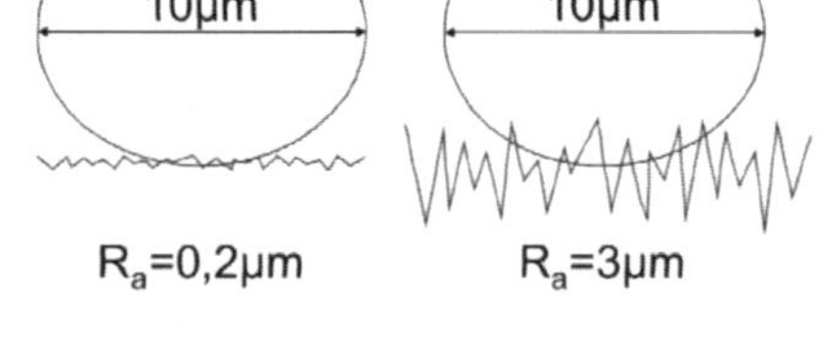

Figure 2. Illustration of a carbon fibre on a polished surface (a) and a rougher surface (b).

Table 1. General requirements for surfaces of metallic bipolar plates [3, 4].

Surface characteristic	2020 Targets
Interfacial contact resistance (ICR)	< 10 mΩcm² at 140 MPa
Corrosion resistance	< 1 μA/cm², with corrosion current at -0.1 V and H_2 purge < 1 μA/cm², with corrosion current at 0.6 V and Air purge
Average roughness R_a	0.2 … 3 μm

3. Testing methods

3.1 Planar interfacial contact resistance measurement

In order to achieve a high current density in the fuel cell, the electrical resistance of a cell must be as low as possible. Primarily, the contact resistance between the individual components of the gas diffusion layer and the bipolar plate influences the entire electrical resistance of a cell. The internal ohmic resistance is significantly smaller. Therefore, there is a lower priority on it. The measuring arrangement is shown in figure 3. The press simulates the surface pressure in the fuel cell stack. In between are two gold-plated copper plates. Copper is used as plate material due to its high electrical conductivity and the gold coating reduces the contact resistance to the actual measuring object. Gold does not form a passivating corrosion layer. In the center, the layer-like structure consists of two gas diffusion layers on both bipolar plate sides. The sample has an area of 1 cm². It is not converted since the resistance per area (mΩ / cm²) is always given for the comparison [4].

A constant current of 1 A is applied and the voltage drop across the measuring arrangement is measured. With the area of 1 cm² and one ampere, a similar current density results as in a PEM fuel cell. The resistance can then be determined by using Ohm's law. Surface pressures between 10 N/mm² and 300 N/mm² are specified in order to obtain a spectrum around the value of N/mm² applied in the fuel cell stack [4].

At first, an arrangement is measured with only one gas diffusion layer and no bipolar plate (see figure 3 – left side). This results in the sum of the contact resistance between the GDL and the gold-coated copper electrode and the ohmic resistance of the measuring arrangement R_1. Afterwards, the actual measurement setup (see figure 3 – right side) is measured.

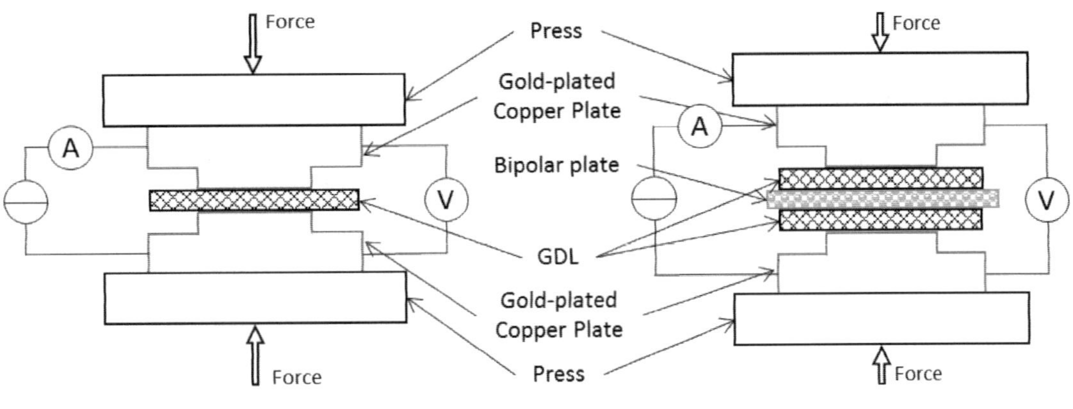

Figure 3. Schematic diagram of interfacial contact resistance (ICR) measurement.

The resistance R_2 determined in this setup is corrected with the aid of the value from the first measurement R_1. The division by 2 results from two contact surfaces in the construction, but only one is of particular interest [4].

$$ICR = (R_2 - R_1)/2 \tag{1}$$

The disadvantage of this measuring method, however, is that only semi-finished material and no bipolar plate can be measured. As a result, production steps cannot investigate possible influences.

3.2 Accelerated corrosion resistance test

The arrangement for measuring the corrosion resistance consists of a container with sulfuric acid solution. Three electrodes are attached: the counter electrode, the reference electrode and the working electrode. The counter electrode is made of conductive corrosion-resistant material such as platinum or graphite. The reference electrode is designed as a calomel electrode or a standard hydrogen electrode. The working electrode is the material sample pressed from the outside to the measuring cell. An aperture forms a connection to the electrolyte [4].

The three electrodes are connected to a potentiostat. A voltage is applied between the working electrode and counter electrode. By the reference electrode a regulation and thus a constant potential becomes possible. The voltage generates a current, namely a corrosive flow [6].

In order to simulate the conditions in a PEM fuel cell, the measurements are carried out at 80 °C. Atmospheric or hydrogen gas is passed through the electrolyte. The gases can act as reaction partners for corrosive and reducing processes. The solution of 0.5 M H_2SO_4 and 2 ppm F- or 2 ppm HF ensures an accelerated aging process [4].

Both potentiodynamic and potentiostatic measurements are applied. The dynamic method uses a potential between 0.5 V below and 1 V above the free corrosion potential and a sampling rate of 1 mV/s. Under static conditions, a potential of 0.6 V is applied over 8 hours. This potentiostatic test allows researchers to simulate the material behavior after a long period of use without having to build up a fuel cell and to operate it over a long period of time. The corrosion behavior is to be read from the potentiodynamic polarization curves. Kumar [5] describes the procedure in detail.

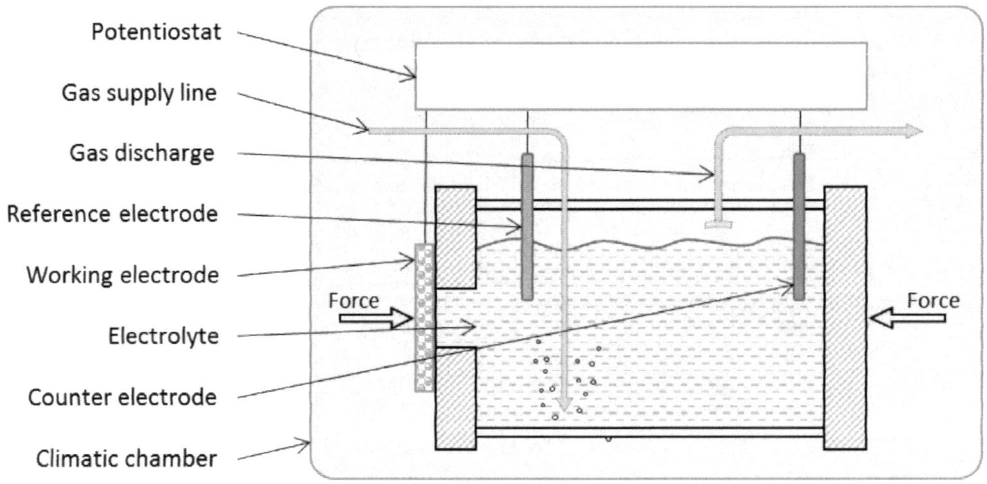

Figure 4. Schematic diagram of corrosion test cell and potentiostat.

IOP Conf. Series: Materials Science and Engineering **181** (2017) 012018 doi:10.1088/1757-899X/181/1/012018

3.3 Roughness measurement

The roughness is influenced by the production of semi-dies, the forming and further production process steps. This, in turn, has influence on contact resistance, corrosion behavior, flow behavior, etc. Hence, this characteristic of the surface is measured by the optical and tactile approach.

3.3.1 Tactile scanning of roughness

During tactile measurement, the surface is tracked over a distance of 4 cm. The probe is provided with a diamond tip, which is only a few nanometers thick and is able to measure surface topographies of this magnitude. The exact procedure is described in DIN EN 4287. The recorded measurement signal is cleaned from general noise and surface deviations of a different magnitude, such as ripples, are calculated.

The disadvantage is that the entire surface of a bipolar plate cannot be examined due to the size of the measuring probe. Only overlying sections, such as webs, can be measured.

3.3.2 Optical roughness measurement

In the case of optical methods, the surface is scanned by means of a laser. The reflection is recorded by a sensor and converted into an image for evaluation. In a following step, the roughness is deduced from this. This method allows scanning of the entire surface. However, the approach is less accurate since unwanted surface reflections can lead to results, which do not result from the low absorption coefficient of the metallic surface instead of the topography.

4. Locally different requirements

The bipolar plate is a component with many different tasks, resulting in a complex requirement spectrum. Therefore, different properties are required locally. The following chapter examines the proper point of property and the measurable variables. Two linked and reshaped sheets form a bipolar plate. Hence, the surface at the attachments must be suitable. This suitability depends mainly on the mating method used. Soldering requires sufficient soldering.

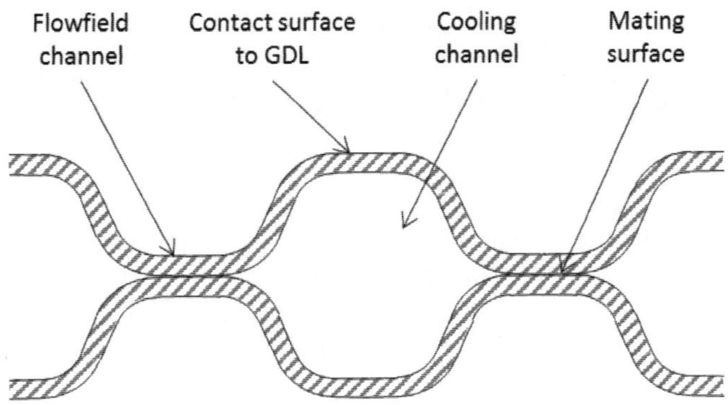

Figure 5. Schematic cross-sectional diagram of locally different requirement zones.

The adjacent gas diffusion layer is located on the web surfaces. Between this component and the bipolar plate, a contact resistance should provide as small as possible. Consequently, the formation of a passive layer has to be prevented. Since it also lacks the corrosion-protective effect of such a layer, is it performed by an actively applied layer system.

The contact resistance in the flowfield channel towards the GDL is not needed. The focus there is on adequate corrosion protection. On the one hand, the corrosion rate must remain low to ensure that the permeation rate of hydrogen through the bipolar plate stays within the limits of the DOE over the entire lifetime (5000 hours in the vehicle). In combination with corrosion, the smallest possible discharge of ions is required, since this effect accelerates membrane degradation. Other effects constraining the design regarding hydrophobicity play a role inside the flow field channel, e.g. low flow resistance is desired in ducts. In the cooling water channels it is also required to minimize the pressure drop. Furthermore, the requirements of the deionized water have to be fulfilled with respect to corrosion.

The requirements for the inactive area are slightly lower. Although corrosion protection requirements as in the ducts are similar to the gas feeds to the cell. However, the contact resistance does not play any role here.

5. Conclusions and future work

The complex requirements for the bipolar plate lead to great efforts to ensure functionality over the required lifetime of the vehicle. Consequently, methods are required to measure the physical quantities. The described methods according to the prior art are limited to the verification of the requirements on the semi-finished product.

Various properties are required at different points of the bipolar plate. A distinction is made between active and passive surfaces. More precise specification means viewing the cooling channel, the flow field channel, the contact surface to the GDL and the joining surface. Not all properties need to be checked. At present, there are still insufficient possibilities for an exact surface test for bipolar plates.

The trend in fuel cell development for vehicles is to use complex 3D structures, which are provided with a functionalizing coating. Therefore, it is necessary to develop measurement methods, which can investigate corrosion resistance, contact resistance and roughness on the filigree structures of the bipolar plate at the desired locations.

Acknowledgements

The authors would like to thank the *Saxon Ministry for Economic Affairs and Labour (SMWA)*, the *European Social Fund (ESF)* and the *Development Bank of Saxony (SAB)* for financially supporting the project *HZwo:BIP - Bipolarplatten aus Sachsen.*

References

[1] Kurzweil P 2013 *Brennstoffzellentechnik* (Wiesbaden: Springer Vieweg)

[2] Vielstich W, Lamm A and Gasteiger H A 2003 *Handbook of Fuel Cells: Fundamentals, Technology, Applications* (New York: Wiley)

[3] DOE Fuel Cell Technologies Office 2016 *Multi-Year Research, Development and Demonstration Plan - 3.4 Fuel Cells* p 22

[4] Tawfik H, Hung Y and Mahajan D 2016 Bipolar Plate Durability and Challenges *Polymer electrolyte fuel cell degradation* ed M M Mench et al (Oxford: Academic Press) pp 249–291

[5] Kumar A, Ricketts M and Hirano S 2010 Ex situ evaluation of nanometer range gold coating on stainless steel substrate for automotive polymer electrolyte membrane fuel cell bipolar plate *J. Power Sources* **195** 1401–7

[6] Asri N F, Husaini T, Sulong A B, Majlan E H and Daud W R W 2016 Coating of stainless steel and titanium bipolar plates for anticorrosion in PEMFC: A review *Int. J. Hydrogen Energy*

Influence of the post-weld surface treatment on the corrosion resistance of the duplex stainless steel 1.4062

P Rosemann[1,2], **C Müller**[2], **O Baumann**[2], **W Modersohn**[3] **and T Halle**[2]

[1]BAM - Bundesanstalt für Materialforschung und –prüfung, Unter den Eichen 87, 12205 Berlin, Germany
[2]Institut für Werkstoff- und Fügetechnik, Otto-von-Guericke University, Universitätsplatz 2, 39106 Magdeburg, Germany
[3]Wilhelm Modersohn GmbH & Co. KG, Industriestraße 23, 32139, Spenge, Germany

Abstract. The duplex stainless steel 1.4062 (X2CrNiN22-2) is used as alternative material to austenitic stainless steels in the construction industry. The corrosion resistance of welded seams is influenced by the base material, the weld filler material, the welding process and also by the final surface treatment. The scale layer next to the weld seam can be removed by grinding, pickling, electro-polished or blasting depending on the application and the requested corrosion resistance. Blasted surfaces are often used in industrial practice due to the easier and cheaper manufacturing process compared to pickled or electro-polished surfaces. Furthermore blasting with corundum-grain is more effective than blasting with glass-beads which also lower the process costs. In recent years, stainless steel surfaces showed an unusually high susceptibility to pitting corrosion after grinding with corundum. For this reason, it is now also questioned critically whether the corrosion resistance is influenced by the applied blasting agent. This question was specifically investigated by comparing grinded, pickled, corundum-grain- and glass-bead-blasted welding seams. Results of the SEM analyses of the blasting agents and the blasted surfaces will be presented and correlated with the different performed corrosion tests (potential measurement, KorroPad-test and pitting potential) on welding seams with different surface treatments.

1. Introduction

The number of applications of duplex stainless steels (DSS) has increased significantly over the last 10 years in the construction industry. DSS combine the high fracture toughness of austenitic stainless steels (ASS) with the high stress corrosion resistance of ferritic stainless steels (FSS) [1]. The higher mechanical toughness of DSS compared to ASS allows furthermore the reduction of component dimensions. Especially the "lean" DSSs (1.4062, 1.4162, 1.4362 and 1.4662) with lower alloying contents of Ni and Mo reduce the materials costs in many applications [2-3]. Scientific investigations prove the equal or better corrosion resistance of DSSs compared to ASSs of the same corrosion resistance class (CRC) [2, 4]. These positive properties initiate the application of DSS in various applications requesting high corrosion resistance [1-3, 5-12]. There are only a few long-term investigations on the performance of DSSs in the construction industry [4]. Furthermore, production, processing and welding are still challenging and require optimization regarding the corrosion resistance [2, 8, 11].

The corrosion resistance of welding seams is influenced by the used base material, the weld filler material, the welding process and the final surface treatment [13, 14]. The aim of this study is to characterize the influence of different surface treatments of welding seams on the corrosion resistance of the "lean" DSS 1.4062. Investigations in natural climates would be the best approach to characterize this issue, but the investigations would require several years [2]. For this reason mainly electrochemical methods were used to characterize the corrosion resistance of industrial welding seams after different surface treatments. The scale layer next to the weld seam is known to reduce the corrosion resistance and has to be removed [2, 4, 14]. Depending on application and requested corrosion resistance, this could be achieved by grinding, pickling, electro-polishing or blasting [4].

It is reported that blasting with corundum is more effective than blasting with glass-beads. In recent years, stainless steel surfaces showed an unusually high susceptibility to pitting corrosion after grinding with corundum [15]. For this reason, it is now questioned whether the corrosion resistance of DSSs is influenced by the used blasting agent. This question was investigated specifically by comparing grinded, pickled, corundum-grain-blasted and glass-bead-blasted surfaces of the DSS 1.4062 and its welding seams. Results of the SEM analyses of the blasting agents and the blasted surfaces will be presented and correlated with the results of various performed corrosion tests.

2. Materials and methods

2.1 Materials, welding process and microstructure characterization

The welding of cold rolled sheets of 1.4062 was performed at Wilhelm Modersohn GmbH & Co KG with gas shielded metal arc welding using a root layer of 1.4062 and higher alloyed top layers of 1.4343. The structure of the welding seams was examined by metallographic investigations. The samples were wet ground with increasing grain size, polished with 3 µm diamond suspension and etched with beraha's reagent (800 ml H_2O, 400 ml HCl, 48 g $(NH_4)HF_2$ and 0,1 g/l $K_2S_2O_5$) at room temperature for 20 s. The microstructure was investigated with optical microscopy. The macroscopic appearance of the welding seams after etching is presented in figure 1. The chemical composition of the alloys (1.4062 and 1.4343) used to perform the welding process were analyzed by optical emission spectrometry. Table 1 shows the major alloying elements of the different alloys.

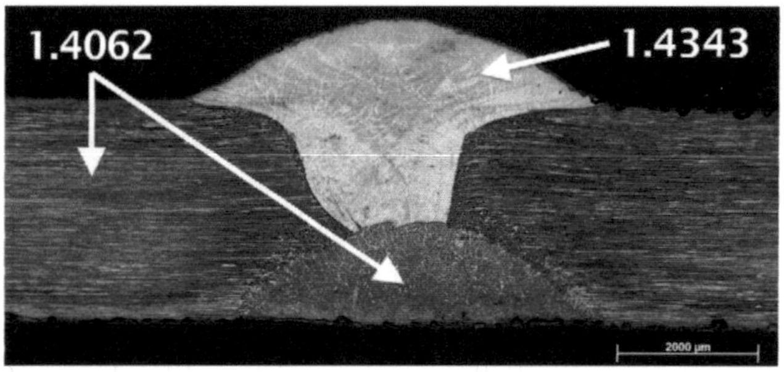

Figure 1. Macrostructure of welding seams used for this study.

Table 1. Alloy composition (mayor elements) of the welding seam in wt.-%.

	alloy	Cr	Ni	Mo	Mn	Si	N	C	P	S
base material	**1.4062**	**24.44**	**2.54**	**0.21**	**1.39**	**0.42**	**0.137**	**0.033**	**0.015**	**<0.001**
weld filler material (root)	**1.4062**	**24.28**	**2.64**	**0.23**	**1.21**	**0.30**	**0.143**	**0.029**	**0.017**	**<0.001**
weld filler material (top layer)	**1.4343**	**22.34**	**8.92**	**2.89**	**1.19**	**0.38**	**0.128**	**0.027**	**0.019**	**<0.001**

2.2 Electrochemical potentiodynamic reactivation

The passivation and reactivation behavior of the welding seams was studied with the double loop electrochemical potentiodynamic reactivation (DL-EPR). These measurements were performed to characterize only the effect of welding on the corrosion resistance on different parts of the welding seam. The weld reinforcement was removed by wet grinding to compare the cold rolled base alloy 1.4062 with the higher alloyed part of the welding seam (1.4343) and the lower alloyed part of the welding seam (1.4062). The surface was initially wet ground with grit 1000, rinsed with deionized water, subsequently cleaned with ethanol and then dried under an air flow before the EPR investigations. An area of 1.54 cm^2 was exposed to 2.3 ml 0.5 M H_2SO_4 with 0.01 M KSCN as activator using a mini test cell with a classical three electrode setup consisting of a saturated Ag/AgCl reference electrode (+198 mV_{SHE}) and a platinum sheet as counter electrode. A Gamry potentiostat was used to polarize the working electrode from -500 $mV_{Ag/AgCl}$ to +100 $mV_{Ag/AgCl}$ (passivation) followed by a reversal of the polarization direction polarizing back to -500 $mV_{Ag/AgCl}$ (reactivation) under a constant polarization rate of 2 mV/s. The maximum current density measured during the forward scan (i_P) characterizes the passivation ability. The maximum current density during the backward scan (i_R) is the result of local reactivation and dissolution of sensitized areas. The degree of sensitization (C) was calculated as the ratio of i_R / i_P.

2.3 Surface treatment and SEM-analysis

Different industrial processes (pickling, blasting with glass bead, blasting with corundum, electro-polishing and plasma-polishing) were used to remove the scaling layer from the welding seams. The effect of these post surface treatments on the corrosion resistance of the welding seams was studied with different analytical and electrochemical methods.

The corundum-grains and the glass-beads were studied using scanning electron microscopy (SEM). Unused and used blasting agents were compared to characterize the effect of surface blasting on the blasting agents. All samples were sputtered with gold to decrease the electric charge of the blasting agents. Furthermore, the corundum-grain and the glass-bead blasted steel surfaces were characterized with SEM using topographic (SE) and elemental contrast (BSE) at the same surface area. An EDX-analysis was performed to identify the elements of detectable surface irregularities.

2.4 Potential measurements and the KorroPad-test

The corrosion behavior of the pickled, corundum-grain and the glass-bead blasted steel surfaces (without welding seam) were characterized by potential measurements. A test solution with 3 g/l Cl^- and pH 4.5 was applied on the surface using a mini test cell (1.54 cm^2, 4 ml). The electrochemical potential was measured against saturated Ag/AgCl reference electrode (+198 mV_{SHE}) with a Gamry potentiostat.

The KorroPad indicator test [16] was used to investigate the pitting corrosion resistance of pickled and blasted (glass and corundum) surfaces (without welding seam). This test detects the susceptibility of stainless steel surfaces to pitting corrosion by means of blue indications [17, 18]. The KorroPad consists of an indicator solution mixed with an activator embedded in a gel pad. The KorroPad has a test area of approximately 300 mm^2 and a nearly constant redox potential of 240 $mV_{Ag/AgCl}$. The KorroPad was originally designed for lower alloyed austenitic stainless steels (1.4301, 1.4404).

At the initiation state of pitting corrosion small dots are generated depending on the stability of the passive film. The steel surfaces were exposed to air with 95 % relative humidity for 24 h before the KorroPad-test to give each state the same opportunity to build up a passive film. The KorroPads were placed on the prepared surface and the electrochemical potential was measured during the reaction time of 15 min against a saturated Ag/AgCl reference electrode (+198 mV_{SHE}) with a Gamry potentiostat. The blue coloration of the KP was evaluated after the test using a high resolution scanner.

2.5 Potentiodynamic polarization test

The pitting corrosion behavior was investigated by electrochemical dynamic polarization in all generated surface states on the higher alloyed welding seam (1.4343). The samples were exposed to air with 95 % relative humidity for 24 h before the experiment to give each state the same opportunity to build up a passive film. The crevice between the sealing ring of the mini test cell and the sample surface was covered with an inert lacquer to prevent crevice corrosion. A sample area of 1.54 cm^2 was exposed to the test solution (3 g/l Cl^-, pH 4.5) and the polarization was performed with a classical three electrode setup consisting of a saturated Ag/AgCl reference electrode (+198 mV_{SHE}) and a platinum sheet as counter electrode. A Gamry potentiostat was used to polarize the working electrode from the open circuit potential to the critical pitting potential at a constant polarization rate of 2 mV/s. The critical pitting potential was determined at a current density of 100 $\mu A/cm^2$.

3. Results and discussion

The surface treatment has a strong influence on the corrosion behavior of the welding seam. Initial metallographic investigations (2.1.) and the EPR-test (2.2.) were performed to describe the quality of the welding seam prior to the effect of surface treatment.

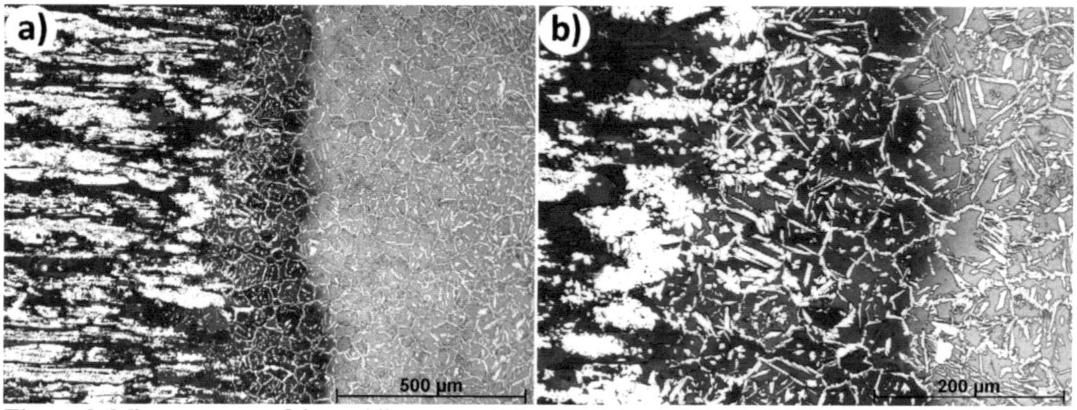

Figure 2. Microstructure of the welding seam: a) metallurgical bonding and b) heat affected zone.

Figure 2 shows the typical duplex microstructure with austenitic (white) and ferritic (dark) phases in all parts of the welding seam. The effect of the tint etching with beraha's reagent is more effective on the cold rolled 1.4062 (left) and its heat affected zone (middle), which is approximately 200 µm thick. The higher alloyed 1.4343 (right) shows a less etched ferritic matrix, which can be attributed to the higher alloying content of Mo and Ni.

The results of the EPR-test performed on the different parts of the welding seam are presented in figure 3. The cold rolled base alloy (1.4062) and the welding seam of the same alloy showed higher passivation current densities which are the result of the lower Mo and Ni-content compared to the dissimilar welding seam with alloy 1.4343. The reactivation at all three parts of the welding seam is low, but the base alloy is less sensitized compared to both parts of the welding seam. The degree of sensitization is below 0.05 for all three parts and the slightly increased reactivation of the welding seams should have no dominant effect on the pitting corrosion resistance. This was confirmed with the KorroPad-test showing no indication of pitting corrosion for ground surfaces. Furthermore, the critical pitting potential of the ground welding seam was as high as for the ground cold rolled base alloy (1.4062). It can be concluded that the welding seam itself has no dominant negative effect on the pitting corrosion resistance.

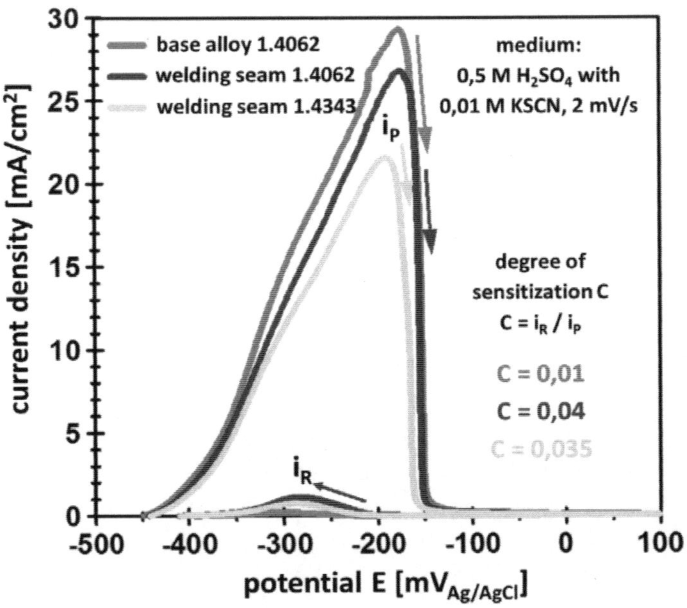

Figure 3. Results of the electrochemical potentiody-namic reactivation (EPR) performed on different locations of the welding seam.

The further investigations focus on the effect of the surface treatment of the welding seam which is necessary to remove the corrosion-prone scale layer. The typical appearance of four different surface states investigated is shown in figure 4. The corundum-blasted and glass-bead-blasted surfaces as well as the electro-polished and plasma-polished surfaces have a similar appearance.

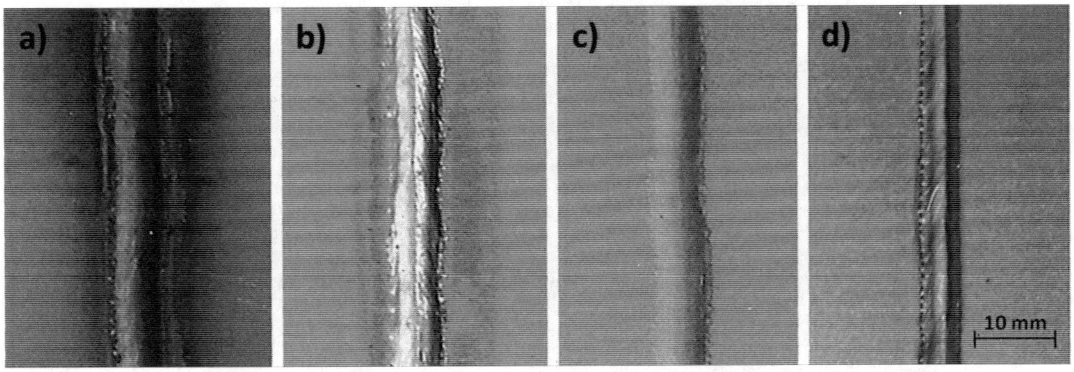

Figure 4. Typical surfaces of welding seams: a) scale layer, b) pickled, c) blasted and d) polished.

The blasting agents and blasted surfaces were studied in order to evaluate the effect of the blasting agents (glass vs. corundum) on the pitting corrosion resistance. Figure 5 shows SEM images of unused and used blasting agents which were as well used for blasting of the welding seam. The grain size of the blasting agents decreases due to the blasting process. The investigation of the blasted surfaces in the SEM with SE- and BSE-contrast is presented in figure 6 for the same surface area. The blasted surfaces are heavily rugged and the roughness is comparable (corundum: Rz 26 ± 2, Ra 3,8 ± 0,2 and glass: Rz 29 ± 8, Ra 4,1 ± 0,5) for both blasting agents. The dark particles were analyzed by EDX-spot-analysis and it was confirmed that the small grained blasting agents were incorporated in the blasted surface.

IOP Conf. Series: Materials Science and Engineering **181** (2017) 012019 doi:10.1088/1757-899X/181/1/012019

Figure 5. SEM images of blasting agents: a) unused glass bead, b) used glass bead, c) unused corundum-grains and d) used corundum-grains.

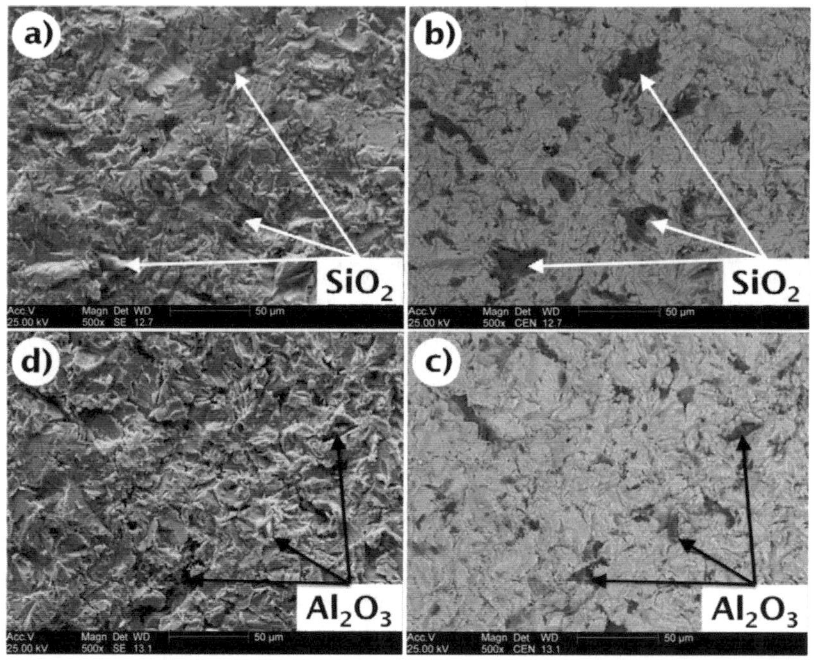

Figure 6. SEM images of blasted surfaces: a) glass bead with SE-contrast, b) glass bead with BSE-contrast, c) corundum-grain with SE-contrast, d) corundum-grain with BSE-contrast.

The effect of these microscopic defects in the surface of the DSS 1.4062 on the corrosion resistance was studied with the KorroPad-test, which was instrumented with an additional potential measurement. Furthermore, a conventional potential measurement in a electrochemical test solution with 3 g/l Cl^-, pH 4.5 (2.4.) was performed for comparison. These investigations were performed in some distance to the welding seam to detect only the effect of the surface treatment. The blasted surfaces showed lower initial potentials during the KorroPad-test compared to the pickled surfaces, see figure 7 a).

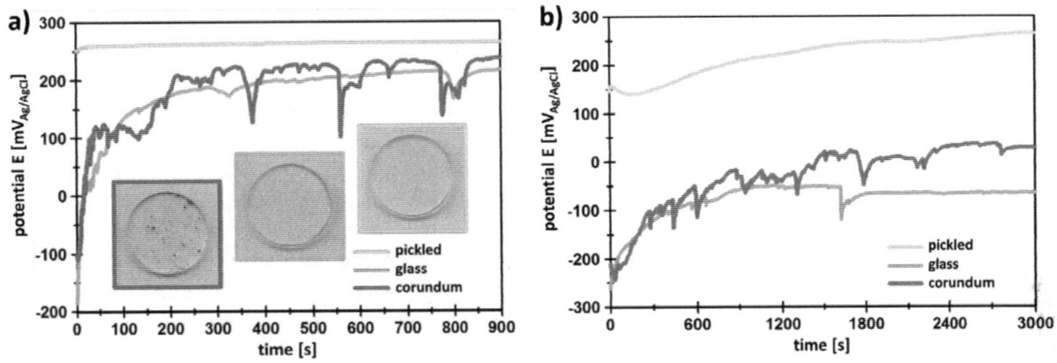

Figure 7. Corrosion potential a) during the KorroPad-test as well as the appearance of the KorroPads b) in 3 g/l Cl-, pH 4.5 for different surface states (without welding seam).

Especially the corundum-blasted surface shows a high number of potential drops. These are the result of metastable pitting corrosion, which leads to many blue indications in the KorroPad. The potential development of the glass-bead-blasted surface shows only a few potential drops, which initiate only isolated indication in the KorroPad. The potential measurement in the test solution with 3 g/l Cl^- and a pH 4.5 shows a similar tendency, (figure 7 b). These results confirm the weaker passive layer stability of corundum-blasted compared to glass-bead-blasted surfaces by means of more potential drops. The pickled surface shows a constant noble potential and therefore a high passive layer stability.

Finally, the critical pitting potentials were evaluated on the welding seams with the higher alloyed weld filler material (1.4343) in all different surfaces states, see figure 8.

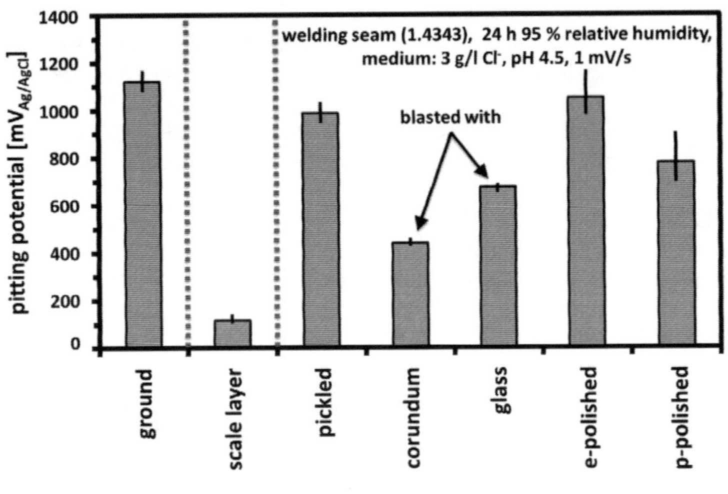

Figure 8. Critical pitting potentials of different surface modifications of the higher alloyed part of the welding seam (1.4343).

The surface treatment determines the critical pitting potential significantly. The welding seam with scale layer shows the weakest pitting corrosion resistance, which is the reason for the typically performed surface treatments. The pitting corrosion resistance of blasted surfaces is higher compared to the scale layer, but it is much lower compared to ground, pickled or polished welding seams. Moreover, blasting with corundum result in a lower critical pitting potential compared to blasting with glass-bead.

4. Conclusions

The corrosion resistance of the "lean" duplex stainless steel 1.4062 (X2CrNiN22-2) was studied on welding seams with different surface treatments. The corrosion resistance of the welding seam itself was comparable to the cold rolled base alloy. The scale layer shows the weakest pitting corrosions resistance, which was estimated. The surface treatments, which were used to remove the scale layer, result in major differences in the pitting corrosion resistance. Blasting leads to highly unstable passive layers, as could be detected by the KorroPad-test and the potential measurements. The blasting process reduces the grain size of the blasting agents and EDX-spot-analysis of the blasted surfaces confirmed the incorporation of the small grained blasting agents in the steels surface. These surface defects hinder the formation of a stable passive layer and reduce the pitting corrosion resistance. In case of blasting with corundum this effect is more pronounced compared to blasting with glass-bead and the susceptibility to pitting corrosion is significantly higher. This was proven by KorroPad-test, the potential measurement and the higher critical pitting potentials of glass bead blasted surfaces compared to corundum grain blasted surfaces. The pitting potential measurements showed a pronounced differentiation of the pitting corrosion resistance in the following order: scale layer < blasting with corundum < blasting with glass-bead < plasma-polished < ground, pickled and electro-polished. Future work will focus on the optimization of the blasting process and the post-treatment of corundum blasted surfaces to increase the passive layer stability.

References

[1] Gunn R N 1997 *Duplex Stainless Steels, Microstructure, Properties and Applications* (Abbington Publishing)

[2] Burkert A Modersohn W Müller T and Busch W B 2016 Innovation und Sicherheit im Bauwesen durch Einsatz von Duplexstahl *Duplex-Stähle – Nische oder Zukunft?* 82

[3] Eich G Burkert A Burkert A and Mietz A 2007 Einsatz des nichtrostenden Duplexstahles 1.4362 als preisgünstige Alternative im Bauwesen *Stahlbau* **76** 898

[4] Burkert A Gümpel P Lehmann J Burkert A Zietelmann C Leu F and Hoffmann C 2011 Technische und wirtschaftliche Alternativen zu den klassischen nichtrostenden, austenitischen Stählen unter atmosphärischen Einsatzbedingungen *final report of the IGF project 16049 N / 1*

[5] Rauscher T and Nürnberger U 2016 Hochfeste nichtrostende austenitische Stähle und Duplex-Stähle im Hinblick auf eine Anwendung im Betonbau *Duplex-Stähle – Nische oder Zukunft?* 70

[6] Venkataraman S Overbeck G W Jakobi D 2016 Einsatz von Duplex-Gusslegierungen *Duplex-Stähle Nische oder Zukunft ?* 60

[7] Schmitt G Kuhl L and Siegmund G 2016 Duplex-Werkstoffe im Sauergaseinsatz - beständiger als ihr Ruf *Duplex-Stähle Nische oder Zukunft ?* 39

[8] Wilke F 2016 Fertigungsprozesse für Duplexstähle beim Stahlhersteller *Duplex-Stähle – Nische oder Zukunft?* 13

[9] Ladwein T 2016 Einführung in die Welt der ferritisch-austenitischen Duplexstähle *Duplex-Stähle – Nische oder Zukunft?* 7

[10] Branner W 2016 Einsatz von Duplexstählen im Bereich Wasser/Abwasser *Duplex-Stähle – Nische oder Zukunft?* 96

[11] Pohl M 2016 Einsatz von Duplex-Gusslegierungen *Duplex-Stähle – Nische oder Zukunft?* 106

[12] Morach R 2016 Einsatz von Duplexstählen in der Chemie *Duplex-Stähle – Nische oder Zukunft?* 110
[13] Schultze S 1999 *Selektive Korrosion von Duplexstahl*, Magdeburg, (Shaker Verlag)
[14] Vollmer T Gümpel P Blaise M and Racky W 1995 *Materials and Corrosion* **46** 92
[15] Burkert A Burkert A Lehmann J Müller T and Bohlmann T 2014 Detektion korrosions empfindlicher Oberflächen nichtrostender Stähle durch die Verarbeiter final report of the IGF project 17136 N / 1
[16] Lehmann J Burkert A and Steinhoff U M 2012 Auflage für den Nachweis von korrosions-empfindlichen Metalloberflächen und Verfahren zum Nachweis von korrosionsempfindlichen Metalloberflächen *German patent* Nr. DE 10 2010 037 775 A1
[17] Burkert A Klapper H S and Lehmann J 2013 Novel strategies for assessing the pitting corrosion resistance of stainless steel surfaces *Materials and Corrosion* **64** 675
[18] Lehmann J Burkert A and Mietz J 2016 Investigations proofing the passive layer stability of stainless steels *Materials and Corrosion* **67** 84

IOP Conf. Series: Materials Science and Engineering **181** (2017) 012020 doi:10.1088/1757-899X/181/1/012020

Analysis of the ductility dip cracking in the nickel-base alloy 617mod

A Eilers[1], J Nellesen[2], R Zielke[2] and W Tillmann[1]

[1]Institute of Materials Engineering, TU Dortmund, Leonhard-Euler-Str. 2, 44227 Dortmund, Germany
[2]RIF e.V., Dortmund, Germany

Abstract. While testing steam leading power plant components made of the nickel-base alloy A617mod at elevated temperatures (700 °C), ductility dip cracking (DDC) was observed in welding seams and their surroundings. In order to clarify the mechanism of crack formation, investigations were carried out on welded specimens made of A617mod. Interrupted tensile tests were performed on tensile specimens taken from the area of the welding seam. To simulate the conditions, the tensile tests were conducted at a temperature of 700 °C and with a low strain rate. Local strain fields at grain boundaries and inside single grains were determined at different deformation states by means of two-dimensional digital image correlation (DIC). Besides the strain fields, local hardnesses (nanoindentation), energy dispersive X-Ray spectroscopy (EDX), and electron backscatter diffraction (EBSD) measurements were performed. Besides information concerning the grain orientation, the EBSD measurement provides information on the coincidence site lattice (CSL) at grain boundaries as well as the Schmid factor of single grains. All results of the analysis methods mentioned above were correlated and compared to each other and related to the crack formation. Among other things, correlations between strain fields and Schmid factors were determined. The investigations show that the following influences affect the crack formation: orientation of the grain boundaries to the direction of the loading, the orientation of the grains to each other (CSL), and grain boundary sliding.

1. Introduction

The nickel-base alloy 617mod is planned to be used for steam leading pipes in power plants with steam temperatures of 700 °C. In a test run under operation conditions at an elevated temperature of 700 °C ductility dip cracking (DDC) occurred in the area of the welding seam [1-3].

DDC occurs in nickel-base alloys and austenitic steels during solidification or reheating (i.e. during the welding process). The ductility of these materials shows a sharp drop at temperatures above approximately half the melting temperature. DDC is described as a grain boundary sliding, creep-like phenomenon. The grain boundary sliding (mainly at 45° to stress direction) leads to higher strain concentrations at triple points and other irregularities, where a grain boundary changes its direction. The high strains result in micro-voids leading to intergranular crack formation [4-6].

In this study, investigations regarding the mechanism of crack formation in the welded area of the alloy 617mod were carried out. Tensile specimens were taken from a welded pipe and deformed in interrupted tensile tests at an elevated temperature of 700 °C. The strain distribution at the specimen surface were determined using scanning electron microscopy (SEM) images and by means of digital image correlation (DIC) after every stage of the interrupted tensile test. The strain fields at grain boundaries and in single grains could be analysed due to the high resolution of the SEM images.

Besides the strain fields, the grain orientation (electron backscatter diffraction), the hardness (nanoindentation), and the Schmid factor were determined and compared to each other.

2. Experimental methods

2.1 Sample manufacturing and preparation
The tensile specimens were cut from a welded pipe segment made of the alloy 617mod using electrical discharge machining. The specimen gauge (length 5 mm, width 2 mm, thickness 1.85 mm) consists in equal parts of the welding seam and the base material (figure 1 left) so that during tensile testing both sections were loaded.

The grain orientations of the specimen gauge were observed by means of electron backscatter diffraction (EBSD) after polishing. The grain orientation is shown in figure 1 (right) with the Miller indices. Furthermore, the coincidence site lattices (CSL) were determined with the help of EBSD measurements. The CSL describes the orientation of the grains to one another using the Σ value [7, 8]. By specifying the stress direction and with the knowledge of the grain orientation, the Schmid factor could be determined, giving information about the orientation of the slip planes and slip directions.

After the EBSD measurements, titanium dots (diameter: approx. 2 µm, distance: 5 µm, thickness: 50 nm) were applied on the surface of the specimen by using photolithography and sputtering. These titanium dots (figure 2 left) created grey-scaled edges which were used for the SEM-based DIC to calculate the strain fields.

Figure 1. Left: Position of the two polished samples in a welded area (between the dashed lines) of a pipe segment; right: Grain orientation map with colour scale (Miller indices).

2.2 Selection of areas for the DIC in reference condition
The strain fields were determined with the help of the digital image correlation based on SEM images. A single image was magnified 650 times (image size approx. 190 x 140 µm). In order to cover a larger area with the images, nine single images were assembled to one image with a size of approx. 500 x 400 µm. Three of these assembled areas were selected for each specimen (figure 2 right).

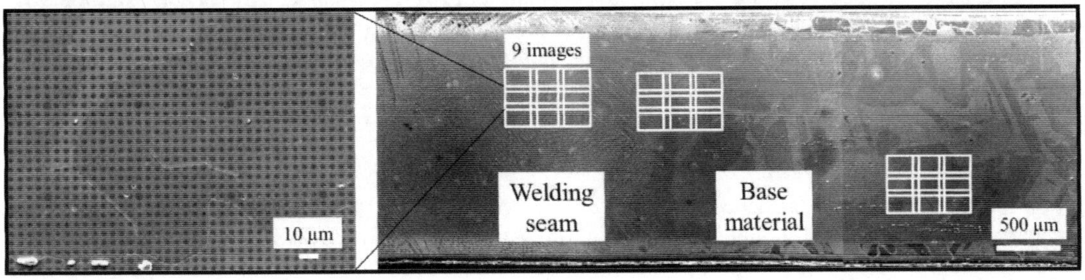

Figure 2. Left: Titanium dots (diameter: approx. 2 µm, distance: 5 µm, thickness: 50 nm) on the specimen surface used for the DIC; right: Gauge section with three areas (nine images each).

2.3 Interrupted tensile test

After the preparation, EBSD measurements, and taking SEM images the specimens were inserted into the experimental setup. The specimens were deformed at an elevated temperature of 700 °C in several stages. After every stage, the specimens were cooled down, removed from the setup, and SEM images were taken of the same areas as in the reference condition. The experiments were conducted with the test machine RMK 100 (Schenk) using a deformation speed of 3.2 µm/min ($\dot{\varepsilon} = 10^{-5}\ s^{-1}$).

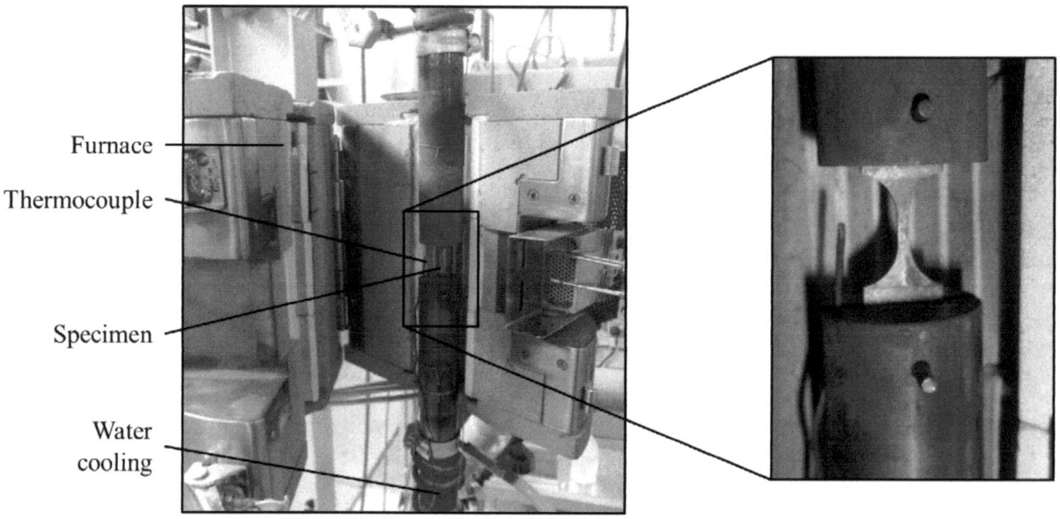

Figure 3. Experimental setup of the tensile test at 700 °C.

2.4 Digital image correlation (DIC)

SEM images in different deformation stages were generated of the same areas as in the reference condition after the interrupted tensile test (see 2.2.). The strain fields were calculated for every SEM image with the help of the DIC. The software used for the calculations was developed by the institute. The reference image and the deformed image were locally mapped to each other. Based on this local mapping a strain tensor was determined, that was used to calculate the scalar equivalent strain ε_{equ}.

2.5 Nanoindentation

After the tensile tests, the oxide layer, which was formed at the surface during these tests, were grinded off and hardness tests were carried out at room temperature by means of nanoindentation with a Nanoindenter G200 (Agilent). The depth of the indents was 500 nm and the distance between the indents was 20 µm. To evaluate if the hardness near the grain boundary deviates from the hardness inside the grains, measurements were conducted at grain boundaries showing DDC. Additionally, the hardness at grain boundaries without DDC was compared to the previous results.

3. Results and discussion

In the following different influences on DDC are described. The influence of the precipitate condition plays an important role [1, 6], but is not a subject of this investigation and will be focused on in future investigations.

3.1 Influence of the stress direction and grain boundary sliding on DDC

The SEM images of different deformation stages were at first visually investigated for cracks. Figure 4a shows a representative grain where DDC occurred. As shown in other investigations, the mechanism of DDC is related to grain boundary sliding [6]. Because of the maximum shear stress at angles of 45° to the stress direction (cf. Schmid´s law), grain boundary sliding mainly occurs at these angles. If a sliding grain boundary changes its direction to higher angles (i.e. 90° to the stress direction), DDC occurs as a result of a higher strain concentration [6].

The above mentioned influence of the stress direction on grain boundary sliding and DDC was also observed in this work. Figure 4a shows the mechanism of grain boundary sliding at grain boundaries that are orientated 45 degrees in respect to the loading direction and DDC at higher angles. Grain boundary sliding could be proven and illustrated with the help of the strain distribution (figure 4b). The strain value is increased along the grain boundary with the highest values at the triple point where DDC will develop in further processes of deformation. Furthermore, the titanium dots are shifted along the grain boundary as visible in the SEM image.

Figure 5a shows a SEM image of the entire specimen gauge. It can be seen that DDC occurs in the welding seam as well as in the base material. The cracked grain boundaries are orientated at higher angles (45° to 90°) to the stress direction.

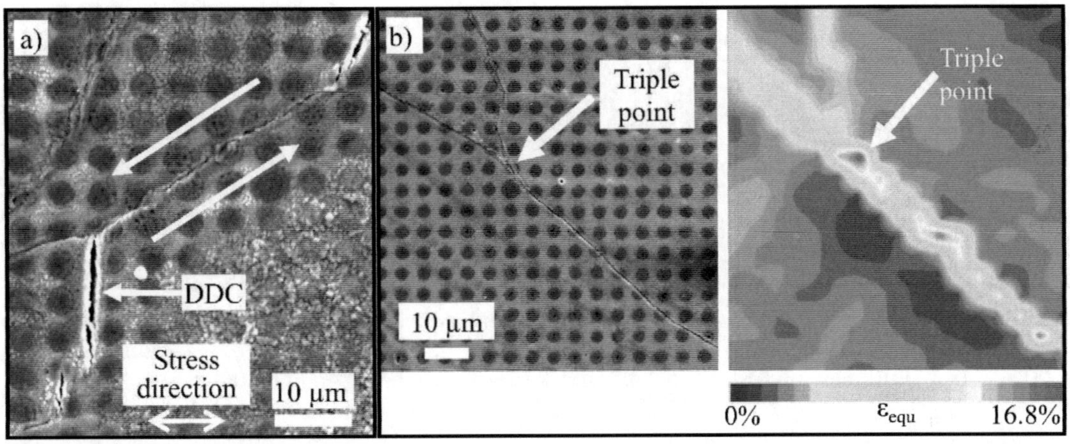

Figure 4. a) SEM image of sliding grain boundary (45° to stress direction) and resulting DDC (often 90° to stress direction); b) SEM image (left) and strain field (right) of a grain boundary (45° to stress direction) in the welding seam after the third deformation stage.

3.2 Influence of CSL on DDC

The CSL describes the orientation of the grains to each other. Figure 5b shows the grain boundaries of the specimen gauge. The Σ3 grain boundaries (red lines) are usually twin boundaries which have a lower grain boundary energy and therefore are more resistant to damages such as cracking and creep [7, 9]. The black lines indicate random grain boundaries divided in low angle (LAGB) and high angle grain boundaries (HAGB). A large number of twin boundaries (parallel red lines) are visible in the base material, while almost no red lines were detected in the welding seam.

The ductility dip cracks, which can be seen in figure 5a, are also marked in figure 5b. Almost all grain boundaries where DDC occurs, are random high angle grain boundaries (bold black lines) with a high angle to the stress direction (see 3.1). Many short cracks occur in the welding seam at wound grain boundaries with a lot of direction changes. These direction changes, along with the grain boundary sliding and the small number of Σ3 grain boundaries in the welding seam, explain the high number of cracks in these areas.

In order to verify that Σ3 grain boundaries are more resistant to DDC, local strain fields were investigated. Figure 6 shows a SEM Image (a), a SEM image with the CSL grain boundaries (b), and the strain field (c) of the same area. Figure 6b shows that a Σ3 grain boundary and a HAGB have the same (high) angle to the stress direction. The SEM image (a) as well as the strain field (b) show that deformation (and DDC) only occur at the HAGB (black line). The parallel Σ3 grain boundary (red line) as well as the inside of the grain show no conspicuous deformation. These observations, together with the observations in figure 5b, confirm the assumption that Σ3 grain boundaries are more resistant to DDC than HAGB.

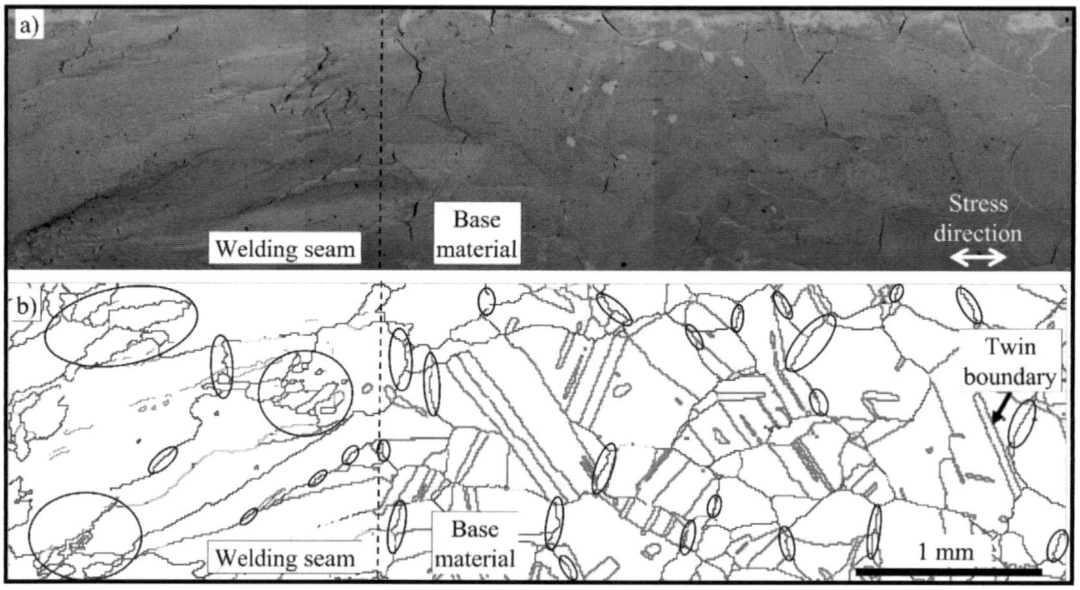

Figure 5. a) SEM image of the A617mod specimen after the sixth deformation stage with a large number of ductility dip cracks; b) grain boundaries of the same specimen: Red: Σ3 grain boundaries (CSL); coloured: Remaining CSL (Σ5 - Σ49); black (thin): Low angle grain boundaries (LAGB); black (bold): High angle grain boundaries (HAGB). The cracked grain boundaries are marked.

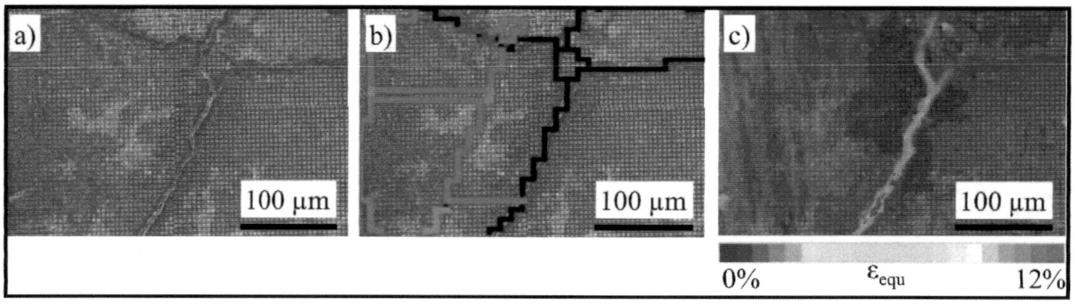

Figure 6. DDC in the base material of the A617mod after the third deformation stage: a) SEM image; b) grain boundaries (black: HAGB; red: Σ3 grain boundary; pink: Σ9 grain boundary; green: Σ25 grain boundary); c) strain field.

3.3 Influence of the hardness

As mentioned above (see 2.5.), hardness measurements were conducted in different areas of the specimen. All average values of the nanoindentation hardness measurements are between 5 and 5.5 GPa. No significant influence of the observed areas could be identified. This includes areas of grain boundaries with and without DDC as well as the inside of grains. In figure 7 an area in the welding seam with cracked grain boundaries is shown. The average hardness value is 5.46 GPa and the standard deviation is 0.46 GPa. For future investigations, hardness measurements at elevated temperatures are planned.

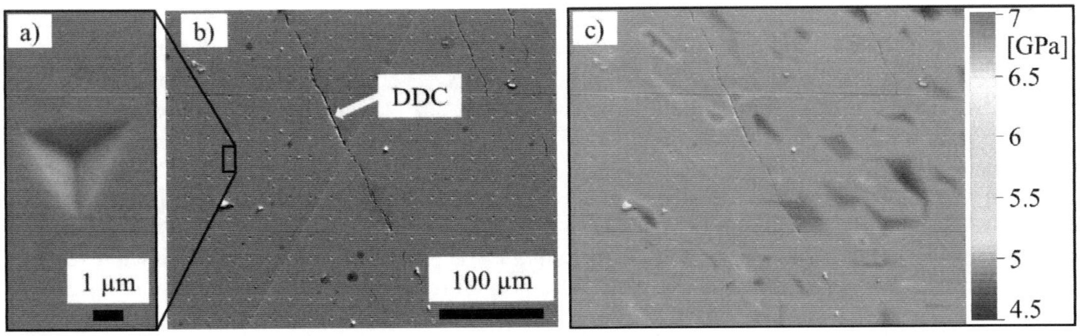

Figure 7. a) SEM image of a single indent; b) SEM image of an area with indents and DDC; c) SEM image of the same area with overlaid hardness values in [GPa].

3.4 Influence of the Schmid factor on the deformation within the grains

With the information of grain orientation and the stress direction, the Schmid factors were determined for every single grain using the software Channel 5 (Oxford). Nickel has a face centred cubic (fcc) crystal system. Thus, the slip systems with slip planes {111} and slip directions [$\bar{1}$10] were selected. The higher the Schmid factor the higher the shear stresses are inside of the grains in the selected slip systems. The highest value (0.5) occurs for an angle of 45° to the stress direction [10]. If the critical resolved shear stress (CRSS) is reached, dislocation motion and, therefore, plastic deformation occurs. These deformations were proven with help of the strain field at grains with a higher Schmid factor (figure 8). The grain with a high Schmid factor (white) shows higher strain values with an angle of 45° to the stress direction. The grain with a low Schmid factor (black) shows no high strain values.

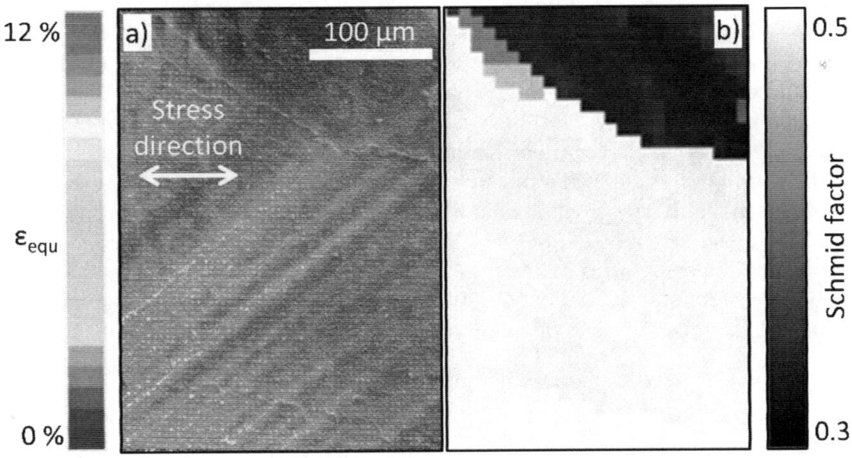

Figure 8. a) SEM image with overlaid strain field; b) Schmid factor of the same area.

4. Summary

Within the scope of this work, the welded nickel base alloy 617mod was investigated with interrupted tensile tests at an elevated temperature of 700 °C. Due to this thermomechanical stress DDC occurred. By means of DIC local strain fields were calculated. The grain orientation, CSL, and the Schmid factor were determined with the help of the EBSD. Furthermore, local hardness measurements were performed using nanoindentation.

DDC occurred at grain boundaries with a high angle (45° to 90°) to the stress direction. Grain boundary sliding appeared at grain boundaries with an angle of approx. 45° to the stress direction, leading to DDC in the further course of the grain boundary. The grain boundary sliding was proven with help of the strain field. Furthermore, it was shown that the orientation of the grains to each other (CSL) has an influence on DDC. Σ3 grain boundaries are more resistant to DDC than other grain boundaries.

Besides DDC, plastic deformation occurs especially in grains with a high Schmid factor, which could be proven with the help of the strain fields. Further investigations address the precipitate condition at the grain boundaries as well as heat treatments, that influence the precipitate condition. Additionally, hardness measurements at elevated temperatures are planned to determine the material behaviour during the formation of DDC.

Acknowledgments
The authors thank the "Deutsche Forschungsgemeinschaft" (DFG) for its financial support within the scope of the project TI 343 / 43-1.

References

[1] Ullrich C, Gierschner G, Stolzenberger C and Tschaffon H 2011 Erfahrungen beim Testbetrieb von Komponenten für das 700 °C Kraftwerk *VDI Wissensforum*
[2] Kuhn B, Ullrich C, Tschaffon H, Beck T and Singheiser L 2013 Stress-relaxation cracking test for welded joints *J. Test. Eval.* **41** 207–16
[3] Van Wortel H 2007 Control of relaxation cracking in austenitic high temperature components *Nace Corrosion Conference*
[4] Lippold J 2005 Recent developments in weldability testing for advanced materials *Joining of Advanced and Specialty Materials* **VII** 1–7
[5] Jang A Y, Lee D J, Shim J H, Kang S W and Lee H W 2011 Effect of Cr/Ni equivalent ratio on ductility-dip cracking in AISI 316L weld metals *Mater. Design* **32** 371–6
[6] Ramirez A J and Lippold J C 2004 High temperature behavior of Ni-base weld metal: Part II – Insight into the mechanism for ductility dip cracking *Mater. Sci. Eng. A* **380** 245–58
[7] Wen X 2014 Creep Behavior of high temperature alloys for generation IV nuclear energy systems *PhD Thesis* University of Cincinnati
[8] Föll H Defects in crystals *Hyperscript Christian-Albrechts-Universität Kiel* http://www.tf.uni-kiel.de/matwis/amat/def_en/index.html
[9] Gertsman V Y and Bruemmer S M 2001 Study of grain boundary character along intergranular stress corrosion crack paths in austenitic alloys *Acta Mater.* **49** 1589–98
[10] Roesler J, Harders H and Baeker M 2007 *Mechanical Behaviour of Engineering Materials: Metals, Ceramics, Polymers, and Composites* (Berlin: Springer) pp 180–182

IOP Conf. Series: Materials Science and Engineering **181** (2017) 012021 doi:10.1088/1757-899X/181/1/012021

Dynamic tafel factor adaption for the evaluation of instantaneous corrosion rates on zinc by using gel-type electrolytes

M Babutzka[1] **and A Heyn**[2]

[1]Bundesanstalt für Materialforschung und –prüfung, Division 7.6 Corrosion and Corrosion Protection, Berlin, Germany
[2]Korrosionsdiagnostik Dr. Andreas Heyn, Magdeburg, Germany

Abstract. Electrochemical corrosion measurements allow calculation of the instantaneous zinc corrosion rate via polarization resistances by using tafel factors. However, the determination of the actual tafel factor is problematic since it depends on the state of the formed zinc layers and the corrosion reactions taking place. Therefore, valid tafel factors are either determined in additional experiments via dynamic polarization or estimated by calculation. In most cases a constant value for tafel factors is assumed for simplification, without regard to the real conditions of the corroding system. Since naturally formed zinc layers are unstable using conventional test electrolyte solutions determination of tafel factors is hindered additionally and inaccurate interpretations can result. For some time now, the use of gel-type electrolytes in corrosion research has enabled minimally invasive investigation of zinc surface layers and thus offers new approaches to a scientifically proven determination of tafel factors. The paper presents a new method for the determination and evaluation of tafel factors using gel-type electrolytes and electrochemical frequency modulation technique (EFM). This relatively new electrochemical method offers the possibility to determine both polarization resistances and tafel factors within one measurement and in short measuring intervals. Starting from a comprehensive parameter study it is shown that a direct relationship between the two values exists that can be described mathematically. This contribution presents the determined tafel factors for the system gel-type electrolyte/zinc and discusses their applicability and their limits.

1. Introduction

When the tafel factor (B value, Stern-Geary constant) is known, the instantaneous corrosion rate (i_{cor}) of an actively corroding system can be described by means of the polarization resistance (R_p). This is carried out using the following relationship:

$$i_{cor} = \frac{B}{R_p} \tag{1}$$

This value can then be used for further calculations to determine mass loss and corrosion loss. Thus, the lifetime of zinc and zinc coatings under atmospheric conditions can be estimated. However, the tafel factor currently valid for the investigated system is not readily accessible. In many cases values for tafel factors are determined by additional experiments (mass loss measurements, tafel relationship) or are taken from the literature. A B value in the range of 5 to 30 mV is described, for example, by Metikos-Hukovic and Zevnik [1] for the system "electrolytically deposited zinc in a neutral 5 % sodium chloride solution". This value determined by means of dynamic polarization was dependent on the parameters of the zinc deposition and the temperature of the test solution. The investigations proved that the B value is not a constant value, but is caused by the corrosion processes taking place. It was shown that, as the polarization resistance increases, the B value decreases statistically. However, in the case of monitoring studies for zinc-coated iron under atmospheric conditions, they used a statistically determined constant B value of 13.8 mV [2] for simplification.

In the case of the specific determination of B values on zinc, zinc alloys and zinc coatings limits may be encountered in certain circumstances. The high reactivity and solubility of the protective layers formed on zinc surfaces under atmospheric conditions are problematic in contact with aqueous test media. As a result, naturally formed protective layers on zinc are stable only for a limited time during the corrosion test with typical aqueous test electrolytes and tend to rebuild or dissolve during the measurement. For this reason, B values from literature should be considered critically concerning atmospheric corrosion reactions. In order to counter this problem, however, two approaches can be utilized. On the one hand, attempts can be made to shorten the test period by selecting a suitable examination procedure and by adapting the examination parameters. On the other hand, the influence on the naturally formed layers on zinc can be minimized by selecting a suitable test electrolyte (minimally invasive test). New scientific knowledge and results from recent research allow new alternatives for both approaches, which appear promising for the determination of tafel factors on zinc under atmospheric conditions.

In the last 15 years the electrochemical frequency modulation (EFM) was introduced as a monitoring technique in the field of electrochemical investigation methods. Bosch et al. introduced the technique for the first time in [3] and extended it in [4]. It is based on the assumption that a corrosion system is a non-linear system. Thus, a current response at more frequencies (at zero, intermodulation, harmonic frequencies) than the perturbation frequencies is obtained by a potential perturbation with two sine waves of different frequency. The simplified principle of the EFM technique is shown in figure 1.

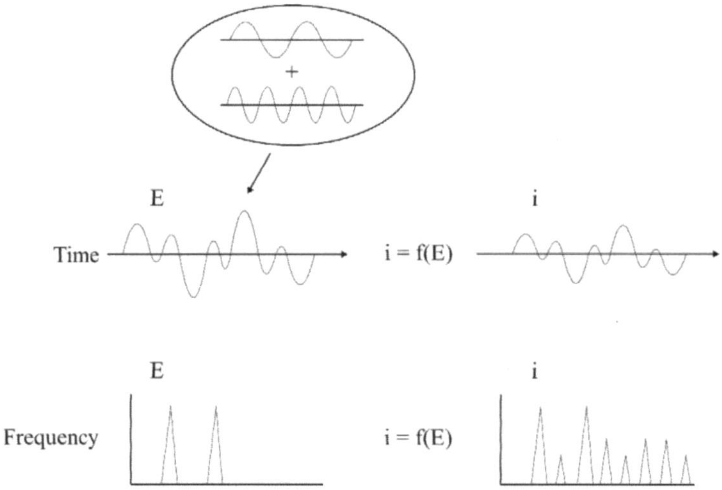

Figure 1. Principle of EFM technique according to [4].

The frequency-dependent current response is modeled (e.g. with Gamry software) by means of a Taylor sequence and the instantaneous corrosion current density as well as the anodic (b_a) and cathodic (b_c) tafel factors are calculated. A typical measuring routine usually only takes about 1 minute if the suitable examination parameters are chosen. Thus, B values from b_a and b_c and corrosion current densities can be determined in a short period of time. The knowledge of tafel factors is not necessary beforehand, since they are linked to the frequency-dependent current responses. A deri-

IOP Conf. Series: Materials Science and Engineering **181** (2017) 012021 doi:10.1088/1757-899X/181/1/012021

vation of this relationship can be found in [3] and will not be explained in this article. In the EFM, the perturbation amplitude and the perturbation frequencies can be varied. The amplitude of the perturbation (E_0) should not exceed ± 20 mV and is subject to the following relationship from [3]:

$$\frac{E_0}{b_{a,c}} \ll 1 \tag{2}$$

Accordingly, the perturbation amplitude should be smaller than the measured tafel factors. However, it is pointed out in [3] that the effective error at a ratio of 1 lies at ≤ 10% and thus in an acceptable range. According to [4] the choice of perturbation frequencies underlies the following three aspects:

- harmonic and intermodulation frequencies should not influence each other,
- base frequency should be as low as possible to avoid capacitive behavior,
- base frequency should be as high as possible to minimize measurement duration.

These relationships show that the choice of optimal frequencies is not possible and a compromise has to be made. In their studies on steel in sulfuric acid and in a sodium chloride solution with adapted test parameters, Bosch et al. proved that the EFM technique is in good correlation with mass loss, impedance and polarization measurements. In contrast, Mansfeld and Kus [5] offer a critical view of the EFM technique. They utilized EFM measurements for copper, aluminum and stainless steel and were unable to find a satisfactory correlation with established methods. They noted that the accuracy of the technique depends on the accuracy of the current measurement and reaches its limits especially for systems with low corrosion currents. Rauf and Bogaerts, on the other hand, were able to observe the onset of pitting corrosion on stainless steel using EFM [6]. The EFM technique has not been utilized for zinc and zinc coatings so far. However, it appears to be promising as it generates a high amount of information within a short period of time in a minimally invasive way (by small perturbation amplitudes).

The problem that determination of B values on zinc and zinc coatings becomes more difficult due to the high solubility of the formed cover layers under the influence of an aqueous test electrolyte can be countered by using of gel-type electrolytes for electrochemical investigations. This electrolyte variant is a conductive polymerized mass, usually made of agar that can be shaped as desired. Due to the syneresis effect (phase separation of a two-phase system), gel-type electrolytes form a thin moisture film at the gel/metal interface. Thus, an electrochemical instrumentation allows electro-chemical corrosion investigations on zinc and zinc coatings. Testing with gel-type electrolytes is minimally invasive since the naturally formed corrosion product layers on zinc coatings are only slightly influenced by the moisture film of the gel-type electrolyte. It has already been demonstrated, that gel-type electrolytes influence zinc corrosion product layers much less than a comparable aqueous test electrolyte [7-9]. Cano et al. could also show this effect for bronze [10]. Due to difficult diffusion conditions in the gel-type electrolyte, dissolution of the corrosion product layers is almost completely inhibited in many applications.

In this paper, the EFM technique is linked to a gel-type test electrolyte and applied to pure zinc. At first, suitable test parameters had to be worked out through means of a parameter study, since no EFM parameters were available in literature for the system "pure zinc/gel-type electrolyte". Subsequently, the interdependency between B values and polarization resistances determined by means of EFM was investigated, mathematically described and verified. For this, pure zinc samples were stored for a defined period of time under atmospheric conditions (city atmosphere) in order to form differently stable zinc corrosion product layers and thus to generate a wide range of samples. The interdependency shall allow to calculate the appropriate B value only by means of a measured polarization resistance (from impedance or dynamic polarization) if EFM measurements are not possible.

2. Experimental

2.1 Sample preparation and test electrolyte

Sample sheets of pure zinc (99.5 % Zn) with a sheet thickness of 0.8 mm were used as specimen for all investigations. At first samples were cleaned and degreased. Then, the samples were pickled with 1 M sodium hydroxide solution for 5 minutes in an ultrasonic bath, cleaned with deionized water and ethanol and dried under a warm airflow. Pickling was conducted to activate the sample surfaces by removing the existing corrosion product layers that formed under undefined atmospheric conditions. Thereby, all samples had the same initial surface conditions. Afterwards, the specimens were stored for different durations with a maximum of 180 days to form a natural zinc corrosion product layer by means of outdoor weathering at city atmosphere (Berlin). During this period, the samples were taken at regular intervals, examined electrochemically and then subjected to the city atmosphere again.

A gel-type electrolyte based on agar was used as test electrolyte. The gel-type electrolyte was shaped as a circular gel pad with a diameter of 19 mm and a pad thickness of 3 mm. From the size of the gel pad, a theoretical electrolyte volume of 0.9 ml was obtained. The measuring area was 2.84 cm^2. High purity agar powder was dissolved in demineralized water and then polymerized at elevated temperature to produce the gel pads. Afterwards, the gel pads had a gelatinous consistency, a dense surface with no visible porosity and could be brought in the dimensions desired for testing. In the gel pads, a chloride concentration of 260 ppm was determined.

2.2 Electrochemical investigations

The gel-type electrolyte in the form of gel pads was first electrochemically instrumented for electrochemical investigations. The typical three-electrode arrangement was used for this purpose. A mixed titanium oxide net was used as counter electrode (CE) and a saturated Ag/AgCl electrode as reference electrode (RE). The samples served as working electrodes (WE). A Potentiostat/Galvanostat Interface 1000 from Gamry Instruments was used for all electrochemical experiments.

The correlations between perturbation frequencies and perturbation amplitude were investigated at two sample states (activated surface before exposure to city atmosphere, after 180 days of exposure) by means of a parameter study. From this the most fitting EFM parameters for the corrosion system zinc/gel pad were derived. Base frequencies of 0.05, 0.1, 0.5 und 1 Hz were varied. Multipliers of the base frequencies were 2 and 5. Thus, for a base frequency of 0.1 Hz a superimposed perturbation with the frequencies 0.2 Hz and 0.5 Hz was obtained in contrast to a perturbation with 2 Hz and 5 Hz for a base frequency of 1 Hz. A measurement included four cycles. In addition, the perturbation amplitude was varied in the range from 5 to 20 mV. After completion of the parameter study, a base frequency of 0.1 Hz (perturbation frequencies 0.2 Hz and 0.5 Hz) with a perturbation amplitude of 5 mV was selected for the remaining sample conditions. Using the Gamry software EchemAnalyst, anodic and cathodic tafel factors as well as corrosion current densities were determined from the EFM data. B values were calculated from the tafel factors using the following relationship:

$$B = \frac{b_a\, b_c}{2.3\,(b_a + b_c)} \tag{3}$$

Polarization resistances R_p were calculated from the determined corrosion current densities i_{cor} using the following relationship:

$$R_p = \frac{B}{i_{cor}} \tag{4}$$

Current density-potential curves were determined by means of dynamic polarization at the two sample states of the EFM parameter study in order to evaluate the EFM results. After 10 minutes of potential measurement, the samples were polarized in anodic direction in a potential range from -50 mV to +50 mV vs. E_{cor}. The sweep rate was 1 mV/s. In a potential range with only slight distance from the free corrosion potential (±5 mV vs. E_{cor}), linear polarization resistances (LPR) were subsequently determined as a characteristic value for the instantaneous protection ability of the zinc layer. In addition, the corrosion current density at the free corrosion potential was determined via tafel relationship. Anodic and cathodic tafel factors were determined at a distance of 5, 10, 15 and 20 mV from the free corrosion potential by using the Gamry software EchemAnalyst.

3. Results

3.1 Dynamic polarization for evaluation of EFM results

By means of an EFM parameter study at two sample states, the best fitting parameters for the selected corrosion system should be derived. In order to be able to evaluate the EFM results, current density-potential curves and their specific characteristic values were determined using dynamic polarization. In the following figure 2, the current density-potential curves for the activated surface state and for a sample after 180 days of exposure to city atmosphere are shown.

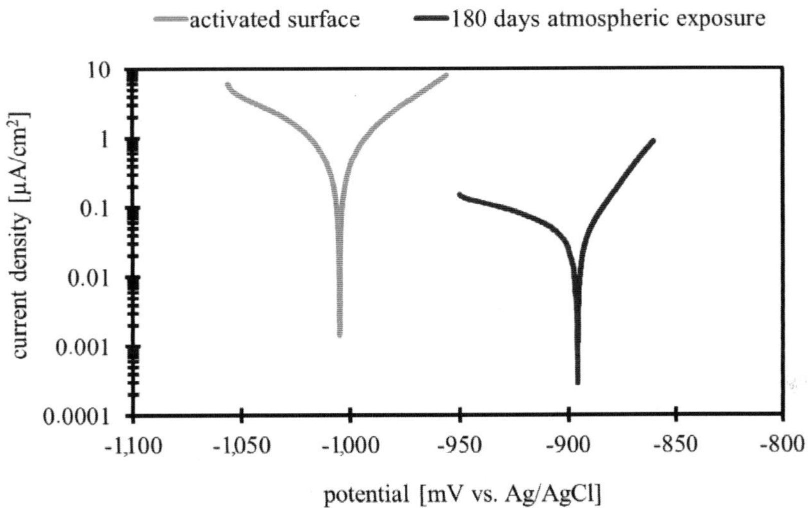

Figure 2. Current density-potential curves on pure zinc with pickled (activated) surface and after 180 days of exposure to city atmosphere, determined by using gel-type electrolyte.

The curves and the position of the free corrosion potentials differ depending on the sample state. The free corrosion potential is approximately 100 mV more negative when the surface is activated than after 180 days of exposure to city atmosphere during which a surface layer of corrosion products could form on the surface. This inhibits the partial reactions so that after 180 days of exposure lower current densities are measured overall. However, typical potential ranges for a zinc surface are measured for both states. The following table 1 compiles the corrosion-relevant tafel factors and the characteristic kinetic values determined by using the current density-potential curves. The determined linear polarization resistances (LPR) and corrosion current densities (i_{cor}) as characteristic values provide an overview of the state of the zinc surface. A stable zinc corrosion product layer was formed after 180 days of exposure to city atmosphere. A low polarization resistance and thus a high corrosion current density were measured on the activated surface. The anodic and cathodic tafel factors determined at defined distances from the free corrosion potential become larger with increasing distance to the free corrosion potential for both sample states. The cathodic tafel factors are larger than the anodic. The tafel factors for the sample exposed to city atmosphere for 180 days tend to be smaller than when the surface is activated.

Table 1. Corrosion-relevant tafel factors and kinetic values on pure zinc determined by means of dynamic polarization and gel-type electrolyte.

Sample	Tafel factors calculated at … vs. E_{cor}	Tafel factors			corrosion relevant values	
		b_c (mV/dec)	b_a (mV/dec)	B (mV)	i_{cor} (µAcm^{-2})	LPR (Ωcm^2)
activated surface	5 mV	18	17	4	0.7	12,936
	10 mV	44	36	9		
	15 mV	53	43	10		
	20 mV	59	47	11		
180 days of atmospheric exposure	5 mV	12	11	3	0.03	158,580
	10 mV	37	26	7		
	15 mV	54	34	9		
	20 mV	59	36	10		

3.2 EFM parameter studies for determination of test parameters

The results of the EFM parameter study are presented below. In the following figure 3, the B values determined on the activated (pickled) sample by means of EFM are depicted as a function of the base frequency and the perturbation amplitude. In addition, B values from the current density-potential curves from table 1 are shown for comparison purposes.

The B values show a strong dependence on the perturbation amplitude and a lower dependence on the base frequency of the perturbation. With higher perturbation amplitude, the B-value is also increased. With a higher base frequency, the B value tends to increase slightly. Compared to the B values from the dynamic polarization, the B values from the EFM measurements at perturbation amplitudes greater than 10 mV are in some cases significantly higher than the comparative values. The same dependencies can be also seen at the sample state after 180 days of exposure to city atmosphere. The results are summarized in figure 4. The B values also show a strong dependence on the perturbation amplitude and a dependence on the base frequency, especially at higher perturbation frequencies. Compared to the B values from the dynamic polarization, there is an even higher deviation to the B values from the EFM measurements at perturbation amplitudes greater than 10 mV than when the surface is activated. In addition, the B values tend to be higher than when the surface is activated. The interdependencies between polarization resistances determined by means of EFM and the chosen test parameters are shown in figure 5 at the example of the activated (pickled) surface.

The polarization resistances show an indirect dependence on the base frequency of the perturbation. If a higher base frequency is chosen, a lower polarization resistance is measured. The perturbation amplitude has no significant influence on the magnitude of the polarization resistance. The polarization resistance of the EFM measurements corresponds to the determined linear polarization resistance from table 1 at a base frequency of 0.1 Hz. The polarization resistances of the sample after 180 days of exposure to city atmosphere are shown in figure 6 as a function of the test parameters.

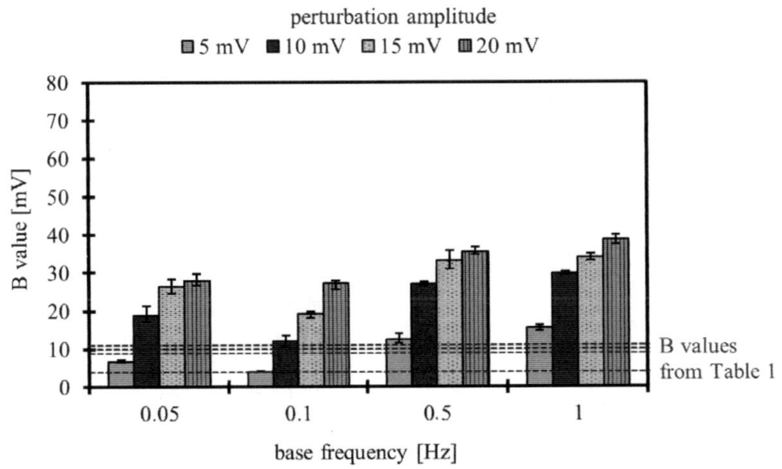

Figure 3. B values in function of base frequency and perturbation amplitude determined by EFM and B values from dynamic polarization (Table 1), pure zinc with activated surface, gel-type electrolyte.

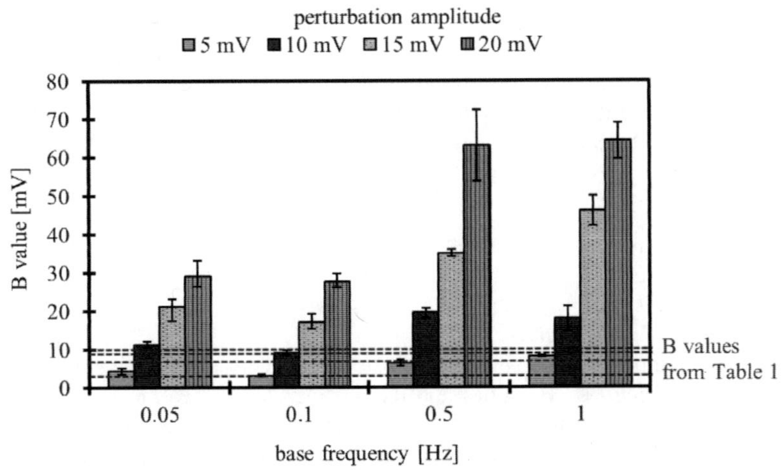

Figure 4. B values in function of base frequency and perturbation amplitude determined by EFM and B values from dynamic polarization (table 1), pure zinc after 180 days of exposure to city atmosphere, gel-type electrolyte.

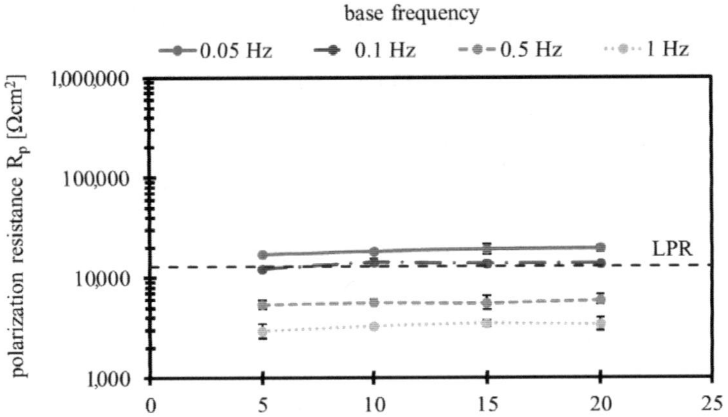

Figure 5. Polarization resistances in function of base frequency and perturbation amplitude determined by EFM, pure zinc with activated surface, gel-type electrolyte.

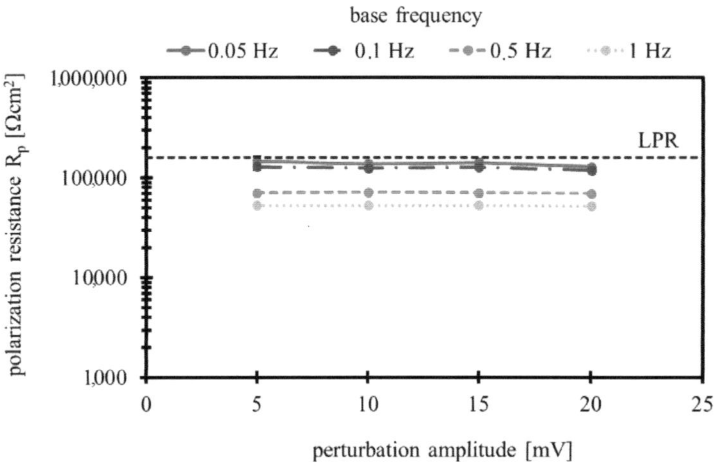

Figure 6. Polarization resistances in function of base frequency and perturbation amplitude determined by EFM, pure zinc after 180 days of exposure to city atmosphere, gel-type electrolyte.

The polarization resistances comprise the same interdependencies between the test parameters as with the activated surface. The perturbation amplitude also has no significant influence on the magnitude of the polarization resistance. The polarization resistances are higher than with the activated surface caused by the protective zinc corrosion product layer. At a base frequency of 0.1 and 0.05 Hz, the polarization resistance of the EFM measurements corresponds sufficiently to the determined linear polarization resistance from table 1. From the results of the EFM parameter study with regard to the results of dynamic polarization, the following test parameters are obtained for subsequent studies:

- a base frequency of 0.1 Hz (= perturbation frequencies 0.2 Hz and 0.5 Hz) in correlation with the LPR,
- a perturbation amplitude of 5 mV in good correlation with results of dynamic polarization and in accordance to equation (2).

3.3 Interdependency between B value and polarization resistance

A large number of sample states after atmospheric exposure were examined by means of a gel type electrolyte to investigate the correlation between B values and the calculated polarization resistances via EFM measurements on pure zinc. These samples differed in the duration they were subjected to city atmosphere and thus in the state of the formed corrosion product layer. Thereby, it was possible to obtain a wide spectrum of different polarization resistances and B values on pure zinc samples. The determined B values as well as the corresponding polarization resistances are shown in figure 7. In addition, the figure shows two regressions with their corresponding equation and the coefficient of determination (R^2), respectively.

It turns out that the B-value decreases with increasing polarization resistance. This interdependency can be described adequately by a logarithmic function or via a power regression. With high polarization resistances above 200,000 Ωcm^2, only a small decrease in the B value is observed. In contrast, with low polarization resistances below 10,000 Ωcm^2 there is a strong dependency between polarization resistance and B value. The interdependency can be described by the following regressions:

$$B = -1.9 \cdot \ln R_p + 26.4 \tag{5}$$

$$B = 116 \cdot R_p^{-0.293} \tag{6}$$

The B values calculated by means of equations (5) and (6) are compiled for predetermined polarization resistances in table 2. Subsequently, the corrosion current density was calculated by means of the B value via the equation (4). Corrosion current densities calculated with a constant B value from literature [2] are also shown.

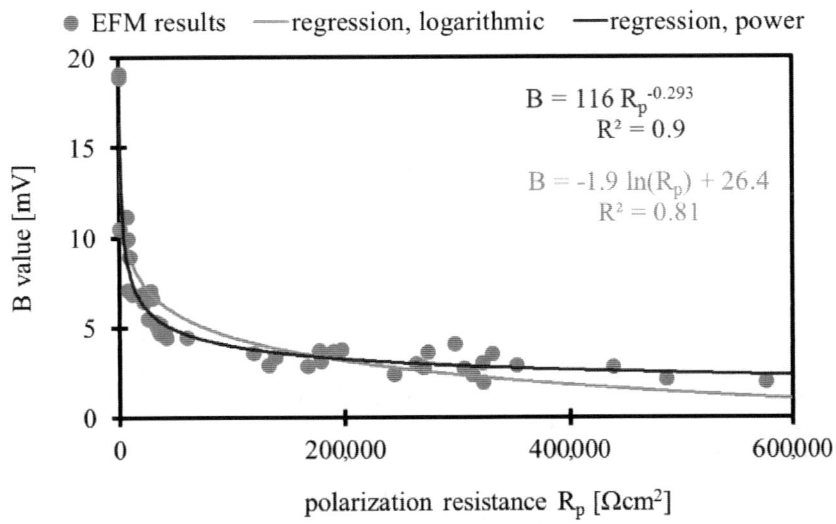

Figure 7. Interdependency between B value and polarization resistance in EFM measurements on pure zinc with different exposure times to city atmosphere, gel-type electrolyte.

Table 2. Calculated B values from regression and literature and corrosion current densities for pure zinc, gel-type electrolyte.

	regression, logarithmic (5)		regression, power (6)		constant B value	
R_p (Ωcm^2)	B (mV)	i_{cor} (μAcm^{-2})	B (mV)	i_{cor} (μAcm^{-2})	B (mV)	i_{cor} (μAcm^{-2})
750	14	18	17	22	13.8	18
1,000	13	13	15	15	13.8	14
2,500	12	5	12	5	13.8	6
5,000	10	2	10	2	13.8	3
7,500	9	1	8	1	13.8	2
10,000	9	0.9	8	0.8	13.8	1
25,000	7	0.3	6	0.2	13.8	0.6
50,000	6	0.1	5	0.1	13.8	0.3
75,000	5	0.07	4	0.06	13.8	0.2
100,000	5	0.05	4	0.04	13.8	0.1
250,000	3	0.01	3	0.01	13.8	0.06
500,000	1	0.003	2	0.005	13.8	0.03
750,000	1	0.001	2	0.003	13.8	0.02

The regression based on a power (6) gives a good agreement with the measured values at polarization resistances below 1,000 Ωcm^2 (figure 7). For higher polarization resistances with a value above 500,000 Ωcm^2 current densities calculated by means of both regressions differ slightly from each other. In comparison to the regressions the effect of a constant B value on the calculated corrosion current density is shown. Being in good agreement with the regressions for very low polarization resistances, a higher corrosion current density is determined for polarization resistances above 2,500 Ωcm^2.

4. Discussion

The investigated parameters from the EFM parameter study are in good correlation with the results from dynamic polarization. An adequate correlation between polarization resistance and B value was possible by means of the selected test parameters. It was possible to confirm the relation described in [1] that the B value decreases with increasing polarization resistance. This relation can be described mathematically by means of regressions. Figure 8 visualizes the values from table 2 and the results from EFM measurements in contrast to the results of dynamic polarization (from table 1) in order to evaluate whether the regressions based on the EFM results or a constant value of 13.8 mV from [2] describe the investigated corrosion system more adequate. The regression on the basis of a power describes the studied system the most accurate, for the EFM results as well as for the kinetic parameters from the dynamic polarization. The regression on the basis of a logarithm describes the system sufficiently accurate up to a polarization resistance of 250,000 Ωcm^2. For higher polarization resistances a lower corrosion current density is calculated.

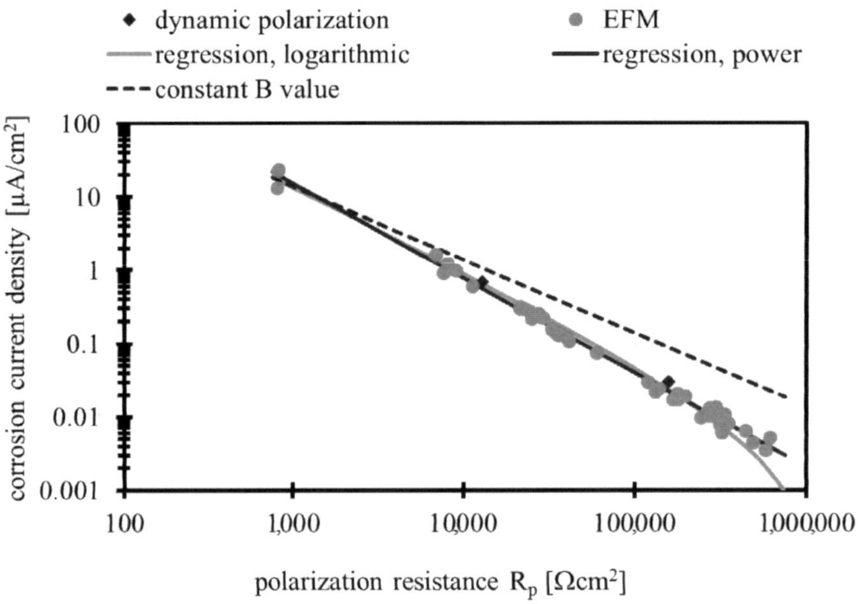

Figure 8. Interdependency between polarization resistance and corrosion current density from EFM measurements on pure zinc with different exposure times to city atmosphere, gel-type electrolyte, verification of results by comparison to values from dynamic polarization (table 1).

If a constant B value is used, the calculated corrosion current densities deviate too much from the measured values and no agreement can be found with values from dynamic polarization.

5. Conclusion

The results of the electrochemical investigations using EFM technique and gel-type electrolyte show that determination of B values and polarization resistances on pure zinc samples with differently stable corrosion product layers is possible. It has to be considered that these B values are only valid for the described corrosion system. These values have to be confirmed and adapted for other zinc coatings and variants as well as gel-type electrolytes with a different composition (e.g. higher chloride content in the pad). It was shown that the B value is not a constant value, but is strongly dependent on the corrosion system, the condition of the surface layer on the surface, and the corresponding polarization resistance. However, a determination of a valid B value is possible by means of the presented mathematical equations and regressions. These provide the possibility to calculate a B value by means of a measured polarization resistance, e.g. from impedance measurements or linear polarization resistances, without additional measurements. The B value can be used for the calculation of instantaneous corrosion current densities via the polarization resistance.

References

[1] Metikos-Hukovic M and Zevnik C 1982 *Werkst. Korros.* **33** 661–8
[2] Metikos-Hukovic M and Zevnik C 1984 *Werkst. Korros.* **35** 116–20
[3] Bosch R W and Bogaerts W F 1996 *Corrosion* **52** 204–12
[4] Bosch R W, Hubrecht J, Bogaerts W F and Syrett B C 2001 *Corrosion* **57** 60–70
[5] Kus E and Mansfeld F 2006 *Corros. Sci.* **48** 965–79
[6] Rauf A and Bogaerts W F 2010 *Corrosi. Sci.* **52** 2773–85
[7] Babutzka M and Heyn A 2016 *Tagungsband zum 18. Werkstofftechnisches Kolloquium in Chemnitz* pp 313–23
[8] Babutzka M and Heyn A 2016 *EUROCORR 2016 - Conference Proceedings (Montpellier)* paper P-56097
[9] Babutzka M, Burkert A and Heyn A 2016 *Tagung Werkstoffprüfung 2016 – Fortschritte in der Werkstoffprüfung für Forschung und Praxis (Neu-Ulm)* pp 125–30
[10] Cano E, Crespo A, Lafuente D and Ramirez Barat B 2014 *Electrochem. Commun.* **41** 16–9

IOP Conf. Series: Materials Science and Engineering **181** (2017) 012022 doi:10.1088/1757-899X/181/1/012022

A numerical and experimental study of temperature effects on deformation behavior of carbon steels at high strain rates

M Pouya, S Winter, S Fritsch and M F-X Wagner

Chemnitz University of Technology, Institute of Materials Science and Engineering, 09125 Chemnitz, Germany

Abstract. Both in research and in the light of industrial applications, there is a growing interest in methods to characterize the mechanical behavior of materials at high strain rates. This is particularly true for steels (the most important structural materials), where often the strain rate-dependent material behavior also needs to be characterized in a wide temperature range. In this study, we use the Finite Element Method (FEM), first, to model the compressive deformation behavior of carbon steels under quasi-static loading conditions. The results are then compared to experimental data (for a simple C75 steel) at room temperature, and up to testing temperatures of 1000 °C. Second, an explicit FEM model that captures wave propagation phenomena during dynamic loading is developed to closely reflect the complex loading conditions in a Split-Hopkinson Pressure Bar (SHPB) – an experimental setup that allows loading of compression samples with strain rates up to 10^4 s^{-1}The dynamic simulations provide a useful basis for an accurate analysis of dynamically measured experimental data, which considers reflected elastic waves. By combining numerical and experimental investigations, we derive material parameters that capture the strain rate- and temperature-dependent behavior of the C75 steel from room temperature to 1000 °C, and from quasi-static to dynamic loading.

1. Introduction

It is well known that during dynamic loading the mechanical behavior of many materials (like carbon steels, the most important structural materials) differs substantially from that under quasi-static conditions. Due to various industrial applications, e.g. in crash-absorbing structural members, the interest to accurately determine the material behavior at high strain rates has strongly increased during recent decades. Combined with high temperatures, a better understanding of the mechanical behavior is also required to determine the material properties during various manufacturing and production processes. For dynamic testing, different experimental techniques for uniaxial compression are well established. In contrast to the conventional uniaxial testing machines used for quasi-static testing, these techniques (including the drop weight testers and the well-known Split-Hopkinson Pressure Bar setup) enable the material properties to be determined at high strain rates. The Split-Hopkinson Pressure Bar (SHPB) is an ideal tool to apply a controlled impulse of energy on a specimen and thus to determine the dynamic mechanical behavior at strain rates of up to 10^4 s^{-1}. This experimental technique has been used widely, typically at room temperature (RT) [1]. Fewer results have been reported on SHPB testing at higher temperatures [2]. The experimental results at high temperatures appear to be more complicated to analyze because of various experimental effects that occur during the sample heating process and subsequent dynamic loading. Numerical simulations, and a careful comparison between the experimental and numerical results [3, 4], are essential to separate different

IOP Conf. Series: Materials Science and Engineering **181** (2017) 012022 doi:10.1088/1757-899X/181/1/012022

effects caused by e.g. measurement techniques, testing machine setup, sample geometry and the material behavior both on the macro- and micro-scale [5, 6].

In this study, a simulation using the Finite Element Method (FEM) is performed to model the compressive deformation behavior of the high strength carbon spring steel C75 under quasi-static loading. The numerical results are compared to those of experiments at RT and at higher temperatures. To capture the complex dynamic loading conditions, an explicit FEM model of the SHPB is developed. The simulations are thereafter expanded to study the effect of higher temperatures under dynamic loading. The dynamic simulations provide an improved theoretical basis for an accurate analysis of dynamically measured experimental data.

2. Experimental and theoretical methods

To study the mechanical behavior of the high strength carbon spring steel C75, cylindrical specimens with a height of 6 mm and a diameter of 4 mm (provided by Risse + Wilke Kaltband GmbH & Co.KG, Iserlohn, Germany) in a hardened and tempered condition were used. The chemical composition was determined by spectroscopic analysis as 0.72 wt.-% C, 0.29 wt.-% Si and 0.74 wt.-% Mn. The material samples were first uniaxially deformed under quasi-static (initial strain rate: 10^{-3} s^{-1}) and dynamic compressive loading conditions. The measured displacements and loads were related to initial sample height and cross section in order to determine engineering strains ε_{eng} and stresses σ_{eng}, respectively (see [7] for further experimental details). Then, true (logarithmic) strains and stresses were calculated from the engineering data. Quasi-static and dynamic compression tests were performed at RT and also at higher temperatures up to 1000 °C. The dynamic compressive tests were conducted using a SHPB with strain rates up to 10^{4} s^{-1}.

The well-known SHPB setup is shown in figure 1. The SHPB system contains a gas launcher, the striker, incident and transmitter bars, and a shock absorber, all of them mounted on a flat base (figure 1a). The test specimen is positioned between the incident and transmitter bars. The gas launcher is charged with a pressurized gas to accelerate the striker bar at a measured velocity and therefore to create an elastic stress pulse. The generated elastic wave pulse (a longitudinal wave) propagates through the incident bar at the speed of sound as determined by the bar material; part of this pulse is reflected at the end of the bar due to impedance mismatches while the other part propagates through the material sample and into the transmitter bar. To prevent signal oscillations, a "pulse shaper" (shown in figure 1b) is embedded between the striker and incident bars. To perform high temperature mechanical testing, the specimen is protected using two ceramic pieces and heated quickly up to a specific testing temperature. Strain gauges (SGs) are positioned on the surfaces of the incident and transmitter bars to measure the strains and the amplitudes of the elastic wave pulse (input and reflected signals in addition to the transmitted signal). Stress and strain values can then be calculated from the data collected by these SGs.

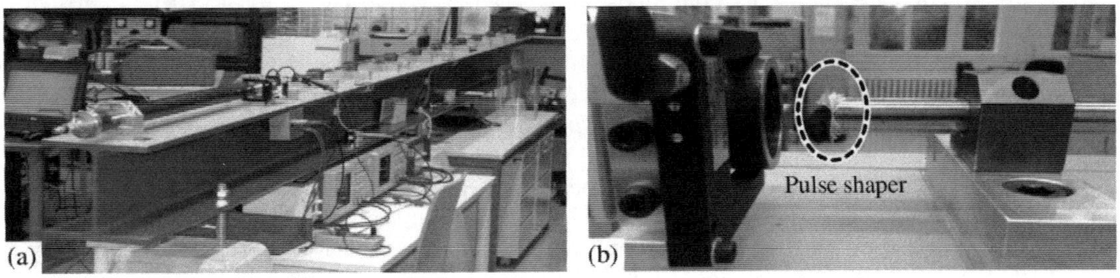

Figure 1. Experimental setup of the SHPB for dynamic compressive loading at high strain rates up to 10^{4} s^{-1} (a). A pulse shaper is embedded between the striker and incident bars to prevent signal oscillations during experimental mechanical testing (b).

The experimental results from quasi-static loading were used as hardening curves in a simple isotropic elastic-plastic material model to simulate the quasi-static thermo-mechanical behavior of C75 using the Abaqus software package. Standard linearly coupled temperature-displacement elements (C3D8T) were employed. Separate simulations were performed, allowing a predefined temperature field (e.g. RT, 200 °C, 400 °C, 600 °C and 800 °C) to reflect the coupled thermo-mechanical behavior of C75 during compressive loading. To capture the dynamic compressive deformation behavior of C75 numerically, the exact experimental setup of the SHPB was carefully defined in Abaqus dynamic explicit, as shown schematically in figure 2.

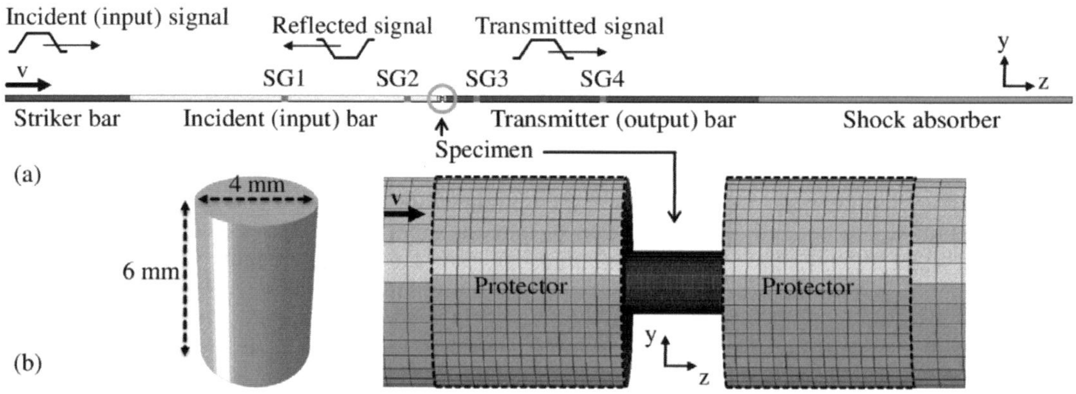

Figure 2. (a) Schematic representation of the simulated SHPB setup. (b) Geometry, mesh and boundary conditions of C75 specimen for the simulation of dynamic compressive loading in the SHPB setup.

The simulated SHPB setup (figure 2a) is a combination of four bars with the same diameter D = 16 mm. All bars consist of the same material, since no wave reflections alone through the SHPB setup is desired. The total length of shock absorber beside the incident and transmitter bars is 1250 mm, while the striker bar has a total length of 500 mm. That ratio of lengths is chosen because the input signal, which contains a double total length of the striker bar, must be able to fit in and travel through the incident bar entirely [8]. The end/front faces of the incident/transmitter bars are protected using two protectors (figure 2b, right; marked by dashed lines), which are replaced by two ceramic pieces at higher temperatures (up to 1000 °C) according to the experimental setup. The effect of the pulse shaper is ignored in our finite element simulations for the sake of simplicity. The striker bar is accelerated dynamically (boundary conditions are defined in terms of velocity) to strike the incident bar. The compression specimen is positioned between the incident and transmitter bar (figure 2b). Based on the real-life experimental setup, the strain gauges number 1 and 4 (SG1 & 4) are positioned in the middle of the incident and transmitter bars (625 mm from end/front faces of incident/transmitter bars) and SG2 and 3 are located closer to the specimen (150 mm from the end/front faces).

Various coupled thermo-mechanical simulations at different test temperatures (RT, 200 °C, 400 °C, 600 °C, 800 °C and 1000 °C) were carried out. Note that the effect of high strain rates was incorporated in the simulations by iteratively adapting the strain rate sensitivity for the RT data (see also figure 3) for a good fit of experimental and simulation results; a constant strain rate sensitivity of $m = 0.0062$ was assumed for all temperatures and included in the simulations by changing the hardening curves for high strain rate deformation accordingly.

3. Results and Discussion

The experimental and numerically fitted results (true stress as a function of true plastic strain) for quasi-static loading are presented in figure 3. The true stress-strain curves are shown at different temperatures of RT, 200 °C, 400 °C, 600 °C and 800 °C. The flow stresses decrease at higher temperatures during quasi-static loading. The resulting numerical quasi-static material model was used to describe the mechanical behavior of the C75 specimen (positioned in SHPB setup) in the dynamic simulations.

To validate the simulated setup of the SHPB, a "test shot" without a specimen was performed experimentally. The resulting elastic wave pulses are compared to those of a corresponding simulation in figure 4a. Because no specimen is placed in the SHPB system, no reflection of the elastic wave pulse is observed. Therefore the experimental data obtained using a pulse shaper (figure 4a, upper part) and the numerical results (figure 4a, below, ignoring the effect of a pulse shaper) represent only the input and output signals during the test shot. Because of the different positions of the SGs, the elastic pulse data occur at different times in each individual SG.

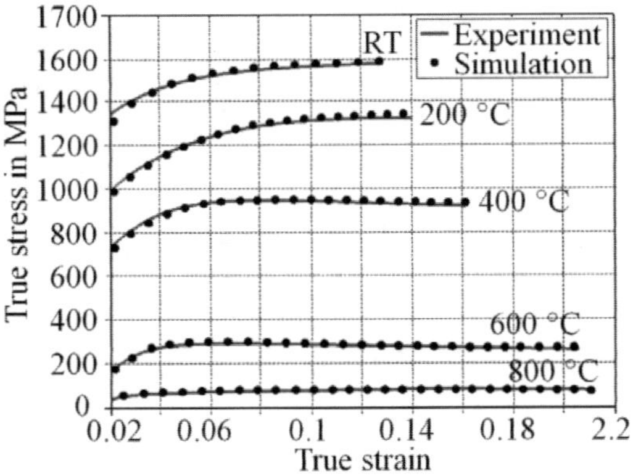

Figure 3. Experimental and numerical true stress-strain curves of quasi-static compression tests at different temperatures (RT, 200 °C, 400 °C, 600 °C and 800 °C). The simulation has been fitted to the experimental (engineering) stress-strain curves.

The propagation of the elastic wave pulse, measured during a regular test and therefore including the compressive dynamic response of the C75 specimen at RT, is presented in figure 4b. The resulting input, reflected and output experimental signals (figure 4b, left) are compared to those of the simulation (figure 4b, right) at the four SGs. With excellent agreement between experiment and simulation, it can be seen that compared to the test shot, by positioning the C75 compression specimen in the SHPB setup, a part of the elastic wave pulse is reflected at the specimen front face. The resulting signal at SG1 consists of both the input and the reflected pulses, while the signal measured in SG2 is a combination of these signals and shifted in time according to its location in the SHPB setup. Therefore the reflected signal at SG1 can be used to calculate the true strains in the specimen. Using the signals measured in SGs 3 and 4, the transmitted mechanical behavior of the specimen is recorded and finally used to calculate true stresses.

The dynamic compressive response of the C75 specimen at RT and higher temperatures (200 °C, 400 °C, 600 °C, 800 °C and 1000 °C) is presented in figure 5, where true stress is plotted as a function of true strain. Since a clear determination of the transition from elastic to plastic behavior is difficult because of the signal oscillations, strains above 0.2 percent are considered to represent plastic deformation. In both experimental and numerical investigations, the true stress-strain curves exhibit strain hardening up to a maximum true stress and afterwards decrease to a steady state response (i.e., negligible hardening). The stress peak marked in the experimental curves (highlighted with an arrow for the curve measured at 800 °C) is not observed in the simulation results. Possible explanations for this experimental effect are geometrical imperfections and/ or misalignment of the compression specimens. Further experimental and theoretical work may be required to fully clarify this effect, which is frequently observed experimentally at increased temperatures.

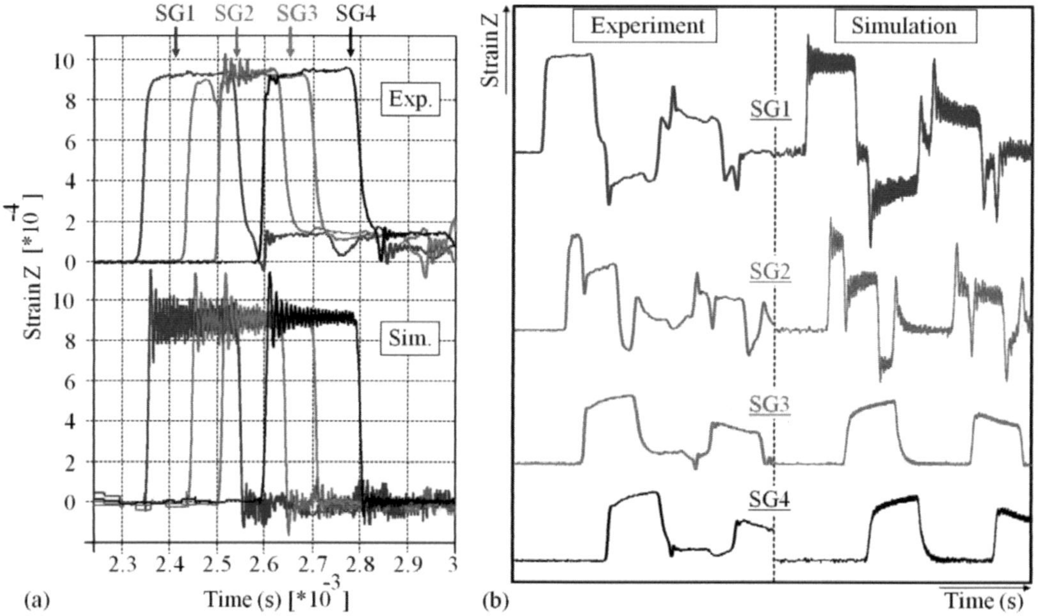

Figure 4. (a) Propagation of elastic wave pulses during a test shot. The experimental results (top, measured while using a pulse shaper) and of the simulation (bottom) present only the input and output signals in the SHPB setup. (b) The experimental (left) and simulated (right) input, reflected and transmitted signals of a SHPB test with a C75 compression specimen shown at four different SGs, are in excellent agreement.

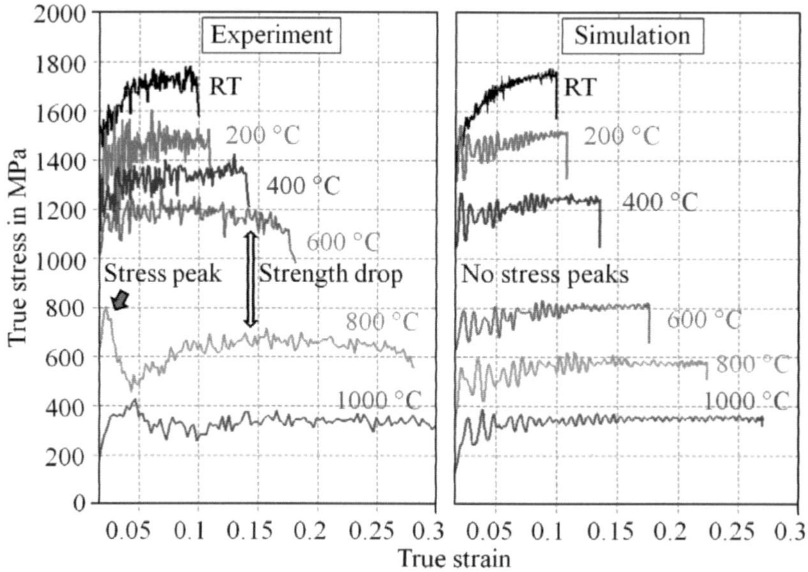

Figure 5. Experimental (left) and numerical (right) true stress-strain curves of the C75 compression specimen under dynamic compressive loading at high strain rates in a SHPB setup at RT and at higher temperatures of 200 °C, 400 °C, 600 °C, 800 °C and 1000 °C.

While the experimental and numerical results are generally in good agreement, a significant strength drop between 600 and 800 °C is observed in the experiments. This strength drop, however, is not included in the simulations as our rather the simple elastic-plastic material model based on the quasi-static results was used for the dynamic simulations. While our material model does predict a gradual decrease of material strength with increasing temperature, the abrupt drop specifically in the temperature range between 600 and 800 °C warrants further microstructural investigation. In the C75 steel, temperatures up to 600 °C are not expected to lead to significant microstructural changes compared to the initial material state at RT. The pronounced strength drop between 600 and 800 °C, however, indicates the onset of the diffusion-controlled phase transformation to austenite at higher temperatures. The micrographs (taken from SHBP specimens after mechanical testing at the corresponding elevated temperatures) shown in figure 6 provide direct microstructural evidence for this phase transition. While the material at 600 °C exhibits the typical features of a tempered martensitic/ bainitic steel, the short heating process (5 s) to 800 °C prior to the dynamic mechanical testing leads to an intermediate bainitic/ austenitic microstructure. In contrast, after fast heating to 1000 °C, the microstructure is austenitic (figure 6c, brighter areas), with additional features of hard pearlite (figure 6c, darker areas).

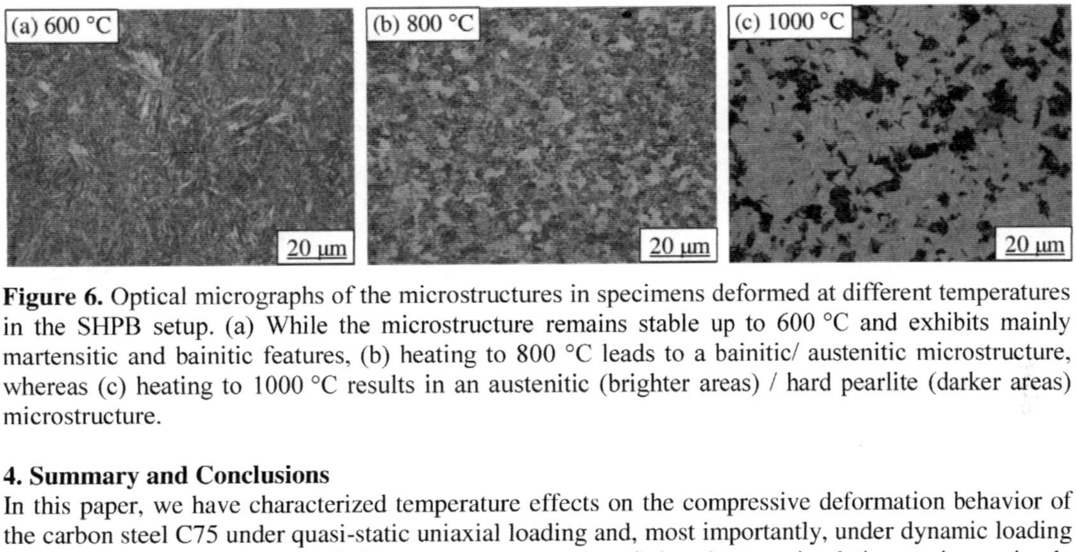

Figure 6. Optical micrographs of the microstructures in specimens deformed at different temperatures in the SHPB setup. (a) While the microstructure remains stable up to 600 °C and exhibits mainly martensitic and bainitic features, (b) heating to 800 °C leads to a bainitic/ austenitic microstructure, whereas (c) heating to 1000 °C results in an austenitic (brighter areas) / hard pearlite (darker areas) microstructure.

4. Summary and Conclusions

In this paper, we have characterized temperature effects on the compressive deformation behavior of the carbon steel C75 under quasi-static uniaxial loading and, most importantly, under dynamic loading conditions. The experimental findings were compared to finite element simulations using a simple elastic-plastic material model. Our main results can be summarized as follows:

1. When a "test shot" without a specimen is performed, it is possible to verify the simulation of wave propagation in the SHPB setup, and experimental and numerical results are in excellent agreement. The simulated input, reflected and output signals including deformation of a C75 specimen in the SHPB also agree well with those of actual experiments.
2. Different parameters, i.e., the effect of geometry and detailed setup of the SHPB (e.g. the use of ceramic plates) can be also considered during finite element simulations. Thus, the true dynamic mechanical response of the C75 specimen at room temperature and at higher temperatures up to 1000 °C can be studied both experimentally and numerically.
3. A strength drop in the true stress-strain curves (between 600 and 800 °C) and at high strain rates is observed experimentally. This effect can be related to the diffusion-controlled phase transformation to austenite at higher temperatures, which is not included in our finite element simulations.

References

[1] Meyer L W, Weise A and Hahn F 1997 Comparison of constitutive flow curve relations in cold and hot forming *J. Physique IV* **7(C3)** 13–20

[2] Lee Y, Kim B M, Park S W and Min O 2002 A study for the constitutive equation of carbon steel subjected to large strains, high temperatures and high strain rates *J. Mater. Process. Technol.* **130–131** 181–8

[3] Sellars C M 1990 Modelling microstructural development during hot rolling *Mater. Sci. Technol.* **6** 1072–81

[4] Wang M, Li X, Du F and Zheng Y 2005 A coupled thermal–mechanical and microstructural simulation of the cross wedge rolling process and experimental verification *Mater. Sci. Eng.* A **391** 305–12

[5] Pinheiro I P, Barbosa R and Cetlin P R 2007 The relevance of dynamic recrystallization in the hot deformation of IF steel at high strain rates *Mater. Sci. Eng.* A **457** 90–3

[6] Dehghan-Manshadi A, Barnett M R and Hodgson P D 2008 Hot deformation and recrystallization of austenitic stainless steel. Part I: Dynamic recrystallization. *Metall. Mater. Trans.* A **39** 1359–70

[7] Meyer L W, Kuprin C and Halle T 2009 High rate material behaviour at hot forming conditions *J. Mech. Time-Depend. Mater.* **13** 49–62

[8] Halle T 2009 Ermittlung und Modellierung von Werkstoffkenndaten metallischer Werkstoffe bei hohen Dehnungsgeschwindigkeiten *Habilitation Thesis* TU Chemnitz

IOP Conf. Series: Materials Science and Engineering **181** (2017) 012023 doi:10.1088/1757-899X/181/1/012023

Experimental and numerical investigation of the residual yield strength of aluminium alloy EN AW-2024-T3 affected by artificially produced pitting corrosion

R Pippig[1], E Schmidl[1], P Steinert[2], A Schubert[2] and T Lampke[1]

[1]Institute of Materials Science and Engineering, Chemnitz University of Technology, 09125 Chemnitz, Germany
[2]Professorship Micromanufacturing Technology, Institute for Machine Tools and Production Processes, Chemnitz University of Technology, 09126 Chemnitz, Germany

Abstract. In this study, the behaviour of the residual yield strength of aluminium alloy EN AW-2024-T3 affected by the morphology and numbers of corrosion pits (defects) is presented. Since specific defect structures are not reproducible during experimental corrosion tests, metal sheets with different numbers of pits and pit shapes are produced using laser micro structuring. The defect structures are measured using laser scanning microscopy. To compare the stress states of the micro structured and real corroded metal sheets, FE-analysis is used. Afterwards, uniaxial tensile tests are carried out and critical defect parameters in terms of yield strength reduction of the investigated aluminium alloy are detected.

1. Introduction

High strength aluminium alloys are susceptive to local corrosion phenomena, especially pitting corrosion [1]. The susceptibility to local corrosion is caused by damage of the passive layer, e.g. by chloride ions [2, 3]. When the passive layer is destroyed, the propagation of the local corrosion begins. The process is accelerated by local differences in electrochemical potentials caused by precipitates [1, 4, 5]. As a result, deep penetrating pits grow from the surface into the bulk material. This pitting corrosion leads to a loss in mechanical properties due to a reduced cross section, local dealloying [6] and notch stresses [7, 8].

High-strength aluminium alloys are widely used in the automotive and aircraft industry, where the most components are stressed by oscillating forces. Hence, common research focuses on the impact of pitting corrosion on the fatigue strength [9-11]. But also the static behaviour of high-strength aluminium alloys damaged by pitting corrosion is a subject of present investigations [12-15]. Pantelakis et al. showed a significant decrease in yield strength of the aluminium alloy EN AW 2024-T3 after being exposed to the so called EXCO test [13, 15] and a 3.5% NaCl solution [13]. In further investigations Setsika et al. [14] used finite element simulations (FE) to calculate this decrease of yield strength. Thereby, they approximate the complex local corrosion damage with conical frustums. The geometry of the frustums is obtained from corrosion damage obtained from metallographic specimens [15].

In order to apply the founded correlations to component parts in service, the pit shape should be determined by a non-destructive method. For modelling the strength loss, the geometry of the pits should be preferred to the corrosion duration. The same corrosion parameters, e.g. exposure time, result in different degrees of corrosion damage. Due to this reason it is not possible to achieve samples with equal damage by corrosion experiments in order to test parallel samples.

The aim of this work is to correlate the geometry and amount of corrosion damage (shape of pits, approximated by defect parameters) to the resulting quasi-static yield strength of the aluminium alloy EN AW 2024-T3. The natural pitting corrosion damage was imitated by artificial pits produced by laser micromachining. Metallographic and numerical results verify the suitability of this method. Using this approach, specimens with artificial pits varying in geometry and numbers are produced and uniaxial, quasi-static tensile tests are carried out to determine the residual yield strength.

2. Experimental procedure

The experiments were carried out using standard commercial aluminium alloy EN AW-2024-T3 sheets with dimensions of 100 x 25 x 1.6 mm^3 (rolling direction (RD) parallel to the specimen length). All specimens were grinded along the transversal direction (TD) using 1000 mesh sizing SiC paper to achieve a uniform and comparable surface condition over all samples.

To recreate real pitting corrosion caused by electrochemical corrosion, laser micromachining was used. Therefore, defects with different machining parameters were generated on the surface of a metal sheet using a $ND:YVO_4$ laser by Spectra-Physics. Afterwards, the artificial defects were scanned using laser scanning microscopy (Keyence VK-X200) and metallographic cuttings were recorded to identify the best set of machining parameters, see table 1, in terms of reproducing real pitting corrosion. The metallographic cuttings were prepared using SiC grinding paper with a mesh size of 600, 1000, 2400 and 4000. After grinding, the samples were polished using diamond paste with a grain size of 3 µm and finally OPS-suspension. To quantify the morphology of pitting corrosion based on laser scanning microscopy (LSM), three parameters of a pit were defined. These are the maximal depth z, the width in RD x and the width in TD y. The widths x and y are measured through the coordinates of the maximal depth in RD and TD, respectively.

Beside the optical measurements, microhardness measurements (KB 250 BVRZ) around the artificial pits were performed to observe unexpected changes in the microstructure caused by the laser beam. Five measurements per area of interest were done.

Table 1. Set of layer machining parameters to produce the artificial pits.

Parameter	Value
Wave length	532 nm
Pulse duration	10 ns
Focus diameter	15 µm
Mean power	10.4 W
Pulse repetition frequency	150 kHz
Pulse and line distance	8 µm

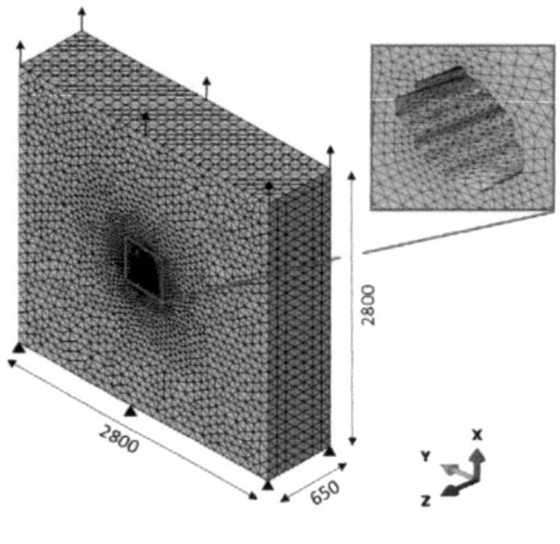

Figure 1. Geometry and mesh refinement for the FE-Analysis of a real corroded pit.

To compare the stress distributions induced by real corroded and artificial pits, the finite element analysis is used. Therefore, 3D profiles of the individual pits were measured using laser scanning microscopy and imported into the 3D surface modeller software Rhinozeros to create a solid CAD model. The pits were scaled to a size of 350 µm in x- and y-direction and a depth of 230 µm. Around the pits, a large enough solid block is defined to prevent interactions between the stress field of a pit and the boundaries of the simulation area, see figure 1. In the literature the shape of the pits are often approximated using simple geometries, e.g. ellipsoids [16, 17] or conical frustums [14]. Therefore, the stress distribution around a conical frustum is also analysed and compared to the other pit structures. To calculate the stress distributions around the pits, one side of the block is fixed in x-direction and a load of 100 MPa is applied on the other side, see figure 1. The material is isotropic with a Young's Modulus of 77.4 GPa, a Possion's ratio of 0.33 and a density of 2.78 g/cm³. The plastic material behaviour is modelled using the experimental measured stress strain curve of the aluminium alloy EN AW 2024-T3 without pitting corrosion. The mesh size around and within the pits was refined until constant stress distributions were noted, see figure 1. All FE simulations were performed using the dynamic implicit solver with a quasi-static strain rate of 10^{-3} s^{-1}.

In order to determine the quasi-static strength of the alloy damaged by pitting, an unstandardized tensile specimen was used. The alternating dimensions for the tensile specimen were necessary because of the dimensions of the metal sheets. The tensile specimens had a measuring length of 15 mm and a width of 4 mm and were taken parallel to RD. Afterwards, pits with different defect parameters were generated on each tensile specimen after statistical considerations using the machining parameters presented in table 1. In order to minimize the reduction of the cross section, no more than one defect was generated per cross section. After laser micromachining, the generated pits were measured using laser scanning microscopy and the corresponding defect parameters were calculated. At the end, uniaxial tensile tests at room temperature (295 K) and a constant strain rate of 10^{-3} s^{-1} were carried out using a ZWICK Allround-Line 20 kN testing machine.

3. Results and discussion

3.1 Comparison of the pit shapes

Figure 2 illustrates the cross section of a typical pit resulting from electrochemical corrosion, see figure 2 (a), and an artificially produced pit, see figure 2 (b). The geometrical shapes are very similar, including the rough topography at the bottom of the defects. However, the micro structured pits have a convex shape at the flanks of the defects and no undercuts which can be obtained in case of real pitting corrosion.

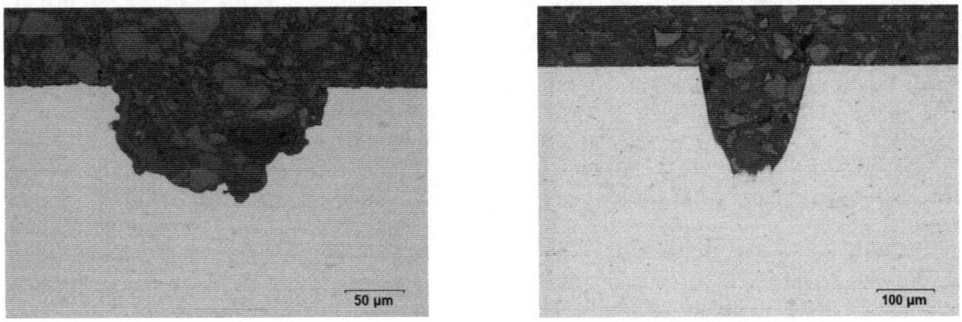

(a) (b)

Figure 2. Metallurgical specimens of a pit caused by (a) electrochemical corrosion (5 days salt spray test, 3.5 % NaCl, 40 °C) and (b) laser micromachining.

3.2 Comparison of hardness around the pits

Due to the laser beam, a heat impact into the material is seen when using laser microstructuring. This impact may result in microstructural changes of the heat treated aluminium alloy EN AW 2024-T3, influencing the microhardness and the yield strength, respectively. Therefore, the microhardness of the bulk material and the material around the pit were compared. For high-strength aluminium alloys, the hardness is usually measured in Brinell. However, because of the very limited space around the pits, the microhardness measurements in Vickers hardness (HV) are more convenient for comparing the areas around the pits with the bulk material. The hardness of the bulk material is 162±8 HV 0.1 and the hardness around the area of the artificial pits is 151±11 HV 0.1, showing no significant differences. The fluctuations of the values are caused by the heterogeneous microstructure of the aluminium alloy, whereas changes in microstructure due to a heat treatment would result in much larger differences. An influence of the hardness caused by the laser could not be observed.

3.3 Numerical FE-Analysis of the stress distribution around the pits

In figure 3 (a), a representative overview over the total von Mises stresses inside and outside of the individual pits is shown. The stress distribution outside of a real corroded pit, see figure 3 (a), shows nearly zero stresses in tensile direction and maximum stresses of around 150 MPa perpendicular to it. This stresses are qualitatively and quantitatively comparable to the stresses outside of an artificially generated pit, see figure 3 (b), and even of a conical frustum pit, see figure 3 (c). In contrast to this, the stresses inside the conical frustum pit are totally different compared to the real and artificially generated defects. The maximum von Mises stress for the real corroded pit when applying a load of 100 MPa is 348.9 MPa, meanwhile the maximum stress of the conical frustum is 282.7 MPa (19.0 % lower). While the conical frustum pit shows a continuous decrease from the middle to the edge of the total von Mises stress in tensile direction, the stress inside of a real corroded pit is inhomogeneous due to its rough and rugged morphology. At the flanks of real pits, higher and lower stresses alternate from the middle to the edge in tensile direction. It can be assumed that due to the presented large stress state differences inside a simplified pit structure, the predicted strength behavior will depict insufficient results when using simple geometries.

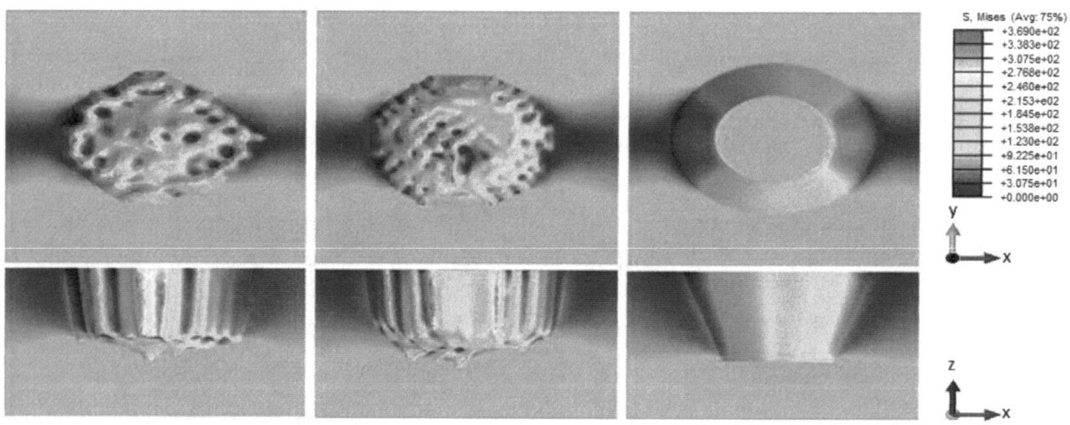

(a) (b) (c)

Figure 3. Numerical FE-Analysis of the total von Mises stress distribution around the (a) electrochemical corroded, (b) artificially generated and (c) conical frustum pit.

The artificially laser microstructured pits show a more accurate representation of real pits. The maximum von Mises stress with 368.3 MPa are only 5.6 % higher than the maximum stress of the representative real corroded pit. Also, the inhomogeneous stress distribution within the pit and the alternating stress profiles at the flanks of real pits is reproduced.

The finite element stress analysis as well as the experimental shape and hardness investigations show that micro laser machining is a sufficient method to imitate real corroded pits in terms of reproducing the strength behavior of the corroded aluminum alloy EN AW 2024-T3.

3.4 Mechanical properties

The influence of pitting on the mechanical properties in form of the yield strength reduction $\Delta R_{p0.2}$ is presented in figure 4. The relatively large standard deviations in yield strength between some parallel samples, see the bars in figure 4, are caused by the slightly different defect parameters of these samples. Especially, the pit depths can vary up to 21 μm because of the heterogeneous microstructure due to the precipitations within the aluminum matrix. This leads to a different amount of material removal during the laser micro machining process, resulting in slightly different depths and widths of the pits.

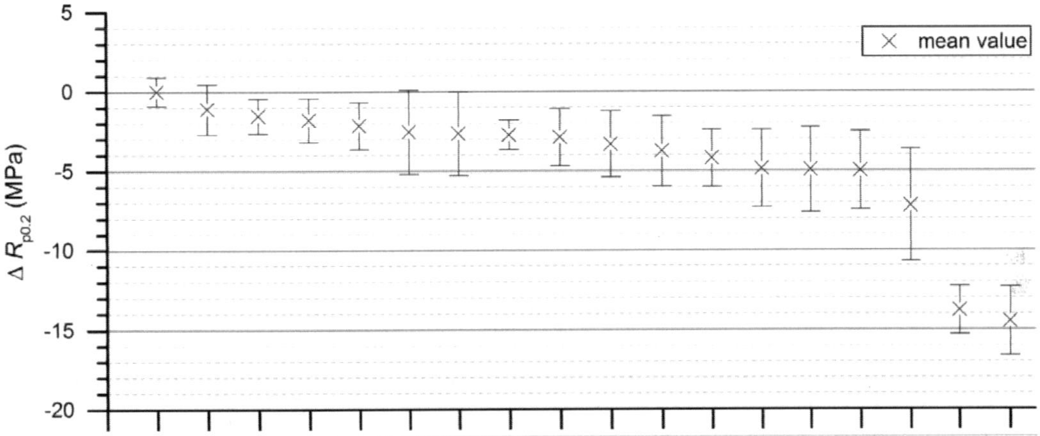

z (μm)	0	188±12	133±12	83±6	130±13	89±4	261±6	164±7	75±14	252±1	56±11	274±18	301±2	253±4	95±4	296±21	291±19	225±9
x (μm)	0	194±7	138±8	110±6	131±10	85±4	236±21	244±2	47±5	353±1	49±8	242±16	341±4	336±8	139±6	433±7	433±10	345±9
y/x (-)	0	0.89	0.77	1	0.53	0.72	0.81	0.76	1.28	0.51	0.84	1.04	1.27	1.02	0.70	0.71	1.01	1.29
n (-)	0	25	48	48	17	1	1	1	1	1	33	17	1	1	33	17	33	48

Figure 4. Differences of the yield strength $R_{p0.2}$ of samples with artificial pits and the uncorroded aluminium alloy EN AW 2024-T3 with a yield strength of 350 MPa. The x-label shows the LSM measured average defect parameters and corresponding standard deviations.

As presented in figure 4, a large number of the investigated samples show a small reduction in yield strength of up to 5 MPa. For stationary loading applications, this reduction is compensated by the used safety factor during the dimensioning process. The yield strength of the uncorroded aluminium alloy EN AW 2024-T3 is 350 MPa. The strongest yield reduction of 14 MPa occurs for the samples with 48 defects, an aspect ratio of 1.29 and a depth of 225 μm. Only three samples show a yield strength reduction greater than 5 MPa. The pit depths for all three samples are greater than 225 μm, the widths in tensile direction are greater than 345 μm and the aspect ratios are greater than 0.7. Considering only these samples, the yield strength decreases if the aspect ratio *y/x* of the pits increases even if the pit depth decreases. Moreover, the yield strength decreases if the number of pits increases. Therefore, not only the penetration depths influences the yield strength reduction, but the aspect ratio and the number of pits seems to be the dominating factors when it comes to strength reduction after reaching the critical pitting corrosion parameters.

However, the differences in $R_{p0.2}$ are not large enough in this work to model the resulting yield strength in dependence on the defect parameters.

4. Conclusions

Electrochemical pitting corrosion was imitated using laser micro machining. The resulting uniaxial yield strength of the aluminium alloy EN AW-2024-T3 was investigated. It was observed that the hardness of the alloy is not significantly affected by the heat impact of the laser beam. The stress field around and inside the real and artificially produced pits are similar. Simple defect structures, e.g. conical frustums, show larger differences. Overall, it was shown that laser micro machining is a suitable technique to reproduce electrochemical pitting corrosion (excluding undercuts). In this work, a wide range of the investigated pitting corrosion states are within a uniaxial yield strength reduction of 5 MPa. A strong influence of pitting corrosion could be observed just for pits with a depth greater than 225 µm, a width in tensile direction greater than 345 µm and an aspect ratio of 0.7.

In further investigations, the number of tested samples with individual defect parameters could be raised. This would lead to more data for modelling the resulting yield strength in dependence of the defect parameters. After that, the constitutive relation for corroded materials could be applied on component parts in FE simulations.

Acknowledgments

Financial support was granted by the Deutsche Forschungsgemeinschaft (German Research Foundation) via the project LA 1274/36-1. The technical assistance of C. Gläser, K. Muhr, D. Spieler, H. Böhme and the numerical assistance of Mehadi Ziko is gratefully acknowledged.

References

[1] Campestrini P, van Westing E, van Rooijen H and de Wit J 2000 *Corros. Sci.* **42** 1853-61
[2] Strehblow H H 1984 *Mater. Corros.* **35** 437–48
[3] Blanc C and Mankowski G 1998 *Corros. Sci.* **40** 411–29
[4] Suter T and Alkire R C 2001 *J. Electrochem. Soc.* **148** B36–42
[5] Birbilis N and Buchheit R G 2005 *J. Electrochem. Soc.* **152** B140–51
[6] Obispo H, Murr L, Arrowood R and Trillo E 2000 *J. Mater. Sci.* **35** 3479-95
[7] Cerit M, Genel K and Eksi S 2009 *Eng. Fail. Anal.* **16** 2467–72
[8] Pidaparti R M and Rao A S 2008 *Corros. Sci.* **50** 1932–38
[9] Turnbull A, Wright L and Crocker L 2010 *Corros. Sci.* **52** 1492–98
[10] Sankaran K, Perez R and Jata K 2001 *Mater. Sci. Eng. A* **297** 223–29
[11] Wang Q, Kawagoishi N and Chen Q 2003 *Scripta Mater.* **49** 711–16
[12] Pantelakis S and Kermanidis A 2009 *Corrosion Control in the Aerospace Industry* 67 – 108
[13] Pantelakis S, Chamos A and Kermanidis A 2012 *Theor. Appl. Fract. Mec.* **57** 36–42
[14] Setsika D, Tserpes K and Pantelakis S 2015 *Int. J. Struct. Integ.* **6** 451–67
[15] Pantelakis S, Setsika D, Chamos A and Zervaki A 2016 *Int. J. Struct. Integ.* **7** 25–46
[16] van der Walde K, Brockenbrough J, Craig B and Hillberry B 2005 *Int. J. Fatigue* **27** 1509–18
[17] van der Walde K and Hillberry B 2008 *Int. J. Fatigue* **30** 106–18

IOP Conf. Series: Materials Science and Engineering **181** (2017) 012024 doi:10.1088/1757-899X/181/1/012024

Microstructural evolution during tension-compression in-plane deformation of a pure aluminum sheet

M Härtel, B Bohne and M F-X Wagner

Institute of Materials Science and Engineering, Chemnitz University of Technology, 09125 Chemnitz, Germany

Abstract. Classically, the Bauschinger effect refers to a reduction of yield strength after a load path change. In this contribution, we present results of an experimental and microstructural investigation on Bauschinger effects in an AA1050 sheet metal (with 1 mm thickness) subjected to in-plane uniaxial loading. We performed tension-compression tests with different values of maximum tensile strains in a novel tool that was specifically designed to avoid buckling under compressive loading. Our experimental results show that the sheet material exhibits distinct Bauschinger effects. At different stages of deformation, we interrupted the tests and prepared samples for transmission electron microscopy. Our microstructural observations allow rationalizing the occurrence and magnitude of the observed Bauschinger effects.

1. Introduction

The well-known Bauschinger effect typically refers to a reduction of yield strength after a load path change [1]. Different reasons for this material behavior have been discussed in a variety of materials: According to Masing, different strengths in adjacent grains and developing residual stresses during the initial deformation are the main reason for macroscopic Bauschinger effects [2] after load reversal; this model certainly holds for simple, polycrystalline materials. Sachs and Shoji [3], as well as Buckley and Entwistle [4], however, also observed Bauschinger effects in single crystals. Margolin explained this phenomenon by considering the formation of substructures, and the corresponding anisotropic mechanical properties, in individual grains [5]. Brown [6] and various other groups [7-9] identified the interaction of free dislocations with grain boundaries, precipitates, impurities and other dislocations, and the resulting back stress after load removal, as important mechanisms affecting the Bauschinger effect in metals. Clearly, a detailed analysis of microstructural evolution during the different stages of mechanical loading is required to fully rationalize Bauschinger effects in different materials. For example, Lewandowska [10], Hasegawa et al. [11], Caceres et al. [12] and Vincze et al. [13] carefully documented the formation of substructures (and their decomposition after a load path change) and related these observations to the macroscopic Bauschinger effect. Previous experimental work made use of tension and subsequent compression tests of bulk materials [11-12]. Sheet materials have been studied in shear tests (by shearing in one direction, load removal, and subsequent shearing in the opposite direction, [10, 13]); evaluation of mechanical data from testing with different shear specimen geometries, however, is not trivial [14, 15]. In the present study, the goal is to perform tension-compression tests on a sheet material. This experimental approach is quite challenging as well because sheet materials are prone to buckling under compressive loading [16]. Different groups have overcome this issue by using devices that provide additional support from the samples' sides to prevent buckling [17-23]. Based on their reports [17-23], we designed and established a comb-shaped support that has been described in [16]. The present paper primarily reports the first results of the tension-compression tests on a pure aluminum alloy AA1050 that were performed using our new experimental setup. We evaluate the magnitude of the Bauschinger effect based on experimental stress-strain data [7, 16]. Moreover, we document the development of the microstructure at different stages of the deformation by transmission electron microscopy (TEM).

2. Experimental

In the experiments presented in this paper, an aluminum sheet material with a thickness of 1 mm was used. The pure aluminum alloy (AA1050) consists of 99.5 (wt.-)% Al, 0.08 % Si, 0.341 % Fe, 0.002 % Cu, 0.003 % Mn, 0.05 % Mg, 0.005 % Zn and 0.013 % Ti. Prior to testing, the material was annealed for 120 min at 500 °C to ensure a homogeneous microstructure with a low dislocation density.

For all mechanical tests, our new tension-compression setup [16] was used, see figure 1. The setup consists of four comb-shaped parts (two of them with a notch of 1 mm) that can move relative to each other, and clamps to hold the sample and to avoid buckling under compression. This setup was installed in a conventional testing machine (ZWICK/ROELL ZMART PRO UNIVERSAL 100 kN). A standardized tensile specimen (following the design specified in [24]) with a gauge length of 50 mm and a width of 12.5 mm was used. Force measurements, force data correction and strain measurements with a digital image correlation system were performed as described in [16]. All mechanical tests were performed at room temperature and at a strain rate of 10^{-3} s^{-1}.

For TEM analysis, samples were prepared at different stages of the deformation during the tension-compression tests, i.e., (i) at the stage of maximum tensile strain; (ii) immediately after yielding after load reversal, and (iii) at 4 % compression. All samples were ground, mechanically thinned and finally electro-polished to produce the TEM samples. TEM analysis was performed using a TEM Hitachi H8100 with an accelerating voltage of 200 kV (see [25] for further details).

Figure 1. Tension-compression setup, [16]: comb-shaped fingers (a) installed between a die set in a conventional testing machine. The standardized tensile specimen is placed between the two front and the two back parts of the tool (b). For sufficient illumination between the combs' fingers, a ring of lights is placed around the DIC camera (c); the illuminated surface of the specimen is covered with speckle patterns for DIC measurement (d).

3. Results and Discussion

3.1 Mechanical behavior during tension-compression testing

Figure 2 shows the results of the tension-compression tests with different maximum tensile strains (0.5, 1, 1.5 and 2 %). The tensile stress-strain curves do not fall exactly on top of each other, similar to our observations with the same set up for steel sheet DC06 presented in [16]. Furthermore, we observe a deviation of about 15 MPa in the initial yield strength compared with conventional tensile test data of the AA1050 sheet material. It is therefore difficult to quantify the decrease in the yield strength after load reversal. The Bauschinger effect factor α [7, 16], with values about 0.3, is uncharacteristically low (for such small maximum tensile strains, values of $\alpha = 0.8\text{-}0.9$ are commonly observed [7, 16]), which may be a consequence of the scatter in the experimental data. This error in our measurements is probably related to internal friction in the experimental setup during the deformation: For very soft materials such as AA1050 (initial yield strength: 25 MPa), even small amounts of internal friction affect the measurement, whereas for steel sheets (even for soft steels), this influence is negligible [16]. Despite this drawback in terms of stress data, the surface strain measurements with DIC are accurate, and this ensures that different stages of deformation can be quantified and related to the results of our TEM analysis discussed below.

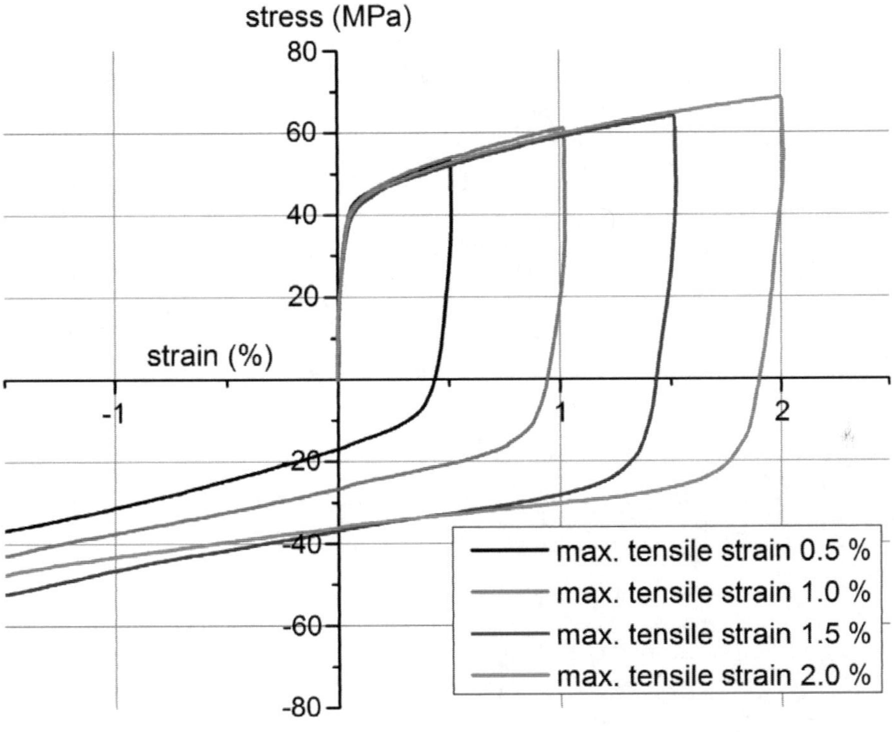

Figure 2. Stress-strain curves of tension-compression experiments with different maximum tensile strains of 0.5, 1, 1.5 and 2 %, respectively. The initial yield strength under tensile load deviates by about 15 MPa from data determined by conventional tensile testing, which likely results from friction in the experimental setup. After load reversal, all specimens exhibit a reduced yield strength under compression, but an accurate quantification of the Bauschinger effect is unreliable.

3.2 Microstructural analysis

In figure 3, microstructures at different stages of the tension-compression tests with a low maximum tensile strain are shown. Figure 3a shows a representative microstructure after 0.5 % maximum tensile strain, figure 3b shows the microstructure immediately after yielding under compression, and figure 3c shows the microstructure after a strain of 4 % in compression. Figure 3d shows a stress-strain curve which indicates where the samples corresponding to figures. 3a, b and c were taken out for TEM analysis. The microstructure shown in the bright field image in figure 3a is characterized by a very low dislocation density. Slip seems to primarily occur on a family of favorably oriented, parallel slip systems. In the vicinity of impurities or precipitates, we observe slightly increased dislocation densities, while other parts of the microstructure are mostly free of dislocations. After load reversal and subsequent yielding under compression, the dislocation density decreases, figure 3b. Even in the vicinity of impurities or precipitates, the dislocation density seems to be smaller than in figure 3a, which indicates the occurrence of recovery processes; this interpretation is in line with previous reports [10, 11, 13]. Further deformation under compression leads to the onset of the early formation of substructures, figure 3c.

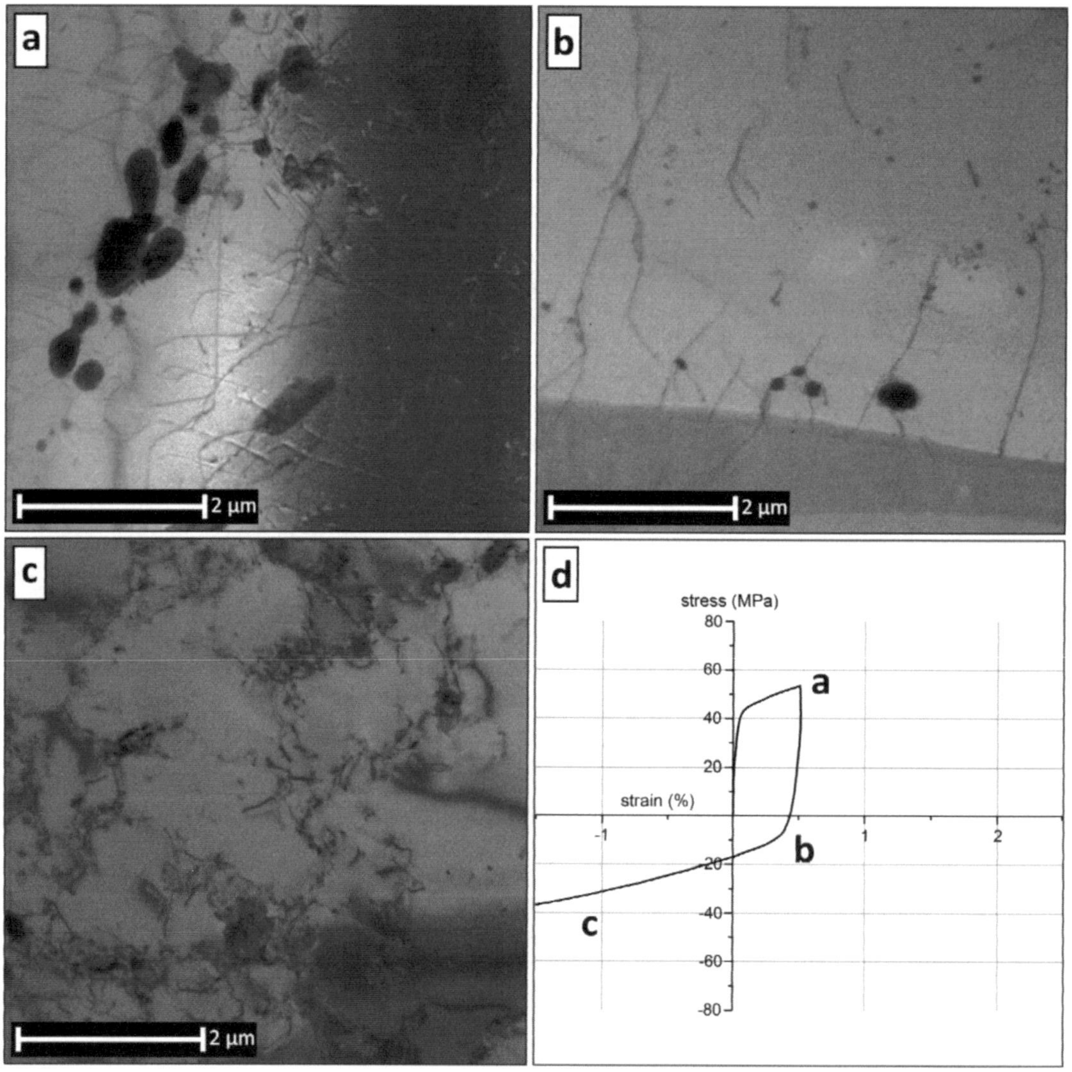

Figure 3. TEM bright field images of the AA1050 sheet material at different stages of the tension-compression test (maximum tensile strain: 0.5 %). After loading to the maximum tensile strain, we observe a low dislocation density (a). A decrease in the dislocation density due to annihilation is observed after the load reversal (b). After reaching a strain of 4 % in compression, we observe the beginning of the formation of substructures (c). The stress-strain-curve in (d) highlights the stages of deformation where the corresponding samples for TEM analysis were taken out.

Such observations have also been reported by other groups [10, 11, 13, 26], and the annihilation and/ or interaction of the dislocations with precipitates and impurities are typically discussed as the main microstructural processes that lead to the Bauschinger effect. More specifically, Vucko et al. discussed how the deformation of dislocation cells leads to a decrease of residual stresses which in turn allows to reduce strain energy [26].

In figure 4, the microstructures at different stages of deformation in the tension-compression test with an increased maximum tensile strain are shown.

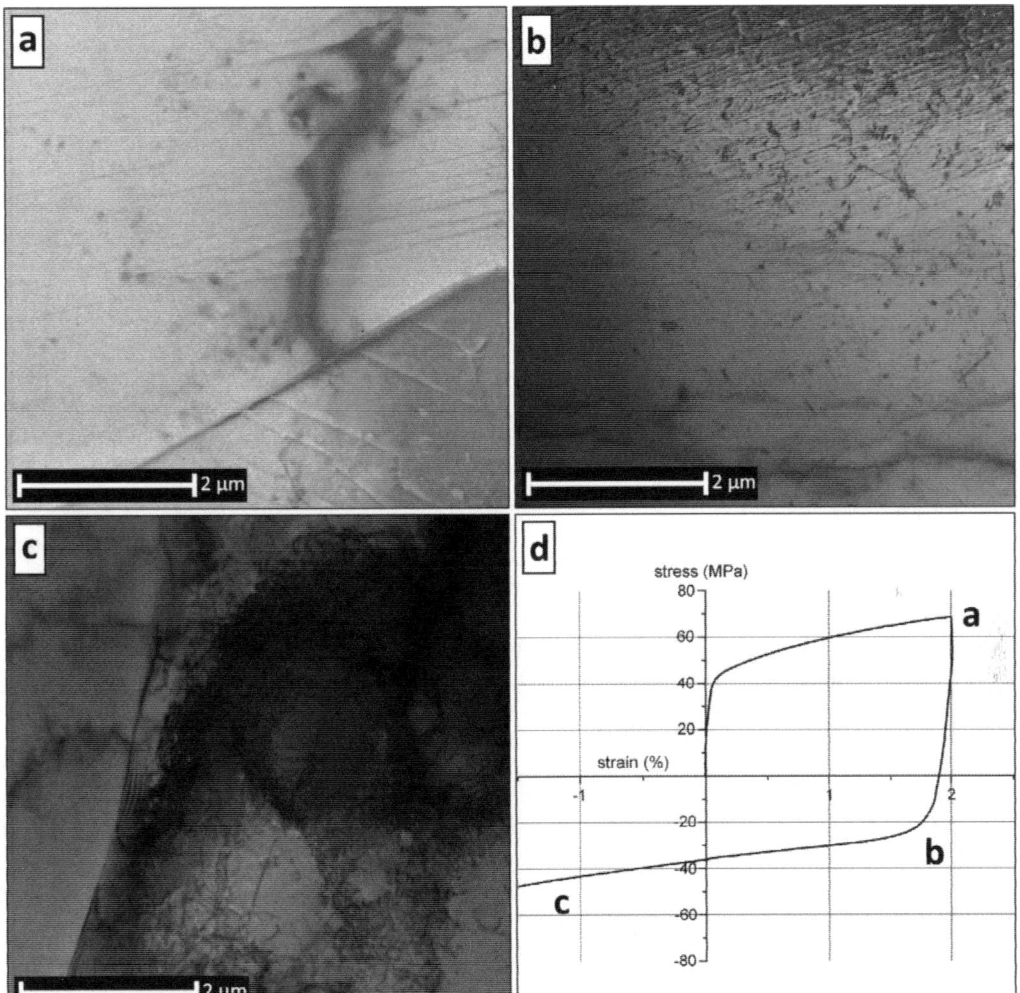

Figure 4. TEM micrographs of the AA1050 sheet material at different stages of a tension-compression test with a maximum tensile strain of 2 %. After tensile straining to 2 %, we observe a higher dislocation density on parallel slip planes (a) compared to smaller tensile strains (see figure 3a). Additional slip systems are activated after load reversal (b). After strains of 4 % in compression, we observe the beginning of the formation of substructures (c). In the cell interiors, some of the dislocation segments formed earlier can still be detected. The stress-strain-curve in (d) highlights the stages of deformation where samples for TEM analysis were taken out.

Figure 4a shows a representative microstructure after loading to a maximum tensile strain of 2 %. The dislocation density appears to be increased compared to figure 3a, and dislocations can be observed throughout the entire grain. In the vicinity of impurities or precipitates, dislocation loops, where the dislocations are not able to pass the obstacles, can be observed. The stress fields of these Orowan loops can also be directly related to the occurrence of a Bauschinger effect. Figure 4b shows the microstructure after load reversal and early yielding under compression. Instead of an annihilation of dislocations (as observed for smaller maximum tensile strains, see figure 3b), the bright field image in figure 4b indicates the activation of a new family of parallel slip systems, where straight dislocation segments seem to cross the other sets of dislocations (formed earlier during tensile deformation).This leads to an increase in the dislocation density already immediately after load reversal and may also lead to an increase of internal stresses that further promote Bauschinger effects. Subsequent compression leads to a further increase in dislocation density, and to the formation of substructures that are more favorable concerning strain energy, figure 4c. These microstructural observations represent only the beginning of the formation of substructures; in the interior of several cell structures we still observe some conventional dislocations. Such remains of the earlier dislocation network are likely to be reduced during the ongoing formation of dislocation cells when the material is strained even further.

4. Summary and conclusions

This investigation focused on both mechanical and microstructural investigations of Bauschinger effects in an AA1050 sheet material. Our study forms the basis for ongoing research on the effect of uni- and biaxial loading on Bauschinger effects in different sheet materials [27, 28]. We performed uniaxial tension-compression tests with an experimental setup with comb-shaped support plates to avoid buckling of the thin sheet specimens after load reversal. The AA1050 material exhibits distinct Bauschinger effects for different maximum tensile strains in the range of relatively small strains from 0.5 % to 2 % investigated here.

Using transmission electron microscopy, we analyzed the development of the microstructures at different stages of the tension-compression tests for experiments with 0.5 % and 2 % maximum tensile strains, respectively. For the 0.5 % maximum tensile strain condition, we observed annihilation of dislocations after load reversal and the formation of dislocation substructures during subsequent compression. These processes clearly contribute to the Bauschinger effect observed in the corresponding mechanical data. For 2 % maximum tensile strain, we observed the activation of a dominant slip system on parallel slip planes after tensile deformation. After load reversal, additional slip systems are activated, and interaction of dislocations on different slip planes is more pronounced than the annihilation processes observed for smaller maximum tensile strains. The developing stress fields also contribute to the macroscopic Bauschinger effect observed in the sheet material. Further compressive straining results in the formation of dislocation cells. The microstructural observations reported here confirm earlier studies on pure aluminum, and they highlight the dominant microstructural processes that also most likely affect Bauschinger effects during more complex, multi-axial load cases.

Acknowledgements

The authors gratefully acknowledge funding by the German Research Foundation (Deutsche Forschungsgemeinschaft, DFG) through project WA 2602/12-1 and through the Collaborative Research Center (Sonderforschungsbereich) 692.

References

[1] Bauschinger J 1886 *Mitteilung XV aus dem Mechanisch-technischen Laboratorium der Königlichen Technischen Hochschule in München* **13** 115
[2] Masing G 1926 *Wiss. Veröff. Siemens Konzern* **3** 231
[3] Sachs G and Shoji H 1927 *Zeitschrift für Physik* **45** 755
[4] Buckley S N and Entwistle K M 1956 *Acta Metall.* **4** 352
[5] Margolin H, Hazaveh F and Yaguchi H 1978 *Scr. Metall.* **12** 1141
[6] Brown L M 1977 *Scr. Metall.* **11** 127
[7] Abel A and Muir H 1972 *Philos. Mag.* **26** 489
[8] Kostryzhev A G, Strangwood M and Davis C L 2010 *Metall. Mater. Trans A* **6** 1399
[9] Sleeswyk A W and Kemerink G J 1984 *Scr. Metall.* **19** 471
[10] Lewandowska M 2003 *Mat. Chem. Phys.* **81** 555
[11] Hasegawa T and Yakou T 1975 *Mat. Sci. Eng.* **20** 276
[12] Caceres C H, Griffiths J R and Reiner P 1996 *Acta Mater.* **44** 15
[13] Vincze G, Rauch E F, Gracio J J, Barlat F and Lopes A B 2005 *Acta Mater.* **53** 1005
[14] Yin Q, Zillmann B, Suttner S, Gerstein G, Biasutti M, Tekkaya A E, Wagner M F-W, Merklein M, Schaper M, Halle T and Brosius A 2014 *Int J Solids Struct* ***51*** 1066
[15] Zillmann B, Clausmeyer T, Bargmann S, Lampke T, Wagner M F-W and Halle T 2012 *Technische Mechanik* **32** 622
[16] Härtel M, Illgen C and Wagner M F-W 2016 *IOP Conf. Series: Material Sciences and Engineering* **118** 012018
[17] Kuwabara T, Nagata K and Nakako T 2001 *Proc. of Int. Conf. Advanced Materials and Processing Technol.* 407
[18] Tan Z and Magnusson C 1995 *Mater. Sci. Eng. A* **183** 7926
[19] Boger R K, Wagoner R H, Barlat F, Lee M G and Chung K 2005 *Int. J. Plast.* **21** 2319
[20] Balakrishnan V 1999 *Measurement of in-plane Bauschinger effect in metal sheet* (Columbus, Ohio: The Ohio State University, Master Thesis)
[21] Piao K, Lee J K, Kim J H, Kim H Y, Chung K, Barlat F and Wagoner R H 2012 *Int. J. Plast.* **38** 27
[22] Lee M G, Kim J H, Kim D, Seo O S, Nguyen N T and Kim H Y 2013 *Expe. Mech.* **53** 1039
[23] Knoerr L, Sever N, McKune P and Faath T 2014 *Proc. of AIP Conference* **1567** 654
[24] Norm DIN 50125: Prüfung metallischer Werkstoffe – Zugproben. **2009**. Berlin: Beuth Verlag
[25] Härtel M, Wagner S, Frint P and Wagner M F-X 2014 *IOP Conf. Series: Materials Science and Engineering* **63** 012080
[26] Vucko F, Bosch C and Delafosse D *2014 Mater. Sci. Eng. A* **597** 381
[27] Zillmann B, Härtel M, Halle T, Lampke T and Wagner M F-W 2011 *Proc. ICTP 2011 Aachen* 691
[28] Zillmann B, Wagner M F-W, Schmaltz S, Schmidl E, Lampke T, Willner K and Halle T 2015 *Int. J. Solids Struct.* **66** 111

Analysis of the complex stress state during early loading in cylindrical compression-shear specimens

S Pfeiffer, P Frint and M F-X Wagner

Institute of Materials Science and Engineering, Chemnitz University of Technology, 09125 Chemnitz, Germany

Abstract. In most engineering applications, materials are subjected to complex load cases rather than the simple uniaxial ones typically used for material characterization. To experimentally study the material behavior under a combination of compression and shear, an inclined compression specimen can be used. This specimen has been applied in various earlier experimental studies, typically to investigate shear localization under quasi-static or impact loading. In this contribution, we analyze the stress state in a compression-shear specimen in detail using an elastic-ideal plastic finite element simulation. The effects of specimen aspect ratio (height/diameter), inclination angle, and friction conditions between specimen and tool plates are investigated using the material parameters of different conventional steels as input. Shear stress distributions in characteristic shear directions on specific planes in the specimen that control the subsequent plastic deformation behavior are evaluated. Our results show that, even in the absence of friction, shear stresses are distributed heterogeneously in the inclined specimen, which differs from the stress distribution in a conventional compression specimen. Moreover, the highest shear and equivalent stresses always occur at the edges of the short diagonal plane of the specimen, independent of the investigated parameters. This study contributes to a more detailed understanding of the elasto-plastic mechanics in compression-shear specimens, and it specifically provides information for the analysis of the onset of early plastic deformation.

1. Introduction

Structural components face complex stress states in most applications. Industrial forming processes, impact loading and ductile fracture are just a few examples. One load case of special interest is the combination of compressive and shear loading. Different tests have been developed to investigate material behavior under such specific loading conditions. Rittel et al. [1] developed the shear-compression specimen (SCS) for large strains and a wide range of strain rates. The SCS was for instance used by Sang et al. [2] investigating a 7050 aluminum alloy, and by Moemi et al. [3] to study an AZ31 magnesium alloy, both at high temperatures. This cylindrical specimen has two notches, machined under an angle of 45° with respect to the longitudinal axis that defines the gage section. Finite element (FE) simulations show that stresses and strains are both distributed relatively uniformly within the gage section [1]. Ashab et al. [4] presented special fixtures, consisting of a support and a loading block, that can be used in universal testing machines to apply a combined compression-shear loading. Three different sets of fixtures with loading angles of 15°, 30° and 45° were used to investigate aluminum honeycomb structures, [4]. Meyer et al. [5] introduced another variation of a compression-shear specimen: The specimen (with the shape of an inclined cylinder, see figure 1) was originally developed to investigate the material behavior under combined compression and shear loading at high strain rates, with a special focus on adiabatic shear failure [6]. Recently Elibol and

Wagner [7, 8] applied this specimen geometry in static tests to study the localized phase transformation in NiTi shape memory alloys. The compression-shear specimen also allows the characterization of anisotropic material behavior under combined shear and compressive loading especially in textured materials [9]. It can be used in conventional testing machines, where uniaxial compressive loading results in a combined compression-shear stress state within the specimen, and the ratio of shear stresses to compressive stresses depends on the inclination angle of the specimen. A first analytical, but considerably simplified assessment of shear and compressive load was presented in [9], demonstrating that an inclination angle of 6° results in a shear to compressive load ratio of about 0.1. FE simulations of a similar specimen, in comparison to a conventional (i.e., non-inclined) compression specimen showed that the highest equivalent stresses occur at the edges of the shortest connecting line between the top and bottom faces of the specimen [5]. In this study, shear strains in different orientations with respect to a global coordinate system were evaluated. For an inclination of 6°, maximum shear strains of 26 % were found. Moreover, these simulations predict that the highest shear strains do not occur in a shear plane at an inclination of 45° with respect to the compressive load axis, which would be the case in simple compression.

While the results presented in [9] provide important information on the special stress state in compression-shear specimens, they are only valid for one specimen geometry (with an angle of inclination of 6°); moreover, friction between the tool plates and the faces of the specimen (which are known to play an important role in simple compression testing) have not yet been addressed. In the present study, we perform a more detailed analysis of the stress state in compression-shear specimens. We investigate different specimen inclination angles, different specimen aspect ratios (height/diameter), friction conditions and material parameters. A fully three-dimensional FE simulation, considering elastic–ideal plastic material behavior as well as contact and friction between the tool plates and the specimen, enables an in-depth analysis of the present stress state. The goal of our study is to contribute to a more detailed understanding of the mechanics of the compression-shear specimen. Based on the results of our elastic FE simulations, an assessment of the subsequent onset of early plastic deformation can be performed.

2. Specimen geometry and material parameters

The geometry of a compression-shear specimen can be fully described by the height of the specimen h, its radius r (or diameter d), and inclination angle α (see figure 1a). The geometry corresponds to a cylinder with a longitudinal axis that is inclined to the parallel, circular faces of the sample and the uniaxial (compressive) loading direction, respectively. We note that there are two characteristic planes of ellipsoidal shape within the specimen that can be constructed by cutting the specimen along a plane through points A and C in the case of the "long diagonal plane", and through points B and D for the "short diagonal plane" (figures 1b and c).

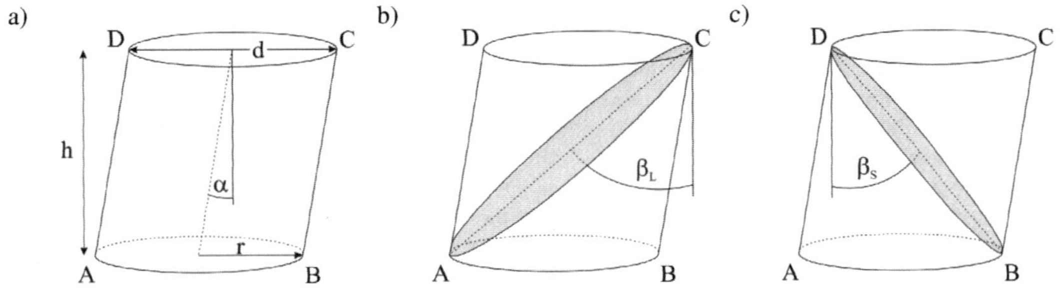

Figure 1. Geometry of the compression-shear specimen (a). Long (b) and short (c) diagonal planes of the specimen can be described by their angles to the vertical axis, β_L and β_S, respectively.

These characteristic planes can also be defined by their angles to the vertical axis, β_L and β_S, respectively. As discussed below, these angles are a function of α. In this study, FE simulations were used to analyze the characteristic stress states in compression-shear experiments for three different (elastic, ideal-plastic) materials, roughly corresponding to three conventional steels: a mild steel S235 with a yield strength of 235 MPa and two tempering steels C45 and 42CrMo4 with yield strengths of 490 MPa and 900 MPa. The elastic constants were assumed to be constant for all three materials, with a Young's modulus of 205 GPa and a Poisson's ratio of 0.3.

3. Analytical assessment of shear stresses

The inclination of the compression-shear specimen leads to a compressive stress state superimposed with a shear stress state when the sample is loaded by a compressive force on its two circular faces. A first approximation of the ratio of shear stresses (in the short diagonal plane) to compressive stresses was derived by Meyer et al. [9, 6], which is simply a function of α and therefore naturally only holds for the undeformed specimen (i.e., during early loading):

$$\tau/\sigma = \tan(\alpha) \qquad (1)$$

Equation (1) is, moreover, only valid for frictionless conditions, and it is based on the assumption of constant shear and normal stresses within any plane parallel to the faces of the compression-shear specimen. With similar assumptions, the shear and normal stresses in an arbitrary plane within the specimen (inclined by an angle φ to the vertical axis) can be determined using Mohr's circle [10]. This calculation corresponds to a simple coordinate transformation of the 2D stress state:

$$\sigma_\varphi = -{}^1\!/_2\left(\sigma_{xx}+\sigma_{yy}\right) - {}^1\!/_2\left(\sigma_{yy}-\sigma_{xx}\right)\cos(2\varphi) + \sigma_{xy}\sin(2\varphi) \qquad (2)$$

$$\tau_\varphi = {}^1\!/_2\left(\sigma_{yy}-\sigma_{xx}\right)\sin(2\varphi) + \sigma_{xy}\cos(2\varphi) \qquad (3)$$

Using the tilt angles of the short and long diagonal planes of the compression-shear specimen, the amount of superimposed shear stress can be estimated for different specimen geometries. Figure 2 shows the tilt angles of short and long diagonal plane for three different specimen aspect ratios and for the experimentally relevant range of inclination angles and d/h ratios. Empty circles represent specimen geometries that were also analyzed using the finite element method in this study (see below). In case of a conventional compression specimen (α = 0°) with an aspect ratio of 1 (blue curves), both tilt angles are 45°. Thus, maximum shear stresses occur in these planes during loading. Inclining the specimen leads to an increase of the tilt angle of the long diagonal plane while the tilt angle of the short diagonal plane is decreased. Shear stresses decrease within both planes with increasing angle of inclination due to their increasing deviation from the 45° inclination. An increase of the specimen aspect ratio to 1.25 (orange curves) and 1.5 (pink curves) results in a decrease of both tilt angles. Shear stresses are expected to decrease with increasing specimen aspect ratio for a given angle of inclination. Note that for aspect ratios larger than 1 (and for α > 0°), shear stresses in the long diagonal plane are higher than those in the short diagonal plane, because the corresponding tilt angles are closer to 45°. Only small angles of inclination and negligible amounts of friction allow for such a direct estimation of shear stresses with the tilt angle approach according to Mohr's circle. As is discussed in the remainder of this paper, a more accurate analysis of the stress state, also accounting for the effect of different material parameters and friction conditions, can be performed by means of FE simulations.

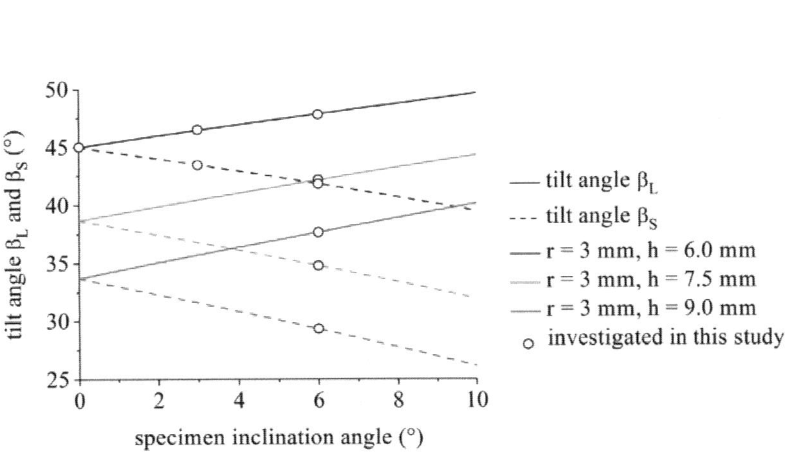

Figure 2. Tilt angles of the short (dashed lines) and long (solid lines) diagonal planes as a function of the inclination angle for compression-shear specimens with different height/diameter ratios. Empty circles mark orientations of the characteristic short and long diagonal planes of specimens that were further investigated in this study.

4. Finite element analysis

In our FE simulations, we systematically considered different specimen geometries with different inclination angles, specimen aspect ratios (height/diameter), friction conditions between tool plates and specimens, and material parameters (yield stresses). All simulations consider an elastic-ideal plastic material behavior with a Young's modulus of 205 GPa and a Poisson's ratio of 0.3. Table 1 provides a summary of the parameter variations used in the different simulations.

Table 1. Overview of the parameters used in the different FE simulations.

μ (–)	d (mm)	h (mm)	α (°)	σ_Y (MPa)	investigated parameter
0	6	6	**0**	490	
0	6	6	**3**	490	angle of inclination
0	6	6	**6**	490	
0	6	**6**	6	490	
0	6	**7.5**	6	490	specimen aspect ratio
0	6	**9**	6	490	
0.05	6	6	6	490	
0.10	6	6	6	490	friction
0.15	6	6	6	490	
0	6	6	6	**225**	
0	6	6	6	**490**	yield strength
0	6	6	6	**900**	

To investigate the multiaxial stress state in a compression-shear specimen subjected to uniaxial load in detail, 3D FE simulations were performed with the FE software package Abaqus, using 26.000 brick elements (C3D8R type) in a static analysis. The tool plates were considered as two rigid surfaces, using node to surface contact with a hard contact in normal direction and a constant friction coefficient in tangential direction. All simulations were stopped when the von Mises stress at any integration point met the static yield strength of the investigated material; our simulation results therefore represent the stress at the onset of yielding in a compression-shear test.

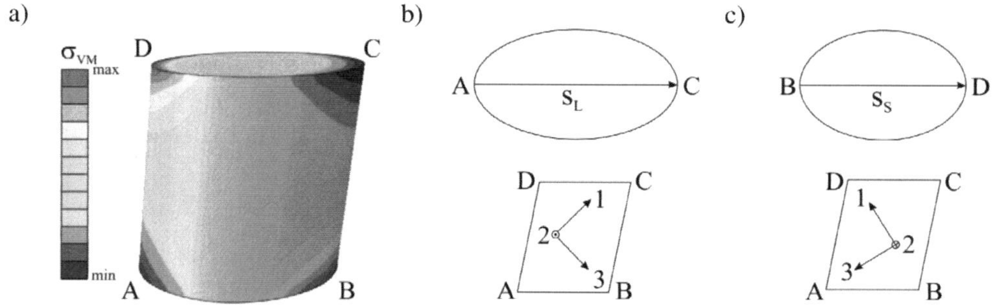

Figure 3. Equivalent von Mises stress distribution on the surface of the inclined compression-shear specimen (a). Long diagonal plane with path s_L and corresponding local coordinate system for shear stress evaluation (b). Short diagonal plane of the compression-shear specimen, path s_S and corresponding local coordinate system (c).

Figure 3a qualitatively shows a representative result valid for all inclined compression-shear specimens (for general validity, the color scale does not indicate distinct stress values since these are a function of the material parameters used in each simulation). The maximum von Mises stress occurs in the edges of the short diagonal plane, whereas minimum von Mises stress values are observed in the edges of the long diagonal plane. Obviously, both planes are interesting for further analysis of the stress state since they represent the limiting cases for the stress states throughout the entire specimen. Figures 3b and 3c schematically show top-down views on the long and short diagonal planes of elliptic shape, with characteristic paths along the long axes of the ellipses s_L and s_S, respectively. Note that s_L (path from point A to C) is longer compared to s_S (point B to D) because of the specimen inclination. The corresponding local coordinate systems are oriented (lower parts of figures 3b and 3c) with respect to the respective paths: 1-directions are parallel to the direction of the investigated path, 3-directions are perpendicular to the path and the 2-directions are oriented following the conventions for right-hand Cartesian coordinate systems.

5. Results and Discussion

In our FE simulations shear stresses τ_{31} were calculated along the two paths (s_L and s_S) of each long and short diagonal plane. Different simulations were performed to investigate the effect of four parameters (see table 1): specimen inclination angle, aspect ratio, friction coefficient between specimen and the tool plates and yield strength of the material. In the following discussion, path lengths are normalized and shear stresses are always related to the corresponding local coordinate system (as defined in figure 3).

5.1 Inclination angle and specimen aspect ratio

Loading of the compression-shear specimen leads to a combined stress state of compression and shear. The inclination angle as well as the specimen aspect ratio control the geometry and therefore also affect the imposed stress state. Figure 4 shows how the shear stresses τ_{31} is varied along the paths in the long and short diagonal planes for different angles of inclination.

It is well-known that in a conventional compression specimen (dotted line) the highest shear stresses occur in planes that are inclined by 45° relative to the longitudinal axis. In this case, the shear stresses are equal to half of the applied normal stress. In the simulations considered here, the normal stress was limited by the yield strength of the material (490 MPa). As a consequence, the shear stress is constant with τ_{31} = 245 MPa along both paths. A specimen inclination of 3° (dashed lines in figure 4) results in a decrease of shear stress along the long diagonal path (figure 4a). Along path s_L shear stresses are low (148.3 MPa) at the edges and increase towards their maximum (198.1 MPa) at the center of the specimen. A similar distribution of shear stresses is observed for 6° of inclination (solid line), with a maximum of 163.3 MPa in the center and a minimum of 87.7 MPa at the edges. A different shear stress distribution is observed in the short diagonal plane (figure 4b) along path s_S: At the edges, high shear stresses of 250.1 MPa (3° inclination) and 251.7 MPa (6° inclination) arise due to the geometry of the specimen – these stresses are even somewhat higher than the constant shear stress in a conventional compression specimen. These results show that Mohr's circle approach cannot be used to estimate shear stresses, particularly at the edges, of compression-shear specimens. Towards the center of the specimen the shear stresses decrease to minima of 198.0 MPa (3° inclination) and 162.9 MPa (6° inclination), respectively. There is a much higher gradient of the shear stress for the 6° inclination, although both compression-shear specimens exhibit nearly similar shear stresses at the edges of the short diagonal path. We highlight that particularly in less ductile or brittle materials, the formation of cracks, and thus the onset of shear failure, typically occurs at the edges of the short diagonal plane – our results demonstrate that this is a natural consequence of the geometry-imposed increase of stresses in these regions. The shear stresses distributions along both paths are symmetric, with a local maximum/ minimum at the center of the specimen. These shear stress distributions are characteristic for the compression-shear specimen. Clearly, calculating a simple ratio of shear to compressive stresses using the analytical model (see equation 1) cannot accurately represent such stress distributions even though our FE simulations are still limited to the stress states at the onset of plastic deformation only.

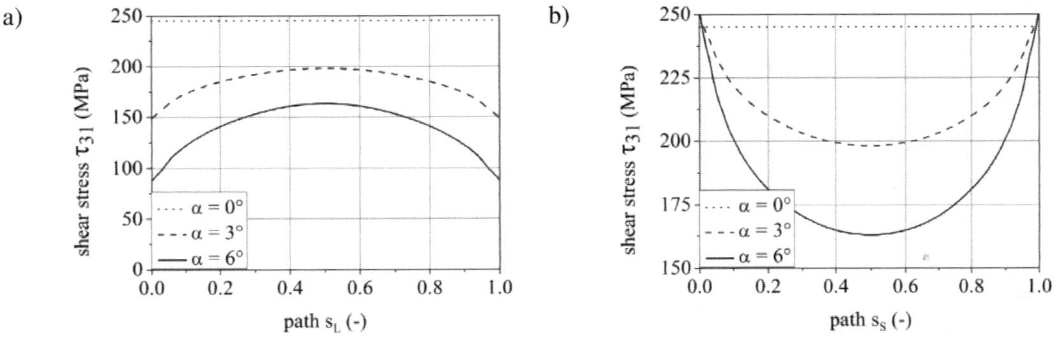

Figure 4. Shear stresses τ_{31} along paths s_L and s_S for a non-inclined compression specimen with inclination 0° (dotted lines) and for two compression-shear specimens with inclinations of 3° (dashed lines) and 6° (solid lines). Aspect ratio h/d: 1, no friction.

An increase of the inclination angle results in an overall decrease of shear stresses along s_L, again with highest shear stresses at the center of the specimen. In contrast, the highest shear stresses arise at the edges of the short diagonal path s_S and decrease towards the center. Moreover, an increasing inclination angle results in an increase of the shear stress gradient. Both paths intersect at the center of the compression-shear specimen, but the shear stresses at this point are not equal. Shear stresses referring to the local coordinate system of the long diagonal plane are slightly higher (198.1 MPa and 163.3 MPa) compared to the shear stresses referring to the local coordinate system of the short diagonal plane (198.0 MPa and 162.9 MPa). This is, of course, in line with Mohr's circle: the tilt angles β_L (46.5° and 47.9°) are closer to 45° (highest shear stresses expected for small inclinations and negligible friction) compared to β_S (43.4° and 41.8°, see figure 2).

Besides the inclination angle, the specimen aspect ratio controls the orientation of long and short diagonal planes and therefore affects the distribution of shear stresses. Figure 5 shows the shear stresses τ_{31} along both long and short diagonal paths for specimen aspect ratios (h/d) of 1 (solid line), 1.25 (dashed line) and 1.5 (dotted line). A similar progress of shear stress curves is visible compared to the inclination angle parameter study along both paths. With increasing specimen aspect ratio, shear stress curves are shifted towards smaller values. This can be documented regarding the maxima and minima at the center of the specimen (intersection of paths s_L and s_S). For the long diagonal path (figure 5a) maxima are shifted from 163.3 MPa (h/d = 1) to 151.9 MPa (h/d = 1.25) and 139.5 MPa (h/d = 1.5); similarly, minima of τ_{31} on the short diagonal path (figure 5b) change from 162.9 MPa to 137.7 MPa to 115.0 MPa with increasing h/d. Again shear stresses at the center are higher for the long diagonal path, and this observation is in line with Mohr's circle and the corresponding tilt angles (figure 2).

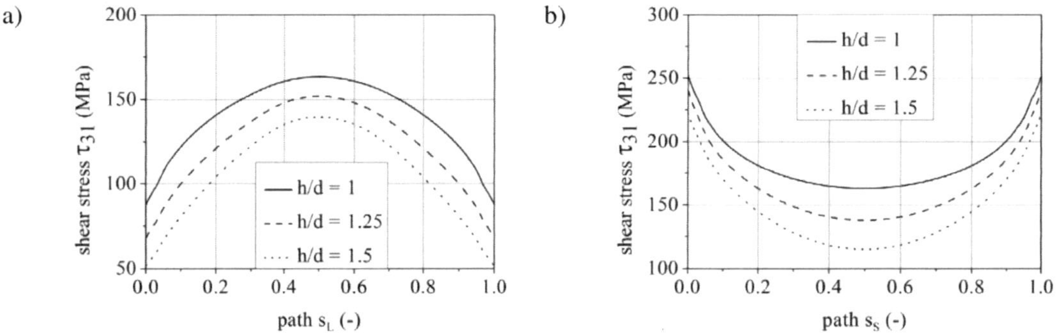

Figure 5. Shear stresses τ_{31} corresponding to the local coordinate systems within long and short diagonal planes for different specimen aspect ratios (h/d), a constant inclination angle of 6° and no friction. Solid lines refer to the specimen aspect ratio of 1, dashed and dotted lines to 1.25 and 1.5, respectively.

5.2 Friction

Friction between the tool plates and the specimen surfaces was neglected in all numerical investigations discussed above. In the FE analysis considered next, the tool plates have been defined as rigid surfaces with hard contact in normal and a constant Coulomb friction coefficient μ in tangential direction. Figure 6 shows the shear stresses τ_{31} along the two paths s_L and s_S within the long and short diagonal planes for different coefficients and a constant inclination angle of 6° as well as a specimen aspect ratio of 1. Again, shear stresses correspond to the local coordinate system of each plane (figure 3).

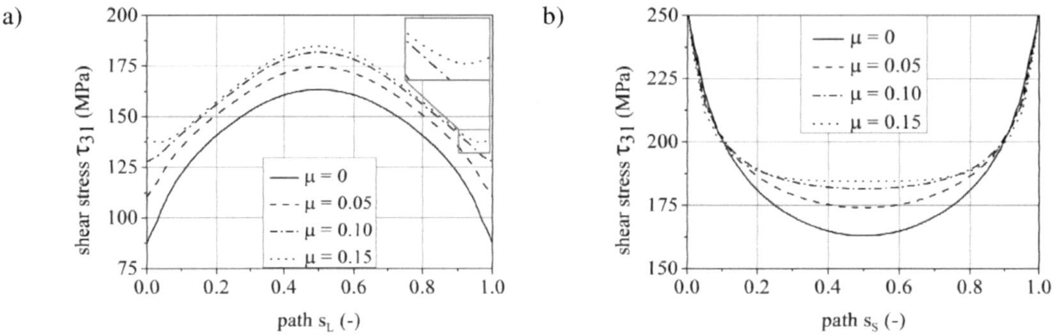

Figure 6. Shear stresses τ_{31} according to the local coordinate system of long and short diagonal plane along the corresponding paths. Results with different friction coefficients in tangential direction $\mu = 0$ (solid line), $\mu = 0.05$ (dashed line), $\mu = 0.10$ (dash-dotted line) and $\mu = 0.15$ (dotted line) are shown.

A similar trend of the curves compared to the results discussed in subsection 5.1 is observed. An increase of the friction coefficient ($\mu = 0$, solid line; $\mu = 0.05$, dashed line; $\mu = 0.10$, dash-dotted line

IOP Conf. Series: Materials Science and Engineering **181** (2017) 012025 doi:10.1088/1757-899X/181/1/012025

and $\mu = 0.15$, dotted line) results in an increase of the absolute shear stress values because a higher normal stress is required to overcome tangential friction. Within the long diagonal path (figure 6a) shear stresses increase at the edges from 87.7 MPa ($\mu = 0$) to 139.7 MPa ($\mu = 0.15$). Note that there is a local minimum for the highest friction coefficient $\mu = 0.15$ (inset in figure 6a). At the center of the specimen, the shear stresses also increase from 163.3 MPa ($\mu = 0$) to 184.8 MPa ($\mu = 0.15$). Since the maximum values in the center for high friction coefficients are nearly similar, it is not expected that a further increase of the friction coefficient considerably changes the shear stresses near the center of the specimen. For the short diagonal path (figure 6b) the shear stresses are very similar at the edges of the path s_S. In the center region of the path (0.2 – 0.8), however, the curves differ: an increase of the friction coefficient results in an increase of shear stress values and the shear stress gradient decreases. Shear stresses change from 162.9 MPa ($\mu = 0$) to 184.6 MPa ($\mu = 0.15$) in the center. It is again expected that a further increase does not change the shear stress in the center of the specimen. These findings demonstrate that the friction coefficient has a strong impact on the shear stress gradient along both paths. Moreover, shear stresses are generally increased compared to a simple scenario where friction is neglected.

5.3 Yield strength

Figure 7 shows the shear stresses along both path s_L and s_S for different (constant) yield strengths σ_Y. A mild steel with $\sigma_Y = 235$ MPa (solid line) and two tempering steels C45 with $\sigma_Y = 490$ MPa (dashed line) and a 42CrMo4 with $\sigma_Y = 900$ MPa (dotted line) were considered as model materials.

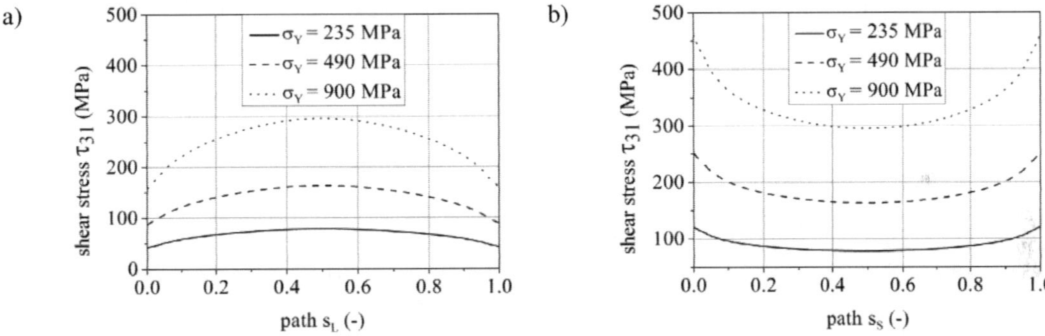

Figure 7. Shear stresses τ_{31} along the paths within the long and short diagonal plane for different yield strength values; solid line (σ_Y = 235 MPa), dashed line (σ_Y = 490 MPa), dotted line (σ_Y = 900 MPa).

Shear stresses along the long diagonal path (figure 7a) are still lower at the edges with a maximum in the center of the specimen. In contrast, shear stresses along the short diagonal path (figure 7b) are lowest at the center of the specimen and increase towards the edges of the path. An increase of the yield strength results in an increase of shear stresses. This is of course due to the need for higher normal stresses as our simulations were stopped when the yield strength was reached at any integration node in the specimen. We also find that an increase of yield strength results in an increase of the shear stress gradients along both paths.

6. Summary and conclusions

Based on simple 3D elastic-ideal plastic FE simulations, we performed an analysis of the stress state at the beginning of plastic deformation in the compression-shear specimen introduced by Meyer [5]. We systematically investigated the impact of specimen inclination angles, specimen aspect ratios, and friction coefficients between the tool plates and the specimen surfaces, as well as the effect of different yield strengths of the material. Since they play a key role in deforming the material in a compression-shear test, shear stress distributions were analyzed in detail taking local coordinate systems oriented parallel to short vs. long diagonal planes into account. The results provide detailed information on the stress state during early loading; they can be summarized as follows:

- During uniaxial, compressive loading of the compression-shear specimen, a geometry imposed combined stress state of superimposed compression and shear loading occurs. The highest equivalent von Mises stresses are observed at the edges of the short diagonal plane, whereas the lowest von Mises stresses occur at the edges of the long diagonal plane.
- In the center of the specimen, where both planes intersect, different shear stresses are determined in each plane. Shear stresses in the long diagonal plane are always higher compared to those in the short diagonal plane. This can be explained by the different tilt angles of these planes with respect to the loading axis.
- An increase of the inclination angle, and/ or an increase of the specimen aspect ratio results in an increase of shear stresses in both planes. Moreover, the shear stress gradients increase with increasing inclination angle.
- Larger amounts of friction result in higher shear stresses and lower gradients in the center region of the short diagonal path and also in the long diagonal path. Moreover, the yield strength of the material has no significant influence on the resulting shear stress distributions, but an increase of the yield strength leads to an increase of shear stress gradients along both paths.

References

[1] Rittel D, Lee S and Ravichandran G 2002 *Exp. Mech.* **42** 58
[2] Sang D-L, Fu R-D and Li Y-J 2016 *Mater. Charact.* **122** 154
[3] Moemeni S, Zarei-Hanzaki A, Abedi H R and Torabinejad V 2012 *Exp. Mech.* **52** 629
[4] Ashab A S M , Ruan D, Lu G and Wong Y C 2016 *Mater. Design* **97** 183
[5] Meyer L W, Staskewitsch E and Burblies A 1994 *Mech. Mater.* **17** 203
[6] Meyer L W and Pursche F 2011 *Engineering Transactions* **59** 21
[7] Elibol C and Wagner M F-X 2015 *Mat. Sci. Eng.* A **621** 76
[8] Elibol C and Wagner M F-X 2015 *Mat. Sci. Eng.* A **643** 194
[9] Meyer L W and Krüger L 2007 *ASM Handbook* vol 8 (Ohio: ASM International) pp 452–4
[10] Salençon J 2001 *Handbook of continuum mechanics* (Berlin: Springer) p 266

High temperature and dynamic testing of AHSS for an analytical description of the adiabatic cutting process

S Winter[1], F Schmitz[2], T Clausmeyer[2], A E Tekkaya[2] and M F-X Wagner[1]

[1]Institute of Materials Science and Engineering, Chemnitz University of Technology, 09125 Chemnitz, Germany
[2]Institute of Forming Technology and Lightweight Construction, Technical University of Dortmund, 44227 Dortmund, Germany

E-Mail: sven.winter@mb.tu-chemnitz.de, fabian.schmitz@iul.tu-dortmund.de

Abstract. In the automotive industry, advanced high strength steels (AHSS) are widely used as sheet part components to reduce weight, even though this leads to several challenges. The demand for high-quality shear cutting surfaces that do not require reworking can be fulfilled by adiabatic shear cutting: High strain rates and local temperatures lead to the formation of adiabatic shear bands (ASB). While this process is well suited to produce AHSS parts with excellent cutting surface quality, a fundamental understanding of the process is still missing today. In this study, compression tests in a Split-Hopkinson Pressure Bar with an initial strain rate of 1000 s^{-1} were performed in a temperature range between 200 °C and 1000 °C. The experimental results show that high strength steels with nearly the same mechanical properties at RT may possess a considerably different behavior at higher temperatures. The resulting microstructures after testing at different temperatures were analyzed by optical microscopy. The thermo-mechanical material behavior was then considered in an analytical model. To predict the local temperature increase that occurs during the adiabatic blanking process, experimentally determined flow curves were used. Furthermore, the influence of temperature evolution with respect to phase transformation is discussed. This study contributes to a more complete understanding of the relevant microstructural and thermo-mechanical mechanisms leading to the evolution of ASB during cutting of AHSS.

1. Introduction

The quality of the cutting edge of sheet metal parts manufactured by adiabatic cutting processes is excellent [1]. Moreover, adiabatic cutting processes have a great potential for the processing of advanced high strength steels (AHSS) [2]. A fundamental understanding of the process and the availability of efficient simulation strategies are necessary requirements to establish this technology for the industrial cutting of sheet materials. The demand for high-quality shear cutting surfaces that do not require reworking can be fulfilled by adiabatic shear cutting: High strain rates and local temperatures lead to the formation of adiabatic shear bands (ASB) [3 - 5]. This process is well suited to produce AHSS parts with excellent cutting surface quality, but both a fundamental understanding of the process and efficient simulation strategies are still missing today.

The key goal of this study is an analytical approach for the prediction of the relevant technological and microstructural processes. First, material characterization of two high strength steels is performed with a special focus on the temperature dependent and strain-rate dependent behavior. Based on this information an analytical model is then used to calculate the local temperature increase in shear bands for such high deformation rates. These results help to validate subsequent simulations for the adiabatic cutting process and increase the process understanding of adiabatic cutting itself. Finally, this provides the basic knowledge for manufacturing sheet metal parts with high quality functional faces and to estimate the quality of the cut surface. Furthermore, the experimental and numerical results help to determine process parameters for the adiabatic cutting of AHSS.

2. Experimental

2.1 Materials and experimental setup

For this study two high strength steels, obtained as sheet metals from different suppliers, were used. The first material was the manganese boron steel 20MnB5 provided by *BILSTEIN GmbH & Co. KG* (Hagen, Germany) in 4.7 mm thick sheets in a hardened condition. The second one was the carbon-spring steel C75S provided by *Risse + Wilke Kaltband GmbH & Co. KG* (Iserlohn, Germany) in 4.5 mm thick sheets in a hardened and tempered condition. The chemical composition was determined by spectroscopic analysis; the results are shown in Table 1. Both steels meet the required specifications of the given values.

Table 1. Chemical composition of the two used high strength steels, determined by spectroscopic analysis.

in wt.-%	C	Si	Mn	Cr	(ppm) B
20MnB5	0.20	0.18	1.20	0.21	19.5
C75S	0.72	0.29	0.74	0.32	2.3

Compression specimens with a diameter of 4 mm and a height of 6 mm were extracted out of the sheet metals. These specimens were used for all experiments discussed below. In order to estimate the anisotropy of mechanical behavior, the samples were taken out both in rolling (RD) and transversal directions (TD), figure 1. The initial microstructures (prior to mechanical testing) of both steels were analyzed by optical microscopy.

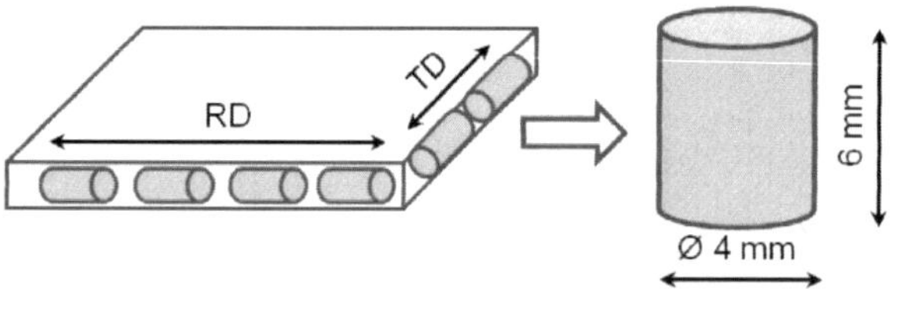

Figure 1. Schematic illustration: Extraction of the compression specimens out of the sheet metal in rolling and transversal direction. The samples have a diameter of 4 mm and a height of 6 mm.

For the characterization of the quasi-static stress-strain behavior, compression tests were performed at room temperature (RT) in a *UPM 1475 Zwick/Roell* universal testing machine with an initial strain rate of 10^{-3} s^{-1}. Due to friction at the faces of the samples, the experiments were stopped at 30 % applied (engineering) strain. The strain was measured with strain gauges applied on the specimens and using an additional inductive distance measuring system. Furthermore, measurements of the Rockwell hardness (HRC) were performed in a *Mitutoyo HTM ARK600* Rockwell hardness tester. The indentation time was 15 s and the force was 150 kp (1471 N).

The main part of the experimental investigations was the dynamic characterization of the steels with a strain rate up to 1000 s^{-1} in a temperature range from RT to 1000 °C. For these experiments, a Split Hopkinson Pressure Bar (SHPB) was used (setup based on the original design of Kolsky [6], figure 2). The SHPB consists of four bars: striker, incident, transmitter and damping bar. During the experiment, the striker bar is accelerated by compressed air and then hits the incident bar that subsequently deforms the specimen. Force data is determined from the strain gauge data measured at four locations on the incident and transmitter bars. The deformation of the specimen is documented with a high-speed camera, and an optical digital image correlation system (Aramis by *GOM*) is used to determine surface strain fields. For the RT experiments, this technique tracks the displacement fields of a speckle pattern on the specimen surface. At higher temperatures, the pattern was applied on the end of the incident bar and the beginning of the transmitter bar in order to allow for an indirect optical measurement of global strain in the specimens.

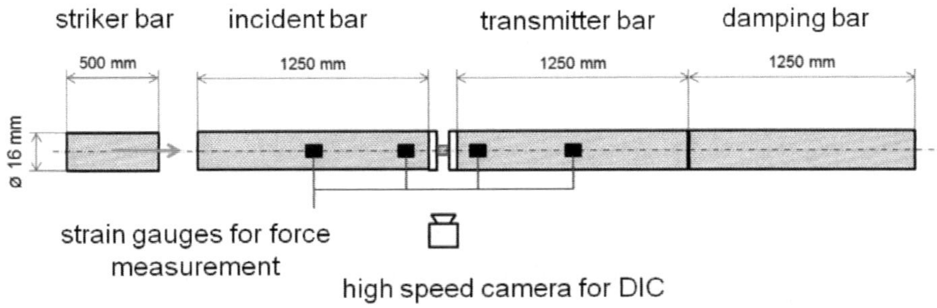

Figure 2. Schematic illustration of the Split Hopkinson Pressure Bar (SHPB) setup used in this study: The SHPB consists of four bars. The force is measured by considering the signals from four strain gauges. The deformation of the specimen is measured with a high-speed camera and analyzed using an optical digital image correlation system.

To perform the compression tests at high strain rates and at temperatures up to 1000 °C, local induction heating was used in the SHPB. For temperature isolation, two ceramic plates were applied between the sample and the bars of the SHPB (figure 3). The experiments at elevated and high temperatures were performed in steps of 200 °C from RT up to 1000 °C. Heating up to the testing temperature was achieved in 5 s, followed by the triggering of the striker bar. The temperature measurement was realized using a pyrometer focused on the specimen. After the dynamic tests, the microstructures of both steels were again analyzed by optical microscopy.

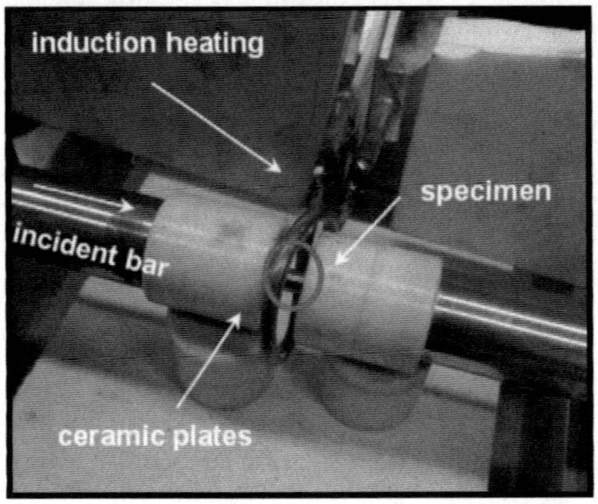

Figure 3. Setup in the SHPB for experiments at high temperatures up to 1000 °C. Two ceramic plates are applied for temperature isolation between the specimen and the incident and transmitter bar, respectively. Heating is realized by local induction heating.

2.2 Analytical description

The aim of the analytical model is to enable the analysis of local process and material effects which cannot be directly determined by experiments. The formation and growth of ASB can be used for technological processes like shear cutting. In order to define a feasible set of process parameters, the material response has to be known. Since both materials considered in this study possess temperature dependent flow stresses, the locally generated heat and thus the temperature is a crucial factor for the process. Due to the locality of the adiabatic shear band (~40 μm) and short process times (~1 ms), the temperature in the area of interest cannot be measured for a nonstationary process like shear cutting. Here, the area of interest is also difficult to access with a pyrometer and it is thus challenging to directly measure the temperature. Therefore, the focus of the analytical model will be the prediction of the temperature in this region.

Zener and Hollomon [7] observed that the yield strength and the heat capacity each have an influence on the adiabatic shear behavior. In this work, only the influence of the yield strength is taken into account, which underlines the necessity for experimental information on the flow stress evolution of the materials at high strain rates (~10^3 s^{-1}), see section 2.

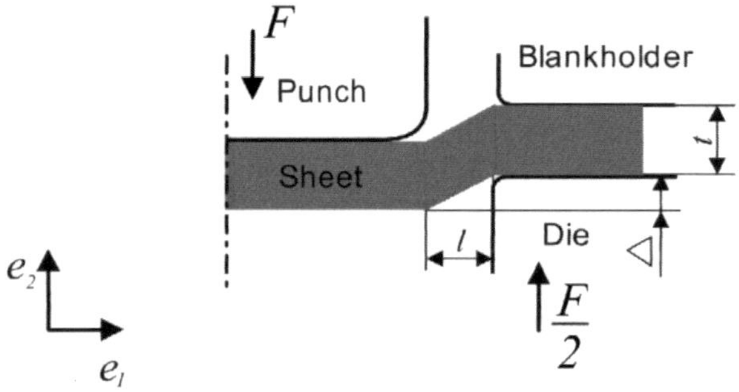

Figure 4. Schematic setup of the axisymmetric shear cutting process. Deformation only occurs in the shear zone (red) with the characteristic length l and shear amplitude Δ.

Finding the link between plastic deformation and temperature evolution is based on the internal dissipation equation. The internal dissipation due to plastic power $\dot{W}$ is coupled with the increase of temperature in the following manner:

$$\dot{T}\, c_\mathrm{p}(T)\rho V = \dot{W} = \beta\, k_\mathrm{f}(T)\, \dot{\bar{\varepsilon}}^{\,p} V, \qquad (1)$$

where c_p and ρ are the specific heat capacity and density, respectively. It is assumed that the deformation only takes place in the volume V (see figure 4, shown in red). Simultaneously, the temperature is assumed to be homogeneous and constant in this area. Since only a part of the plastic work is transformed into heat, the Taylor-Quinney factor β is utilized to incorporate this specific effect. Considering a typical amount of energy converted into heat, the parameter is set to $\beta \sim 0.9$ [8].

To further simplify the model, it is assumed that plastic deformation solely occurs in the shear zone whereas this zone itself exhibits a uniform deformation that can be described as an amplitude Δ, of simple shear deformation. This amplitude of deformation is normalized with respect to the shear width l, such that the shear deformation is given by $\frac{\Delta}{\mathrm{l}}$. Hence, the deformation gradient $\mathbf{F} = \frac{\partial x}{\partial X}$ takes the form

$$\mathbf{F} = \begin{pmatrix} 1 & \gamma & 0 \\ 0 & 1 & 0 \\ 0 & 0 & 1 \end{pmatrix}. \qquad (2)$$

Based on **F,** and the right Cauchy-Green tensor $\mathbf{C} = \mathbf{F}'\mathbf{F}$, the principal stretches λ_i and equivalent plastic strain $\bar{\varepsilon}^p = \sqrt{\frac{2}{3}\sum_{i=1}^{3}\ln(\lambda_i^2)}$ under the assumption of von Mises plasticity can be calculated. It should be noted that for such high deformations, particularly this formulation of equivalent strain represents a feasible measure [9]. The width of the adiabatic shear band (~ 40 μm) was determined by experiments analyzing adiabatic blanking of an axisymmetric coin with diameter of 30 mm. The used materials and thickness corresponds to those given in section 2. Figure 4 shows a schematic sketch of plastic deformation. It is assumed that the crack is initialized after a stroke of 25% of the material thickness t, which corresponds to $\bar{\varepsilon}^{p} = 3.8543 = \bar{\varepsilon}^{p}_{crack}$.

The temperature at $\bar{\varepsilon}^{p}_{crack}$ can be calculated using the differential equation (1), which was solved by means of the explicit Euler forward integration scheme. Room temperature $T_0 = 30°C$ and a virgin material with no plastic deformation $\bar{\varepsilon}^{p} = 0$ was considered as initial condition, while the plastic strain rate was assumed to be constant with $\dot{\bar{\varepsilon}}^{p} = 10^3\, s^{-1}$. For all calculations, the temperature dependent tabulated heat capacity of a commercial software, JMatPro, was used. For a first verification, the heat capacity at room temperature was measured. Experimental and numerical data were in good agreement, thus justifying the use of the JMatPro data ($c_p^{C75S} \sim 460$ [J/(kg K)]). Since the explicit integration needs a sufficiently small time step size, the equation was evaluated in incremental steps of $\Delta t = 10^{-6} s$. Parameter studies of the step size proved no further beneficial effect for shorter step sizes.

3. Results and discussion

3.1 Quasi-static characterization

First we consider the initial microstructure of the two steels. Figure 5a shows the expected fully martensitic microstructure of the manganese-boron steel after hardening. Large martensite plates with a length of 50 µm can be observed. In contrast, the microstructure of the carbon-spring steel consists of both bainite and martensite, which is indicated by many smaller plates in figure 5b.

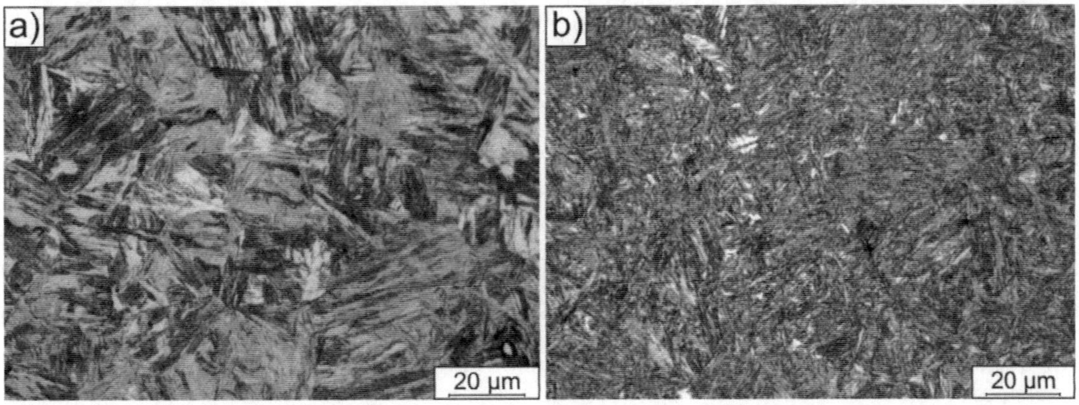

Figure 5. Initial microstructures of the two high strength steels, investigated by optical microscopy. a) Fully martensitic microstructure of 20MnB5. b) Bainitic and martensitic microstructure of C75S.

The Rockwell hardness of the 20MnB5 material was determined as 43 ± 0.5 HRC, in comparison to 44 ± 0.5 HRC for C75S. Both steels possess nearly the same hardness at RT in the as-delivered conditions. This trend is also evident in the quasi-static compression tests. The results for both materials, in rolling direction and in transversal direction, are plotted in figure 6. It is obvious that the manganese-boron steel has nearly the same yield strength (YS) as the carbon-spring steel. The initial yield strengths for both materials are approximately 1200 MPa and the Young's moduli are 212–213 GPa, respectively. Strain hardening in the manganese-boron steel up to engineering strains of 2 % is significantly higher, and then similar to that of the carbon-spring steel. Therefore, beyond strains of 2 %, the strength is about 200 MPa higher than the strength of the carbon-spring steel, which is the only difference in the stress-strain behavior between the two steels. We also observe that there is no significant anisotropy between the rolling and transversal directions: Both RD and TD curves are nearly equal for each material. In summary, these results show that both high strength steels exhibit a very similar stress-strain behavior at room temperature under quasi-static compressive loading.

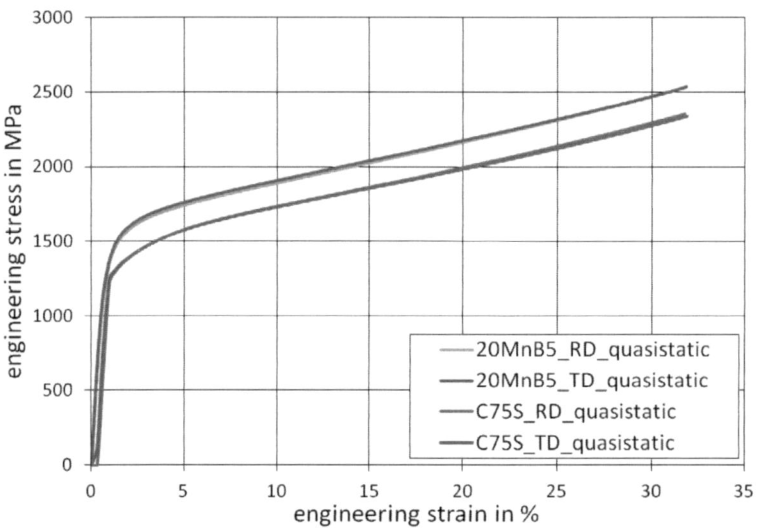

Figure 6. Engineering stress-strain behavior of 20MnB5 and C75S at RT and a strain rate of 10^{-3} s^{-1}. The compression tests were stopped at strains of 30 %. There is no significant anisotropy between rolling and transversal direction (RD and TD, respectively).

3.2 Dynamic characterization

The temperature dependent flow stress - true strain curves of the manganese-boron steel, measured at a strain rate of 1000 s^{-1}, are shown in figure 7. For comparison, the quasi-static stress-strain behavior at RT is also plotted in the diagram. A very high stress peak at the beginning of the curves in all dynamic tests can be observed. This peak does not result from the material behavior, but it is a measurement artifact related to the experimental setup: The incident bar, the ceramic plates for the temperature isolation and the small specimens have different diameters and elastic stiffnesses. This can lead to complex wave reflection phenomena [10], such that elastic waves are superimposed on the "true" measuring signal and thus form these pseudo stress peaks in all curves. While it is therefore not possible to accurately determine the critical stress for the onset of plastic flow (initial yield stress), yield strengths can be compared reasonably well in the range of larger strains (i.e., beyond the initial stress peaks).

As a first result, we observe that the strain rate sensitivity of 20MnB5 is not very high at RT: the stress strain curve measured during dynamic loading exhibits only a slightly increased strength (by about 80 MPa) level compared to the quasi-static one. This is a relatively small effect given that the strain rate is increased by 6 orders of magnitude. Furthermore strength significantly decreases and ductility increases with increasing temperatures up to 1000 °C. We also note that there is a significant strength drop between the stress strain curves measured at 400 °C and 600 °C, respectively. In this temperature range the strength loss is about 500 MPa and simultaneously the ductility is increased considerably. This observation may well have important consequences for the industrial cutting process because only a small amount of impact energy is needed to reach the thermal softening range between 400 °C and 600 °C in the 20MnB5 sheet metal. This softening then results in shear band initiation in the adiabatic cutting process, and can finally lead to a good cutting result (e.g., in terms of surface quality) of the steel.

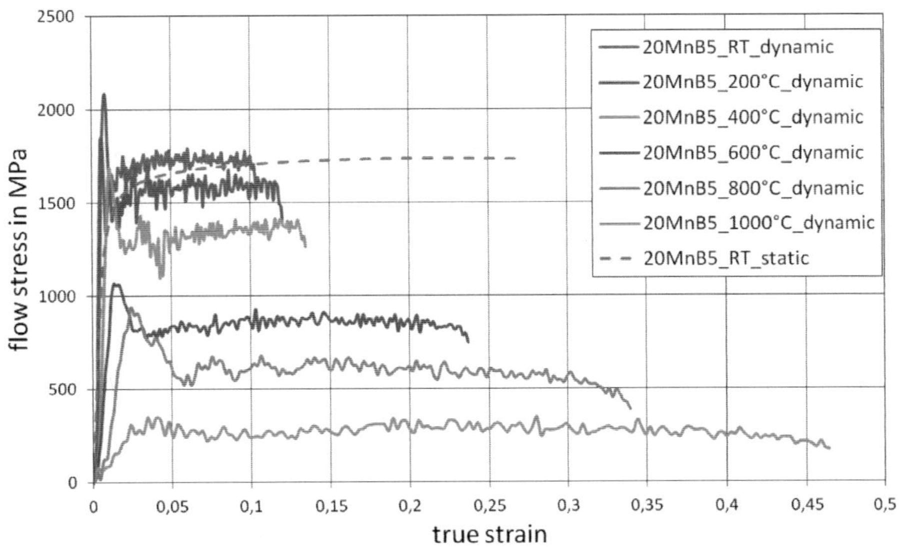

Figure 7. Stress-true strain behavior of 20MnB5 from RT up to 1000 °C (measured at a strain rate of 1000 s^{-1}). For comparison, the quasi-static material behavior at RT is also plotted (dashed line). Flow stresses significantly decrease and ductility increases with increasing temperatures. There is a significant strength drop between 400 °C and 600 °C.

To rationalize the strength drop observed in the temperature range between 400 °C and 600 °C, we analyzed the microstructure of the corresponding specimens after the dynamic experiments, figure 8. The specimen tested at 400 °C is characterized by large martensite plates. While, in comparison to the initial microstructure at RT (see figure 5), the plate size is slightly reduced, the material still exhibits a microstructure that is dominated by martensitic features. After heating to 600 °C, we also observe small martensite needles, but compared to the sample heated to 400 °C the ferrite phase volume fraction (brighter areas in figure 8) has been increased. This indicates that relaxation of the martensite phase by carbon diffusion has taken place. When the carbon content in the martensite is reduced, it transforms back to ferrite, which in turn leads to the lower strength observed in the corresponding mechanical test. Similar microstructural features of so-called high tempered martensite have been reported before, e.g. in [11, 12] for the manganese-boron steel 22MnB5, which has nearly the same chemical composition as the 20MnB5 steel used in the present study. These previous studies documented the temperature dependence of the yield strength from RT up to 450 °C. At RT, the 22MnB5 material had a yield strength of 1500 MPa; at 300 °C, it was decreased to 1400 MPa, and at 450 °C, the yield strength dropped significantly to 950 MPa. The authors of both studies rationalized this material behavior with tempering effects, particularly the transformation from martensite to ferrite; obviously, the new results reported here for 20MnB5 confirm these earlier observations.

The results of the dynamic testing of the carbon-spring steel are shown in figure 9. The grey stress-strain curve again represents the quasi-static behavior at RT. As a first result we observe that the strain rate sensitivity of the C75S material is higher than for the manganese-boron steel: under dynamic loading conditions at RT, the yield strength is higher by about 150 MPa compared to quasi-static testing. Similar to the manganese-boron steel, the flow stresses also significantly decrease (and ductility increases) with increasing temperatures. The carbon-spring steel also exhibits a significant strength drop, but this reduction of flow stresses only occurs between 600 °C and 800 °C. Comparing the two materials, obviously the necessary thermal softening for the initiation of a shear band is only reached at higher temperatures for the C75S materials. This results in a need for higher amounts of energy in the adiabatic cutting process to achieve similar cutting surface qualities in both materials.

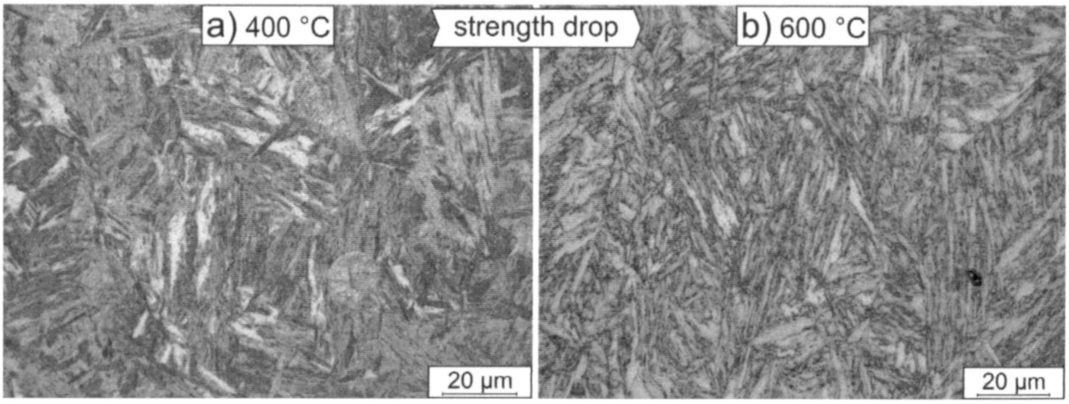

Figure 8. Microstructure of the 20MnB5 steel, investigated (after the experiments at increased temperatures in the SHPB) by optical microscopy. a) Predominantly martensitic microstructure after testing at 400 °C. b) Smaller martensite plates are observed after testing at 600 °C, and in comparison to the sample deformed at 400 °C, the amount of ferrite (brighter areas) is increased. This indicates relaxation of the martensite due to carbon diffusion, which results in a lower flow stress of the material at 600 °C.

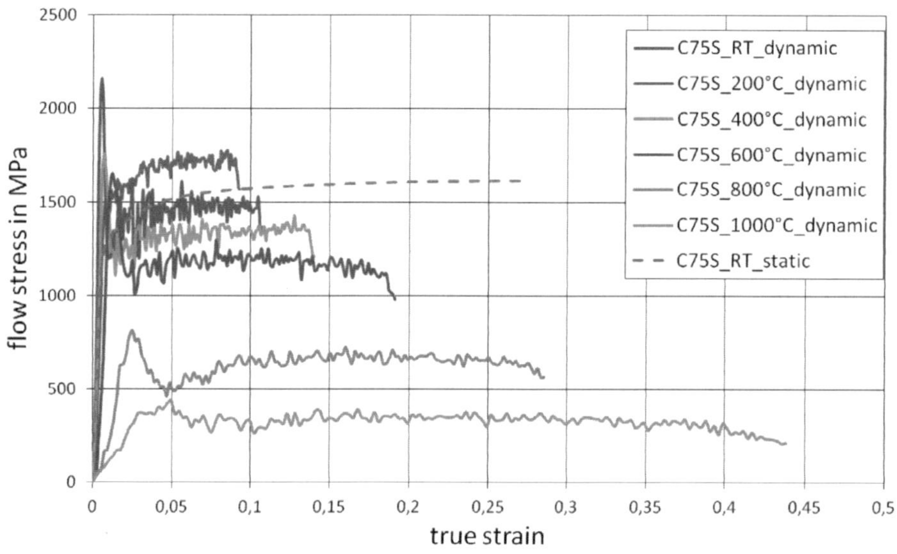

Figure 9. Flow stress-true strain behavior of C75S from RT up to 1000 °C and at a strain rate of 1000 s^{-1}. Strength significantly decreases and ductility increases with increasing temperatures up to 1000 °C. A strength drop occurs between 600 °C and 800 °C.

Figure 10 shows optical micrographs taken from samples that were deformed under dynamic loading conditions at 600 °C and 800 °C, respectively. The sample deformed at 600 °C (figure 10a) is characterized by a bainitic and martensitic microstructure, which is similar to the microstructure observed at RT. After dynamic testing at 800 °C, we instead observe a small amount of bainite, and no more martensite (figure 10b). The microstructure mainly consists of austenite (brighter areas) and hard pearlite (darker areas). This microstructure results from the onset of the phase transformation from α-Fe to γ-Fe followed by air-cooling after the experiment. The A_{c3} temperature of the carbon-spring steel is about 730 °C, and due to the carbon amount of 0.72 %, the steel is very close to the eutectic point (0.80 % C). At the testing temperature of 800 °C, a substantial volume of the initial bainitic and martensitic microstructure is transformed to γ-Fe, and this in turn leads to a significantly lower strength.

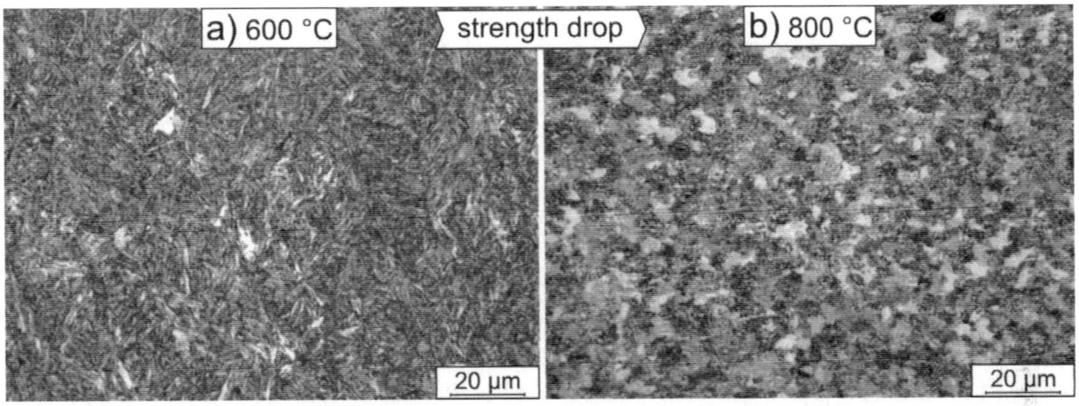

Figure 10. Microstructures of the C75S steel, investigated by optical microscopy after the dynamic experiments in the SHPB at different temperatures. a) Bainitic and martensitic microstructure after testing at 600 °C. b) After testing at 800 °C, the microstructure primarily consists of austenite (brighter areas) and hard pearlite (darker areas), and small amounts of residual bainite.

3.3 Evaluation of the analytical model

Given the information of the flow curves determined by experimental analysis, the next step is to incorporate the findings into the analytical approach presented in section 2.2. The analytical model is used to estimate the temperature evolution in the shear zone as described in section 2.2. For an appropriate comparison between the analytically calculated temperatures, the flow curves from the experiments first had to be modified in the sense that the initial pseudo stress peaks were excluded from the data sets. The reasoning to do so is given in section 3.2. Furthermore, based on a similar line of reasoning, the noise occurring in the experimental data was reduced using a simple filter algorithm. Since the solution of the differential equation (1) requires information about a continuous temperature set of flow curves, but the experiments were conducted at only six discrete temperatures, the remaining data was obtained by means of linear interpolation. With the appropriately processed data at hand, the differential equation for the temperature evolution could then be solved. In figure 11a, the temperature evolution is plotted as a function of the equivalent plastic strain for both materials, 20MnB5 (yellow) and C75S (blue). The final temperature between 766 °C (20MnB5) and 822 °C (C75S), which is predicted by the analytical model, is in good agreement with the literature [13]. As can be seen, the temperature increases monotonically with increasing strain. However, for C75S, based on the experimental data, a short plateau-like behavior can be observed at strains around $\bar{\varepsilon}^p = 3$. Due to their similar flow curves, the temperature evolution between both materials is in good agreement, because a similar amount of energy is dissipated during the deformation process. As pointed out in the analysis of the materials' flow curves, at a specific temperature the flow curves drop significantly. This temperature is different for the two materials (strength drop for 20MnB5 between 400-600 °C and for C75S between 600-800 °C). With this information the increasing difference with increasing equivalent plastic strain (starting at about $\bar{\varepsilon}^p = 1.5$) between the two temperature curves based on experimental data (yellow and blue) can be explained. The aforementioned plateau can, in contrast, be explained by a phase transformation of the material occurring at around 714 °C. The energy necessary for this type of transformation is considered by increasing the heat capacity at this temperature. The absorbed energy between $\bar{\varepsilon}^p = 2.5$ to 3 causes the temperature to remain constant until the material is fully transformed. Since afterwards the heat capacity returns to its actual value, the temperature gradient evolves again according to a decreasing flow stress.

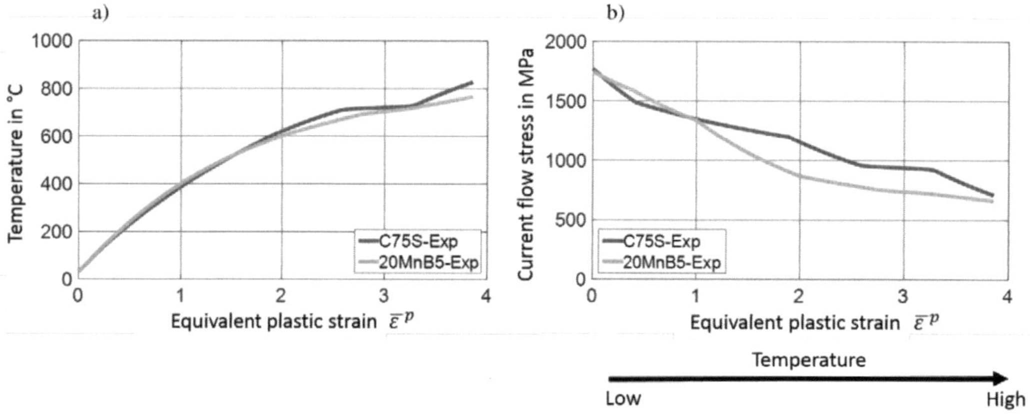

Figure 11. Analytical description of the sheared area by the use of experimentally and numerically determined flow curves. a) Monotonic increase of temperature; a plateau-like behavior can be observed when phase transformation takes place. b) Softening behavior due to the evolution of process temperatures.

A different approach, which is more appropriate for application purpose, is to look at the current flow stress as a function of the current, process-related equivalent plastic strain. This is illustrated in figure 11b. The seemingly softening behavior is a result not of such a material behavior itself but rather of the increasing temperature during the process, which has the effect of reducing the flow stresses of the (still hardening) material. It is therefore useful to think of the temperature evolution, as seen in the left figure, superimposed to the developing equivalent plastic strain in the right figure, for each material respectively. In order to assess the general work required for the entire cutting process, the areas below the curves are analyzed. This provides information about the deformation energy and, hence, the process forces if the distance covered by the tools during the process is known. Since the linkage between the left and right figure is given by the flow curves, see for instance figure 9 for C75S, the curvatures can be explained as follows. At the point, where the absolute value of the gradient increases significantly (for instance: C75S, blue curve, $\bar{\varepsilon}^p = 1.8$) the temperature is known from the left figure (600 °C). Compared with the flow curves, it relates to the experimental data point (temperature) at which the distinctive drop in flow stress due to the phase transformation occurs (flow curves: 600 to 800 °C). However, it should be noted that the linear interpolation between 600 °C ($\bar{\varepsilon}^p = 1.8$) and 800 °C ($\bar{\varepsilon}^p = 3.7$) causes the gradient to decrease prematurely, i.e. the temperature band in which the phase transformation and thus the significant flow stress drop occurs, is overestimated. Further investigation is required to determine more accurate flow curves in this particular temperature band of interest (C75S: 600-800 °C).

4. Summary and conclusions

The mechanical behavior of two high strength steels (20MnB5 and C75S) was characterized in dependence of temperature and strain rate. In addition to conventional quasi-static compression tests at room temperature, compression tests in a Split Hopkinson Pressure Bar with an initial strain rate of 1000 s^{-1} were performed in a temperature range between 200 °C and 1000 °C. The experimental results show that high strength steels with nearly the same mechanical properties at RT may possess a considerably different behavior at higher temperatures, which can lead to different cutting results in the adiabatic cutting process. The manganese-boron steel exhibits a significant strength drop between 400 °C and 600 °C. In contrast, the strength of the C75S steel decreases between 600 °C and 800 °C. This means that the necessary thermal softening for the initiation of a shear band over the whole cross section of a sheet will only start at considerably higher temperatures. This in turn results in a higher need of energy in the adiabatic cutting process for cutting the carbon-spring steel with a comparable cutting surface quality to the 20MnB5 steel. The strength drops observed in both materials can be directly related to microstructural changes in the corresponding temperature ranges. The analytical model presented here shows good agreements with the temperature increase during deformation given in the literature [13]. In order to do obtain a more detailed understanding of the thermo-mechanical behavior of both materials, further microstructural analyses may provide additional information on temperature gradients and grain sizes. Moreover, a more detailed data set (i.e., flow curves) particularly in the temperature range between 600 and 800°C is needed to identify the accurate temperature and amplitude where the strength drop occurs. This will also lead to an enhancement of the results of the analytical model and thus also improve the process simulation, which is necessary for the process design itself.

Acknowledgements

The authors would like to thank the "Arbeitsgemeinschaft industrieller Forschungsvereinigungen (AiF)", the Federal Ministry for Economic Affairs and Energy (BMWi) and the "Forschungsvereinigung Stahlanwendung e. V. (FOSTA)" for support of the research project "Material science-based development of simulation strategies for the application of adiabatic cutting in sheet metal part manufacturing", P 1127/28/2015 / IGF 18865 BG.

References

[1] Neugebauer R, Bouzakis K D, Denkena B, Klocke F, Sterzing A, Tekkaya A E and Wertheim R 2011 *CIRP Annals.-Manu. Tech.* **60 (2)** 627
[2] Lazzarotto L and Michon R 2008 *Tagungsband Chemnitzer Karosseriekolloquium CBC* **5** 242
[3] Meyers M A and Wittmann C L 1990 *Metall. Trans.* **21** 3153
[4] Bai Y and Dodd B 1992 *Theories and Applications* (Oxford: Pergamon Press)
[5] Odeshi A G and Bassim M N 2008 *Mater. Sci. Eng.* A **488** 235
[6] Kolsky H 1949 *Proc. Phys. Soc. B* **62** 676
[7] Zener C and Hollomon J H 1944 *J. Appl. Phys.* **15** 22
[8] Taylor G I and Quinney H 1934 *Proc. Royal Soc. London* **143** 307
[9] Onaka S 2012 *Mater. Trans.* **53** 1547
[10] Meyers M A 1994 *Dynamic Behavior of Materials* (New York: Wiley)
[11] Knezar K, Manzenreiter T, Faderl J and Radlmayr K M 2007 *Tagungsband Erlanger Workshop Warmblechumformung* **2** 93
[12] Laumann T and Pfestorf T 2007 *Tagungsband Erlanger Workshop Warmblechumformung* **2** 149
[13] Lins J F C 2007 *Mater. Sci. Eng.* A **457** 205

Investigation of the effect of contaminations and cleaning processes on the surface properties of brazing surfaces

K Bobzin[1], M Öte[1] and S Wiesner[1,2]

[1]Surface Engineering Institute, RWTH Aachen University, Aachen, Germany
[2]wiesner@iot.rwth-aachen.de

Abstract. The quality of brazed joints is determined by different factors such as the atmosphere and the parameters during brazing as well as the condition of the brazing surfaces. Residues of lubricants used during machining of the components and the subsequent cleaning processes can contaminate the faying surfaces and can hence influence the flow ability of the molten filler metals. Besides their influence on the filler metal flow, the residues can result in the formation of carbonic phases in the joint leading to a possible reduction of the corrosion resistance and the mechanical properties. The first step of the current study with the aim of avoiding these defects is to identify the influence of critical contaminations and cleaning methods on the quality of the brazed joints. In a first step, contaminations on AISI304 and Inconel alloy 625 due to different cooling lubricants and the effect of several cleaning methods, in particular plasma cleaning, have been investigated. Information about the surface energy of contaminated and cleaned surfaces was gained by measuring contact angle of testing fluids. Additionally, the lubricants and the resulting contamination products have been analyzed considering the influence of a heat treatment.

1. Introduction

Technical components processed by brazing are usually subjected to different machining processes prior to brazing. On the faying surfaces contaminations from the machining process such as machining oils and emulsions, swarfs, drawing greases, polishes and degradation products can be found [1, 2]. These contaminations are critical for the wetting behavior of the filler metal during the brazing process. If the surface is not sufficiently free from oxides and contaminations, the molten filler metal cannot wet the complete brazing surface. The result of the insufficient wetting is the formation of voids and porosity in the brazed joint, reducing its mechanical properties and also the corrosion resistant of the whole compound.

Therefore, the components to be brazed pass through a cleaning process consisting of different steps and different mechanical, thermal and chemical cleaning processes depending on the type of contamination. Basic information about possible cleaning processes can be found in DIN 8592 [3]. Often, chemical cleaning agents based on organic solvents or water-based cleaning agents are used in ultrasonic baths to dissociate the contamination particles from the surfaces and to transfer the energy from the cleaning bath in terms of heat or ultrasonic waves to the components. Additionally, treatments with cold atmospheric plasmas recently gain more significance for cleaning processes as this method can be applied for nearly all types of materials. As only small amounts of organic contaminations $m < 0,5$ g/m² [4] can be efficiently removed from the surfaces by means of cold atmospheric plasmas, the plasma treatment is conducted usually as the last step in the cleaning process resulting in very clean surfaces [5]. The carbonic contaminations on the surface are converted by means of the plasma into CO_2 and thus removed from the surface.

While the importance of the pre-cleaning of the components for the quality of the brazed joint is known, there is neither systematic information about the efficiency of the cleaning processes nor a recommended practice for an acceptable type and degree of contamination. Additionally, the behavior of the contaminations during the brazing process is unknown. Therefore, this study focuses on the investigation of the change of surface energy of the common base materials for brazing, X15CrNi18-10 and IN 625, with respect to the surface roughness and the cleaning process. The cleaning process includes pre-cleaning with chemical agents and a subsequent plasma treatment. Additionally, the behavior of contaminations from mineral oil-containing and synthetic lubricants is analyzed, considering the heat treatment conditions, which are comparable with those occurring during a brazing process.

2. Materials and Experimental Procedure

Two base materials, the stainless steel X5CrNi18-10 (1.4301, AISI 304) and the nickel-based super alloy INCONEL® 625 (IN 625) were used in this study. X5CrNi18-10 is an austenitic corrosion-resistant steel that has a wide application range for example in automotive industry and mechanical engineering. IN 625 is a corrosion-resistant nickel-based alloy that is used for components operating at elevated temperatures such as turbine blades due to its high oxidation resistance and good mechanical properties. Both materials are often brazed in industrial practice. The composition of the base materials is listed in table 1.

Table 1. Composition of the base materials X5CrNi18-10 and IN 625 in wt.-%.

	Fe	Ni	Cr	Mo	Mn	Si	C	P	S	N	Al
X5CrNi18-10	≥ 66,8	8-10,5	17-19,5	-	≤ 2	≤ 1	≤ 0,07	≤ 0,045	≤ 0,015	≤ 0,11	-
IN 625	≤ 5	≥ 58	20-23	8-10	≤ 0,5	≤ 0,5	≤ 0,1	≤ 0,015	≤ 0,015	≤ 0,4	≤ 0,4

For this study, substrates were cut from metal sheets made of X15CrNi18-10 and IN 625 in the dimensions of V = 20 mm x 25 mm x 5 mm. After cutting, the samples were milled or manually ground with SiC papers with grit sizes between 180 and 1,200 to generate a variety of the substrate roughnesses. The roughness of the sample surfaces was measured by a confocal laser scanning microscope of the type VKX 210, Keyence (Osaka, Japan) according to DIN EN ISO 25178 [6]. The area-related roughness values Sz and Sa were measured, corresponding with Rz and Ra from profile measurements. For each sample type, at least five measurements were used to calculate average roughness values.

In order to achieve samples with a surface contamination comparable to industrial practice, the cooling lubricants Avialub V3352 and Avia Metacool Syntogrind were used for milling the samples. The content of mineral oil in Avialub, which is a common lubricant for various machining operations, amounts to around 17 %. Avia Metacool, which is a synthetic, mineral oil-free lubricant, is mostly used for grinding and machining steel components. A mixture of 10 vol.-% lubricant and 90 vol.-% demineralized water was applied on the samples to create contaminated surfaces. As a more complex reference contamination, a variation of the Shell standard contamination [1] was used with the following composition:

- 35 wt.-% lanoline
- 25 wt.-% lithium stearate ($C_{18}H_{35}LiO_2$) as a typical compound of lithium soap used as lubricant grease
- 20 % corrosion protection liquid BRANOtect Export
- 10 % polydimethylsiloxane, a member of silicon oils used as lubricants
- 10 % Fe(III) oxide representing chips or shaving from the machining process.

The lubricants and the reference were applied on the milled and ground samples until a homogeneous layer of contamination forms on the surface. Subsequently, the samples were dried for $t_{drying} = 30$ min in ambient atmosphere. Some samples were analyzed in contaminated state and the rest subsequent to the respective cleaning operation.

As chemical cleaning agents Aceton and IBS-Spezialreiniger EL/Extra, referred to as "IBS" henceforth, were used in this study. IBS is a cleaning agent based on non-halogenated, isoparaffinic, aliphatic hydrocarbons. The cleaning process for the contaminated samples includes a cleaning step in an ultrasonic bath with the respective cleaning agent for $t_{cleaning} = 15$ min at $T_{cleaning} = 30$ °C, followed by a rinse cycle in demineralized water resulting in a homogeneous clean surface.

For plasma cleaning, CAT Compact device of Tigres GmbH (Marschacht, Germany) was used. This device generates an almost potential-free atmospheric plasma that can degrease, clean and activated metallic surfaces. The plasma is operated using compressed air as process gas at a net power of P = 600 W. The cleaning device was installed on an industrial robot head moving according to a programmed meandering profile with a velocity of v = 5 mm/s. Based on the plasma nozzle diameter of $\square_{nozzle} = 4$ mm, the distance between the meanders was chosen to be s = 4 mm. Thus, a homogenous treatment of the substrate surface could be ensured. The gap between the nozzle and the sample surface was about d = 5 mm. In order to achieve a stronger cleaning effect, the plasma treatment was conducted twice on the same surface.

The heat treatment of the samples was performed with a temperature-time profile comparable to the brazing process. It was realized in high vacuum with a pressure of $p < 5 \cdot 10^{-5}$ mbar in a PVA MOV 553 T furnace of PVA TePla AG (Wettenberg, Germany). The heating rate was $\dot{T} = 10$ K/min up to T = 1.100 °C for t = 10 min.

The cleanliness of the sample surfaces was at first roughly analyzed by the CleanoSpector from SITA Messtechnik GmbH (Dresden, Germany), a device measuring the fluorescence of surface contaminations. The substances on the surface are activated by ultraviolet light in order to bring these to exhibit fluorescence, which in turn is detected by the device and compared with the intensity of emissions of a standard clean surface. The result is given in percentage cleanliness compared to the cleanliness of the standard surface. A lower intensity of fluorescence compared to the clean surface results in a cleanliness value above 100 %. This measurement method is generally suitable for detecting organic and carbon-based substances, but has also proven to be efficient for perceiving mineral oil based contaminations [7, 8]. Due to the rapidity and facility of the measurement, it is used in industrial practice to evaluate roughly, if the cleanliness of the components delivered by the supplier is suitable for the following brazing process.

In order to determine the surface energy of the cooling lubricants and the sample surfaces, contact angle measurements were performed by means of a DSA 10 video device of KRÜSS GmbH (Hamburg, Germany). The surface energy was calculated using the translational angle of several test liquids that were dynamically measured. As test liquids, demineralized water, diiodomethane, ethylene glycol and glycerin were chosen due to their differing polar and disperse components of the surface energy [9, 10]. For every surface condition and every test liquid, at least five measurements were conducted.

The samples with fresh and heat-treated lubricants were analyzed by means of Micro-Raman spectroscopy. For this analysis, an inVia Reflex spectroscope by Renishaw GmbH (Pliezhausen, Germany) was used. The employed YAG laser had a wavelength of $\lambda = 532$ nm and the Raman shift in the range of $500\ \text{cm}^{-1} \leq \tilde{\upsilon} \leq 3{,}200\ \text{cm}^{-1}$ was determined. The scans were performed five times to achieve statistically reliable results.

Additionally, combined differential scanning calorimetric (DSC) and thermogravimetric (TG) measurements were conducted to analyze the behaviour of the lubricants during the heat treatment process. The focus of this analysis was mainly the investigation of the weight loss by evaporation or decomposition of compounds in the lubricants. For the DSC/TG measurements, the thermoanalytic system Setaram SetSys Evolution 16/18 of Setaram Instrumentation (Caluire, Frankreich) was used. Around m = 20 mg of the lubricants was placed in Al_2O_3 crucibles and heated up in an argon

IOP Conf. Series: Materials Science and Engineering **181** (2017) 012027 doi:10.1088/1757-899X/181/1/012027

atmosphere with a heating rate of $\dot{T}$ = 10 K/min up to temperature of T = 800 °C. The heat flow as a result of the DSC analysis is recorded and the weight change in % is calculated from a tared weight signal.

3. Results and Discussion

First, the influence of the roughness on the surface energy of the substrate materials was determined by contact angle measurements. Additionally, the effect of plasma treatment of contaminated and pre-cleaned surfaces on the cleanliness was determined. It can be expected that the heat treatment, for example during brazing process, significantly influences the chemical composition of the lubricants and the formation of residues on the sample surfaces. Hence, contaminated samples were heat treated and analyzed by combined DSC/TG measurements and Micro-Raman spectroscopy.

3.1 Influence of Surface Roughness on Surface Energy

The surface roughness of the machined samples was measured by means of the confocal laser scanning microscope and the resulting roughness values Sz and Sa are listed in table 2. The roughness values of the X5CrNi18-10 samples have been determined earlier [11]. It can be observed that the roughness of the surfaces of both materials increases from the finely ground surface to the milled surface. Compared with the roughness of the steel samples ranging between Sz = 33.6 µm - 43.6 µm, the roughness of the Ni-based samples is notably lower with values in the range of Sz = 5.1 - 7.1 µm. Also a high deviation from the mean Sz value can be observed for the IN 625 samples. While this deviation is already high for the ground samples, it is still significant after milling. The similar trend can also be observed for the Sa values. While the Sa values of the steel samples were on one level around Sa = 2.5 ± 0.3 µm [11], the roughness of the IN 625 samples ranges between Sa = 0.033 µm - 0.264 µm.

Table 2. Roughness values Sz and Sa of the milled and ground IN 625 and X15CrNi18-10 samples.

		ground 1,200	ground 180	milled
IN 625	Sz	7.5 ± 1.3 µm	6.6 ± 1.6 µm	5.1 ± 1.9 µm
	Sa	0.033 ± 0.003 µm	0.064 ± 0.005 µm	0.264 ± 0.024 µm
X5CrNi18-10 [11]	Sz	33.6 ± 2.1 µm	38.6 ± 4.0 µm	43.6 ± 11.6 µm
	Sa	2.6 ± 0.3 µm	2.3 ± 0.2 µm	2.6 ± 0.5 µm

Figure 1 shows the surface energy of milled and ground X15CrNi18-10 and IN 625 samples calculated from contact angle measurements. As shown before [11], the surface energy of all steel samples amounts to around σ = 40.8 ± 1.4 mN/m with nearly constant disperse and polar shares of $\sigma_{disperse}$ = 26.1 ± 1.8 mN/m and σ_{polar} = 14.4 ± 2.4 mN/m in the analyzed range of surface roughness. In contrast, the surface energy of the Inconel samples shows a notable dependence on the surface roughness. With increasing roughness, the surface energy decreases from around σ = 45.9 ± 0.7 mN/m for the finely ground sample to σ = 35.9 ± 0.7 mN/m for the milled sample. While the amount of the disperse share increases by around 34 %, comparing the finely ground sample to the milled sample, the amount of polar share decreases more significantly by around 90 % to σ_{polar} = 1.77 ± 0.7 mN/m for the milled surface. A possible explanation for this behavior could be that the apparent contact angle θ_a of a liquid changes when applied on a finely structured or very rough surface as explained by Wenzel in [12]. He proposes that a roughness factor r has to be taken into consideration when determining the actual contact angle of the liquid on a finely structured surface. The use of the actual contact angle θ comprising the roughness correction instead of the apparent contact angle θ_a can lead to significant changes in the calculation of the surface energy. The roughness factor is the ratio of the actual surface and the geometric surface and is used as a proportionality factor in the relation $\cos \theta_a = r \cdot \cos \theta$ [12]. Additionally, in some cases for example the hydrophilic characteristics of the system increase with increasing r [13]. This would result in an increase of the disperse shares of the surface energy as it is observed here.

Additionally, further possible explanations have to be taken into consideration. Possibly the predominant direction of the grooves resulting from milling and grinding results in an heterogeneous formation of the surface profile leading to a distorted contact angle of the drop. Furthermore, the different level of thermal stress during milling compared to that occurring in grinding might influence the oxide layer formation on the surface, which would in turn influence the measured contact angle and thus, the apparent surface energy. Currently, the surfaces of the samples are analyzed to identify the relevant mechanism.

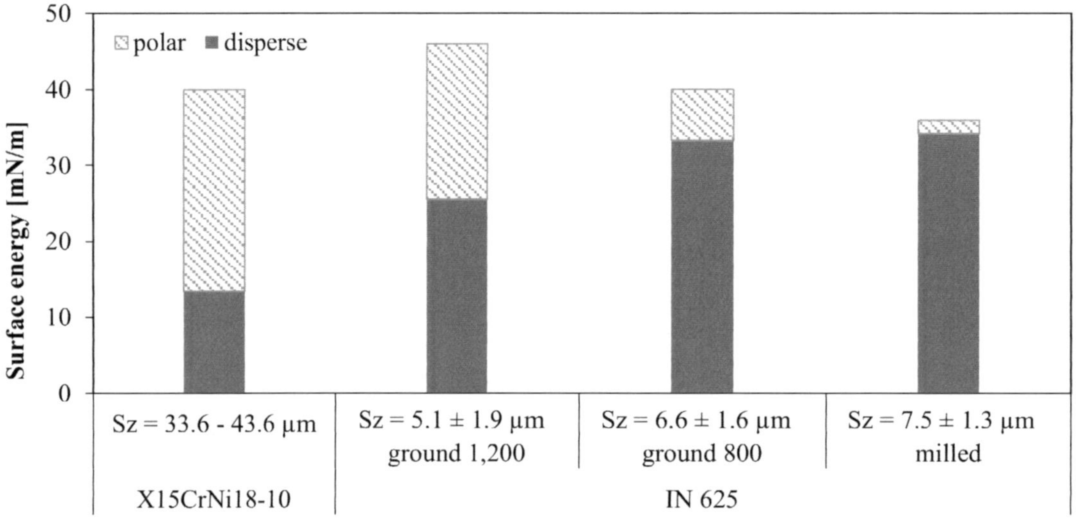

Figure 1. Surface Energy depending on the surface roughness of milled and ground X15CrNi18-10 [11] and IN 625 samples.

3.2 Effect of Plasma Treatment on Cleanliness and Surface Energy

Preliminary tests have shown that plasma treatment alone was not suitable to remove high amount of cooling lubricant residues from the sample surfaces. Using the parameters listed in chapter 2, resulting in short treatment times as known from industrial practice, no effect on the lubricant residues could be optically detected. Even with a high treatment duration of t = 30 s on a fixed position, dark brown, inhomogeneous residues on the surface can be observed. Although higher amounts of carbonic contaminations can be removed from surfaces by repeated plasma treatments, the process is most efficient on pre-cleaned surfaces [14].

Based on the results of the preliminary tests and also based on industrial practice, an alternative cleaning route was chosen. First, the contaminated samples were cleaned by means of the chemical cleaning agents Aceton and IBS respectively, using the cleaning process stated in chapter 2. Afterwards, the plasma treatment was performed. The cleanliness of the samples was determined by fluorescence analysis prior and subsequent to the plasma treatment. Additionally, the samples have been analyzed by Micro-Raman spectroscopy. Figure 2 shows exemplarily the results of the analyses conducted on X15CrNi18-10 samples.

Subsequent to the pre-cleaning with IBS, the measured cleanliness of the samples exhibits already high values of around 96 % (figure 2 a). By the additional plasma treatment, the cleanliness values increase up to 105 % indicating a higher cleanliness level compared to the untreated, clean steel substrate surface used as standard. Additionally, a decrease of the standard deviation of the cleanliness values can be observed leading to the assumption that the plasma treatment results in a surface that is more homogeneous concerning surface composition in comparison to the non-treated surface. Moreover, the Raman spectrograph in figure 2 b) shows the effect of the plasma treatment. The level of Raman intensity after pre-cleaning with IBS is already very low and no distinct peaks can be observed in the curve. After plasma treatment, the intensity level decreases further indicating a very clean metallic surface as most metals don't show characteristic Raman spectra. For a future brazing process, a clean surface is supposed to be favorable to enhance wetting and flowing of the molten filler metal.

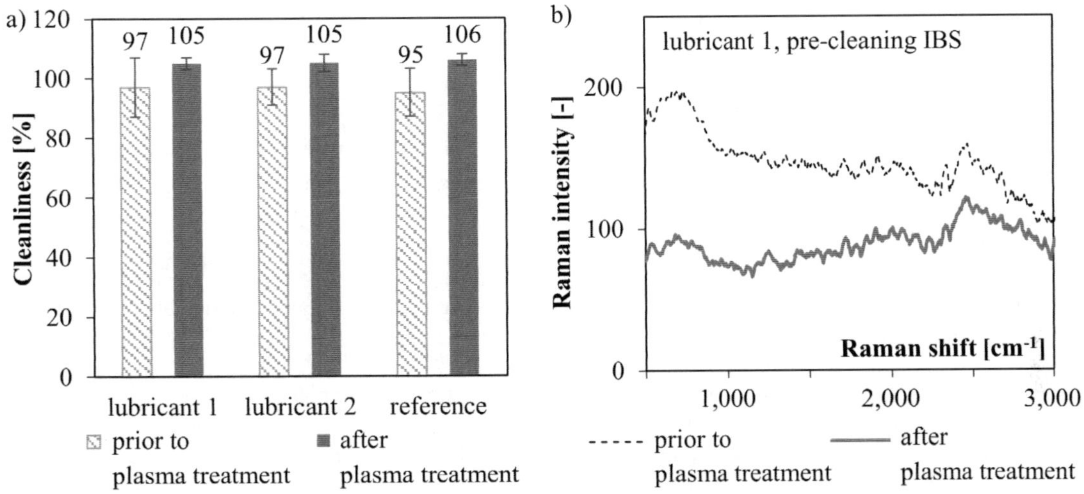

Figure 2. a) Results of fluorescence measurement on pre-cleaned and plasma treated X15CrNi18-10 samples; b) Raman spectrograph pre-cleaned and plasma treated X15CrNi18-10 samples.

In order to investigate the influence of cleanliness on the surface energy of the substrate surface, contact angle measurements were conducted on the pre-cleaned and plasma-treated samples subsequent to a contamination with lubricants. Figure 3 shows exemplarily the results of the IN 625 samples. The results of the plasma-treated samples were first compared with the milled, clean samples from figure 1 as their surfaces were also milled. It can be observed that the surface energy of all plasma-treated samples ranges between 45.7 mN/m $\leq \sigma \leq$ 59.1 mN/m, which is notably higher than that of the originally clean, milled samples with a value of σ = 35.9 mN/m. Furthermore, the amounts of disperse and polar shares differ significantly, as the plasma-treated samples exhibit a notably higher amount of polar shares. However, it can be observed that the level of surface energy and the distribution of the disperse and polar shares of the plasma-treated samples is similar to the clean, finely ground sample from figure 1 with a surface energy of around σ = 45.9 mN/m. This behaviour could probably be partly attributed to the influence of the roughness. As it can be expected that the residues of lubricants accumulate in the grooves and notches of the rough, milled substrate surface, it is very likely that the cavities are filled with residues even after pre-cleaning and plasma-treatment leading to a smoother surface. Thus, the roughness of the surface of the plasma-treated samples on which the contact angle measurements are performed resembles probably more the smooth surface of the finely ground samples. As the roughness is lower, the roughness factor r does not have to be considered for the evaluation of the contact angle measurements, as the apparent and the actual contact angle are similar. Consequently, the surface energy auf the plasma-treated samples is similar to the results of the finely ground sample.

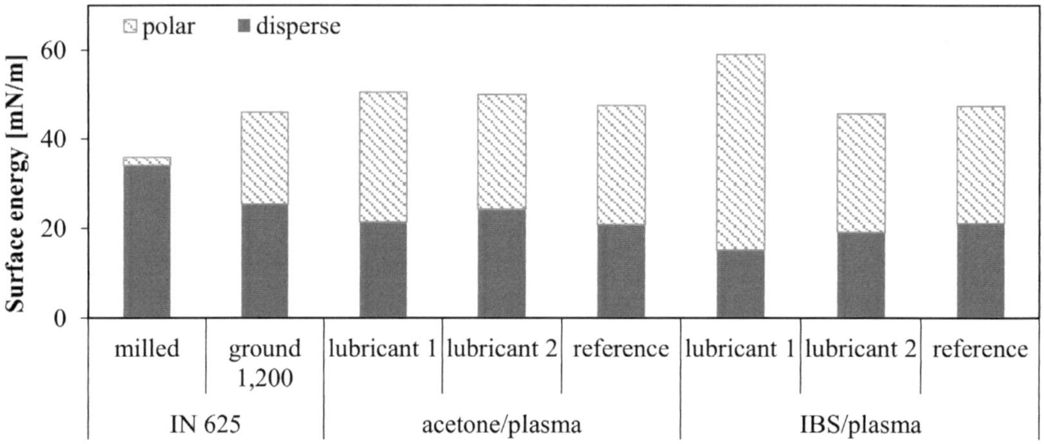

Figure 3. Surface energy of IN 625 samples in machined state as well as in cleaned state following different contaminations.

The disperse shares of the clean, finely ground sample and those of the plasma-treated surfaces are on a similar level of around $\sigma_{disperse} = 22.1 \pm 2.1$ mN /m. However, a higher level of the polar shares of the pre-cleaned and plasma-treated surfaces is reached with values around $\sigma_{polar} = 26.8 \pm 1.2$ mN/m compared to the finely ground surface with a disperse share of around $\sigma_{polar} = 20.5$ mN/m. This is possibly due to the polar residues of the contaminations on the surface based. The sample with the combination of lubricant 1 and IBS pre-cleaning shows the same trend, but more pronounced, with a very low level of disperse share of around $\sigma_{disperse} = 15.2$ mN/m and a high value for the polar share of the surface energy of around $\sigma_{polar} = 43.9$ mN/m.

Previous results on X15CrNi18-10 [11] showed a lower surface energy and a notable reduction of the polar shares after the cleaning process leading to the assumption that the plasma treatment influences the surface energy significantly as well as the amount of polar shares. This is analyzed further on clean samples that are additionally plasma-treated. Furthermore, the influence of the higher surface energy on the wetting and spreading behaviour of the molten filler metal is determined.

3.3 Characterization of the Influence of a Heat Treatment on the Lubricants

The video device for contact angle measurements was used to determine the surface tension of the lubricant. The surface tension of the lubricant 1 containing mineral oil amounts to $\gamma_{lub1} = 31.0 \pm 0.4$ mN/m at room temperature. The surface tension of the second, synthetic lubricants corresponds to $\gamma_{lub2} = 67.1 \pm 0.3$ mN/m. As the reference contamination is rather a paste than a liquid, no surface tension could be measured for it. Due to the relatively low surface tension of the first lubricant, it wets the surface homogeneously, even when mixed with the demineralized water. The surface tension of the second lubricant increases further by the addition of water as water has a surface tension of around $\gamma = 72.8$ mN/m at T = 20 °C [15]. As a result of the use of the mixture, only partial wetting of the surface is achieved on the sample surface.

Additionally, the lubricants and the reference were analyzed in combined DSC/TG analyses up to T = 800 °C. A high weight loss of 80 % for lubricant 1 and of 60 % for lubricant 2 can be observed in the temperature range between 85 °C ≤ T ≤ 115 °C (figure 4 a). This weight loss coincide with an endothermic peak in the heat flow curve in figure 4 b in the temperature range between 20 °C ≤ T ≤ 150 °C. The peak of lubricant 1 is divided in two peaks indicating two different reactions. The first reaction at around T = 100 °C is probably correlated with the evaporation of water and the second one with the evaporation of volatile components of the lubricants. The fact that the second lubricant doesn't exhibits the strong second peak leads to the lower weight loss. Between 115 °C ≤ T ≤ 400 °C the second lubricant shows another but less pronounced step of weight loss resulting in a total weight loss of 77 %. The reference contamination doesn't show neither weight loss nor heat flow peaks at lower temperatures. In the temperature range between 200 °C ≤ T ≤ 500 °C, several small endothermic peaks can be seen in the heat flow curve resulting in a pronounced decrease of weight of around 78 %. It can be concluded that after a heat treatment up to T = 800 ° C, around 10 % - 20 % of the original weight of the lubricants can be found on the surface. These residues can be observed on the sample surfaces in the form of dark, crumbly and voluminous material. Based on the structure and appearance of the residues, it can be assumed that some graphitic material has been formed.

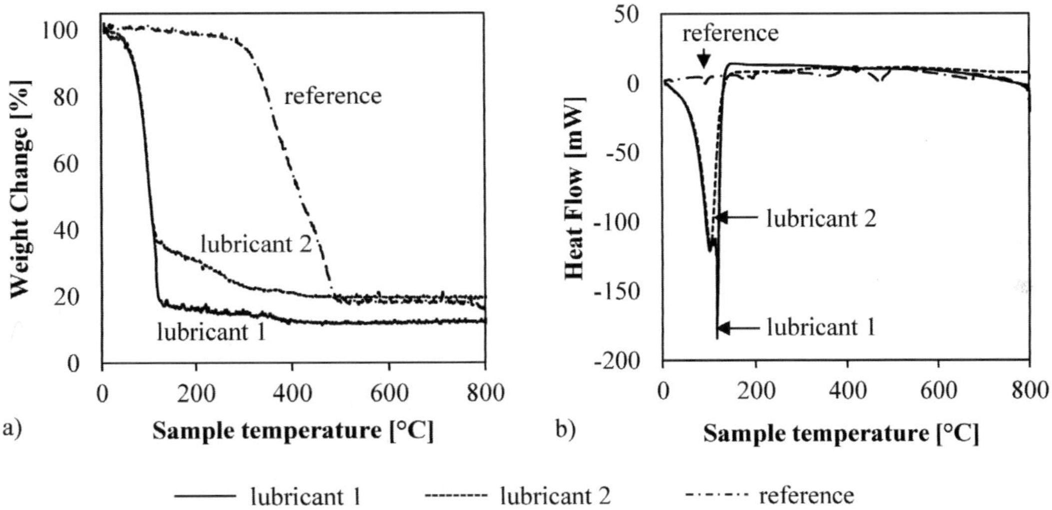

Figure 4. a) Weight change and b) heat flow during combined DSC/TG analyses of lubricants 1, 2 and reference in the temperature range 20 °C ≤ T ≤ 800 °C.

In order to identify the dark, crumbly residues resulting from the heat treatment of the lubricants, Micro-Raman measurements were done on fresh lubricants as well as on heat-treated samples showing the residues on the surface. Figure 5 shows exemplarily the results of the analyses of X15CrNi18-10 samples. Comparable results were observed for IN 625. This was expected because no intense chemical interaction should occur between lubricant and base material. In figure 5 a) hardly any peaks can be detected for the fresh lubricants possibly due to an amount of moisture in the material.

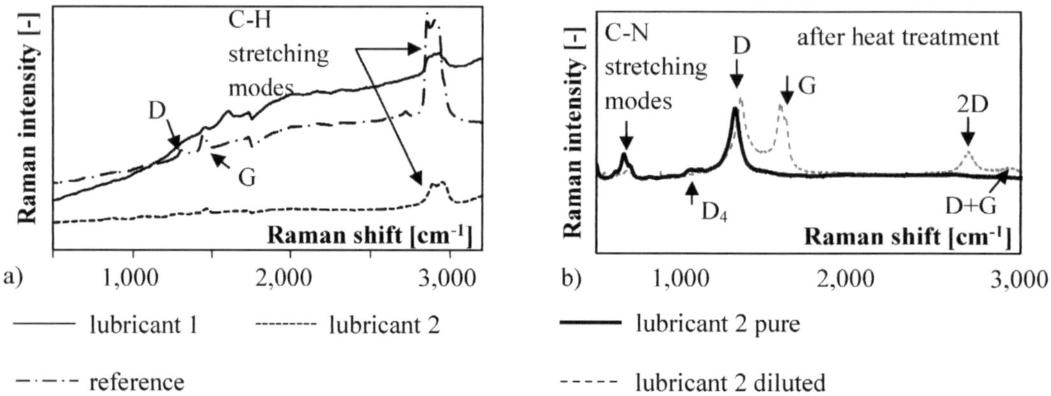

Figure 5. Raman spectrographs of a) fresh lubricant 1, 2 and reference b) heat-treated pure and diluted lubricant 2, both on X15CrNi18-10.

In the range of the Raman shift between $2{,}800\ cm^{-1} \le \tilde{\upsilon} \le 3{,}020\ cm^{-1}$, broad and flat peaks can be found for lubricant 1 and 2 and peaks with higher intensity for the reference. Peaks in this range are due to C-H stretching modes indicating the existence of partly long-chain hydrocarbons containing CH_2 or CH_3 groups [16]. Peaks at Raman shifts of $\tilde{\upsilon} > 3{,}000\ cm^{-1}$, can be assigned to complex nitrogen containing hydrocarbon groups [17]. Additionally, some small peaks can be seen at around $\tilde{\upsilon} = 1{,}367\ cm^{-1}$ and at around $\tilde{\upsilon} = 1{,}540\ cm^{-1}$ indicating sp^3 and sp^2 carbon bonds [18, 19]. Their intensity is comparatively low due to the high intensity of the peaks for the C-H stretching modes. The band correlated to sp^3 is shifted to a higher frequency compared to the diamond sp^3 band at $\tilde{\upsilon} = 1{,}332\ cm^{-1}$. Especially in graphene based material, peaks in the range of $1{,}330\ cm^{-1} \le \tilde{\upsilon} \le 1{,}360\ cm^{-1}$ are called the D band. This peak occurs when the graphene sheets are irregularily stacked in graphite, when defects at the edges or in the plane of the two dimensional graphene stacks exist. The second peak is correlated with the primary phonon G band, typically located at around $\tilde{\upsilon} = 1{,}580\ cm^{-1}$, indicating the carbon lattice stretching in the plane of the graphene sheets. A shift of this peak can be correlated with different interactions between the graphene layers and a change of the peak form can be attributed to a splitting of the peak due to several underlaying band with slightly different energies and frequencies [18].

Figure 5 b) shows the Raman spectrograph of pure and diluted lubricant 2 after heat treatment. It can be noted that the peaks of the spectrographs are relatively broad. This can be attributed to a highly defective structure of the graphene sheets, but also to the amount of sheets that are stacked. It could also be possible that the broading of the peaks is a result of amorphous carbon in the material. But as the cooling rate of the samples is low, the formation of a higher amount of amorphous material is not expected. The position of the peaks and also the relation of the intensities observed here is very similar to another study where the degree of crystallinity and number of the graphene layer increased with decreasing cooling rate leading to broad peaks [20].

While the spectrograph of the different contaminations is characterized only by some C-H stretching modes indicating the existence of long-chained hydrocarbons, more defined peaks can be seen after heat treatment. Especially in the spectrograph of the diluted lubricant, the D band at around $\tilde{\upsilon} = 1{,}360\ cm^{-1}$ and the G band with a doublet at around $\tilde{\upsilon}_1 = 1,\ 600\ cm^{-1}$ and around $\tilde{\upsilon}_2 = 1{,}620\ cm^{-1}$ can be seen. Additionally, the 2D(G') band can be observed at around $\tilde{\upsilon} = 2{,}720\ cm^{-1}$ due to secondary phonons resulting from the same defects as the D band. Furthermore, at higher Ramans shift of around $\tilde{\upsilon} = 2{,}900\ cm^{-1}$ a new band occurs known as the D + G band. It results from a double resonance scattering process and is also induced by the defects in the graphen structure. The spectrograph of the pure, heat-treated lubricants shows fewer and slightly different peaks. It shows only pronounced peak at around $\tilde{\upsilon} \leq 1{,}340\ cm^{-1}$, comparable to the D band of the pure lubricant but slightly shifted to lower frequencies. Furthermore, at around $\tilde{\upsilon} = 1{,}200\ cm^{-1}$, a broad but flat peak can be observed which is attributed to the D_4 band. This band occurs in highly defective carbons and have been related to hydrocarbons or residues of aliphatic compounds connected with the graphitic structure. A small peak at around $\tilde{\upsilon} = 700\ cm^{-1}$ can be attributed to C-N stretching modes [17].

Summarized it can be concluded that the long-chained hydrocarbons in the lubricants react during the heat treatment to graphene-based compounds, for example graphite, especially when the lubricant has been diluted with demineralized water. Due to the higher intensity of the D band, it can be concluded that the structure is highly defective possibly due to distortions during stacking or defects at the edges of the graphene sheets. A difference between the pure and diluted lubricant can be seen as the spectrograph of the pure lubricant shows more indication of residues of aliphatic hydrocarbons and of C-N bonds. These different kinds of residues on the sample surfaces can lead to different wetting behavior by the molten filler metal in a brazing process. It has been shown before that residues of the lubricants and cleaning agents on the surface lead to an enhanced spreading of the filler metal [11].

4. Conclusion and Outlook

The analyses of the sample surfaces by means of confocal laser scanning microscope showed an influence of the roughness of the IN 625 samples on their surface energy. The surface energy and especially the disperse share of the energy increased with increasing roughness. This could be attributed to the necessity to take into account the roughness factor developed by Wenzel for finely structured surfaces. Differences in the flowing behaviour of filler metals on these surfaces can be expected due to the enhanced spreading of the filler metal on substrates with residues of lubricants and cleaning agents shown before [11]. The flowing behaviour of the filler metal on heat-treated samples is currently analysed in detail.

Additionally, it could be seen that a pre-cleaning process combined with a plasma treatment of the samples could further increase the cleanliness of the surfaces. Thus, a reduction of voids and defects and inclusions of lubricant residues in the brazed joint can be expected enhancing the quality of the joints. Furthermore, the surface energy and especially its polar share are increased after the pre-cleaning and plasma treatment. The influence of this increase on the wettability and spreading of a molten filler metal is also currently determined.

The analyses of the behaviour of the lubricants show a strong reaction during a heat treatment similar to that occurring during the actual brazing processes. While a high amount of the former fresh lubricants, consisting mostly of long-chained hydrocarbons, evaporate during the heat treatment, around 20 wt.-% remain on the surface but mostly as a graphene-based compound, probably in the form of graphite. Raman spectrographs indicate that the graphite structure is disordered and defective leading to a voluminous and crumbly structure on the surface. The graphitic structure is a challenge for a brazing process as it is difficult to wet. Additionally, the crumbly structure will be included in the braze material leading to a reduction the mechanical properties. Also the difference between the residues of pure and diluted lubricants has to be further analyzed regarding the wettability of both kinds of residues by a molten filler metal.

Acknowledgement

The project IGF 18387 N of the research association German Welding Society (DVS) is funded by the AiF within the framework of the program to promotion of the IGF (Industrielle Gemeinschaftsforschung) by the German Federal Ministry for Economic Affairs and Energy according to a resolution of the German Bundestag. The authors would also like to thank the project partners from Institute of Materials Engineering for the good cooperation.

References

[1] Hertlein K 2000 *Reinigen mit Kohlenwasserstoff-Lösemitteln und Wasser im metallbearbeitenden Gewerbe, im Druck- und Elektronikbereich* ed W J Bartz and E Wippler (Renningen-Malmsheim: Expert-Verlag)
[2] Grün R 2005 *reprinting STAHL Heft* **2** 35–38
[3] DIN 5892:2003 *Manufacturing processes cleaning – classification, subdivision, terms and definitions*
[4] Kegel B and Schmid H 1999 *Surf. Coat. Technol.* 112(1) 63–66
[5] Adams H-N and Jelinek T W 1999 *Reinigen und Entfetten in der Metallindustrie,* ed T W Jelinek (Bad Saulgau: Eugen G. Leuze Verlag)
[6] DIN EN ISO 28178:2016 *Geometrical product specifications (GPS) - Surface texture: Areal*
[7] Oberhausen M 2007, Dissertation Universität Saarbrücken
[8] Schütz A and Triebert J 2008 *Metalloberfläche* **62** 29–31
[9] Lugscheider E and Bobzin K 2003 *Surf. Coat. Technol.* **165** 51–57
[10] Mantel M and Wightman J P 1994 *Surf. Interface Anal.* **21** 595–605
[11] Bobzin K, Öte M, Wiesner S, Tillmann W, Wojarski L, Manka M and Eilers A 2016 *Proc. of 11thInt. Conf. on Brazing, High Temperature Brazing and Diffusion Bonding LÖT 2016,* 7th - 9th June 2016 (Aachen) in *DVS Berichte* **325** (Düsseldorf: DVS Media GmbH), 128–132
[12] Wenzel R N 1936 *Ind. Eng. Chem.* **28** 988–994
[13] Quéré D 2005 *Rep. Prog. Phys.* **68** 2495-2532
[14] Saito H, Shiki H, Tsjii K, Oke S, Suda Y, Takikawa H and Amanaka S 2009 *Jpn. J. Appl. Phys., Part 1* **48.8S1** (2009) 08HH03
[15] Harkins W D and Brown F E 1919 *J. Am. Chem. Soc.* **4** 499–524
[16] Spiker R C, Levin J R and Levin I W 1975 *Biochim. Biophys. Acta* **338** 361–373
[17] Garber B P and Peticolas W L 1977 *Biochim. Biophys. Acta* **465** 260–274
[18] Hodkiewicz J 2010 *Thermo Scientific Application Note* 51946 (2010)
[19] Bokobza L, Brunel J-L and Couzi M 2015 *C* **1** 77–94
[20] Li C, Li D, Yang J, Zeng X and Yuan W 2011 *J. Nanomat.* 19624

Investigation on the cold rolling and structuring of cold sprayed copper-coated steel sheets

K Bobzin[1], M Öte[1], S Wiesner[1], L Gerdt[1,3], S Senge[2] and G Hirt[2]

[1]Surface Engineering Institute, RWTH Aachen University
[2]Institute of Metal Forming, RWTH Aachen University
[3]gerdt@iot.rwth-aachen.de

Abstract. A current driving force of research is lightweight design. One of the approaches to reduce the weight of a component without causing an overall stiffness decrease is the use of multi-material components. One of the main challenges of this approach is the low bonding strength between different materials. Focusing on steel-aluminum multi-material components, thermally sprayed copper coatings can come into use as a bonding agent between steel sheets and high pressure die cast aluminum to improve the bonding strength. This paper presents a combination of cold gas spraying of copper coatings and their subsequent structuring by rolling as surface pretreatment method of the steel inserts. Therefore, flat rolling experiments are performed with samples in "as sprayed" and heat treated conditions to determine the influence of the rolling process on the bond strength and the formability of the coating. Furthermore, the influence of the rolling on the roughness and the hardness of the coating was examined. In the next step, the coated surface was structured, to create a surface topology suited for a form closure connection in a subsequent high-pressure die casting process. No cracks were observed after the cold rolling process with a thickness reduction of up to ε = 14 % for heat treated samples. Structuring of heat treated samples could be realized without delamination and cracking.

1. Introduction

A current research project at the RWTH Aachen deals with the joining of molten aluminum and a steel sheet by high-pressure die casting. Due to the challenge that no sufficient metallurgical bonding between steel and aluminum occurs, thermal spraying processes shall be used as a pretreatment of steel sheets to overcome the issues occurring during the joining of steel and aluminum. Thermally sprayed coatings can come into use to reduce the formation of brittle intermetallic phases and to combine the advantages of metallurgical bonding with mechanical interlocking. In order to achieve higher bond strength, the thermally sprayed coating shall be structured. This requires a high formability of the coating.

Cold spraying as no other thermal spraying process allows the deposition of coatings with properties similar to those of bulk materials. As the particles of the deposited material remain solid during the spraying process, they are less susceptible to oxidation [1]. Coatings of ductile materials such as copper or aluminum applied by cold gas spraying exhibit a low porosity and oxide content. However, cold gas sprayed coatings have low ductility due to a strong work hardening effect occurring during deposition and are therefore difficult to deform plastically. Recently some studies have investigated the mechanical behavior of cold sprayed coatings indicating the possibility to improve the microstructure and mechanical properties by means of heat treatment. Renzhong et al. have investigated the mechanical properties of detached cold sprayed copper coatings under tensile loading in as sprayed condition and after heat treatment at different temperatures [2]. It was shown that the elongation of the copper coatings can be increased from A = 0.5 % up to A = 35 % after heat treatment at T = 500 °C for t = 5 h at the expense of the tensile strength which decreased from σ_M = 290 N/mm² to σ_M = 240 N/mm². Gärtner et al. compared in their study mechanical properties of cold sprayed, arc sprayed and high velocity oxygen fuel (HVOF) sprayed coatings in as sprayed and annealed conditions [3]. Taking hardness as an indicator for work hardening, the amount of deformation for cold sprayed copper corresponded to that of cold rolling of bulk material with thickness reduction of about ε = 90 %. It was also shown that the formability of cold sprayed coatings can be significantly improved by annealing. Despite the interest in mechanical performance of thermal sprayed coatings, the formability of the coated components by means of cold rolling has not yet been investigated.

Copper coatings often serve as bonding agent for multi-material components. During high pressure casting of aluminum onto steel sheets, copper interlayers can be used as a barrier to prevent the formation of brittle iron aluminides. The binary system of copper and aluminum has a low melting eutectic with a melting temperature of T_m = 548.2 °C implying the possibility of a better metallurgical bonding between these materials in comparison to that between steel and aluminum. Copper possesses a further benefit, as it has a coefficient of thermal expansion in the range between steel and aluminum and thus can reduce the thermal stresses occurring during the solidification process after high pressure die casting. In addition, the bond strength can be optimized by a form closure connection. This can be achieved by forming the coated steel sheet to increase the contact area and to create a suitable surface structure for mechanical interlocking between aluminum cast and steel inserts.

Surface structuring by rolling is an established research area at the Institute of Metal Forming (IBF) of RWTH Aachen University. Some work in the past concentrated on the rolling of channel-like grooves into steel sheets made of steel DC04 using a modular roll-setup. Grooves with a height of up to h = 0.5 mm and aspect ratios, thus the ratio of structures height to structures width, between 0.5 and 1.0 could be realized using different roll setups. Figure 1a) shows a top view of such a structure and the profile measured with a digital microscope. For all investigated structures, the same trend could be determined regarding the resulting size of the structure h_s in dependence of the thickness reduction. For the structuring of steel DC04 by solid roll with a structure height of a roll $h_{s\text{-roll}}$ = 0.5 mm it was shown that the thickness reduction of ε = 28 % was required, to achieve a relative size of 1 ($h_{s\text{-roll}} = h_s$), Figure 1b) [4].

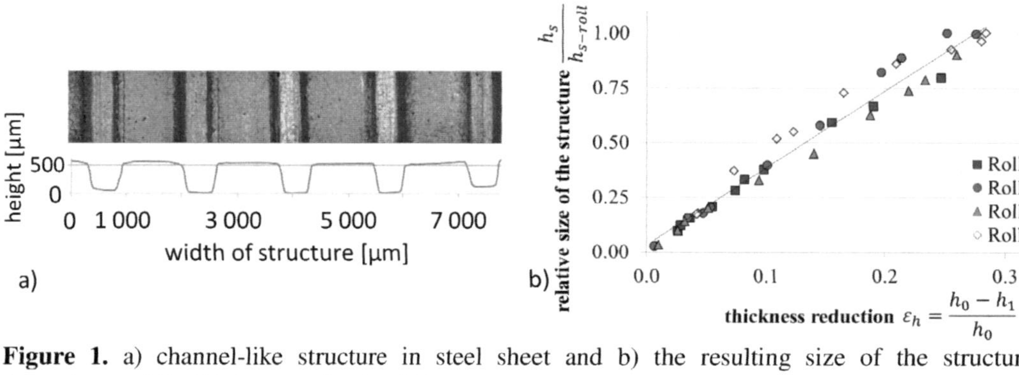

Figure 1. a) channel-like structure in steel sheet and b) the resulting size of the structure in dependence of the thickness reduction [4].

In the current study the process limitations of cold rolling and surface structuring of cold sprayed copper coated steel sheets are investigated. In order to increase the formability of the copper coatings a heat treatment is performed prior to cold rolling. Moreover, the influence of mechanical loads in a range comparable to those occurring during roll bonding on the bond strength of the coating is investigated. For steel sheets, a thickness reduction of at least ε = 50 % is needed to realize an adhesive bond by roll bonding [5], as the oxide layers at the metal surface can only be broken by a high degree of deformation. For coated components, the adhesion between coating and substrate is achieved through the high velocity impact of the particles during the deposition process but an additional influence of the rolling process on the bond strength can be expected.

2. Experimental Setup

The experimental procedure includes cold spraying process, heat treatment, cold rolling and sample analysis. The principle scheme of the experimental setup is shown in Figure 2.

2.1 Cold Gas Spraying

In the present study, the conventional steel DC04 was selected as a substrate material. Prior to spraying process, the steel plates were grit blasted using alumina with the grit size of F20 and cut to rectangular plates with the dimensions of $V = 200 \times 30 \times 2$ mm^3. For the deposition, a 99 % pure copper powder with a size distribution of -45 +15 µm, delivered by ECKA Granules GmbH (Fürth, Germany), was employed. The spray parameters are shown in Table 1. The deposition of the coatings was performed at Surface Engineering Institute of RWTH Aachen University using the Kinetics 8000 Cold spray system from Oerlikon Metco AG (Switzerland) with nitrogen as process and carrier gas.

Table 1. Spray parameters.

Parameter	Value
Gas temperature [°C]	500
Process gas flow [m^3/h]	66
Carrier gas flow [m^3/h]	2.4
Spray distance [mm]	30
Transverse movement velocity [mm/s]	300

In order to improve the formability of the sprayed coating, some samples were heat treated in vacuum at a temperature of T = 500 °C for t = 5 h. The choice of these parameters is based on the analysis of the results presented in the study of Renzhong et al. [2]. The cold sprayed samples were heat treated using a PVA MOV 553 vacuum furnace. The non-heat-treated samples were also cold rolled as a reference for the investigation of the effect of annealing.

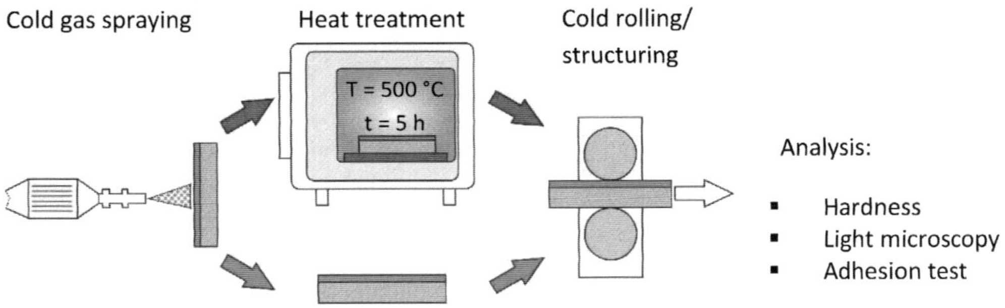

Figure 2. Principle scheme of the experimental procedure.

2.2 Cold Rolling Process

All rolling experiments were performed with a rolling mill at IBF. The two principles for flat rolling and rolling of a channel like structure are shown in Figure 3. In order to avoid difficulties regarding the insufficient stiffness of the tooling in case of a modular roll, observed by Senge et al. in [4], a solid roll with a higher stiffness was used for these experiments. During the experiments, the force and the global thickness of the compound, h_l for flat rolling and $h_{c\text{-}peak}$ for structure rolling, are measured. The dimensions of the structure such as the sheet thickness measured in the valley of a channel $h_{c\text{-}val}$ and the depth of a channel $h_{c\text{-}depth}$, are determined in metallographic cross sections, Figure 3.

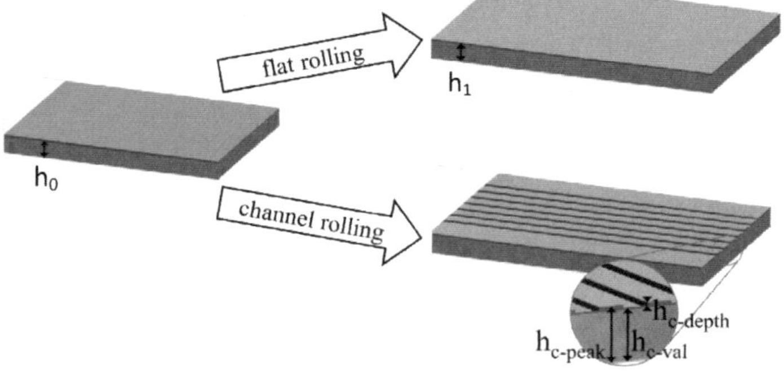

Figure 3. Principle scheme of the two rolling procedures, where $h_{c\text{-}peak}$ is the thickness in the peak of the structure, $h_{c\text{-}val}$ is the thickness in the valley of the structure.

The global thickness reduction performed in these experiments was about ε = 8 % in one rolling pass and about ε = 14 % in two subsequent passes. Table 2 gives an overview of the cold rolled and structured samples from both processing routes with and without heat treatment step. The thickness reduction of the structured samples is defined as the difference between initial thickness h_0 and final thickness measured in the valley of the structure $h_{c\text{-}val}$.

Table 2. Overview of the performed experiments and the thickness of the coated steel sheets (CR - cold rolled, HT - heat treated, S - structured).

Sample	initial thickness, h_0 [mm]	final thickness, h_l [mm]		thickness reduction, ε [%]		heat treatment	rolling process
CR8	2.07	1.91		8.0		no	flat
CR14	2.07	1.78		14.2		no	flat
HT+CR8	2.07	1.89		8.5		yes	flat
HT+CR14	2.08	1.78		14.3		yes	flat
		$h_{c\text{-}peak}$	$h_{c\text{-}val}$	ε_{peak}	ε_{val}		
HT+S	2.26	2.26	2.06	0.0	8.7	yes	channel

2.3 Testing

In order to characterize the coating, the cold sprayed samples were metallographically prepared and examined in terms of their microstructure. The coating thickness was measured on the cross sections using a light microscope by Carl Zeiss with Axio Vision image analysis tool. The roughness of the surfaces in as sprayed and cold rolled condition was measured by a confocal laser microscope (VKX 210, Keyence, Osaka, Japan). The acquired topography images were used to evaluate the roughness parameters according to the evaluation procedure defined in DIN EN ISO 4288:1997 [6]. Three measurements were carried out on each sample at different random positions. The bonding strength of the deposited coatings was determined using the PATHandy tester (PATHandy, DFD Instruments, Kristiansand, Norway). The measurement principle is described in [8]. Even though this method is primarily standardized for lacquer coatings according to DIN EN ISO 4624 [7], it is industrially deployed for thermally sprayed coatings due to its simple handling and robustness. As it was shown by Schlaefer [9], the measurements by PATHandy are in good agreement with the adhesion tests conducted according to DIN 582 [10]. A microhardness tester Fisherscope HM2000 by Helmut Fischer GmbH (Sindelfingen-Maichingen, Germany) and Bühler Micromet 1 by ITW Test & Measurement GmbH (Esslingen am Neckar, Germany) were used for the examination of the Martens and Vickers hardness of the coatings, respectively. In order to identify a possible formation of cracks in the coating during the cold rolling process, a SEM analysis was performed by means of the scanning electron microscope Zeiss Leo 1530.

3. Results

The cold spraying experiments revealed that no adhesion of copper particles to the untreated steel surface could be achieved in the process temperature range of T = 450 °C to T = 500 °C, as the particles bounced off the substrate. After grit blasting the steel surface with alumina F20, a coating with a thickness of about s = 120 µm could be deposited at a temperature of T = 500 °C, yielding a bonding strength of 13.3 ± 0.9 MPa. A further increase of the process temperature did not improve the coating properties significantly and led to particles clogging on the nozzle. The process parameters shown in Table 1 were adopted for further investigations.

3.1 Microstructure

The porosity of the copper coatings after deposition was about 0.5 %, Figure 4a). Copper feedstock exhibits a good plasticity resulting in a tight bonding between particles. The deposition quality was uniform through the whole thickness. Only the upper layer of the coating exhibits a lower density resulting in higher porosity close to the surface. This phenomenon is well known in cold spraying as a tamping effect, describing the process of additional deformation of deposited particles by impacting ones [11].

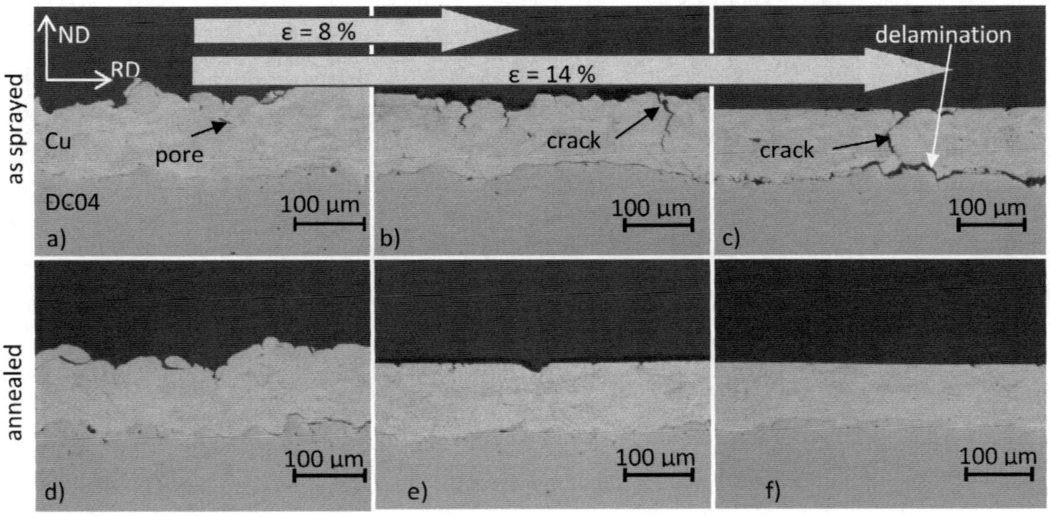

Figure 4. Cross section in plane of normal direction (ND)/rolling direction (RD) of the a) sprayed, b) cold rolled with ε = 8 %, c) cold rolled with ε = 14 %, d) annealed, e) annealed and cold rolled with ε = 8 %, f) annealed and cold rolled ε = 14 % copper coated samples.

When the coated samples were flat rolled without previous annealing, some cracks were observed in normal direction/rolling direction (ND/RD) cross sections. After the first rolling pass, the adhesion of the copper coatings was still maintained. With a higher thickness reduction of ε = 14 %, cracks propagation through the whole coating thickness towards the substrate surface and the coating delamination occurred at different positions, Figure 4c). The SEM images show that the crack propagation occurs mainly at the particles boundaries, Figure 5.

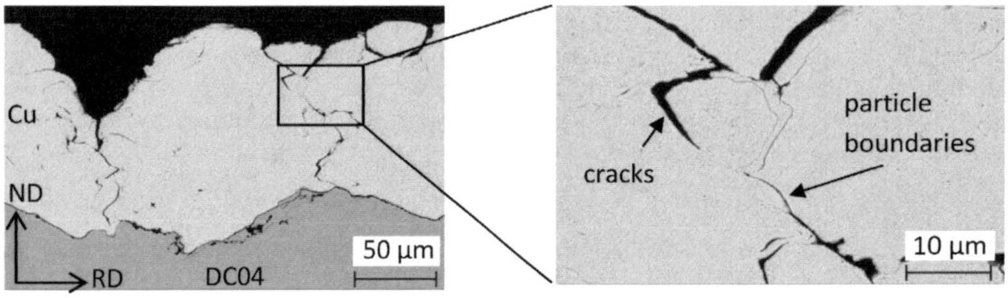

Figure 5. SEM/RBSD images of the coatings cross section, showing the crack propagation after the flat rolling process with a thickness reduction of $\varepsilon = 8$ % without annealing.

After annealing, the interfaces between the particles almost disappear. The EDS analysis showed that mutual diffusion of copper from the coating into the steel and iron from the steel into the copper coating occurs at the interface coating/steel during the annealing. The EDS measurement revealed at point P1 within the steel substrate the presence of copper in an amount of about 1.4 wt.%, and at the measurement point P2 within the copper coating the presence of about 1.8 wt.% iron, Figure 6. No cracks were observed in the annealed coating subsequent to flat rolling with a thickness reduction of $\varepsilon = 8$ % and $\varepsilon = 14$ %, Figure 4f).

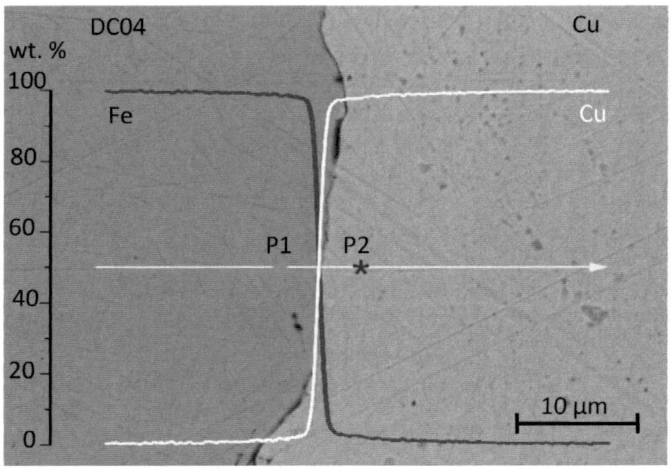

Figure 6. EDS analysis and iron and copper content at the interface steel/coating after annealing.

In the next step, the annealed samples were successfully structured by channel rolling without delamination or cracking. The mean structure height was about $h_{c\text{-}depth} = 120$ µm at a thickness reduction of nearly $\varepsilon_{val} = 9$ %. Cross section of the structure is shown in Figure 7. For the structuring of bare DC04 sheet the same thickness reduction of about $\varepsilon = 9$ % resulted in a structure height of nearly $h_{c\text{-}depth} = 170$ µm [4].

Figure 7. Metallographic cross section in plane of normal direction (ND)/transverse direction (TD) of a structured copper coated sample with a mean height of the structure of $h_{c\text{-}depth}$ = 120 µm.

3.2 Surface roughness

The surface roughness after deposition and after each flat rolling pass was determined to observe the evolution of surface topography, Figure 8. The surface roughness is of interest as it can affect the wetting behavior of the aluminum melt during the high pressure die casting. Directly after deposition, the mean surface roughness was Ra = 13.9 ± 0.3 µm. After subsequent flat rolling with the total cross section reduction ratio of ε = 8 % and ε = 14 %, the roughness decreased to Ra = 3.7 ± 0.3 µm and Ra = 1.9 ± 0.1 µm, respectively.

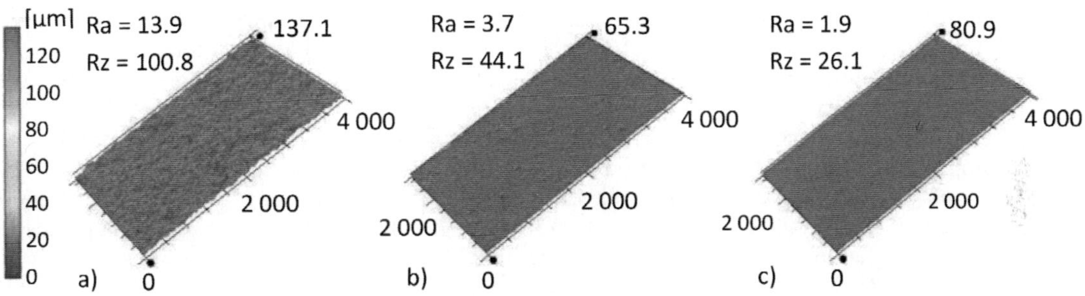

Figure 8. Surface topography of copper coatings after a) deposition, b) flat rolling with cross section reduction of ε = 8 %, c) flat rolling with cross section reduction of ε = 14 %.

The same trend can be seen for the surface roughness depth Rz which decreased from Rz = 100.8 ± 4.2 µm in the as sprayed condition to Rz = 44.1 ± 3.2 µm after the first cold rolling pass and to Rz = 26.1 ± 1.0 µm after the second pass. No significant difference was observed when comparing non-treated and heat treated samples, Figure 9.

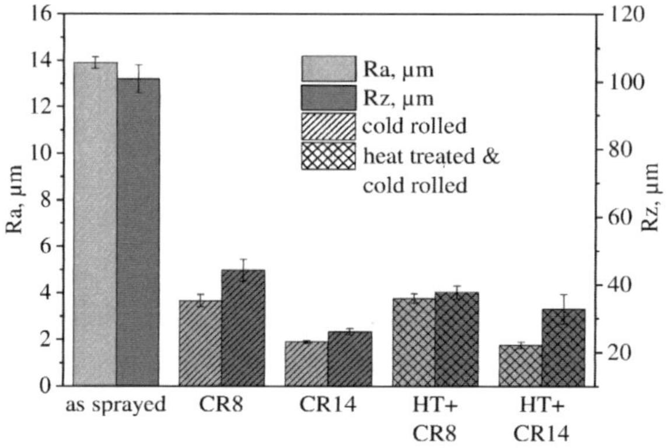

Figure 9. The mean surface roughness Ra and surface roughness depth Rz of the copper coating after deposition and subsequent heat treatment and flat rolling with $\varepsilon = 8$ %, and $\varepsilon = 14$ %.

3.3 Hardness of the coating

The cold spraying as well as the cold rolling process causes a work hardening effect and a respective reduction of formability of the coatings. The hardness measurement provides a simple way to estimate the extent of strengthening after cold spraying and cold rolling. In Figure 10a), the evolution of the Vickers hardness of the copper coating after each process step is depicted. After the coating deposition, the hardness was about 140 HV0.05. The subsequent flat rolling does not contribute to the further work hardening of the copper coating. This can be related to the high degree of the particle deformation during the spraying process where they possibly reached their strain limit. This assumption is in accordance with the coating behavior observed metallographically, indicating the crack propagation already after the flat rolling with a cross section reduction of $\varepsilon = 8$ %. After annealing of the cold sprayed samples, the hardness of the copper coatings recovers to the level comparable with that of bulk copper of about 70 HV0.05. A subsequent cold rolling process leads to an increase of the hardness up to 100 HV0.05. The initial hardness of the copper powder couldn't be determined by Vickers hardness measurement, as the particles were too small for indentation. To observe the evolution of the mechanical properties from initial state through cold spraying process and subsequent annealing, the Martens hardness were measured by means of Fisherscope. As can be seen in Figure 10b), the hardness after the heat treatment has almost reached the level of the bulk material. These results indicate the near-complete recovery of mechanical properties after annealing.

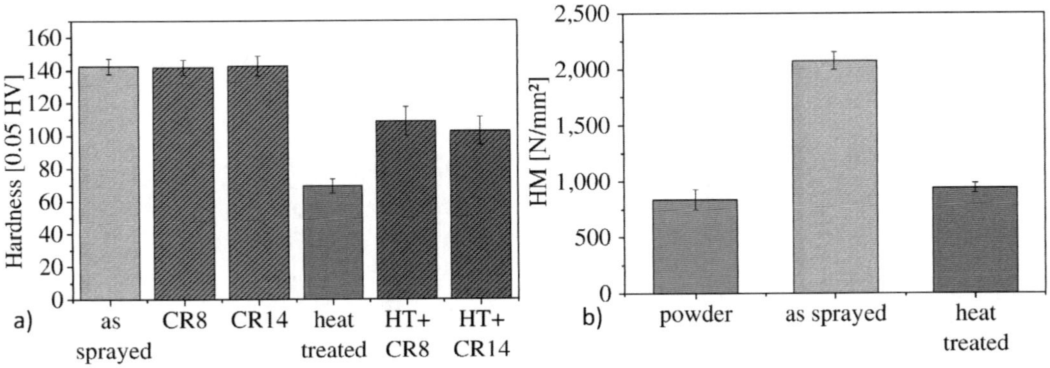

Figure 10. a) Vickers hardness of the copper coatings after deposition, heat treatment and flat rolling and b) Martens hardness of the copper powder, after deposition and subsequent annealing.

For the pure copper coating, the strain hardening is the only mechanism contributing to the hardness increase as the other common mechanisms like precipitation hardening and solid solution hardening can be excluded due to the absence of alloying elements. For this reason, the hardness evolution directly indicates the deformed state of the microstructure. The comparison of the Martens hardness of the copper powder with the hardness in the as sprayed and annealed condition shows that the heat treatment at a temperature of T = 500 °C for t = 5 h is sufficient for reaching almost complete recovery of the mechanical properties.

3.4 Adhesion strength

To determine the influence of the annealing and the cold rolling processes on the adhesion of the coating to the steel substrate, the bond strengths were measured after corresponding process. Recently, some studies have pointed out, that a heat treatment can improve the bonding strength of the coating. Li et al. have observed an almost 100 % increase in the adhesive strength of a cold sprayed titanium coating after annealing at T = 850 °C for t = 4 h [12]. Although the EDS measurement revealed mutual diffusion processes at the interface Cu coating/steel during the annealing, no significant changes of the bonding strength were observed in the current study after the heat treatment process compared to the as sprayed condition, Figure 11. The flat rolling process without heat pretreatment influenced the adhesion of the copper coatings negatively. After the first rolling pass with thickness reduction of about $\varepsilon = 8$ %, the bond strength decreased slightly. However, a number of cracks was observed by light microscopy in the normal direction. After the second rolling pass with the thickness reduction of about $\varepsilon = 14$ %, the bonding strength decreased by approx. 50 %. For annealed samples, the flat rolling process did not lead to deterioration of the adhesive strength even after a thickness reduction of $\varepsilon = 14$ %.

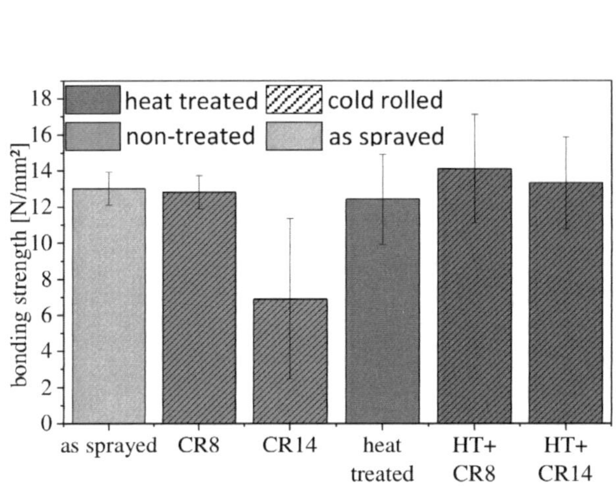

Figure 11. The bonding strength of the copper coatings after deposition and flat rolling.

The substrate material, surface preparation as well as spraying condition are known to have an influence on the tensile adhesion strength of the coatings. For the material combination steel substrate/copper coating applied by cold gas spraying the typical values of adhesion strength are reported to be in the range of 9 to 30 MPa depending on parameters described above [13,14]. In the current study, it was shown that the annealing treatment is an efficient tool to increase the formability of the Cu coating without influencing its adhesion strength negatively.

4. Conclusions

In the present study the formability of cold sprayed copper coating on steel substrates was investigated in as sprayed and annealed conditions. The effect of the heat treatment and cold rolling on the adhesion strength of the coating and its mechanical properties was the main focus of the research. The results have demonstrated the high extent of the work hardening effect during the deposition. Thus, the Martens hardness of the bulk copper has increased by the factor of two after coating application. The subsequent annealing at a temperature of T = 500 °C for t = 5 h resulted in the recovery of the initial hardness values of the feedstock material. The microstructure of the coating also experiences a modification during the heat treatment. Due to the diffusion processes, the boundaries between particles almost disappear, leading to a homogenous microstructure.

Following observations were made about the coating behavior during the cold rolling process. The non-heat-treated samples are prone to crack initiation and propagation in the copper coating due to cold rolling process. For a thickness reduction of ε = 14 %, the bonding strength decreased drastically by a factor of 2. If the sample was annealed prior to cold rolling, no cracking and no coating delamination was observed after rolling. The bonding strength is not influenced negatively. The heat treated samples were structured by the rolling of channel-like grooves in the coating with a maximal structure height of $h_{c\text{-}depth}$ = 120 µm. Concluding, it can be stated that an annealing treatment can drastically increase the formability of cold sprayed coatings by retaining the initial bond strength. The current study showed that metal forming processes like cold rolling are applicable for cold gas sprayed coatings in the annealed condition. Moreover, coatings can be structured to increase the bonding surface and to create a suitable surface topology for high pressure die casting process.

Acknowledgements

The authors gratefully acknowledge the financial support of the German Research Foundation (DFG) received within the research project “Surface pretreatment of steel sheet for compound casting process of low-distortion and low-gap aluminum cast/steel sheet hybrid”.

References

[1] Fauchais P L; Heberlein J V and Boulos M I 2014 *Thermal Spray Fundamentals* (Boston, MA, USA: Springer)

[2] Renzhong H, Michiyoshi S, Wenhua M and Hirotaka F 2015 The effects of heat treatment on the mechanical properties of cold-sprayed coatings *Surface and Coatings Technology* **261** 278–288

[3] Gärtner F, Stoltenhoff T, Schmidt T and Kreye H 2006 The Cold Spray Process and Its Potential for Industrial Applications *J Therm Spray Tech* **15** 223–232

[4] Senge S and Hirt G 2015 Evaluation of Modular Roll-Setup to Roll Grooves into Steel Sheets *AMM* **794** 120–127

[5] McEwan K and Milner D R 1962 Pressure Welding of Dissimilar Metals *British Welding Journal* **9** 406–420

[6] DIN EN ISO4288:1997. Geometrical Product Specifications (GPS) - Surface texture: Profile method - Rules and procedures for the assessment of surface texture

[7] DIN EN ISO 4624:2014-06. Paints and varnishes - Pull-off test for adhesion

[8] Bobzin K, Öte M, Knoch M A, Liao X, Hopmann C and Ochotta P 2016 Investigation of Influencing Factors on the Transfer of Wire Arc Sprayed Zn Coatings for the Metallisation of Plastic Parts *ITSC 2016 International Thermal Spray Conference & Exposition* 24–29

[9] Schläfer T 2014 *Entwicklung von Verfahren zur Bestimmung der Haftzugfestigkeit von thermisch gespritzten Zylinderlaufbahnen* (Herzogenrath: Shaker)

[10] DIN EN 582:1994-01 Thermal spraying; determination of tensile adhesive strength

[11] Suhonen T, Varis T, Dosta S, Torrell M, and Guilemany J M 2013 Residual stress development in cold sprayed Al, Cu and Ti coatings *Acta Materialia* **61** 6329–6337

[12] Li W-Y, Zhang C, Liao H and Coddet C 2009 Effect of heat treatment on microstructure and mechanical properties of cold sprayed Ti coatings with relatively large powder particles *J Coat Technol Res* **6** 401–406

[13] Mäkinen H, Lagerbom P and Vuoristo P 2007 Adhesion of Cold Sprayed Coatings: Effect of Powder, Substrate, and Heat Treatment. *Proceedings of the 2007 International Thermal Spray Conference ; May 14 - 16, 2007, Beijing, People's Republic of China* (Ohio: ASM International)

[14] Fernández R, MacDonald D, Nastić A, Jodoin B, Tieu A and Vijay M 2016 Enhancement and Prediction of Adhesion Strength of Copper Cold Spray Coatings on Steel Substrates for Nuclear Fuel Repository *J Therm Spray Tech.* **25** 1577–1587

IOP Conf. Series: Materials Science and Engineering **181** (2017) 012029 doi:10.1088/1757-899X/181/1/012029

Interfacial strength analyses of Al/Mg compounds using bending tests

T Lehmann, C Kirbach, J Müller, M Stockmann and J Ihlemann

Professorship of Solid Mechanics, Chemnitz University of Technology, Germany

Abstract. In the Collaborative Research Center 692, subproject B3 Al/Mg compounds are investigated. The hydrostatic co-extruded compounds presented in this paper were further manufactured by the forging process Rising. To continue the first investigations of Rising specimens regarding interface strength, a bending test developed in a previous project period was used. The specific load case and the bending specimen geometry considers the requirements concerning the special geometry of the Rising specimen. Based on experimentally determined failure forces (maximum forces), the stress state for the investigation of the interface strength was calculated by means of the elementary bending theory extended with a numerical determined correction factor. The numerical analyses were based on a parametric FE model of the load case. Crack initiation was caused by the maximum interlaminar interfacial tension stress. In the demonstrated investigations co-extruded compounds with different ratio of core material (Mg) in the transversal cross sectional area of the initial billet were analyzed. A particular feature of the investigations is the interfacial strength analysis of a subset of Rising specimens in different areas of the transversal cross section. This was enabled by using compounds with larger sleeve thickness due to a lower Mg ratio. Thus, in this case a more extensive characterization could be performed. The results show higher strength values for Rising specimens with the largest sleeve thickness compared to the other investigated configurations.

1. Introduction

The requirement of weight saving becomes more and more important in several fields of industry, e. g. automotive [1] and aircraft. Therefore, lightweight materials and structures are essential to consider in product development processes. Aluminum and magnesium provide high lightweight potential with advantageous mechanical properties. In the Collaborative Research Center 692, subproject B3 Al/Mg compounds are developed and analyzed regarding production process and the resulting bonding quality, strength of the basic materials, interfacial strength, residual stresses, fracture mechanical properties as well as formability [2, 3, 4, 5, 6]. The material was produced by hydrostatic co-extrusion [7]. The production of Mg/Al macro composites by co-extrusion can also be found in [8]. Special focus of the actual investigations is on further manufacturing processes of the Al/Mg compounds. Within the research project, the fundamental forging processes Upsetting, Spreading and Rising are analyzed. The investigations presented in this paper include the analysis of Rising specimens. The actual standard configuration features a Mg ratio of the transversal cross sectional area of 53 % of the initial billet. Using this configuration, in areas of high strains the flank thickness of the sleeve is reduced [5]. To enlarge the sleeve thickness in these critical sections of the Rising specimen, concentric co-extrusion rods with lower Mg ratios of the transversal cross sectional area were used as semi-finished product for the forging process Rising instead of the eccentric version presented in [6]. Continuing these investigations, the specimens were analyzed concerning interfacial strength.

2. Production process of the analyzed material

2.1 Hydrostatic co-extrusion

The initial rods of the investigated Rising specimens were produced by hydrostatic co-extrusion at CEP GmbH, Freiberg. In figure 1 the general principle of the production process is given. Details of the co-extrusion process of Al/Mg compounds can be found in [2]. The force is transmitted by the pressure medium which is surrounding the bi-material billet. Thus, the billet is pressed through the die, which results in the compound.

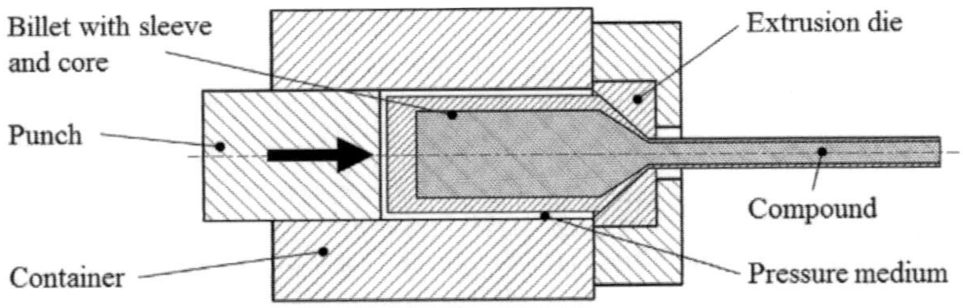

Figure 1. Principle of hydrostatic co-extrusion.

The materials used were the lightweight alloys AZ31 and AA6082. The aluminum alloy was used as sleeve material and assumes function of structure and corrosion protection of the magnesium alloy. In table 1 essential parameters of the rod production by co-extrusion (Mg ratio of the transversal cross sectional area, diameters of the rod d_{sleeve} / d_{core}, degree of deformation φ, punch velocity v_{Punch} and the extrusion temperature T_{extr}) are given. As a new feature, the Mg ratio (core ratio) of the transversal cross sectional area was reduced and varied to achieve a larger sleeve thickness in critical zones of the cross section, see [5]. Furthermore, the sleeve thickness in the other zones of the transversal cross section is enlarged as well using this configuration. Hence, the core diameter was reduced compared to the actual standard configuration of 14.5 mm. Another strategy of improving the sleeve and core geometry is the eccentric co-extrusion of the two materials, see [5,6].

Table 1. Essential extrusion parameters of the initial rods.

Mg ratio [%]	d_{sleeve} / d_{core} [mm]	φ [-]	v_{Punch} [mm/s]	T_{extr} [°C]
47	20 / 13.75	2.77	3.6	300
29	20 / 10.75			

2.2 Forging process Rising

The Rising experiments were performed by the Professorship of Virtual Production Engineering (within the subproject B3). Following the procedure presented in [6] the forging processes were performed with the same parameters as used by FEUERHACK [5]. The co-extruded initial billets with a length of 40 mm were forged by means of a 1000 kN eccentric press (Raster-Zeulenroda) and the upper and lower die, see figure 2 a). The achieved specimen temperature was of about 300 °C and the die temperature of about 200 °C. For lubrication a composition of MoS_2 and graphite was applied to the surfaces of the upper and lower die. An example of the transversal cross section is given in figure 2 b).

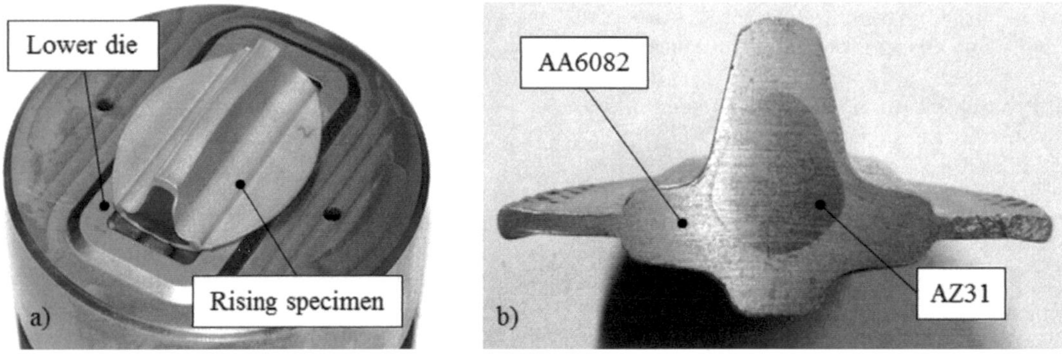

Figure 2. a) Lower die and Rising specimen – photo of Professorship of Virtual Production Engineering, b) Transversal cross section of a Rising specimen, Mg ratio of 29 %.

3. Bending tests

3.1 Specimen separation and preparation

Bending tests were performed to determine the interfacial strength. The separating of the bending specimens was done by wire eroding from the middle part of the Rising specimens, see figure 3 a). At first, 3 slices were separated. In a second step, the bending specimen geometry was manufactured out of the slices. In all Rising components, bending specimens were separated from the upper part of the transversal cross section (bending specimen 1 in figure 3 b)). In case of the configuration with a Mg ratio of 29 %, bending specimens from the lower part were eroded additionally (bending specimen 2 in figure 3 b)) to analyze the strength of the lower interface of the Rising specimens. After wire eroding the specimens were analyzed by dye penetrant test to detect samples with interface damages which were separated out and not mechanical tested. Subsequent to damage detection the specimens were grinded.

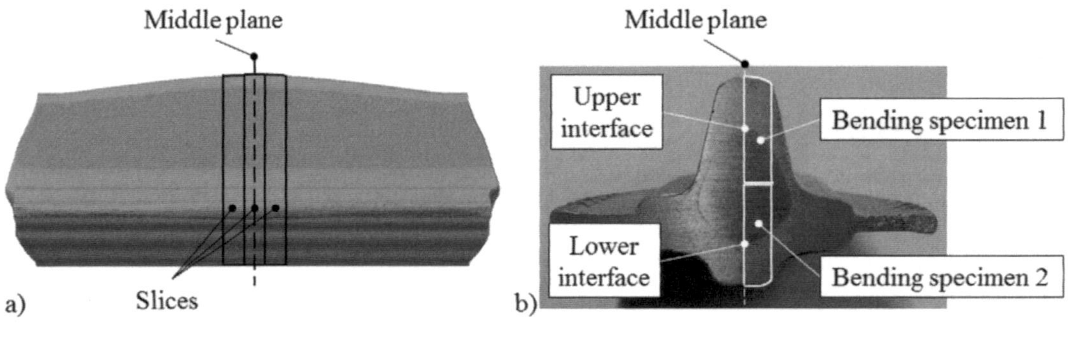

Figure 3. a) Section of slice separation – longitudinal, b) Section of bending specimen separation – transversal.

3.2 Experimental setup and procedure

The bending tests were performed in a 100 kN testing machine of Zwick/Roell. The three point bending load case was developed based on FE analyses already in previous investigations, given in [6], see figure 4 a). Thus, for the experiments the same bending device which is assembled in the testing machine and the same testing procedure as presented in [6] was used, see figure 4 b). The force is applied by means of a punch and measured by a 5 kN GTM load cell. Two prisms which are bonded to linear bearings enable the symmetric floating bearing of the specimen with the degree of freedom of displacement in longitudinal direction. The interface at the lower surface intersects the line of load application. The bending experiments were performed with a punch velocity of 3 mm/min until interface failure occurs at maximum force.

a)
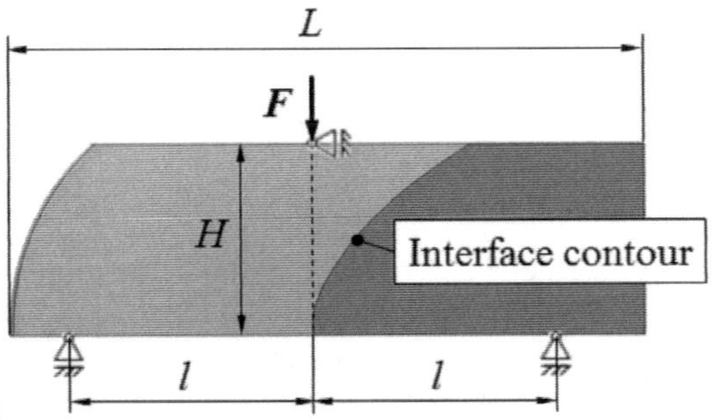

b)
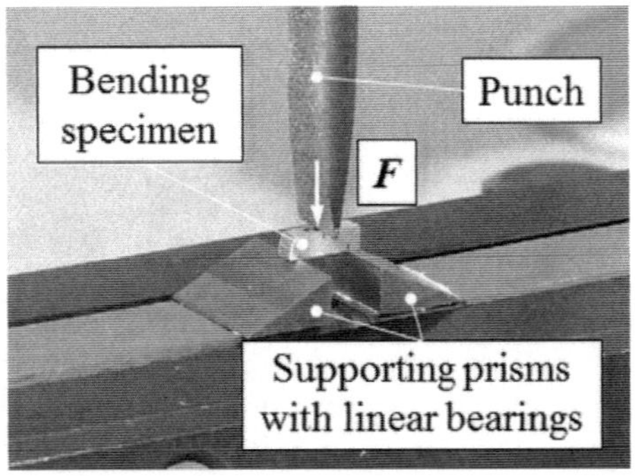

Figure 4. a) Load case, specimen – nominal geometry: W x H x L = 2.5 mm x 2.5 mm x L (width W, height H, length L = 8.5…11.5 mm, depending on the configuration), supporting distance l = 3 mm, b) Bending device and specimen.

3.3 Evaluation of experimental data

As a result of brittle interface cracking, the force F is decreasing abruptly after reaching the maximum value. The interfacial crack initiation is caused by the maximum interlaminar tensile stress $\sigma_{x\,max}$ at the bottom of the specimen which occurs in the middle plane (regarding to the width), see figure 5. Due to the load case there is no superposition of interlaminar shear stress at this position. If the interface strength is reached, the crack is initiated with subsequent instable crack propagation until the whole interface has failed. Therefore, the maximum value of the interlaminar tensile stress in longitudinal direction of the specimen equals to the interface strength σ_{IF} (interfacial tensile strength). Because of deviations to ideal bending beam conditions caused by the special geometry, the calculation of stresses was performed using the equation of the elementary bending theory extended with a correction factor C, see equation (1). For interfacial strength calculation, the maximum forces F_{max} determined in the bending tests were used.

$$\sigma_{x\max}\left(F_{\max}\right)=C\frac{3l}{WH^2}F_{\max} \qquad (1)$$

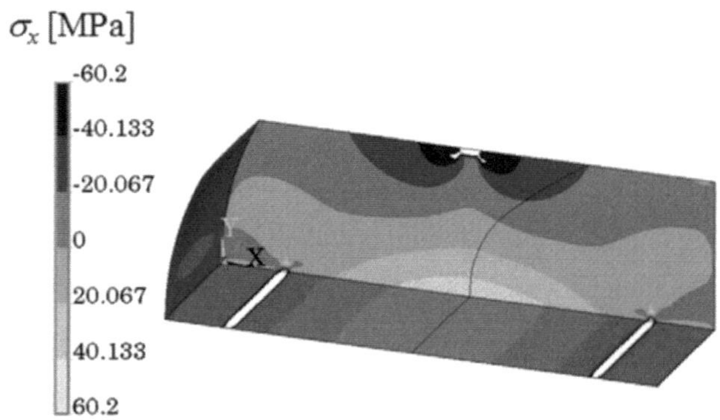

Figure 5. Example of the stress distribution σ_x calculated by means of the FE Model of the bending load case (minimum stress is limited to -60.2 MPa, white areas) [6].

C was calculated using the result of the linear bending theory and the solution of the FE model of the bending load case which was developed and already used in previous investigations. Details to the model are given in [6]. For the specimens geometry the correction factor C equals to 0.937. An example of the stress distribution σ_x is given in figure 5.

3.4 Results and discussion

In table 2 the determined averaged interface strength $\sigma_{IF\,av}$ is given. The upper interface (IF) is associated with bending specimen 1 and the lower interface with bending specimen 2 in figure 3. The specimens, which were separated from the material with a Mg ratio of 47 % show moderate averaged interfacial strength values with good correlation to the results of the Rising specimens with eccentric co-extruded initial rods [6], see figure 6. Furthermore, good correlation can be observed to the interfacial strength of bending specimens which were separated from rods which were further processed by free axial and radial upsetting [4]. For the Rising specimens with lower Mg ratio which causes a larger sleeve thickness, the averaged interfacial strength value is much higher (nearly factor 2). This leads to the hypothesis of an increasing interfacial strength by an increasing sleeve thickness in normal direction of the analyzed interface section. The same trend was observed in the investigations given in [6], see figure 6 (the larger the offset, the larger the sleeve thickness on the upper side). Furthermore, the results show lower averaged strength values of the lower interface compared to the upper interface of the Rising specimens (Mg ratio of 29 %). Nevertheless, the strength of this lower interface is on a high level.

Generally, the scattering of the strength values is high (figure 6), which corresponds to the previous investigations regarding these special Al/Mg interfaces [3, 4, 6].

Table 2. Averaged interfacial strength $\sigma_{IF\,av}$ of the analyzed material.

Mg ratio [%]	Number of bending specimens	Interface	$\sigma_{IF\,av}$ [MPa]
47	8	upper	50
29	6	upper	97
	9	lower	76

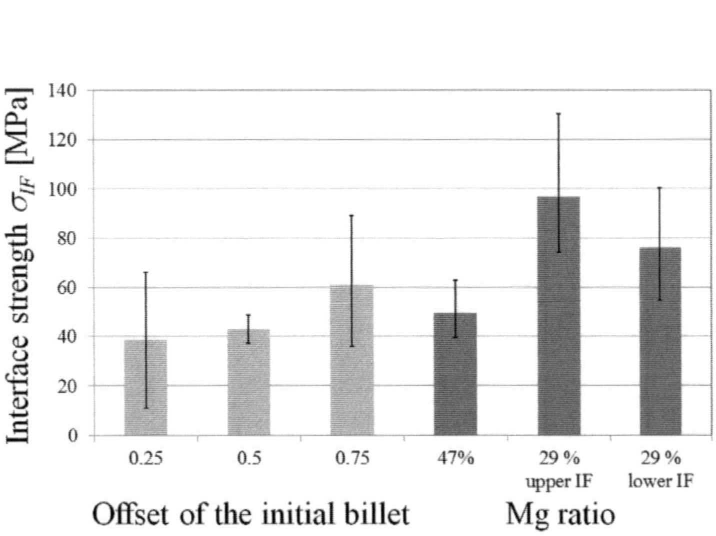

Figure 6. Averaged interface strength and scattering (black lines), left: previous analyses [6], right: actual investigations.

4. Summary

Interfacial strength investigations were performed with Al/Mg compounds which were further processed by the forging process Rising. Continuing previous investigations of the research project, Rising specimens with reduced Mg ratio were used (enlarged sleeve thickness at the critical flanks). For strength determination bending tests were performed. Interfacial strength values were calculated using the experimental maximum bending force and the elementary bending theory extended with a correction factor, which was determined by means of a FE model of the load case.

The material with the lowest Mg ratio (largest sleeve thickness) showed the highest level of interfacial strength. Furthermore, the interfacial strength of the lower interface of this configuration is lower but on a high level as well. In case of the material with Mg ratio of 47 % good correlation to the previous results can be observed. The scattering of the strength values is high. In further investigations concerning the interfacial strength, bending specimens of the initial rods of the configurations of [6] and of the Rising specimens presented in this paper will be analyzed.

Acknowledgments

The authors thank the German Research Foundation (DFG) for its funding within the framework of the Collaborative Research Centre 692 (SFB HALS 692).

References

[1] Fischerworring-Bunk A, Landerl C, Fent A and Wolf J 2005 The new BMW Inline Six-Cylinder Mg/Al Crankcase *Proc. IMA World Magnesium Conf. (Berlin)* pp 49–58

[2] Kittner K, Feuerhack A, Förster W, Binotsch C and Graf M 2015 Recent developments for the production of Al-Mg compounds *Mater. Today: Proc.* 2S pp 225–232

[3] Lehmann T 2012 Experimentell-numerische Analyse mechanischer Eigenschaften von Aluminium/Magnesium-Werkstoffverbunden *PhD-Thesis TU Chemnitz* (online: http://nbn-resolving.de/urn:nbn:de:bsz:ch1-qucosa-100576)

[4] Kirbach C, Stockmann M and Ihlemann J 2014 Experimental and numerical investigations of Al/Mg-compound specimens under bending after mass forming processes *Proc. 31st Danubia Adria Symp. on Advances in Exp. Mechanics* pp 32–33

[5] Feuerhack A 2014 Experimentelle und numerische Untersuchungen von Al-Mg Verbunden mittels Verbundschmieden *PhD-Thesis TU Chemnitz*
(online: http://nbn-resolving.de/urn:nbn:de:bsz:ch1-qucosa-146654)

[6] Förster W, Binotsch C, Awiszus B, Lehmann T, Müller J, Kirbach C, Stockmann M and Ihlemann J 2016 Forging of eccentric co-extruded Al-Mg compounds and analysis of the interface strength *IOP Conf. Series: Mater. Science and Eng.* **118** 012032
(online: http://iopscience.iop.org/article/10.1088/1757-899X/118/1/012032)

[7] Möhler W and Menzel H-U 2004 Strangpressen von Werkstoffverbunden *Proc. MEFORM* pp 53–63

[8] Negendank M, Müller S and Reimers W 2012 Coextrusion of Mg-Al- macro composites *J. Mater. Proc. Techn.* **212** pp 1954–1962

IOP Conf. Series: Materials Science and Engineering **181** (2017) 012030 doi:10.1088/1757-899X/181/1/012030

FE-simulation of the Presta joining process for assembled camshafts – modelling of the joining process

R Scherzer, C B Silbermann, R Landgraf and J Ihlemann

Professorship of Solid Mechanics, Chemnitz University of Technology,
Reichenhainer Straße 70, 09126 Chemnitz, Germany

robert.scherzer@mb.tu-chemnitz.de

Abstract. The Presta joining process is based on the separate manufacturing of internal profiled cams and shafts with previously widened diameters at the intended cam seat through rolling. With regard to further optimization and enhancement, the two substeps are modelled and simulated using FEM. The results of the rolling process simulation are employed as the starting point for the simulation of the joining step, which is presented in this contribution. First, the shaft is widened with a profile oriented orthogonal to the shaft axis at the intended cam seat. Subsequently, the shaft is joined with a cam that has an internal profile oriented parallel to the shaft axis. Due to plastic deformations these perpendicular profiles form a tight fit. Because of these complex plastic deformations, that are related to large local forming degrees, the use of precise material models is essential. Moreover, it is necessary to maintain the deformation history of the rolling simulation in the joining step with regard to effects like kinematic or isotropic hardening. For that purpose, the applied material model's invariance under change of the reference configuration is used by transferring its internal state variables from one simulation step to another. Finally, the results of the simulation are compared to measurements of series production.

1. Introduction

Over the last decades, the use of numerical methods, especially the Finite Element Method (FEM), increased rapidly. The computer simulations basing on these methods allow to save costs by reducing the number of expensive prototypes in the development of new designs and technologies. However, to get reliable simulation results, it is absolutely necessary to understand the relevant processes and the behaviour of the employed materials. Moreover, it is required to apply realistic boundary conditions to the models as wells as to simplify them as much as necessary to reduce computing time. Hence, the Presta joining process will first be explained in detail to make the boundary conditions of the simulation model introduced later in this work comprehensible.

The Presta joining process is a manufacturing method to build shaft-hub connections and is used primarily as the world's leading process in the series production of assembled camshafts. It allows up to 30% reduced weight compared to conventional cast or forged camshafts. Consequently, there is a significant inertia reduction of the rotating camshafts [1, 2].

The manufacturing of Presta camshafts is divided into two substeps. First, the shaft is locally widened by rolling at the position of the intended cam seat. To achieve this widening the rollers have an approximately sinusoidal cross-sectional surface. Hence, due to the rolling of the shaft a circumferentially oriented profile is formed (figure 1a).

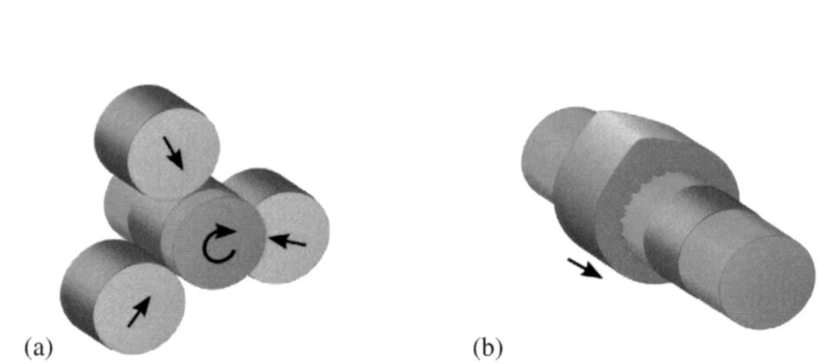

Figure 1. Schematic illustration of the Presta joining process: Local widening of the shaft through rolling (a) and joining of the cam and the shaft (b).

Subsequent to the rolling step, the shaft and the cam are joined, which results in a combination of press fit and form fit. This is achieved by the use of cams with an inner profile oriented parallel to the shaft axis. Consequently, the perpendicular profiles of the shaft and the cam generate a tight fit due to plastic deformations (figure 1b). The inner diameter of the cam is smaller than the outer diameter of the widened cam seat. Thus, during the joining process the rolled profile is squeezed by the cam (figure 2).

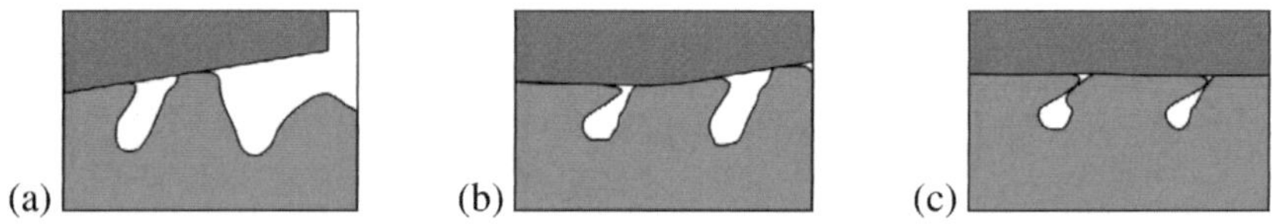

Figure 2. Sectional view derived from a grinding pattern of the deformation of the rolled profile during the joining process: (a) chamfered edge, (b) intermediate stage and (c) final state (cf. [1]).

The simulation of the rolling step of the Presta process has already been presented and discussed before [3]. Within this rolling model two symmetries were used. First, the cam seat has been modelled symmetrically in axial direction such that the model size could be halved. Secondly, a cyclic symmetry due to the arrangement of rollers around the shaft was employed, which reduced the model size to a section of 120 degrees (figure 3).

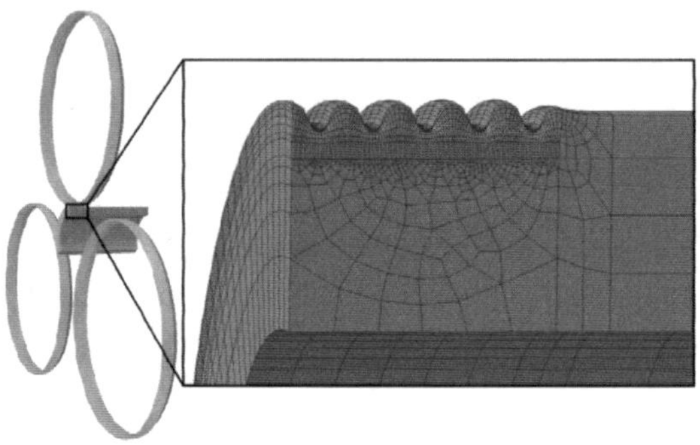

Figure 3. Final state of the rolling simulation with the resulting deformed mesh and the applied symmetries.

The simulation model of the second substep of the Presta joining process simulation is shown in Section 2. At this point, the final state of the rolling simulation represents the initial state of the joining simulation. Thus, despite having new symmetries, load boundary conditions and a new FE mesh, the history of deformations remains in the model. This transfer of the deformation history implies a change of the reference configuration in the employed material model.

In both substeps of the simulation of the Presta joining process, a phenomenological model of finite strain viscoplasticity introduced by Shutov and Kreißig is applied [3, 4]. First, this material formulation is capable of representing the realistic material behaviour even for large deformations. Secondly, this model exactly meets the requirements of multi-stage deformations like the Presta joining process because of its invariance after a change of the reference configuration [5]. Section 3 shows the transfer of the deformation history.

Finally, in Section 4 the results of the forming simulation will be discussed and compared to the experience of the series production.

2. Three-dimensional simulation model of the joining step

2.1 Geometric model

In the second step of the Presta joining process, the cam is forced onto the previously rolled cam seat of the shaft. In analogy to the rolling step the joining process has a very small deformation zone compared to the size of the single parts used in there. Hence, to reduce the model size it is beneficial to utilise symmetries in the assembly of this joining step.

The simulation model of the rolling process included the arrangement of rollers around the shaft to reduce the model to a section of 120 degrees. Moreover, it has been modelled as cyclically symmetric. With regard to its resulting deformation, the joining process model is not axisymmetric (figure 1b). However, the resulting profile of the rolling step has been identified as approximately axisymmetric [3]. Hence, to use the rolled profile as the initial profile of the shaft for the joining model, it is necessary to expand the profile of the shaft in the axial direction. Here, it is sufficient to use one cutting plane to represent the complete shaft. Subsequently, this plane can be expanded to a three-dimensional model.

The size of the shaft sections arc in the joining step depends on the simplifications used in the model. Here, two possible options may be applied: The first one is to keep the original outer contour that controls the kinematics of the valve train in combustion engines (figure 4a). In this model, only one symmetry plane can be employed, which leads to a model that includes a section of 180 degree with regard to the shaft and, consequently, to high computational effort. Another option is to model the outer contour of the shaft as a circular shape (figure 4b).

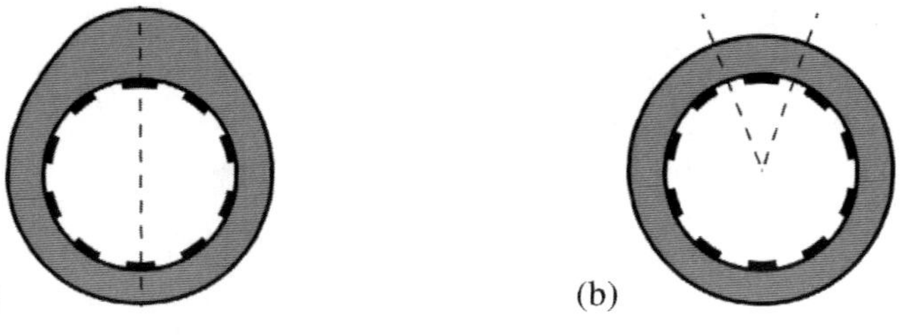

(a) (b)

Figure 4. The cam with its inner profile shown in (a) with the original outer contour and in (b) with the simplified circular shape. The dashed lines show the respectively minimal section needed for the simulation model.

The advantage of this approach is that a cyclic symmetry depending on the number of teeth of the inner profile can be applied. The model discussed in this work contains 36 teeth and, consequently, the assembly of shaft and cam expands over a section of 10 degrees (figure 5a).

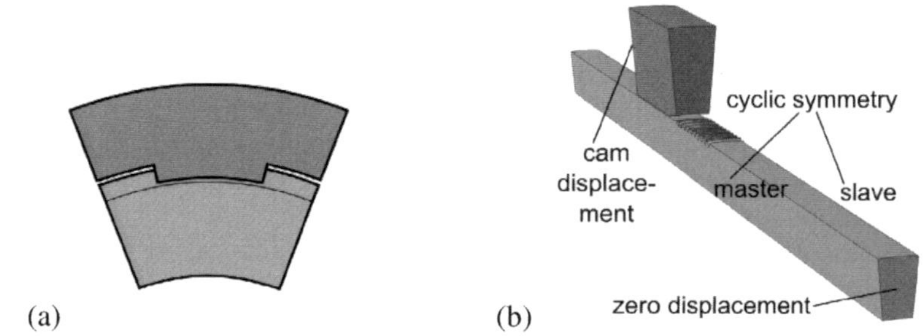

Figure 5. Assembly of the joining simulation: (a) schematic front view of the section and (b) perspective view of the simulation model.

As a result there are three boundary conditions in the joining simulation. First, cyclic symmetry has to be defined at the side surfaces of the model. Furthermore, to force the cam onto the rolled profile there has to be a displacement boundary condition at the cam. In addition, axial displacement of the shaft is set to zero to prevent rigid body motions (figure 5b).

2.2 FE-mesh of the shaft

Due to the change of geometric models between the two simulation steps, a new mesh is required for the simulation of the joining step. Thus, the new mesh can be adjusted to the requirements of the expected deformations in this second model. This in particular applies to the tangential mesh density. The resulting profile of the rolling step is nearly axisymmetric with small circumferential gradients in the displacements and stresses of the rolled profile [3]. Consequently, its tangential mesh is coarse compared to the cross-sectional mesh.

Due to the inner profile of the cam, the deformation of the rolled profile after the joining process cannot be expected as axisymmetric anymore. The teeth's edges cause high circumferential gradients in displacements and in stresses. Thus, in this region a finer mesh is necessary. Furthermore, the interpolation error between the old and the new mesh can be minimized by reusing the cross-sectional mesh of the rolling step (figure 6).

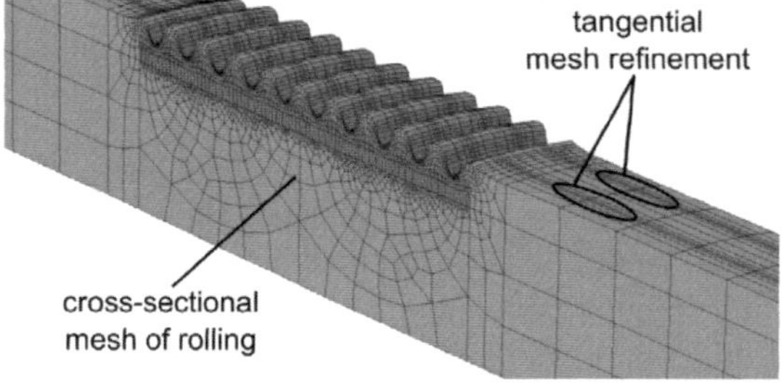

Figure 6. Mesh of the shaft section with the tangential finer mesh in the close to the tooth's edges and the reused cross-sectional mesh of the rolling step.

Likewise, it is possible to generate a new mesh to the shaft and to interpolate the internal state variables at the new integration points. However, a new cross-sectional mesh leads to higher interpolation errors regarding internal state variables. For this purpose, the following section shows the general procedure of transferring the deformation history between the several simulation models.

3. Transfer of deformation history

As already mentioned before, the Presta joining process is a multi-stage manufacturing method consisting of rolling and joining. Within the simulation of both steps, different geometric models with different boundary conditions are required. For the change between the FE-models, the deformed cross-sectional geometry of the rolling process is adopted within the joining step. However, the switch from a 120 degree section to 10 degree results in a new mesh for the second model. Consequently, the internal state variables of the material model would be set to zero by default and there is no deformation history stored at the integration points of the FE-model of the joining step. However, for the applied phenomenological model of finite strain viscoplasticity, a suitable approach to transfer the internal state variables to a new model has been introduced by Shutov et al. [5].

The transfer of internal state variables between the two simulation models has been implemented as follows. First, the integration point coordinates of the joining step simulation model are exported to a data file. Next, these coordinates are rotationally transformed into one plane. Subsequently, the resulting point coordinates are employed to define a path within the final FE-mesh of the rolling step (figure 7).

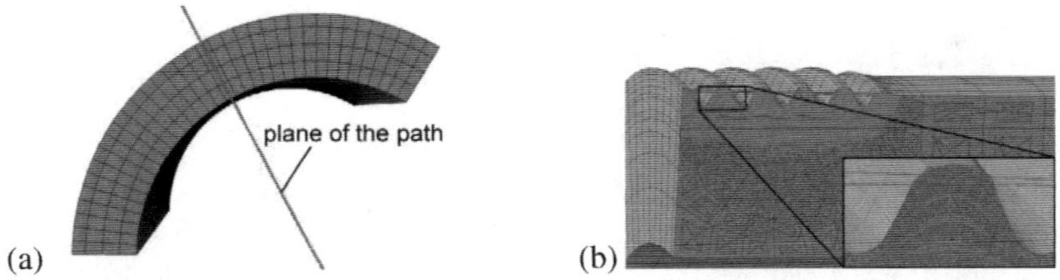

Figure 7. (a) The position of the path's plane in the FE-mesh of the rolling step and (b) the data points of the path connected with lines.

At the data points of the path, the required internal state variables are interpolated by the employed FE-software and passed to a second data file. Hence, in this second file every integration point of the joining step has an associated set of internal state variables representing the deformation history of the rolling step. Finally, at the start of the joining simulation, the state variables are imported and initialized at every integration point.

The applied material model includes the scalar inelastic arc length that can be seen as the degree of deformation since it increases monotonically with plastic deformation [6]. In the rolling step this inelastic arc length reaches its maximum in the grooves of the rolled profile (figure 8).

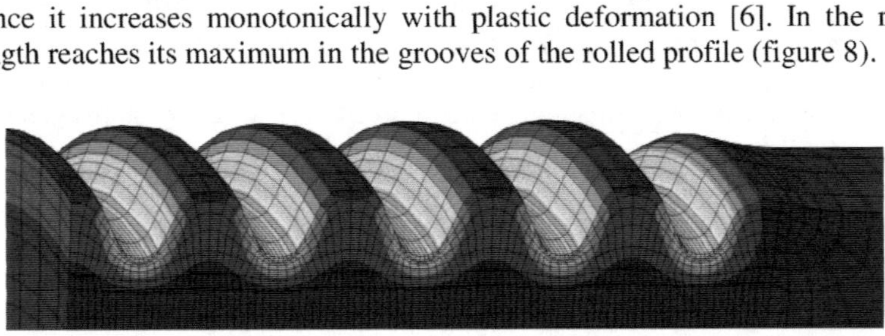

Figure 8. Contour plots of the inelastic arc length at the end of the rolling simulation with its maxima in the grooves of the rolled profile. In addition, the coarse mesh in tangential direction is shown.

A comparison between the plots of the inelastic arc length at the end of the rolling step and at the beginning of the joining step shows almost no differences (figure 9). First, this is due to the identical cross-sectional mesh in both simulations. Secondly, this can be attributed to very small gradients of the physical values in tangential direction.

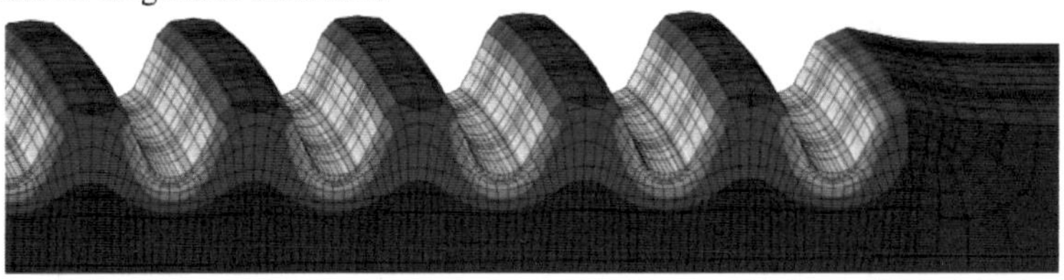

Figure 9. Contour plots of the inelastic arc length at the beginning of the joining simulation with the almost exact reproduction of the contour at the end of the rolling step. Here, the tangential mesh density with its refinement for the specific requirements of the joining simulation is shown.

4. Simulation results and discussion

Once the internal state variables are initialized in the simulation of the joining step, the transfer of the deformation history of the rolling process is completed. Hence, the effects of local isotropic and kinematic hardening due to the deformations of the rolling are still included in the model (figure 10).

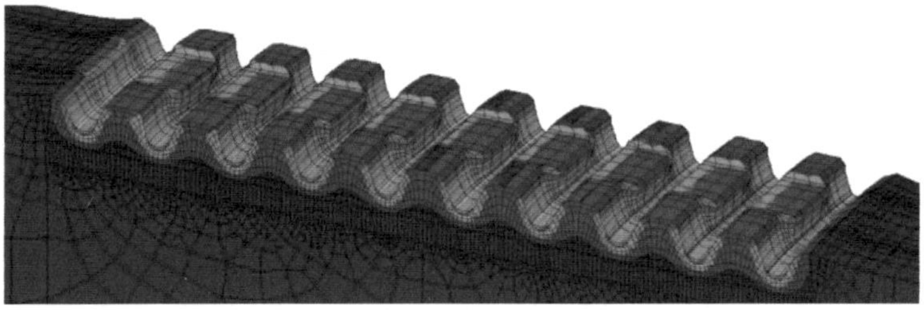

Figure 10. Contour plot of the inelastic arc length after the joining simulation with the cam hidden to get a clear view of the resulting profile.

Furthermore, the final inelastic arc length shows that the rolling of the shaft causes more plastic deformation than the joining step. The degree of deformation increases significantly only at regions, where the inner edges of the cam cut into the rolled profile.

To compare the results of the joining simulation with the series production of Presta camshafts, it is appropriate to examine the joining force. The joining force can be measured easily in the production process. Mathematically, it results from the cam displacement boundary condition as the reaction force. The comparison between experimental and computed data shows suitable agreement (figure 11).

IOP Conf. Series: Materials Science and Engineering **181** (2017) 012030 doi:10.1088/1757-899X/181/1/012030

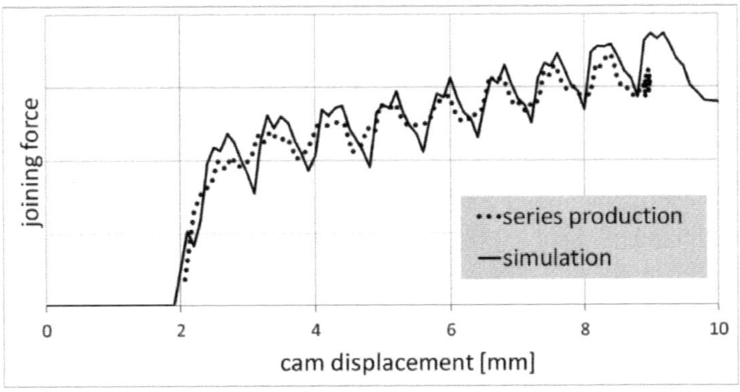

Figure 11. Comparison between the joining force measured in series production (dots) and computed in the joining process simulation (lines).

Both curves of the joining force show similar absolute values. Moreover, nearly constant distances between local maxima and their adjacent minima can be observed. This can be explained by the fact that the rolling profile in the simulation is perfectly symmetric. Manufactured rolling profiles are typically a little bit lopsided and, consequently, some of the rolled grooves are deeper and some of them are rather flat. Hence, the required force to push the cam over the bulges varies with their height.

The general accordance between the measured and computed joining forces can be seen as an indicator for a realistic simulation of the joining process. Nonetheless, more experiments are needed to prove the model's ability to provide reliable predictions for different materials or geometries. For example, to compare the results of the FE-model with manufacturing a possible test would be to cut through a completed camshaft and to measure the deformed profile (cf. figure 2).

5. Conclusions

This work presents a simulation model for the joining step of the Presta process. The model refers to the simulation model of the rolling process introduced by Scherzer et al. [3]. Hence, like in the rolling simulation the parameters and geometries used in the model are taken from the series production of Presta camshafts.

First the geometry, the boundary conditions and the employed symmetries of the model have been explained in detail. Here, the model could be reduced to a section of 10 degree due to the cyclic symmetry based on the simplification to a circular outer profile of the cam. Furthermore, the mesh of the joining model has reused the existing cross-sectional mesh of the rolling simulation to minimize interpolation errors.

The simulation models representing the Presta joining process have been realized using different geometries. Such a change of models is generally associated with a loss of deformation history. However, in multi-stage forming processes like the Presta process the deformation in the first substep is significant. Therefore, in this work an approach to transfer the deformation history has been applied using the invariance under change of the reference configuration of the applied phenomenological model of finite strain viscoplasticity. This is done by transferring the internal state variables of the rolling simulation to the joining model and thus to the new reference configuration.

Ultimately, this work gives an overview of the results of the joining simulation. First, the quality of the transfer of internal state variables is proved by a comparison between the inelastic arc length at the end of the first step and at the beginning of the second step. Based on this, the further increase of the degree of deformation due to the joining process is shown. This reveals the major plastic deformation occurring in the rolling step and shows the importance of maintaining the deformation history in the simulation of the Presta joining process. Most importantly, the comparison between the joining force curves of the series production and of the simulation exhibits suitable accordance.

Perspectively, the two simulation models of the rolling step and the joining step will be extended to a third substep: loading the connection. Here, the Presta camshaft will be loaded similar to the torque in combustion engines. Then, it will be tested how much load the connection endures (figure 12).

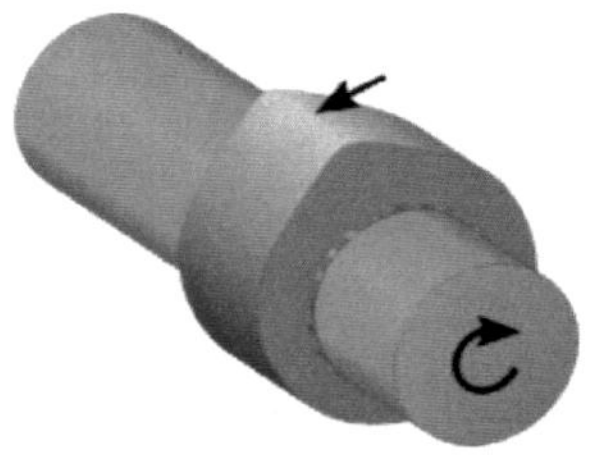

Figure 12. Loading of the camshaft representing the load of the valve train.

Acknowledgements
The research shown in this work originates from the collaborative research center SFB 692 "High-strength aluminum based lightweight materials for safety components", which is supported by German Research Foundation (DFG). Furthermore, this research was carried out in close cooperation with ThyssenKrupp Presta GmbH in Chemnitz.

References

[1] Lengwiler A 2011 *Fehlerfortpflanzung, Simulation und Optimierung von Prozessketten anhand der gebauten Nockenwelle* (Eidgenössische Technische Hochschule ETH Zürich)

[2] Meusburger P 2006 Lightweight design in engine construction by use of assembled camshafts *MTZ worldwide* **67.3** p 10–2

[3] Scherzer R, Silbermann C B and Ihlemann J 2016 FE-simulation of the Presta joining process for assembled camshafts – local widening of shafts through rolling *IOP Conf. Series: Materials Science and Engineering* **118** p 012039

[4] Shutov A V and Kreißig R 2008 Finite strain viscoplasticity with nonlinear kinematic hardening: Phenomenological modeling and time integration *Comp. Methods Appl. Mech. Eng.* **197.21** p 2015–29

[5] Shutov A V, Pfeiffer S and Ihlemann J 2012 On the simulation of multi-stage forming processes: invariance under change of the reference configuration *Materialwiss. Werkstofftech.* **43.7** p 617-25

[6] Shutov A V, Kuprin C, Ihlemann J, Wagner M F X and Silbermann C 2010 Experimental investigation and numerical simulation of the incremental deformation of a 42CrMo4 steel *Materialwiss. Werkstofftech.* **41.9** p 765–75

A numerical and experimental comparison of test methods for the shear strength in hybrid metal/thermoplastic-compounds

E Saborowski, M Scholze, T Lindner and T Lampke

Institute of Materials Science and Engineering, Chemnitz University of Technology, 09125 Chemnitz, Germany

E-mail: erik.saborowski@mb.tu-chemnitz.de

Abstract. The lap shear test is a common method for determining the interlaminar shear strength of material compounds due to its simplicity with regard to specimen production and experimental realization. However, the obtained results strongly depend on the material pairing and the specimen geometry, which have a significant influence to the stress state within the interface during the experiment. A torsion test method using butt-bonded hollow cylinders seems more appropriate due to the expected more homogeneous shear stress distribution. The aim of this work is to compare both testing methods by experimental and numerical investigations. A comparative study with thermally joined aluminum-polyamide hybrid compounds was performed. The corresponding simulations were carried out with the commercial FE-software Abaqus™, using a user defined material model for the thermoplastic joining partner and the built-in cohesive behavior contact property for modelling interfacial damage.

1. Introduction

Current developments in lightweight applications confirm the trend towards multi-material design. The combination of different material classes offers an efficient link of beneficial properties in hybrid material compounds. Especially in the automotive industry, hybrid material compounds of polymer and metal provide an economical way for lightweight design [1]. In addition to that the interest in thermoplastic polymers like polyamide is increasing because of advances like short cycle times and better recyclability compared with thermosetting materials. Furthermore, the application of thermoplastic polymers offers a direct thermal joining without adhesives or joining elements. A suitable way for mechanical interlocking of both components under thermal joining conditions is the surface pretreatment of the metal by abrasive or additive processes prior to the joining process [2].

The mechanical behavior of the compound depends on the properties of each single component as well as on the interface between them. Also, the assisted design by finite element (FE)-simulations in all stages of development requires input data that describes single material properties of the compound and in addition to that the interface between the single materials. The properties of the metallic joining partner can be determined by tensile tests, whereas the thermoplastic joining partner has to be tested under various load conditions and speeds due to strongly strain rate dependent and viscoelastic-viscoplastic behavior. The properties of the interface play a key role under the combination of different material classes in hybrid material compounds. The interface behavior has to be replicated by data determined in test methods of linear elastic fracture methods (LEFM) and values of interface strength at different load directions [3].

IOP Conf. Series: Materials Science and Engineering **181** (2017) 012031 doi:10.1088/1757-899X/181/1/012031

In particular, the lap shear test based on DIN EN 1465 is one of the most used methods for determining the shear strength in multi material compounds reasoned for example by the simple production of test specimens. One disadvantage of the lap shear test is the non-uniform stress distribution in the interface [4], particularly dependent on the difference in stiffness between the joined materials. Especially non-reinforced thermoplastic polymers in combination with metallic layers exhibit an early crack initiation at the interface in lap shear tests. A consequence of this is a premature failure and overlap length-dependent values for the obtained shear strength [4, 5, 6]. In this study, a torsion test method using butt-bonded hollow cylinders based on DIN EN 14869-1 is investigated as an experiment for the determination of interface shear strength in hybrid material compounds and compared to the lap shear test. The advantage of a more homogeneous shear stress distribution and the resulting, more reliable determination of the shear strength value should be shown by FE-simulations and validated by experimental data on a polyamide 6 (PA6)/aluminum compound.

2. Experimental

2.1 Lap shear test

In the lap shear test, two flat plates, each consisting of a partner of the compound to be examined, are bonded together. The specimen geometry used here is shown in Figure 1. The specimen is loaded with a tensile force perpendicular to the joining zone, causing a shear stress within the interface. Assuming an ideal homogeneous stress distribution, the shear strength τ_{max} is calculated from the fracture force F_{max} divided by the overlap area A_o.

$$\tau_{max} = \frac{F_{max}}{A_o} = \frac{F_{max}}{w \bullet l_o} \tag{1}$$

Width:	w = 25 mm
Height plate 1:	h_1 = 3 mm
Height plate 2:	h_2 = 3 mm
Length plate 1:	l_1 = 100 mm
Length plate 2:	l_2 = 100 mm
Overlap length:	l_o = 12.5 mm

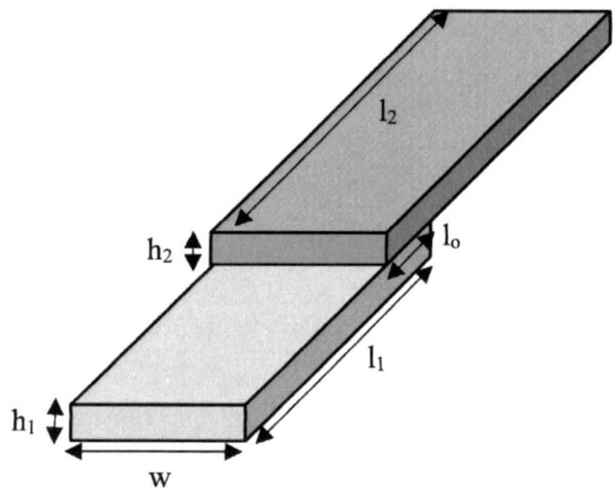

Figure 1. Geometry of lap shear specimen based on[7].

2.2 Butt-bonded hollow cylinder torsion test

In the hollow cylinder torsion test, two thin-walled pipes, each consisting of a partner of the compound to be examined, are butt-bonded together. The specimen geometry used in this work is shown in Figure 2. The compound is loaded with a torsional moment, causing a shear stress parallel to the interface. The specimen is twisted until fracture and the shear strength τ_{max} then calculated from the maximum torque T_{max} divided by the polar section modulus W_P.

IOP Conf. Series: Materials Science and Engineering **181** (2017) 012031 doi:10.1088/1757-899X/181/1/012031

$$\tau_{max} = \frac{T_{max}}{W_P} = \frac{T_{max} \bullet 16 \bullet d_o}{\pi \bullet ({d_o}^4 - {d_i}^4)} \tag{2}$$

Outer diameter: $d_o = 24.5$ mm
Inner diameter: $d_i = 20.5$ mm
Height cylinder 1: $h_1 = 30$ mm
Height cylinder 2: $h_2 = 30$ mm

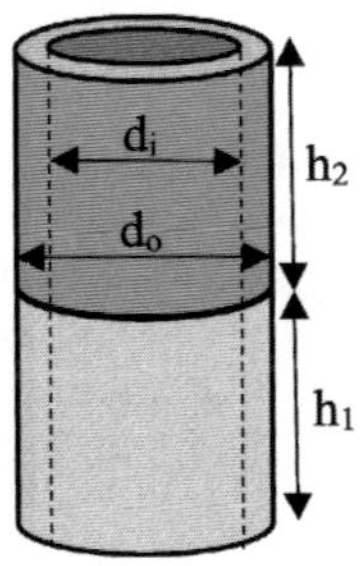

Figure 2. Geometry of butt-bonded hollow cylinder specimen based on[8].

2.3 Specimen production

The tests were carried out at an EN AW 6082/PA6 hybrid material compound. To create a reproducible and well adhering surface, undercuts were produced by coating with thermal spraying. After grit blasting, the aluminum substrate was coated by a high-velocity electric wire arc spraying system developed at Chemnitz University of Technology (CUT) [9]. As feedstock material, a cored wire NiAl95/5 alloy with a diameter of 1.6 mm was used. This is a typical material for bonding layers which provides excellent adhesion to the substrate. The parameters used for producing the rough surface structure are shown in Table 1. According to the production parameters determined in [5], the joining process was done by hot pressing with a joining pressure of 0.3 MPa and a joining temperature of 230 °C, which is approximately 10 °C above the melting temperature of the polyamide. The lap shear specimens shown in Figure 3a have to be reworked after pressing by removing expelled melt around the interface with a metal saw. The rework on the hollow cylinder specimens however causes much more effort because they need to be turned on the inside as well as on the outside to ensure the necessary centricity for testing. The polyamide has been conditioned before manufacturing and testing according to [10] at 70 °C and 62% humidity. 3-6 specimens have been produced and tested per geometry.

Table 1. Parameters for thermal spraying.

Spraying system	High-velocity electric wire arc (modified Visu Arc 350)
Substrate activation	Grit blasting Al_2O_3, 2 bar , 70°, 20 cm, type WFA F 16
Spraying material	Ni - 95% / Al - 95% (DURUM DURMAT AS-756 1.6 mm)
Current	120 A
Voltage	28 V
Process gas	Air, 5.5 bar
Spraying Distance	200 mm
Row spacing	6 mm
Feed speed	1 m/s

IOP Conf. Series: Materials Science and Engineering **181** (2017) 012031 doi:10.1088/1757-899X/181/1/012031

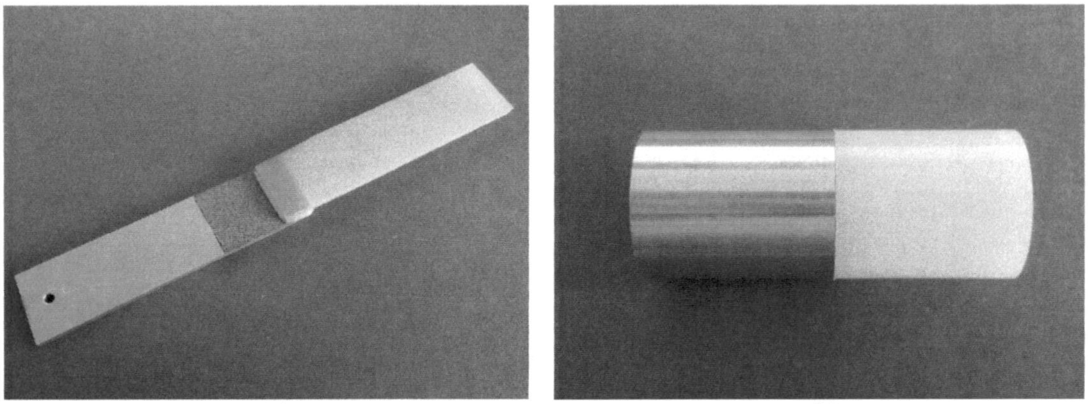

(a) (b)

Figure 3. (a) Lap shear specimen (b) Butt-bonded hollow cylinders specimen.

3. Simulation

To investigate the obtained results depending on the used test method, FEM simulations offer a detailed view into the relevant mechanical quantities. Therefore, a sufficiently exact simulation model is needed. This includes the material data as well as the parameters of the interface. The simulations were carried out with the commercial FE-software Abaqus™ 6.14.

3.1 Material model

The metallic joining partner of the compound can be depicted with a built-in elastic-plastic material model, whereas a user defined material is needed for the thermoplastic partner due to its time- and strain rate-dependent behavior. This user defined material is modelled by a viscoplastic material model by Kießling et al. [3, 11]. This material model is formulated based on a concept of material modeling at large strains by directly connected rheological elements, which has been introduced by Kießling et al. [12]. The parameters for the user defined material have to be determined by solving an optimization problem, which is based on minimizing the deviation between the load-displacement curves produced in material tests (tensile tests, relaxation tests, cyclic tests) and the numerically reproduced load-displacement curves using the material model.

3.2 Interface

The interfacial damage is modeled by the built-in cohesive behavior contact property. The principle of the failure mechanism is the same for the normal (subscript I) as well as for both shear (subscripts II and III) directions. For reasons of comprehensibility, the following explanations do not distinguish between the load directions. The interfacial adhesion can be described as a spring braced between the two contacting surfaces (see Figure 4). When a load is applied, the spring keeps its penalty stiffness K till the damage initiation point at the separation δ_1 is reached. Further increasing separation causes damage evolution and corresponding softening of the spring till complete failure at δ_2 .

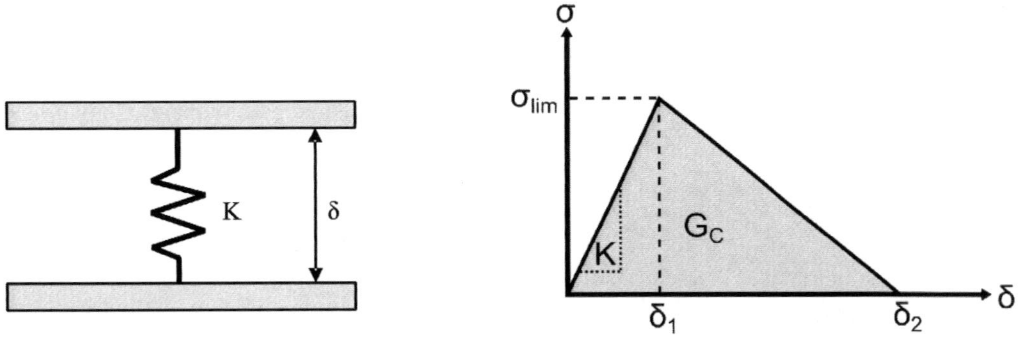

Figure 4. Cohesive behavior contact property, principle and stress-separation curve.

The parametrization is done by the critical stress σ_{lim}, where the damage is initiated, and by the interlaminar fracture toughness G_C, which is the energy per area needed to destroy the interface. Both of these values can be obtained experimentally. The separation at the damage initiation point δ_1 can be calculated from the chosen penalty stiffness of the spring K and the critical stress σ_{lim} (Equation (3)). The separation at the point of complete failure δ_2 is calculated from the interlaminar fracture toughness G_C and the critical stress σ_{lim} (Equation (4)).

$$\delta_1 = \frac{\sigma_{lim}}{K} \tag{3}$$

$$\delta_2 = 2 \bullet \frac{G_C}{\sigma_{lim}} \tag{4}$$

The damage variable d describes how the penalty stiffness is reduced once the initiation criterion is reached. For linear softening, it can by calculated by Equation (5). It has a value of 0 at the damage initiation point δ_1 and reaches 1 at the point of complete delamination δ_2.

$$d = \frac{\delta_2 \bullet (\delta - \delta_1)}{\delta \bullet (\delta_2 - \delta_1)} \tag{5}$$

The resulting stress σ can then be calculated by Equation (6).

$$\sigma = \begin{cases} K \bullet \delta, & \delta < \delta_1 \\ (1-d) \bullet K \bullet \delta, & \delta_1 \leq \delta \leq \delta_2 \\ 0, & \delta > \delta_2 \end{cases} \tag{6}$$

The critical tension as well as the fracture toughness vary depending on the load direction. In the investigated case, the adhesion in shear direction is much higher than the adhesion in normal direction. For this reason, at least 4 parameters have to be obtained experimentally to describe the interface sufficiently. These are the critical stresses in normal and shear direction $\sigma_{I,lim}$ and $\tau_{II,lim}$ as well as the corresponding fracture toughness G_{IC} and G_{IIC}. The critical stresses can be determined with the hollow cylinder specimens described in this article. For the critical tension in normal direction, the hollow cylinders are loaded with a tensile force instead of a torque. For the fracture toughness, modified versions of the Single cantilever beam (SCB) test based on [13] for G_{IC} and End Notched Flexure (ENF) test based on [14] for G_{IIC} are suitable. The behavior under mixed mode loading (combined normal and shear) can be specified by using analytical forms. For the damage initiation, the quadratic stress criterion is used. The damage is initiated, when the quadratic interaction function involving the contact stress ratios reaches the value of 1 (Equation (7)) [15]. For the normal direction, only tensile stress is taken in to consideration, whereas compressive stress does not cause damage.

$$\left(\frac{\langle\sigma_I\rangle}{\sigma_{I,lim}}\right)^2 + \left(\frac{\tau_{II}}{\tau_{II,lim}}\right)^2 + \left(\frac{\tau_{III}}{\tau_{II,lim}}\right)^2 = 1 \tag{7}$$

The damage evolution law describes how the damage evolves under mixed mode loading. The Benzeggagh-Kenane criterion is used in this work because it seems to produce the most reasonable results over a large range of mode ratios [16]. The corresponding fracture toughness can be calculated by Equation (8), where G_I, G_{II} and G_{III} denote the strain energy release rates in normal and both shear directions. The cohesive property parameter η considers the influence of the mode mix to the resulting fracture toughness G_C. The entire calculation of the resulting stiffness and damage under mixed mode loading can be seen in[16].

$$G_C = G_{IC} + (G_{IIC} - G_{IC}) \bullet \left(\frac{G_{II} + G_{III}}{G_I + G_{II} + G_{III}}\right)^{\eta} \tag{8}$$

The values used for the simulation can be seen in Table 2. The critical stresses $\sigma_{I,lim}$ and $\tau_{II,lim}$ were obtained with the specimens shown in Figure 3b. Experiments for determining the fracture toughness G_{IC} and G_{IIC} as well as the corresponding exponent η are currently under investigation and did not produce reasonable results yet, so they have been chosen to fit the experimental data.

Table 2. Simulation parameters.

$\sigma_{I,lim}$ [N/mm²]	$\tau_{II,lim}$ [N/mm²]	G_{IC} [mJ/mm²]	G_{IIC} [mJ/mm²]	K_I [N/mm³]	K_{II} [N/mm³]	η [-]
8	17	0.7	3.5	1000	1000	0.6

4. Results and discussion

4.1 Interface microstructure

The bonding mechanism between substrate and coating system is mainly mechanical interlocking. For NiAl95/5, also micro welding is proven. The coating as an additional element results in the formation of two interfaces in the hybrid compound. Due to the bad polar matching of metal and PA6, the main bonding mechanism between coating and PA6 is also mechanical interlocking. As seen in the cross section of the joined specimens (Figure 5), the coating forms a relatively thin layer which fully covers the aluminum sheet. A good adhesion is achieved by broadly excluded void parts at the interface. The rough surface of the coating with undercuts as well as the porous structure are offering a better potential for mechanical interlocking. Nevertheless, PA6 shows partially delamination along straight tracks of the interface. A reason for this is the mismatch in chemical compatibility as well as the difference in thermal expansion. The shrinkage tension during the cooling in the joining process favors the void formation at straight tracks. Nevertheless, mechanical interlocking and clamping can be provided by the coating roughness, which offers a good adhesion.

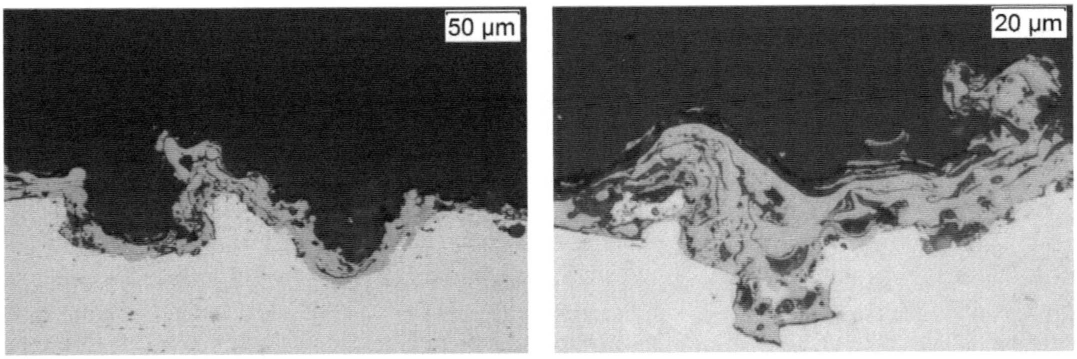

Figure 5. Cross section of a joined specimen (bottom: aluminum, center: NiAl95/5, top: polyamide 6).

4.2 Grey scale correlation

For visualizing the specimen deformation and crack propagation during the lap shear experiment, grey scale correlation data has been recorded. Figure 6 shows the side view of the overlapping area of a specimen with the standard geometry (l_o = 12.5 mm) with aluminum on the top side and polyamide 6 on the bottom side. The evaluated quantity shown is the deformation due it makes the crack visible. Figure **6**a shows that the crack is already emerging at a small cross head displacement of 11% of the total displacement where fracture occurs. The delamination propagates continuously along the interface (Figure 6b). Just before failure at 99% of the fracture displacement, an additional crack emerges at the end of the overlapping area (Figure 6c). Immediately afterwards the specimen breaks.

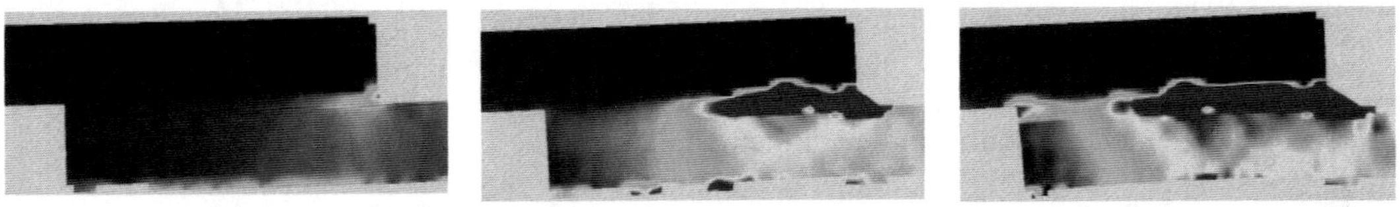

(a) (b) (c)

Figure 6. Grey scale correlation data of the interface (a) 11% fracture displacement (b) 62% fracture displacement (c) 99% fracture displacement.

4.3 Experimental and numerical results

To investigate the overlap length-dependent results for the shear strength τ_{max} obtained by the lap shear test, specimens with an overlap length of l_o = 5 mm respectively l_o = 25 mm were produced in addition to the standard geometry shown in Figure 1 with l_o = 12.5 mm. The tests were carried out with a cross head displacement of 1 mm/min. The experimental as well as the corresponding numerical results can be seen in Table 3.

Table 3. Experimental and numerical results of lap shear tests.

l_o [mm]	$\overline{F_{max}}$ [N]	$\overline{\tau_{max}}$ [N/mm²]	$\tau_{max,Sim}$ [N/mm²]	Δ_{Sim} [%]	$\tau_{max,Sim}/\tau_{II,lim}$ [%]
5	1425	10.59	10.61	0.2	62.4
12.5	2529	7.99	8.27	3.5	48.6
25	2763	4.47	4.29	-4	25.2

It can be seen, that there is indeed a strong influence of the overlap length on the resulting shear strength. An increase of 400% in l_o only causes an increase of 94% in the maximum tensile force F_{max} and a decrease of 58% in the corresponding shear strength τ_{max}. Multiple overlap lengths were also considered in the carried-out simulations to investigate this effect by post processing the numerical results. Figure 7a and Table 3 show the same effect as the experiments, the resulting shear strength $\tau_{max,Sim}$ declines with increasing overlap length. Moreover, the determined tensile force converges to a maximum (Figure 7b). The crack front evolves with nearly constant shape (see Figure 8b) without significantly increasing force (see Figure 10a) if the overlap length is long enough. The stresses concentrate around the crack front, whereas the still adhering interface areas experience only small stresses (see Figure 9a). The partial delamination causes an elongation of the unconstrained polyamide, which reduces the system stiffness. This reduction in stiffness reduces the increase in tensile force caused by further cross head displacement. From a certain delaminated length, both effects compensate and the delamination continues without increasing tensile force.

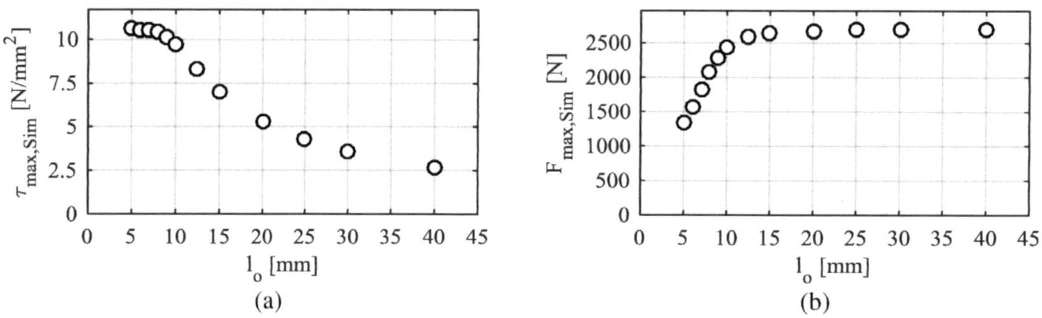

Figure 7. Overlap length-dependent results obtained by simulated lap shear test (a) shear strength (b) maximum tensile force.

Figure 8 shows the interfacial damage variable CSDMG in the joining zone during the simulated lap shear test. It can be seen that the crack already emerges at small loads (a). The crack front evolves continuously (b) until a tensile force causes additional delamination at the end of the overlapping area (c). Immediately afterwards total failure occurs and the polyamide is separated completely from the aluminum (d). The way of crack propagation that can be seen here fits well to the grey scale correlation data shown in Figure 6.

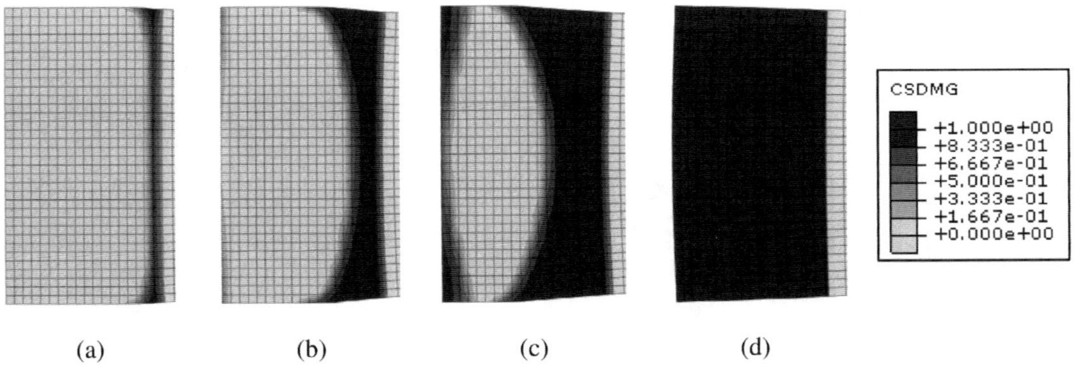

Figure 8. Damage variable CSDMG in the interface during lap shear test with l_o = 12.5 mm (a) 2.5 mm cross head displacement (b) 7.6 mm cross head displacement (c) 9.3 mm cross head displacement (d) 9.5 mm cross head displacement.

The hollow cylinder torsion tests were carried out with an angular velocity of 5 °/min. The experimental and numerical results are shown in Table 4.

Table 4. Experimental and numerical results of hollow cylinder torsion test.

$\overline{T_{max}}$ [Nm]	$\overline{\tau_{max}}$ [N/mm²]	$\tau_{max,Sim}$ [N/mm²]	Δ_{Sim} [%]	$\tau_{max,Sim}/\tau_{II,lim}$ [%]
25.35	17.91	17.76	-0.8	104.5

The shear strength τ_{max} determined by hollow cylinder torsion is, with a value of 17.91 N/mm², about 124 % higher than the 7.99 N/mm² determined by the lap shear test with standard geometry. The corresponding simulation result for the determined shear strength $\tau_{max,Sim}$ is very close to chosen simulation parameter $\tau_{II,lim}$ with an error of only 4.5%, which shows that the shear strength value obtained by hollow cylinder torsion is much more reliable than the value obtained by the lap shear test with l_o = 12.5 mm with an error of 51.4%. This significant difference can be explained with the much more homogenous shear stress distribution in the interface. Figure 9 shows the shear stress distribution just before total failure for both testing methods. In the lap shear test, a significant share of the interface is already delaminated, whereas the shear stress in the still connected spots varies between very low values and the critical stress $\tau_{II,lim}$. In contrast, the shear stress is almost homogenous all over the interface in the hollow cylinder torsion test. The failure occurs nearly instantly after reaching the maximum torsional load without premature crack propagation. The load-displacement curves for both tests can be seen in Figure 10. During the lap shear test, the force increases close to its maximum value and then stays nearly constant despite further increasing displacement due to the crack propagation. During the hollow cylinder torsion test, however the torque increases fitting to the increase of displacement. The declining gradient of the torque curve is caused by the viscoelasticity and -plasticity of the polyamide.

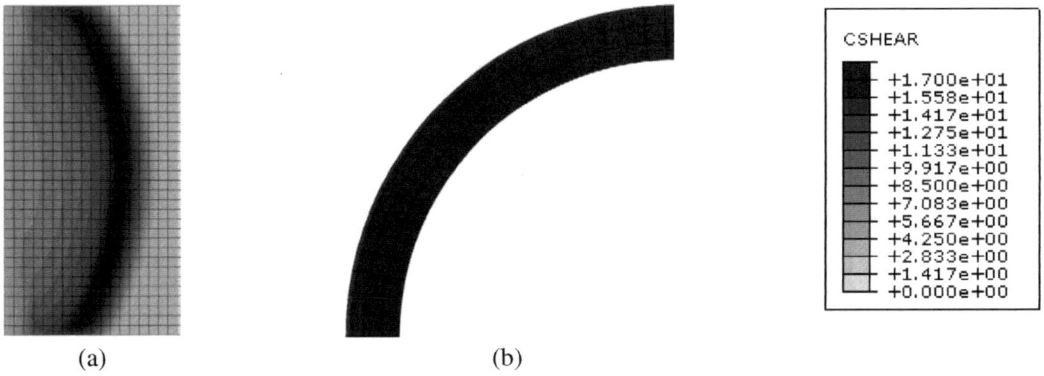

Figure 9. Shear stress distribution in the interface just before failure (a) lap shear test with l_o = 12.5 mm (b) hollow cylinder torsion test.

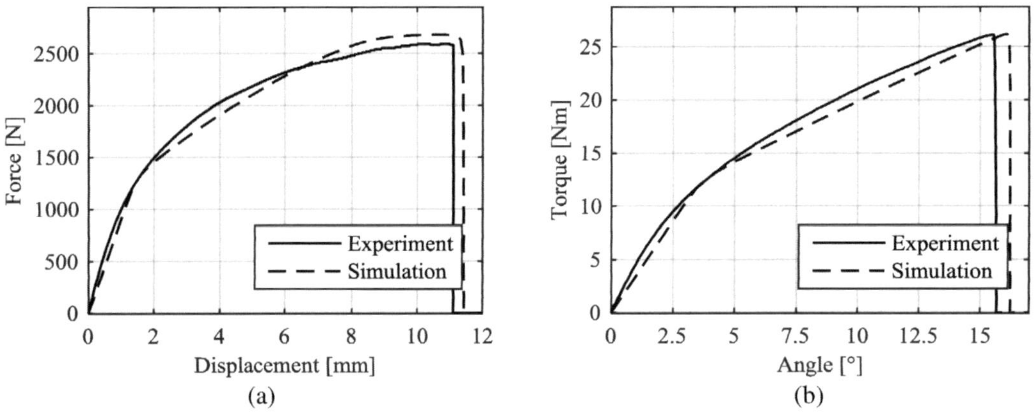

Figure 10. Load-displacement curve (a) Lap shear test with l_o = 25mm (b) hollow cylinder torsion test.

Figure 11 shows the obtained shear strength values from the experiments as well as from the simulations for all examined specimen geometries. It can be seen that there is a good correlation for the obtained shear strength as well as for the load-displacement curves (see Figure 10), which leads to the conclusion that the numerical model based on the viscoelastic-viscoplastic material model from Kießling in connection with the cohesive behavior contact property is a useful approach for simulating hybrid metal/thermoplastic-compounds.

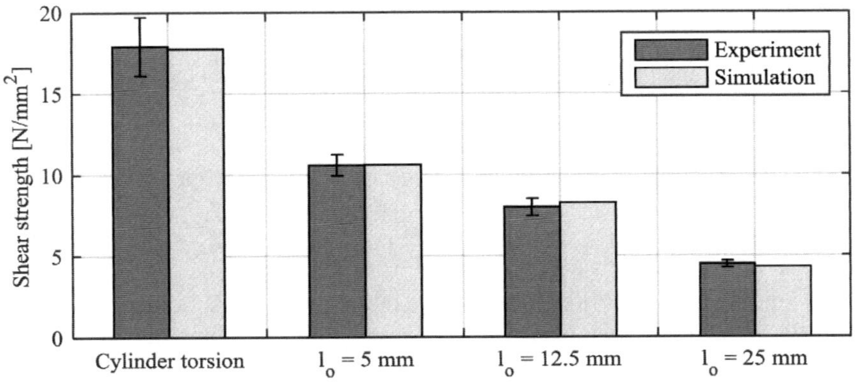

Figure 11. Obtained shear strength depending on specimen geometry.

5. Summary and conclusions

The lap shear test is a commonly used experiment for determining the shear strength of a material compound. In context of other investigations at various metal-thermoplastic compounds within the cluster of excellence MERGE, it showed that the obtained results for the shear strength are varying widely depending on the specimen geometry and the material pairing. Another test, where two hollow cylinders, butt-bonded with the same procedure, are twisted until failure seems more appropriate for determining the shear strength because of a more homogenous shear stress distribution within the interface. To investigate the quality of both testing methods in comparison to each other, a numerical study, validated by experimental results, has been carried out in this work. The numerical models were built up in Abaqus™ 6.14. For the metallic joining partner, a built-in elastic-plastic material behavior could be used, whereas a user defined material model was necessary for the viscoelastic-viscoplastic thermoplastic part. The interfacial damage was modeled with the built-in cohesive behavior contact property. Experiments have been carried out for three lap shear specimen geometries and one hollow cylinder torsion specimen geometry.

The numerical simulations as well as the experiments showed that the results determined by the hollow cylinder torsion test are much more reliable than the results from the lap shear test. The reasons could be seen best in the post processing of the numerical results, where interfacial damage and stresses during the crack propagation were evaluated. During the lap shear test, the crack evolves early before failure with a very inhomogeneous shear stress distribution. In contrast, the failure during the hollow cylinder torsion test occurs nearly instantaneous after the crack evolves. As a result, the critical shear stress obtained from the hollow cylinder torsion test differs only 4.5% from the corresponding simulation parameter, whereas the value obtained from the lap shear test with the standard geometry is 51.4% too low. The results from the lap shear test can be improved by a shorter overlap length, but because of the occurring inhomogeneous stress state, there will always be a significant error.

Consequently, the lap shear test should only be used to qualitatively compare the shear strength of specimens with the same geometry and the same material pairing. It is not appropriate to determine the real shear strength, whereas the hollow cylinder torsion test is well suited. Furthermore, a good correlation between simulation and experiment with regard to crack propagation, the resulting load-displacement curves and the obtained shear strengths shows that the numerical model based on the viscoelastic-viscoplastic material model from Kießling in connection with the cohesive behavior contact property for modeling the interfacial damage is a useful approach for simulating hybrid metal/thermoplastic-compounds.

Acknowledgement

This study was supported by the Federal Cluster of Excellence EXC 1075 "MERGE Technologies for Multifunctional Lightweight Structures". Funding by the German Research Foundation (Deutsche Forschungsgemeinschaft, DFG), Germany is gratefully acknowledged. The support by Felix Schubert, Christoph Wollschläger, Steffen Clauß, Karla Muhr, Elke Benedix, Torsten Rabes, Michael Zychski, Franz Nozicka, Gerd Paczkowski, Martin Schällig (all from the Institute of Materials Science and Engineering) and Robert Kießling (Institute of Mechanics and Thermodynamics) is also gratefully acknowledged.

References

[1] Eckstein L, Eichhorn U, Fischer K, Hackenberg U, Jagt P v D, Ohlsen J, Schöneburg R, Schwenk W, Stracke K F and Weitzenböck J 2015 Multimaterialbauweise - Welche Herausforderungen bringt sie mit sich? *ATZ* **117** no. 06 pp. 8-12

[2] Schuberth A, Göring M, Lindner T, Töberling G, Puschmann M, Riedel F, Scharf I, Schreiter K, Spange S and Lampke T 2016 Effect of new adhesion promoter and mechanical interlocking on bonding strength in metal-polymer composites *IOP Conference Series: Materials Science and Engineering* **118**

[3] Kießling R, Ihlemann J, Pohl M, Stommel M, Dammann C, Mahnken R, Bobbert M, Meschut G, Hirsch F and Kästner M 2016 On the Design, Characterization and Simulation of Hybrid Metal-Composite Interfaces *Applied Composite Materials*

[4] Rasche M 1990 Der Zugscherversuch in der Klebtechnik *Adhäsion* no. 11 pp. 36-43

[5] Flock D 2011 *Heat Conduction Bonding Of Plastic-Metal Hybrid Parts* RWTH Aachen pp. 81-86

[6] Naghipour P, Schulze K, Hausmann J and Bartsch M 2012 Numerical and experimental investigation on lap shear fracture of Al/CFRP laminates *Composites Science and Technology* no. 72 pp. 1718-1724

[7] *DIN EN 1465: Determination of tensile lap-shear strength of bonded assemblies* 2009 Berlin: Beuth Verlag

[8] *DIN EN 14869-1: Determination of shear behaviour of structural bonds – Part 1: Torsion test method using butt-bonded hollow cylinders (ISO 11003-1:2001,modified)* 2011 Berlin: Beuth Verlag

[9] Paczkowski G, Rupprecht C and Wielage B 2012 High-velocity arc spraying - The renaissance of a coating process *Thermal Spray Bulletin* no. 2 pp. 140-147

[10] *DIN EN ISO 1110: Plastics - Polyamides - Accelerated conditioning of test specimens* 1997 Berlin: Beuth Verlag

[11] Kießling R, Ihlemann J, Riemer M, Drossel W G, Scharf I, Lampke T, Sharafiev S, Pouya M and Wagner M F X 2016 On the development of an intrinsic hybrid composite *IOP Conference Series: Materials Science and Engineering* **118**

[12] Kießling R, Landgraf R, Scherzer R and Ihlemann J 2016 Introducing the concept of directly connected rheological elements by reviewing rheological models at large strains *International Journal of Solids and Structures* **97-98** pp. 650–667

[13] *ASTM D5528 - 13: Standard Test Method for Mode I Interlaminar Fracture Toughness of Unidirectional Fiber-Reinforced Polymer Matrix Composites* 2013 West Conshohocken: ASTM International

[14] *ASTM D7905 / D7905M-14: Standard Test Method for Determination of the Mode II Interlaminar Fracture Toughness of Unidirectional Fiber-Reinforced Polymer Matrix Composites* 2014 West Conshohocken: ASTM International

[15] Simulia 2016 *Abaqus Analysis User's Manual 6.14 (37.1.10)*

[16] Camanho P P, Dávila C G and DeMoura M F 2003 Numerical Simulation of Mixed-mode Progressive Delamination in Composite Materials *Journal of Composite Materials* **37** no. 16 pp. 1415-1438

IOP Conf. Series: Materials Science and Engineering **181** (2017) 012032 doi:10.1088/1757-899X/181/1/012032

Reduction of liquid metal embrittlement in copper-brazed stainless steel joints

T Uhlig[1], V Fedorov[1], M Elßner[1], G Wagner[1] and S Weis[2]

[1]Institute of Materials Science and Engineering, Chemnitz University of Technology, 09125 Chemnitz, Germany
[2]Professur Füge- und Beschichtungstechnik, Institut für Produktionstechnik, Westsächsische Hochschule Zwickau, 08012 Zwickau, Germany

E-mail: thomas.uhlig@mb.tu-chemnitz.de, sebastian.weis@fh-zwickau.de

Abstract. Due to its very good formability and the low raw material cost, pure copper in form of foils is commonly used to braze plate heat exchangers made of stainless steel. The difference in the electrochemical potentials of brazing filler and base material leads to corrosion effects in contact with electrolytes. This may lead to leakages, which decrease the reliability of the heat exchanger during service in potable water. The dissolution of the emerging corrosion products of brazing filler and base material induces the migration of heavy metal ions, such as Cu^{2+} and Ni^{2+}, into the potable water. The so-called liquid metal embrittlement, which takes place during the brazing process, may intensify the corrosion. The brazing filler infiltrates the stainless steel along the grain boundaries and causes an embrittlement. This paper deals with the determination of the grain boundary erosion dependent on the degree of deformation and heat treatment of the stainless steel AISI 316L.

1. Introduction

Plate heat exchangers (PHE) are used to transfer thermal energy from one fluid to another without a direct contact of the media. They consist of corrugated stainless steel plates that are arranged diametrically opposed creating channels for the liquids. The resulting surface extension in combination with to the small sheet thickness and high thermal conductivity of the steel result in a very high efficiency. Joining the plates together is carried out by pressing, welding and brazing. In case of pressing there is a need to use rubber sealing gaskets, which limit the applicable pressure. Welding requires lots of process steps due to the many stacks of plates. The advantage of brazing is the possibility to produce all the joints within the PHE in one process step. This offers the possibility to produce big batch sizes in a very cost efficient joining process [1]. The brazing process is carried out in vacuum furnaces using brazing fillers in form of foil or paste. Because of their good formability and the low raw material cost, pure copper foils are used as filler metal [2]. The difference in the electrochemical potentials of copper and stainless steel leads to corrosion effects in contact with electrolytes [3]. Investigations indicate different reasons for leakages and failures: intergranular attack in the braze metal and gap corrosion between braze metal and steel [4]. The chromium oxide layer on the stainless steel surface causes the anodic dissolution of the braze metal [5] [6]. The corrosion behaviour is also influenced by the residual stresses within the heat exchanger plates. During the forming process of the corrugated PHE plates, different degrees of deformation emerge depending on the position at the corrugation. When the plates are brazed, the liquid copper filler infiltrates the grain boundaries. Cracks mostly occur in the high-deformed areas of the plates originating from the eroded grain boundaries [7]. This phenomenon is known as liquid metal embrittlement (LME). Another factor of influence on the corrosion behaviour is the sensitization of the stainless steel during the brazing process. The precipitation of chromium carbides at the grain boundaries of the steel may cause local differences in the chromium content. Especially with regard to the application of PHE in potable water, all the corrosion processes are of interest. The dissolution of corrosion products of brazing filler and base material into the water must be taken into account. The leaching of heavy metal ions may have harmful influence on the health of people in contact. For this reason, there are legally fixed maximum values for the concentrations of these ions. Besides the expectable leaching of copper ions, investigations indicate a nickel enrichment of the potable water [8]. This leads to the assumption that the corrosion of the steel plates and the consequent dissolution of the corrosion products in the potable water cause the leaching. With respect to the influence of LME on the corrosion resistance, a lower tendency to grain boundary erosion should improve the corrosion resistance. There is evidence that LME can be reduced by a thermomechanical treatment of the stainless steel. Investigations on the influence of thermal pretreatments of stainless steels on the LME by copper-based brazing fillers indicate the possibility of an improvement [9]. The determination of the resistance against intergranular corrosion of stainless steels in acids leads so similar results. A thermomechanical treatment of the stainless steel, which leads to the formation of coincident grains is carried out. They form the so-called coincidence site lattice (CSL). The grain boundaries in CSL are regarded as low energy grain boundaries [10] [11] [12].

2. Experimental

The stainless steel AISI 316L (X2CrNiMo17-12-2) in dimensions of 300 x 50 x 1.5 mm is used as base material for the investigations. For the cold rolling experiments, sheets are used in the as delivered as well as the normalised condition. The normalising treatment is carried out at 1100 °C for 1 h in a vacuum furnace (10^{-2} Pa). The cold rolling process is conducted resulting in different reductions of sheet thickness. The values are selected referring to literature data at 0 %, 2 %, 4 % and 6 %. The subsequent heat treatment is done for 1 h in vacuum at 900 °C, 940 °C, 980 °C and 1020 °C. The thermomechanically treated sheets are cut using a metallographic cut-off machine to dimensions of 25 x 25 mm². During the cutting process, the sheets are water cooled to prevent a heat input into the samples. The brazing experiments are carried out in a vacuum furnace (10^{-2} Pa) at 1120 °C with a holding time of 15 min. Pure copper foil (B-Cu100-1085) is used as brazing filler.

The brazed joints are cut in half, ground and polished using grinding consumables and polishing cloths. The metallographic specimens are etched using V2A etchant in order to generate a better contrast between braze metal and base material. Microsections are captured by light microscopy. The tendency to LME is determined by measuring the infiltration depth as well as the number of infiltrations using optical image analysis. Each value corresponds to the average of 9 measurements at different positions of the cross sections. The texture is determined by X-ray diffraction (XRD) and illustrated in pole figures. They represent the orientation distribution of crystallographic lattice planes. In case of a face centered cubic (fcc) lattice it is common to measure three kinds of planes. The surface of the cube (200), the plane along the face diagonal (220) and the plain along the space diagonal (111).

3. Results and Discussion

Before performing the brazing experiments, the influence of the normalizing process on the grain size is determined. The surfaces of the stainless steel sheets are polished and etched in order to contrast the grain boundaries. The grain size increases significantly, approx. form 10 - 50 µm in the as delivered state to 100 - 300 µm after the normalizing treatment. This should not have any negative influence on the mechanical properties of the steel because the brazing process is carried out at almost the same temperature, which should lead to a similar grain growth during brazing. The results of the texture measurements are summarized in the following figures, the rolling direction is horizontal. In the as delivered condition, a typical rolling texture can be observed, figure 1. The (200) and (111) planes show minor maxima at about 22 to 32° transverse to the rolling direction. In the (220) plane a main maximum in the centre is observable. The smooth transitions between the different intensities indicate a small size of crystallites. After the normalising treatment, a uniform distribution of the grain orientations occurs in all measured planes, figure 2. The sharp transitions between the different intensities imply big crystallites, which is in good agreement with the grain size measurements. A subsequent rolling reduction of 2 % causes only minor changes in texture. There is an indication of minor maxima in the (111) plane, figure 3. At a rolling reduction of 4 %, the transition between the intensities gets smoother, which implies a grain refinement. The minor maxima in the (111) plane can be seen more clearly, figure 4. After 6 % reduction in thickness, a rolling texture is observable in all measured planes. The minor maxima in the (200) and (111) plane occur in the same angles as in the as delivered state. The most significant modification can be seen in the (220) plane. Besides the main maximum in the centre, two minor maxima transverse to the rolling direction occur at about 45 to 65°.

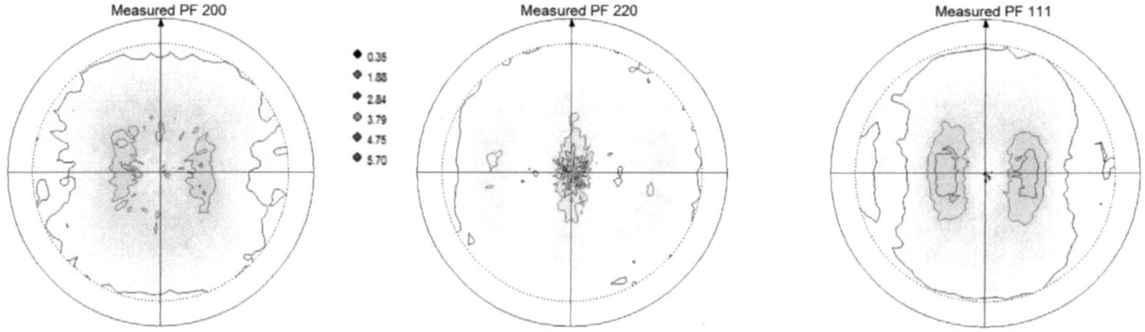

Figure 1. Texture of AISI 316L in the as delivered state.

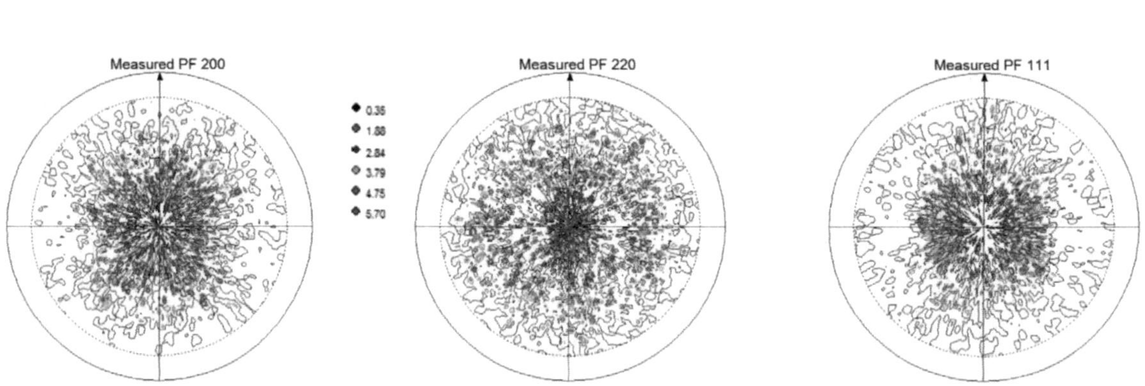

Figure 2. Texture of AISI 316L after the normalising treatment (1 h, 1100 °C, 10^{-2} Pa).

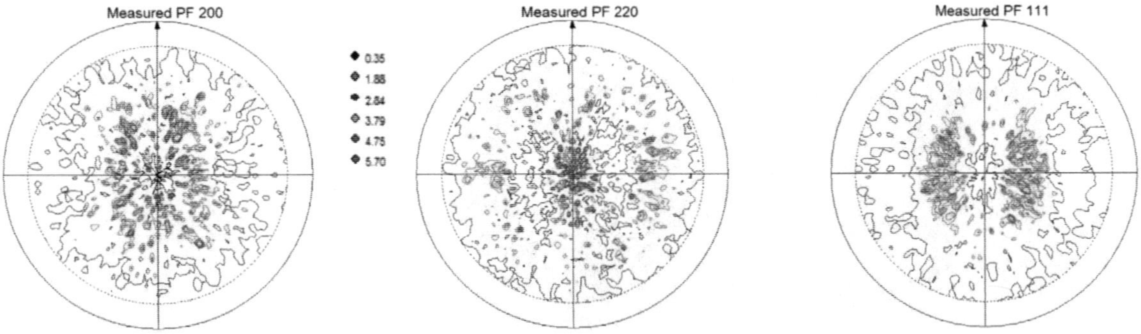

Figure 3. Texture of AISI 316L after the normalising treatment and 2 % rolling reduction.

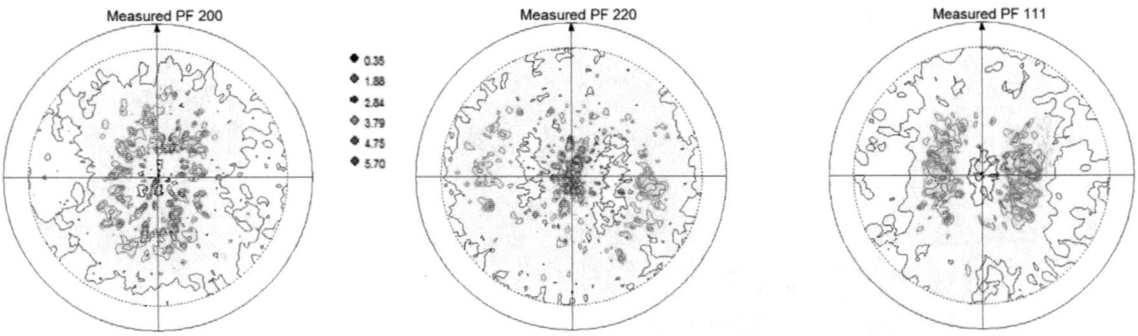

Figure 4. Texture of AISI 316L after the normalising treatment and 4 % rolling reduction.

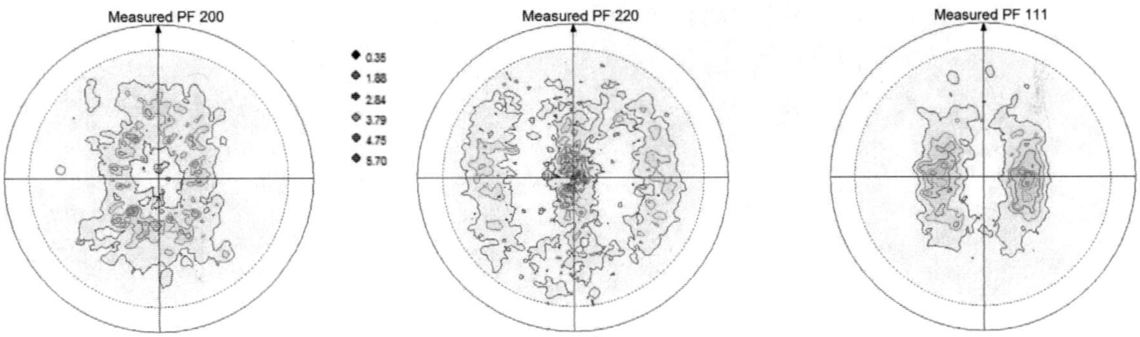

Figure 5. Texture of AISI 316L after the normalising treatment and 6 % rolling reduction.

The results of the metallographic studies on brazed samples are shown in the following. The normalised condition is illustrated in figure 6 and figure 7. At low heat treating temperatures, the rolling reduction rates strongly affect the tendency to LME. The highest number of infiltrations occurs at high deformation rates. With increasing heat treatment temperature, this effect seems to shift to lower deformation rates. As can be seen in figure 6, there are also high values at 980 °C and 2 % deformation. At 1020 °C the lowest values are measured. The rolling reduction does not have a noticeable effect on the number of infiltrations. The lowest values of infiltration depth are measured at low deformation rates, resp. no rolling reduction after normalising, figure 7. Whilst at low heat treating temperatures the values are strongly dependent on the rolling reduction rate, the lowest effect occurs at 1020 °C, too.

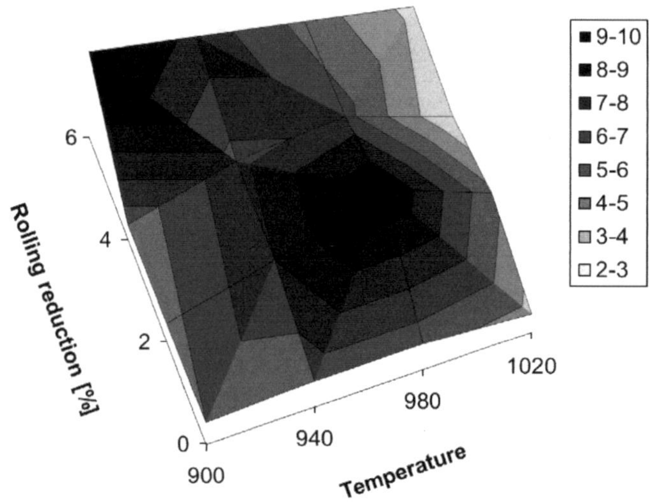

Figure 6. Number of infiltrations of normalised AISI 316L as function of rolling reduction rate and heat treating temperature.

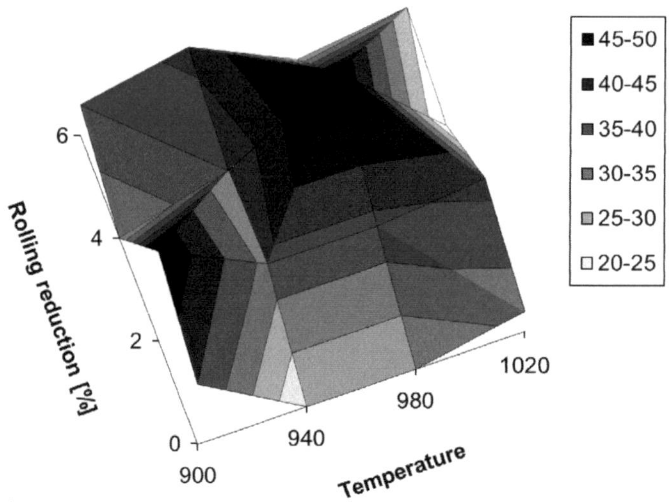

Figure 7. Mean depth of infiltrations of normalised AISI 316L as function of rolling reduction rate and heat treating temperature.

The results of the experiments using samples in the as delivered condition are shown in figure 8 and figure 9. The influence of heat treating temperature and rolling reduction rate on the number of infiltrations is lower than in the normalised state, figure 8. This can be explained by the high deformation rate which is already applied into the stainless steel sheets during the manufacturing process. Hence, the influence of the further rolling reduction is relatively small. As can be seen in figure 9, there is a noticeable influence of the heat treating temperature on the infiltration depth. The minimum values occur after a heat treatment at 940 °C. In contrast to the normalised condition, the lowest values and also lowest deviations are determined at a lower heat treatment temperature. On the assumption of higher residual stresses within the sheets in the as delivered state, the heat treatment leads to a relaxation at lower temperatures which decreases the tendency to LME.

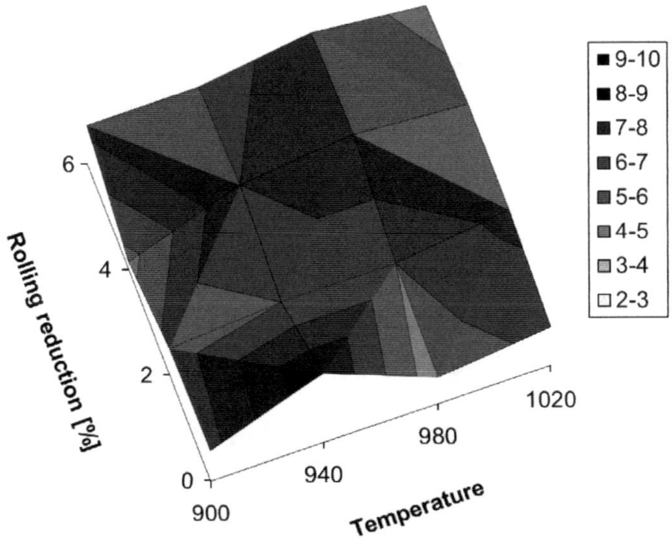

Figure 8. Number of infiltrations of as delivered AISI 316L as function of rolling reduction rate and heat treating temperature.

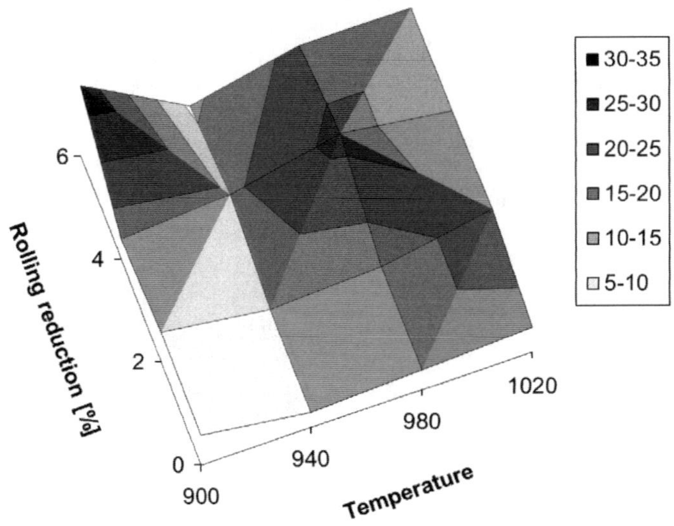

Figure 9. Mean depth of infiltrations of as delivered AISI 316L as function of rolling reduction rate and heat treating temperature.

4. Conclusions

The performed investigations show the possibility to reduce the grain boundary erosion, resp. liquid metal embrittlement (LME) of stainless steel by copper brazing filler. Investigations are carried out using steel sheets in the as delivered as well as the normalised condition. During the normalising process, the grain size increases approximately fivefold. In the normalised state, there is a significant influence of the rolling reduction rate and the heat treatment temperature on the number and the depth of the infiltrations. The lowest values as well as deviations are determined after a heat treatment at 1020 °C. In the as delivered state, the minimum infiltration depth values are determined after heat treating at 940 °C. In comparison to the normalised state, the number of infiltrations is higher in the as delivered state. This can be explained by the smaller grain size, which unavoidably leads to a higher number of grain boundaries. The residual stresses within the sheets in the as delivered state lead to a relaxation at lower temperatures, which decreases the tendency to LME.

Acknowledgements
Part of the research was carried out within the scope of the project # 18507 BR funded by the Bundesministerium für Wirtschaft und Energie (BMWi) via the Arbeitsgemeinschaft industrieller Forschungsvereinigungen "Otto von Guericke" e.V. The authors thank the BMWi for financing this project.

References

[1] Oehmigen H-G 2009 Produktionstechnik in der Wärmeübertragerfertigung *Verlag Publico* 6-18
[2] Wielage B, Weis S, Uhlig T, Fedorov V 2015 Löten von Wärmeübertragern aus Al, Cu und Edelstahl *DVS-Berichte Band* **315** 438-443
[3] Baskutis S 2011 Mechanics and material aspects in serviceability prediction of the heat exchangers *Mechanics* **17** 239-245
[4] Stichel W 1998 Korrosionsprobleme mit gelöteten Plattenwärmetauschern aus nichtrostendem Stahl *IKZ Haustechnik* **15** 27-33
[5] Ehreke J 1998 Messung der Kupfer- Ionenkonzentration in Stagnationswässern aus kupfergelöteten Plattenwärmeübertragern *Materials and Corrosion* **49** 195-197
[6] Pajonk G 2004 Korrosionsschäden an kupfergelöteten Plattenwärmetauschern
[7] Köhler S 1999 Schäden an gelöteten Plattenwärmetauschern bei der Erwärmung von Trinkwasser *Materials and Corrosion* **50** 227-232
[8] DVGW W10/01/05 2009 Schwermetallmigration aus Bauteilen der Trinkwassererwärmung
[9] Pashkov I 2013 Phenomenon of interaction between the solder melts on the basis of copper on the steel detail surface *Welding Technology Review* 36-40
[10] Kurihara K 2011 Grain boundary engineering of titanium-stabilized 321 austenitic stainless steel *Journal of materials science* **46** 4270-4275
[11] Michiuchi M 2006 Twin-induced grain boundary engineering for 316 austenitic stainless steel *Acta Materialia* **54** 5179-5184
[12] Shimada M 2002 Optimization of grain boundary character distribution for intergranular corrosion resistant 304 stainless steel by twin-induced grain boundary engineering *Acta Materialia* **50** 2331-2341

IOP Conf. Series: Materials Science and Engineering **181** (2017) 012033 doi:10.1088/1757-899X/181/1/012033

Prediction of phase distribution pattern in phase field simulations on Mo_5SiB_2-primary areas in near eutectic Mo-Si-B alloy

O Kazemi, G Hasemann, M Krüger and T Halle

Institut für Werkstoff- und Fügetechnik, Otto-von-Guericke-Universität Magdeburg, P.O. Box 4120, D-39016 Magdeburg, Germany

E-Mail: omid.kazemi@ovgu.de

Abstract. A Mo-10.9Si-20.3B (if not stated otherwise all compositions are given in at.%) alloy was modeled using the phase field method with linearized phase diagrams and thermodynamic data. The simulation results showed that there are two specific microstructural constituents. According to the simulations and experimental microstructural investigations the primary Mo_5SiB_2 phase observed in this study was combined by the Mo_{ss}-Mo_3Si-Mo_5SiB_2 eutectics. The overall composition of the both primary and eutectic area was probed and the phase evolution along with the concentration change of the core areas were explored and visualized in the solidification domain. To verify the accuracy of the simulation results, they were comprised with the experimental achievements. In terms of the fraction of phases, the portion of the primary phase and the eutectic constituent and the phase distribution pattern our results were in good agreement with the experimental observations.

1. Introduction

Materials applied in high temperature functions must maintain their mechanical and physical properties at high surface temperatures. Currently used Ni-based superalloys have met their limit near 1100 °C and therefore many efforts have been conducted to find an alternative to meet the high temperature application's requirements and to reach the aim of an optimum use of resources. Regarding these aspects, Mo-Si-B alloys experienced great attention through the last years. The excellent creep behavior and acceptable oxidation resistance of these ternary alloys are favorable for high temperature applications [1-5].

In their recent thermodynamic calculations, Yang and Chang [5, 6] have shown that two-phase and three-phase alloys, which pass through eutectic reactions among the solidification, are located in the Mo-rich portion of the alloy. Liquidus projection conducted by Yang and Chang was considered to figure out the consequence of the reaction pathway with respect to the alloy's composition in this study [5, 6].

In our previous studies [3, 4] we investigated the phase distribution pattern for the alloy Mo-17.5Si-8B which is located close to the ternary eutectic point and has finished in a volume fraction of almost 50 % Mo_3Si and equal amounts of Mo_5SiB_2 and Mo_{ss} phases.

There are a couple of phases containing boron and silicon in the Mo-Si-B ternary system that have ascended the alloy's melting point. However, the only ternary phase Mo_5SiB_2 (the so-called T2-phase) has received high attention because of its excellent mechanical properties at ultra-high temperatures which are superior compared to Ni-based superalloys [1, 2]. High creep strength along with the relatively low density made this ternary phase of the Mo-Si-B system a favorable structural material. Furthermore, two-phase alloys like Mo_{ss}-Mo_5SiB_2 or more complex alloys like Mo_{ss}-Mo_3Si-Mo_5SiB_2 have shown improved toughness [1]. Accordingly, the Mo_5SiB_2 rich portion of the Mo-Si-B system is the focus of the present study. Moreover, because of favorite properties of the eutectic composition a combination of a primarily solidified Mo_5SiB_2 phase and a ternary eutectic constituent is preferred [7]. Therefore, we have modified the previous model with more detailed thermodynamic data [8] so that we could investigate the solidification behavior of an alloy located in the Mo_5SiB_2 primary area.

Microstructural evolution and solidification reactions of Mo-Si-B alloys in different primary solidification were studied using casting experiments by Ha et al. [1]. They reported a dendritic primary Mo_5SiB_2 phase along with a ternary Mo_{ss}-Mo_3Si-Mo_5SiB_2 eutectic phase for alloy Mo-10.9Si-20.3B. Figure 1 shows the location of the alloys investigated in the both the current and the previous studies.

Figure 2 shows a SEM-BSE image of the as-cast microstructure of the assigned alloy. Probing the phase fractions resulted in 70 % Mo_5SiB_2 primary phase with the balance of the ternary eutectic Mo_{ss}-Mo_3Si-Mo_5SiB_2.

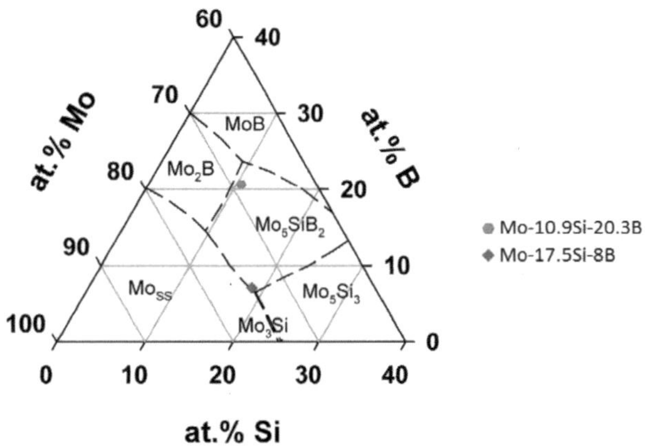

Figure 1. Schematic view of the primary solidification areas according to Yang and Chang [5, 6] and the chemical composition of investigated alloys; alloy shown in red was examined in a previous study [3, 4], the alloy marked in green is subject of the current study.

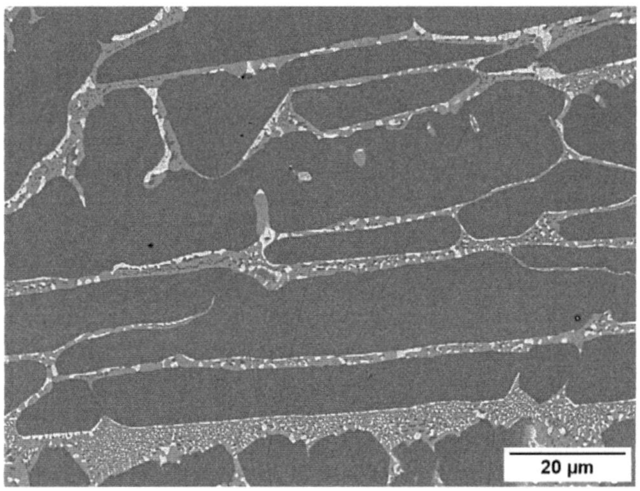

Figure 2. SEM-BSE micrograph of Mo-10.9Si-20.3B, thankfully provided by Ha et al. [1].

First of all we performed the simulation for predicting the solidification of Mo-10.9Si-20.3B alloy within the phase-field method to examine the evolution of the Mo_5SiB_2 primary phase and the Mo_{ss}-Mo_3Si-Mo_5SiB_2 ternary eutectic phase. A comparison between the simulations and experimental observation of Ha et al. [1] proved that the simulations results were in a good agreement with the experimental achievements.

In the following the phase evolution in dependence on the temperature was investigated. In order to probe the dispersion pattern of the elements, their change in the temperature field was examined for both the eutectic and primary area.

2. Simulation procedure

The solidification process was considered to include three solid phases: Mo_{ss}, Mo_3Si and Mo_5SiB_2 [6]. In order to define the correlation between phases and phase fractions, the liquidus projection was performed using the Factsage software [8].

The domain was considered to be a two-dimensional field. The size of the calculation domain along with the grid size has a considerable effect on the calculation time and accuracy. The calculation domain should have an optimal size so that along with generating the most accurate results, the calculation must be performed in a reasonable time. Therefore, after examining various numerical setups, the domain size was adjusted to 24×24 microns containing 90,000 cells with a grid size of 0.08 microns.

In order to apply the actual operational parameter precisely, a thermal gradient was not considered in the model. The temperature trend was applied linearly changing over the time, starting from a temperature in which the whole domain is liquid.

In the simulation domain both of the side walls were adjusted to periodic boundary conditions. Since there is no flux of materials during the solidification, the insulation boundary condition was set for both the top and the bottom wall.

Latent heat release of the produced phases can affect the temperature changes and from a microscopic point of view it may cause some early generated nuclei to melt, which might change the phase distribution pattern. However, the overall phase fractions will be preserved. Due to the serious impact on the computational time, the latent heat was not considered in this study. Nevertheless, the microstructure and phase distribution pattern were acceptably compatible with the experimental observation. It is planned to apply this variable for all phases in prospective studies.

Initial concentrations for the components of the simulation domain were assumed to be at equilibrium situation. At the beginning of the process it was presumed that the composition of the whole domain is unified and equal to Mo-10.9Si-20.3B.

The phase evolution was studied by setting up the nucleation procedure for the phases out of liquid. In order to predict the potential points for the new nuclei of each phase, it was set up into the model to check the possibility of phase generation in each grid cell based on three credentials: the temperature, the composition of the cell and melting temperatures of each phase. The new generated nuclei were supposed to be randomly oriented and it was adjusted to check their stability at every time step to vanish the metastable ones.

The time step was set to automatic and the initial temperature was set to 2370 K. At this temperature, all of the domain must be liquid, based on Factsage prediction. In order to predict the solidification pathway, the correlation between phases and element repartitioning during solidification were taken from the phase diagram. To define the boundary conditions for the used phase field software Micress, two isothermal cross sections of the ternary phase diagram must be incorporated, i.e. one cross section for each phase. The composition changing rate for each element along with the two cross sections should be defined. Table 1 shows the input of the phase diagram for all three phases. It is noteworthy that the phase Mo_3Si is considered as stoichiometric composition. Further it should be noted that all the calculations in this study were performed in the isotropic condition.

Table 1. Applied values for the definition of the phase diagram; reference points 1 and 2 stand for the isothermal section at 2275 K and 2278 K, respectively.

Phase	Mo_3Si	Mo_5SiB_2	Mo_{ss}
Temperature of reference point (K)	2275	2275	2275
Temperature difference of crosses (K)	3	3	3
Si content at reference point 1 (at.%)	19.5	18.6	18.6
Si content at reference point 2 (at.%)	25	10.4	4
B content at reference point 1 (at.%)	5.9	6.8	6.8
B content at reference point 2 (at.%)	0	25	0
dT/dC (K/at.%) for Si at reference point 1	10	-10	-10
dT/dC (K/at.%) for Si at reference point 2	Inf	10	Inf
dT/dC (K/at.%) for B at reference point 1	-7.5	15	15
dT/dC (K/at.%) for B at reference point 2	Inf	15	Inf

The mechanical interaction between phases would affect the phase boundary evolution and consequently the solidification pattern. The surface energy and mobility between different phases are applied from our previous study [3, 4]. Table 2 simplifies the applied values for the required variables.

The diffusion of the elements into different phases was considered by applying the modified values of diffusion coefficients which were used in our previous study. Table 3 shows the diffusion variables of the components in the phases. The simulations were performed with different interface thicknesses and an interface with a width of seven cells proved to result in the most compatible results.

Table 2. Applied values for the operational and thermodynamic parameters.

Cooling rate [K/s]	100
Surface energy between liquid and phases [J/cm^2]	10^{-4}
Entropy of fusion in solid phases [J/cm^{3}K]	0.1
Entropy of fusion in liquid phases [J/cm^{3}K]	0.9
Diffusion coefficient in liquid [cm^2/s]	$2 \cdot 10^{-4}$

Table 3. Applied values for the required diffusion parameters.

Phase	Mo_5SiB_2	Mo_3Si	Mo_{ss}
Diffusion coefficient (cm^2/s)	10^{-8}	10^{-6}	10^{-5}

3. Results and discussion

According to Ha et al.'s research, solidification of the referred alloy composition would result in about 70 % of the domain occupied by the primary Mo_5SiB_2 phase [1]. The balance is occupied with the ternary eutectic composition: Mo_{ss}, Mo_3Si and Mo_5SiB_2. The phase distribution and the distribution of components among the phases for the solidified domain are shown in Figure 3.

The phase distribution pattern as well as the volume fractions of the primary and the eutectic phases were entirely compatible with the Ha et al.'s [1] observations. Figure 4 shows the phase distribution pattern for the Mo-10.9Si-20.3B alloy during the arc melting process. In order to have a thorough comparison of the phase distribution, the phase fraction in both experimental results and simulations are compared to each other in Table 4.

After reproducing Ha et al.'s [1] experimental observations via phase field method, the results were compared to each other and proved to be completely compatible. Figure 5 compares the simulation's prediction of phase distribution and experimental achievements.

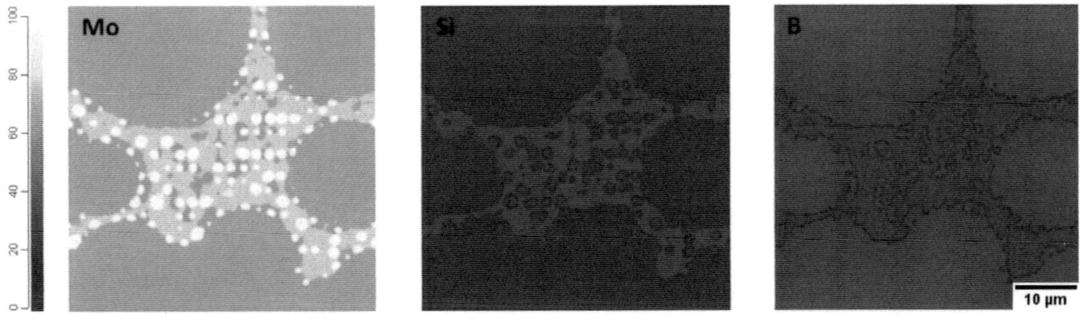

Figure 3. Schematic distribution of components in the solidified Mo-10.9Si-20.3B alloy.

Table 4. Comparison of phase fraction of Mo-10.9Si-20.3B for experiments and simulations.

Phase	Mo_5SiB_2	Mo_3Si	Mo_{ss}
Fraction (%) in experiments	76	17.1	6.9
Fraction (%) in simulations	74	18.6	7.4

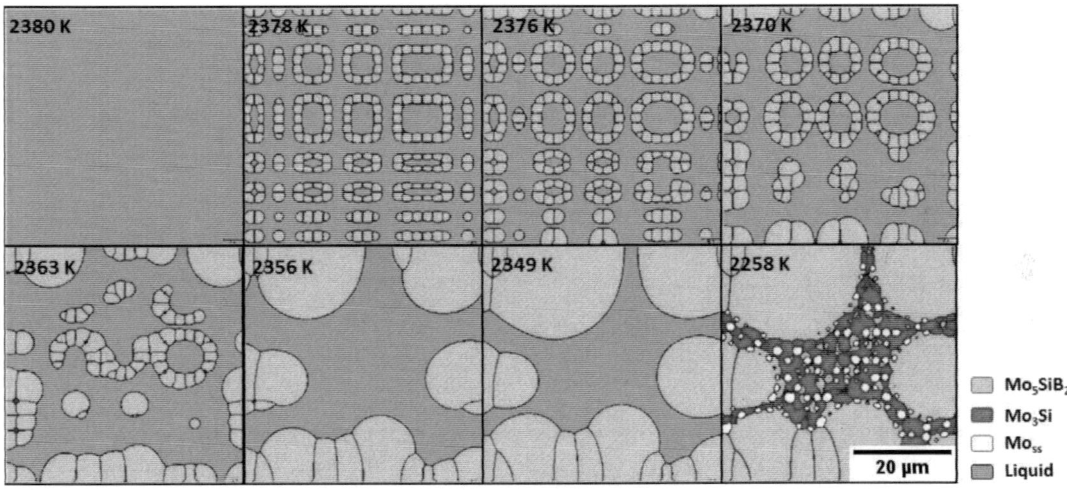

Figure 4. The phase distribution evolution and pattern of Mo-10.9Si-20.3B during cooling down the simulation domain from 2380 K to 2258 K.

The solidification pattern was studied by considering the concentration evolution of each component in the all phases over the time. The visualized results showed that the solidifying paths of all phases meet each other in a ternary point which is supposed to be the ternary eutectic point. Figure 6 shows the solidification path for both the primary and the eutectic phases.

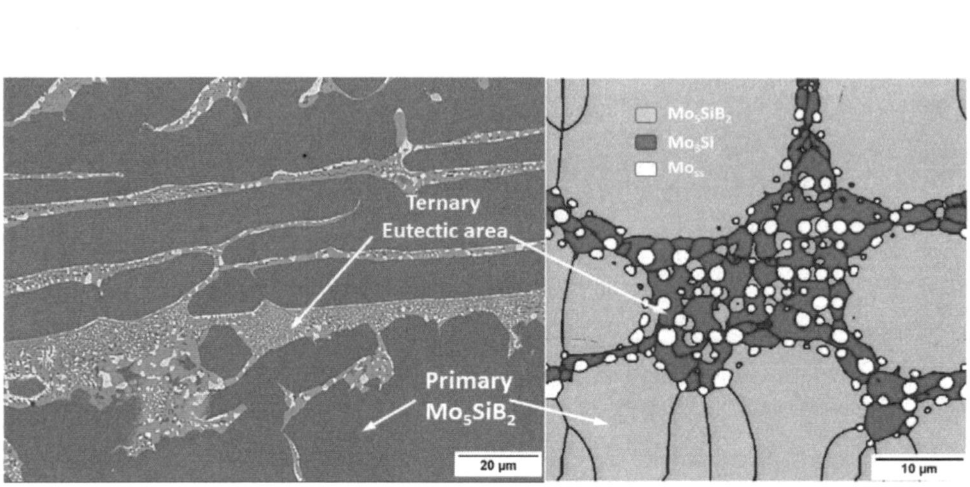

Figure 5. Microstructural evolution of alloy Mo-10.9Si-20.3B in experimental observations vs. simulation.

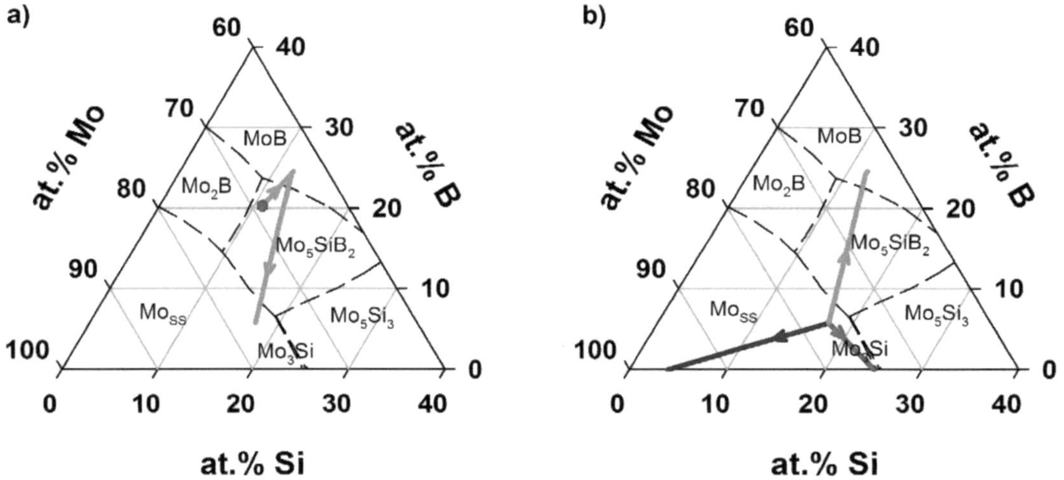

Figure 6. Solidification path derived by simulations for a) primary Mo_5SiB_2 and b) eutectic composition in the context of Yang and Chang´s prediction of the solidification path and primary areas [5]. The orange line refers to primary Mo_5SiB_2, the blue line refers to Mo_{ss} and the red line refers to Mo_3Si. The green point shows the initial composition of the melt. Path a) shows that the liquid start to solidify from the initial composition and by reducing the temperature follows the shown way to eutectic point.

As is apparent in the graphs, the primary Mo_5SiB_2 phase starts to solidify from the initial composition Mo-10.9Si-20.3B and evolve to the stoichiometric composition. Afterwards the residual melt enriches in its Mo concentration until it reaches the ternary eutectic composition at the ternary point. At this point a coexistence of the four phases, namely liquid, Mo_{ss}, Mo_3Si and Mo_5SiB_2, is present and determines the ternary eutectic reaction. By calculation of the Si and B content of each phase it was concluded that Mo_{ss} tends to the assumed maximum solubility of 4 at.% for Si [9]. Also it was found that both Mo_5SiB_2 and Mo_3Si are stoichiometric compositions with respect to the Si content as defined in the model.

As shown in ternary graphs, there is a slight difference between the ternary eutectic point predicted by Yang and Chang and our simulation results. It is worth mentioning that, for their thermodynamic calculations, Yang and Chang assumed a possible variation of the composition of the phase Mo_5SiB_2 by $(Mo)_{0.625}(B,Si,Va)_{0.375}$ due to the presence of vacancies (Va) in the Mo_5SiB_2 lattice. However, in order to generate the phase diagram within the Factsage data along with preserving Yang and Chang´s database we have considered a solubility range for B in Mo_5SiB_2 as $(Mo,B)_{0.625}(B,Si,Va)_{0.375}$. Moreover, Yang and Chang assumed the phase transitions to be in equilibrium conversion but due to the high cooling rate in the simulation setup, which is a typical boundary condition in casting processes, it is not expected to really meet the equilibrium phase transition.

5. Conclusions

The main objective of this research was to investigate the solidification of the alloy Mo-10.9Si-20.3B using the phase field method to extend the already published model [3, 4] which was only able to predict the solidification and microstructure evolution for an eutectic alloy composition. In order to perform this study the Micress software was used. To verify the software's generated phase fractions, micro-structural morphologies and component distribution pattern, the experimental study of Ha et al. [1] was used. The before-mentioned composition was selected for this investigation, because it consists of a considerable amount of primary Mo_5SiB_2 phase, which is supposed to perform an enhanced mechanical behavior in combination with the ternary eutectic Mo_{ss}-Mo_3Si-Mo_5SiB_2 microstructural constituent. The simulation results showed to be in a good agreement with the experimental observations and the presented model is able to predict the observed phase formation patterns and the amount of the individual phases.

The solidification path extracted from the compositional changes of the components in the phases has shown that the melt solidifies from its initial composition and then, after producing a high amount of primary Mo_5SiB_2, the residual melt turns into a ternary composition at the ternary eutectic point. The point varies slightly from Yang and Chang's calculations [5] due to the fast cooling and also slightly different assumptions in thermodynamic calculations. However, the practical and simulation results were in a good agreement.

In order to improve the current study the release of latent heat might be considered. For future studies it is planned to calculate the heat capacity data for the phases and apply them to the model. Also the boundary condition which is derived from the experimental process affects the solidification behavior and the microstructure. It must be reviewed if the simulations could also be performed considering other solidification processes such as zone melting.

References

[1] Ha S-H, Yoshimi K, Nakamura J, Kaneko T, Maruyama K, Tu R and Goto T 2014 *J. Alloys Comp.* **594** 52–9
[2] Hasemann G, Bogomol I, Schliephake D, Loboda P I and Krüger M 2014 *Intermetallics* **48** 25–33
[3] Kazemi O, Hasemann G, Krüger M and Halle T 2016 *IOP Conf. Series: Mater. Sci. Eng.* **118** 012028
[4] Krüger M, Hasemann G, Kazemi O and Halle T 2016 *Mat. Sci. Forum* **879** 1226–32
[5] Yang Y and Chang Y A 2005 *Intermetallics* **13** 121–8
[6] Yang Y 2004 Thermodynamic modeling and experimental investigation of the Mo-Si-B-T quaternary system *PhD Thesis* University of Wisconsin-Madison
[7] Hasemann G, Kaplunenko D, Bogomol I and Krüger M 2016 *JOM* **68** 2847–53
[8] Azim M 2016 University Siegen, personal communication with G Hasemann
[9] Gokhale A B and Abbaschian G J 1991 J *Phase Equilib.* **12** 493–8

IOP Conf. Series: Materials Science and Engineering **181** (2017) 012034 doi:10.1088/1757-899X/181/1/012034

Reversed austenite for enhancing ductility of martensitic stainless steel

S Dieck[1], P Rosemann[1, 2], A Kromm[2] and T Halle[1]

[1]Institute of Materials and Joining Technology, Otto-von-Guericke University, Universitätsplatz 2, 39106 Magdeburg, Germany
[2]BAM – Federal Institute for Materials Research and Testing, Unter den Eichen 87, 12205 Berlin, Germany

Abstract. The novel heat treatment concept, "quenching and partitioning" (Q&P) has been developed for high strength steels with enhanced formability. This heat treatment involves quenching of austenite to a temperature between martensite start and finish, to receive a several amount of retained austenite. During the subsequent annealing treatment, the so called partitioning, the retained austenite is stabilized due to carbon diffusion, which results in enhanced formability and strength regarding strain induced austenite to martensite transformation. In this study a Q&P heat treatment was applied to a Fe-0.45C-0.65Mn-0.34Si-13.95Cr stainless martensite. Thereby the initial quench end temperature and the partitioning time were varied to characterize their influence on microstructural evolution. The microstructural changes were analysed by dilatometer measurements, X-ray diffraction and scanning electron microscopy, including electron back-scatter diffraction. Compression testing was made to examine the mechanical behaviour. It was found that an increasing partitioning time up to 30 min leads to an enhanced formability without loss in strength due to a higher amount of stabilized retained and reversed austenite as well as precipitation hardening.

1. Introduction

Martensitic stainless steels possess high strength and hardness, as well as a cost advantage compared with Ni-containing stainless steel. Thus, they are preferred for application as machine parts that are used in mildly corrosive environments, for example edge tools or pump shafts. The ductility of these steels can be improved by tempering at high temperatures (600 – 650 °C) [1, 2]. This heat treatment is associated with a decreasing strength and corrosion resistance. Furthermore there is the sensibility to temper embrittlement [3, 4]. Recent research focusses on enhancing the ductility without loss in strength by a low temperature heat treatment.

For this purpose SPEER et al. suggest a heat treatment, named quenching and partitioning (Q&P). This two-step heat treatment starts with quenching of homogenized (or partial in the case of austenitizing in the two phase region) austenite to a temperature between martensite start (M_s) and martensite finish (M_f) to remain an amount of austenite within a martensitic (ferritic) matrix. During the following partitioning this microstructure is treated either at, or above, the quench temperature, which results in carbon (C) diffusion from supersaturated martensite into retained austenite. The C-enrichment stabilizes the retained metastable austenite to room temperature [1, 5 – 7].
Q&P steels combine ultrahigh strength (up to 2000 MPa), which results from the martensitic matrix and with a high ductility, attributed to the transformation induced plasticity (TRIP) of austenite to martensite under loading condition [6, 8, 9].

Regarding the alloy composition C has the biggest effect on properties of Q&P steels because of its influence on transformation behaviour, martensite type and austenite stability. Furthermore commonly used steel for Q&P heat treatment include Manganese (Mn) to stabilize austenite and Silicon (Si) to supress carbide forming [10, 11].

Regarding medium carbon stainless martensite YUAN et al. investigated a modified Q&P process with low temperature partitioning at 300 – 500 °C. The alloy composition (Fe-0.44C-13.6Cr) lowers the M_f temperature below room temperature leading to a high fraction of retained austenite without interrupted quenching. Furthermore they found that subsequent low temperature partitioning (400 °C) leads to a stabilization of the retained austenite and also to an increasing fraction of austenite. This is effected by a local carbon enrichment at martensite – martensite and martensite – austenite boundaries, which results in a local reversion of martensite to austenite [12], figure 1.

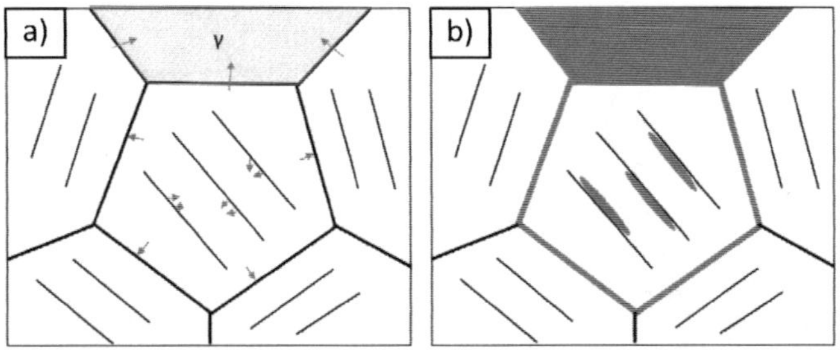

Figure 1. Austenite reversion during low temperature partitioning: a) diffusion paths of C (red arrows); b) microstructure after final quenching (austenite coloured red).

2. Materials and methods

2.1 Material

The chemical composition of the alloy used to apply a Q&P process was analyzed by optical emission spectroscopy (table 1). The heat treatment was performed with cylindrical specimen cut from soft-annealed round bars.

Table 1. Alloy composition of the material in wt.-%.

Material number	C	Mn	Si	Cr	P	S	Fe
1.4034	0.45	0.65	0.34	13.95	0.01	< 0.001	Bal.

The choice of austenitizing parameters for the investigations was based on thermodynamic considerations involving temperature and the resulting microstructure. The thermodynamically stable carbon and chromium content in the austenitic phase solution were calculated as a function of temperature with the software Thermo-Calc (figure 2) [13]. The transformation from bcc to fcc enhances the solubility of carbon. Further increase of temperature leads to a higher degree of carbide dissolution. Thus the carbon and chromium content in the austenitic solid solution is increased until the complete dissolution of carbides is achieved at about 1100°C. The applied temperature was raised up to 1150 °C to dissolve the large carbides within the soft-annealed raw material.

Two initial cooling end temperatures (QT1 – 25 °C and QT2 – 100 °C) were chosen for dilatometer investigations. Room temperature (25 °C) was determined as QT1 for comparability with literature. To retain a higher austenite fraction QT2 was at 100 °C. The quenching end temperature was limited to QT1 for further experiments, because of the results from dilatometer measurements.

After initial quenching, the partitioning treatment was conducted at 400 °C. To investigate the time dependency of the austenite reversion and its influence on mechanical behavior the partitioning time was varied (0 min – as quenched, 1 min and 30 min). The medium for final cooling was water, too.

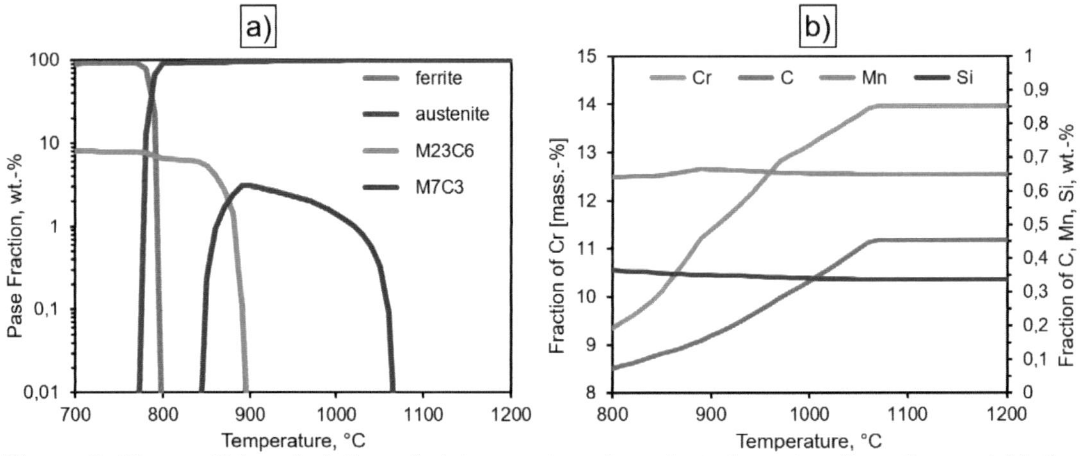

Figure 2. Thermo-Calc calculation of a) temperature dependent phase transformation and b) the accompanied element distribution (right).

2.2 Characterization methods

Mechanical properties were determined by compression testing at room temperature along the rolling direction of the specimen. The samples for compression testing had a height of 9 mm and a diameter of 6 mm. The testing was carried out with a TIRAtest 28100 universal tension and compression testing machine.

Dilatometer measurements were made with DIL805A with sub-zero expansion. The volume fraction of austenite after heat treatment was measured by X-ray diffraction (XRD). The samples were prepared by standard grinding and polishing procedure and finally etched with V2A etchant. The measurements were realized with a PANalytical X'Pert PRO diffractometer using Cu Kα radiation. The data were collected over a 2θ range of 30 – 120° with a step size of 0.017°.

The austenite distribution was measured by electron back-scatter diffraction (EBSD). Therefore the specimen were prepared by standard grinding and polishing with a subsequent vibration polishing using Buehler MasterMet2 polisher slurry to achieve a surface without residual stresses due to preparation. EBSD measurements were performed with EDAX/TSL OIM DC 5 and evaluated with OIM Analysis 5 (clean up via grain dilation; parameters: grain tolerance angle: 5°, minimum grain size: 10 pixel). Scanning electron microscopy (SEM) was conducted with a FEI XL30 ESEM-FEG operating with an accelerating voltage of 25 kV.

3. Results and discussion

3.1 Transformation behaviour

Figure 3 shows a martensite start temperature of ~170 °C. Regarding enhanced formability of Q&P steels a certain volume fraction of stabilized retained austenite is needed [15, 16]. Therefore two initial quench end temperatures (QT), leading to different fractions of retained austenite, are examined concerning phase development during partitioning. Figure 4 shows the results of this investigation. For initial quenching to room temperature (figure 4 a) there are no phase transformation during final quenching from partitioning temperature (400 °C) as expected. On the contrary a sample quenched to 100 °C during the initial cooling exhibits a martensitic transformation during the final cooling (figure 4 b). It is supposed from that data that the stabilizing effect of carbon diffusion during the partitioning treatment is insufficient for such a high potion of austenite remaining when quenching Fe-0.45C-0.65Mn-0.34Si-13.95Cr to elevated temperature. The formation of "fresh" (newly formed) untempered martensite is expected to decrease the formability. Thus the further work on mechanical properties and microstructural evolution will focus the Q&P process with QT1 (25 °C).

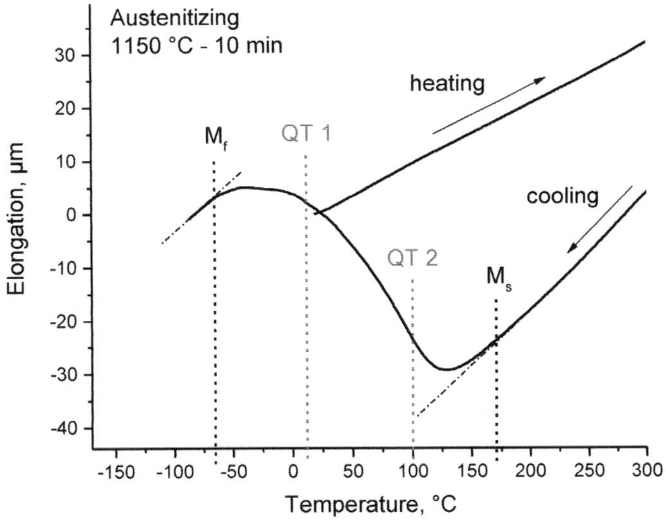

Figure 3. Dilatometer measurement of transformation behavior of Fe-0.45C-0.65Mn-0.34Si-13.59Cr during quenching from 1150 °C to -100 °C; determined initial quench end temperatures marked with red dashed lines.

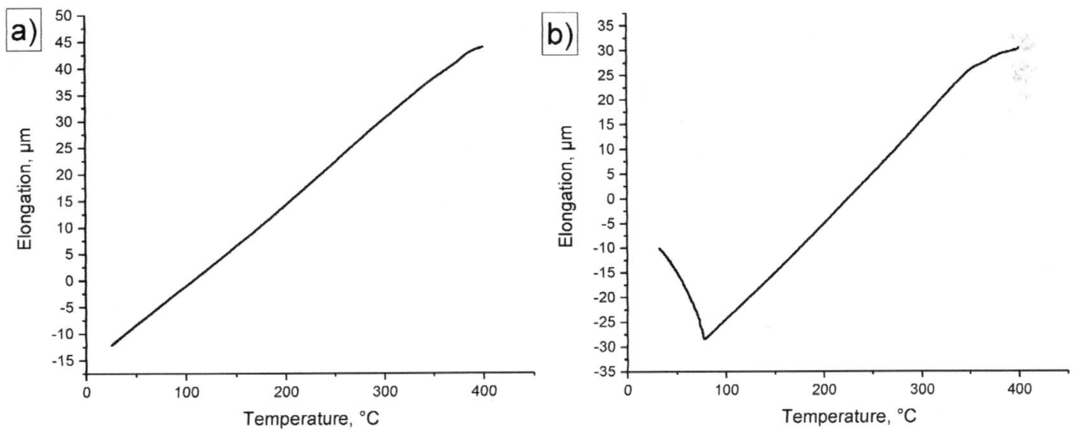

Figure 4. Dilatometer measurements of transformation behaviour during final cooling to room temperature after partitioning: a) austenitizing → quenching to QT1 (25 °C) → partitioning and b) austenitizing → quenching to QT2 (100 °C) → partitioning.

3.2 Mechanical properties

The results of compression testing are summarized in table 2. It has to be considered that all tests stopped at the machine maximum force of 100 kN before fracture of the samples. So an evaluation of total elongation and compression strength is hindered.

IOP Conf. Series: Materials Science and Engineering **181** (2017) 012034 doi:10.1088/1757-899X/181/1/012034

What can be seen from the results is a decrease of compression yield strength with increasing partitioning time. For the as quenched states a yield strength of 1885 MPa is achieved. For partitioned states a loss of ~100 MPa (1 min partitioned) and ~300 MPa (30 min partitioned) is registered. While the maximum compressive strength is nearly the same for all states (~3500 MPa), the elongation when reaching the machine limit increases with increasing partitioning time from 13.7 % (as quenched state) to 21.7 % (1 min partitioned) and 34.1 % (30 min partitioned). The enhancement of elongation behaviour is attributed to the evolution of austenitic phase and the softening of martensite as further investigations will prove.

Table 2. Compression behaviour: comparison of an as quenched with a 1 min and a 30 min partitioned sample.

state	$R_{p0.2}$ (MPa)	σ_{max} (MPa)	ε_{max} (%)
as quenched	1885	3366*	13.7
1 min partitioned	1788	3525*	21.7
30 min partitioned	1602	3602*	34.1

*tension when reaching the force limit of the testing machine

3.3 Phase fraction

Figure 5 shows the results from XRD measurements. For the as quenched state there is a small peak of austenite. This is shifted compared to the reference data (red bars), because of substitutional elements and residual stresses as a consequence of phase transformation. The partitioned states are characterized by a time dependent increasing of integral intensity of austenite peaks and decreased integral intensity of ferrite peaks. Hence, the partitioning treatment results in a higher austenite fraction, which grows with increasing partitioning time. Notable is the significant increasing of austenite fraction even for the 1 min partitioned state. Such a high amount of austenite causes the significant enhanced elongation seen in results of compression testing. The carbon enrichment of austenite during the partitioning process additionally enables the TRIP effect under loading condition. While the peak shift compared with reference data remains unaffected, the width of peaks becomes smaller with increasing partitioning time. This signalizes a decrease of second and third order stresses which is a result of relaxation of martensite, whereas the first order stresses are constant. Because of the used Cu Kα radiation no quantitative analysis of phase fractions was possible.

3.4 SEM investigations

The EBSD investigation of the as quenched state, figure 6 a), proves the existence of retained austenite islands with an area fraction of 9 % within the microstructure. Compared to that the 1 min partitioned sample (figure 6 b)) shows a smaller austenite fraction, which is not confirm with XRD results. This is attributed to circumstance of small size of reversed austenite in the initial state of partitioning. The 30 min partitioned sample (figure 6 c)) is characterised by an austenite fraction of 39 %. Moreover two types of austenite can be found. There are islands of coarse austenite and dispersed needle shaped austenite located on former martensite – martensite boundaries. In agreement with the investigations of YUAN et al. [12] the coarse one can be characterised as stabilized retained austenite and the needle shaped austenite represent reversed austenite. So the EBSD investigations confirm that the growth of austenite fraction due to reversion during the partitioning treatment is a time dependent process. It is expected that the carbon diffusion to austenite during partitioning leads to a softening of remaining martensite which enhances the formability of the specimen further. A correlation between Image Quality and stress level development due to the partitioning couldn't be observed.

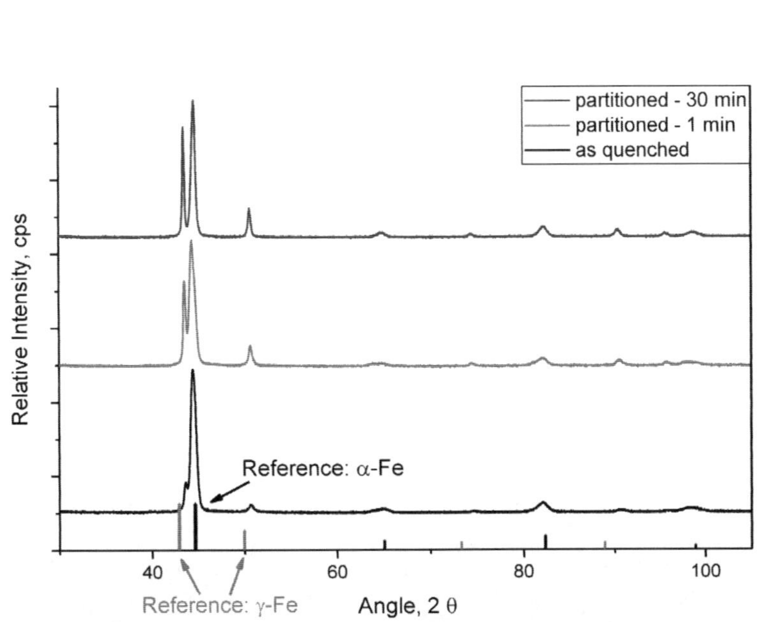

Figure 5. XRD measurements: comparison of an as quenched with a 1 min and 30 min partitioned sample.

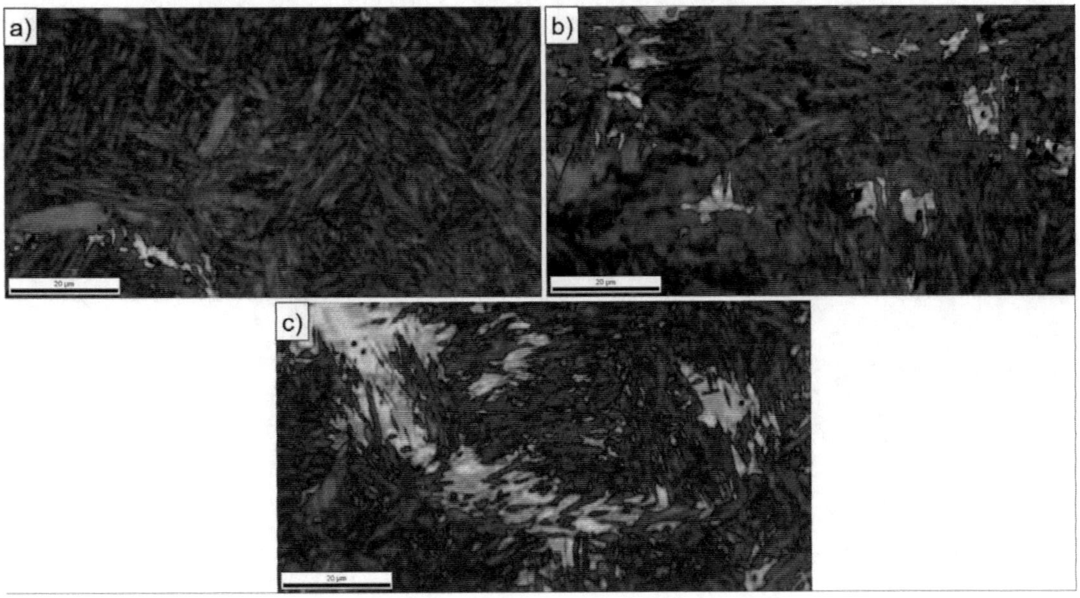

Figure 6. EBSD maps (Image Quality overlain by phase classification - fcc (austenite): green, bcc (martensite): red): comparison of a) an as quenched with b) a 1 min and c) a 30 min partitioned sample.

SEM investigations on the 1 min partitioned state, figure 7 a), are confirm to the EBSD results in such a way that this sample has a microstructure similar to the as quenched state, showing a martensitic matrix and austenitic islands. Additionally globular pores with white surroundings can be found in the SEM pictures. The high Cr content of these white edges, examined with EDX, figure 7 d), indicates the presence of carbides breaking out during sample preparation. These carbides remain from the large carbides of the soft-annealed raw material, which means that the austenitizing time was insufficient. Thus less C is available for the austenite stabilizing and reversion, which is expected to restrict the maximum of austenite phase fraction due to partitioning. Also the 30 min partitioned sample shows these kind of pores resulting from carbide break-out, figure 7 b). When raising the magnitude, figure 7 c), nm-sized white needles within the martensite laths become visible. Similar investigations of LU et al. characterized them as M_3C [14]. The fact, that the earlier stage of partitioning not includes this small carbide needles, define them to be precipitations as a consequence of extended partitioning time. Thus a precipitation hardening should improve the strength of the 30 min partitioned state in case of having an optimum size regarding OROWAN mechanism [17]. The extent of the influence of carbide precipitations on strength development cannot be clarified in this study and has to be investigated in future work.

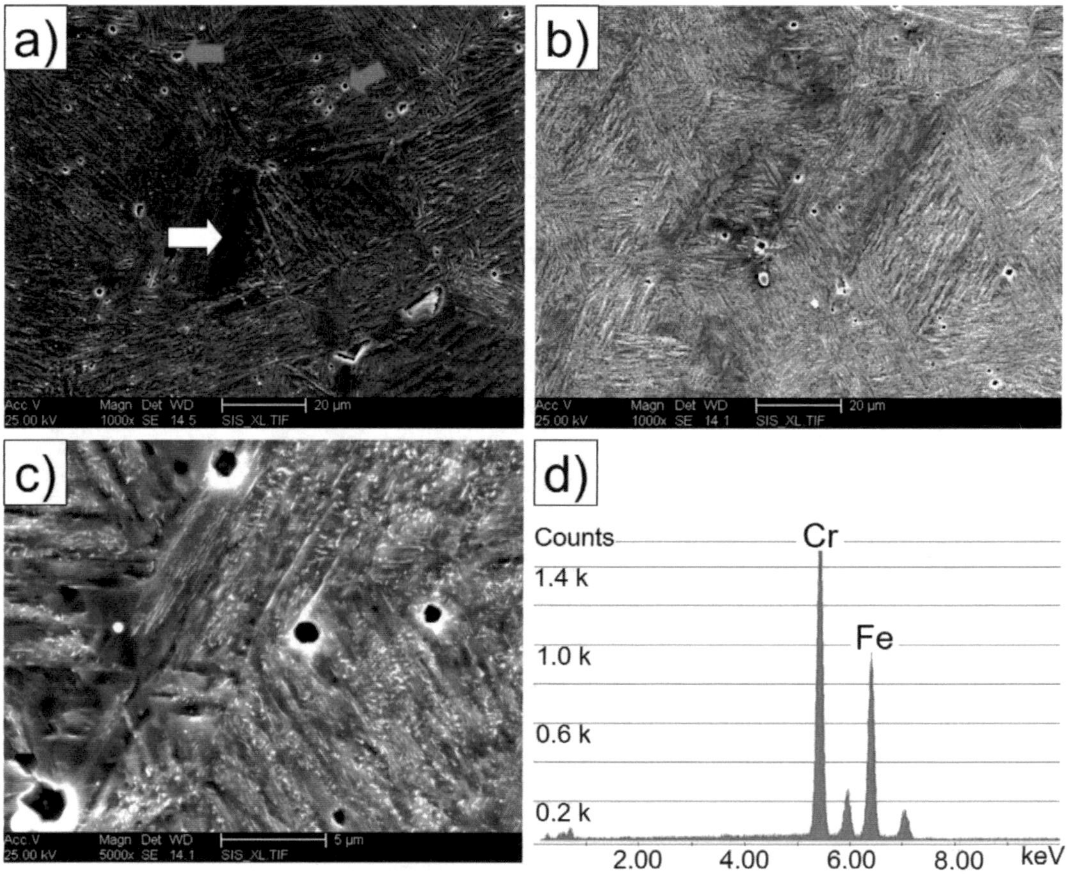

Figure 7. SEM (SE contrast) investigations on a) the 1 min partitioned state, showing austenite (white arrow and former carbide locations (red arrow), and b) the 30 min partitioned state; higher magnification of red window is shown in c). EDX analysis of the surrounding area of a pore is shown in d).

4. Conclusions and outlook

Quenching and partitioning heat treatment was applied to Fe-0.45C-0.65Mn-0.34Si-13.95Cr. The phase development based on different end temperatures of initial quenching was studied by dilatometer measurements. The mechanical properties were determined by compression testing and the microstructural evolution was investigated with XRD and SEM (EBSD). The results were summarized as follows:

(1) When initial quenching to QT2 (100 °C) the carbon partitioning is insufficient to stabilize the elevated volume fraction of retained austenite compared with an initial quenching to 25 °C. As a consequence unstabilized retained austenite transforms into fresh martensite during final cooling.

(2) An additional partitioning at 400 °C of the QT1 (25 °C) state leads to an enhanced elongation behaviour in compression testing. However, there is a loss in compression yield strength. Although the enhancement of elongation increases when increasing the partitioning time up to 30 min.

(3) The austenite fraction increases with the time of partitioning treatment. This effect is based on the reversion of martensite to austenite due to carbon diffusion out of supersaturated martensite and local enrichment on martensite-martensite and martensite-austenite boundaries. A softening of the remaining martensite was connected with this process.

(4) A partitioning at 400 °C leads to a time dependent precipitation of carbides, which has an additional strengthening effect beside TRIP, i.e. compensation of strength loss by austenite formation (ductility gain).

Acknowledgment

The work was financially supported by Graduate School (No. 1554) "Micro-Macro-Interactions in Structured Media and Particle Systems". Further we thank the Federal Institute for Materials Research and Testing for dilatometer measurements.

References

[1] Edmonds D V, He K, Rizzo F C, De Cooman B C, Matlock D K and Speer J G 2006 *Mater. Sci. Eng.* A **438-440** 25–34
[2] Krauss G and Repas P E 1992 *Fundamentals of Aging and Tempering in Bainitic and Martensitic Steel Products* (Warrendale: Iron and Steel Society)
[3] Irvine K J and Crowe D J and 1979 *The Metallurgical Evolution of Stainless Steels* ed F B Pickering (Ohio: American Society for Metals) pp 43–62
[4] Gaunkar G V P, Huntz A M and Lacombe P 1980 *Met. Sci.* **14** 241–52
[5] Speer J G, Rizzo Assunção F C, Matlock D K and Edmonds D V 2005 *Mater. Res.* **8** 417–23
[6] Wang L and Speer J G 2013 *Metallogr. Microstruct. Anal.* **2** 268–81
[7] Santofimia M J, Zhao L and Sietsma J 2011 *Metall. Mater. Trans.* A **42** 3620–6
[8] Torralba J M, Navarro A and Campos M 2013 *Mater. Sci. Eng.* A **537** 253–6
[9] De Moor E, Lacriox S, Clarke A J, Penning J and Speer J G 2008 *Metall. Mater. Trans.* A **39** 2586–9
[10] Hsu T Y and Jin X 2011 Ultra-high Strength Steel Treated by Using Quenching–Partitioning–Tempering Process *Advanced Steels* ed Y Weng et al. (Beijing: Springer) pp 67–73
[11] Fonstein N 2015 *Advanced High Strength Sheet Steels* (Chicago: Springer) pp 350–4
[12] Yuan L, Ponge D, Wittig J, Choi P, Jiménez J A and Raabe D 2012 *Act. Mat.* **60** 2790–804
[13] Thermo-Calc Software with limited academic version of Thermo-Calc database version 3.1 (accessed on March 2014)
[14] Lu S-Y, Yao K-F, Chen Y-B, Wang M-H, Chen N and Ge X-Y 2016 *Corr. Sci.* **103** 95–104
[15] Hao Q, Qin S, Liu Y, Zou X, Chen N and Rong Y 2016 *Mater. Sci. Eng.* A **671** 135–46
[16] Lee C G, Kim S-J, Lee T-H and Lee S 2004 *Mater. Sci. Eng.* A **371** 16–23
[17] Rösler J, Harders H and Bäker M 2012 *Mechanisches Verhalten der Werkstoffe* (Wiesbaden: Springer Vieweg) pp 206–11

IOP Conf. Series: Materials Science and Engineering **181** (2017) 012035 doi:10.1088/1757-899X/181/1/012035

Evolution of microstructure and mechanical properties during Q&P processing of medium-carbon steels with different silicon levels

S Jenicek, I Vorel, J Kana, K Opatova, K Rubesova, V Kotesovec and B Masek

University of West Bohemia, Faculty of Mechanical Engineering, Univerzitní 22, Plzeň, Czech Republic

E-mail: frost@rti.zcu.cz

Abstract. Evolution of microstructure during heat treatment plays a fundamental role in the resulting mechanical properties of steel. Today, mechanical properties in conjunction with technological properties, such as weldability, formability, and machinability, and their optimum combinations, are widely discussed in a number of mechanical engineering disciplines. In this manner, requirements arise for developing steels which could offer high strength and good formability, and which could be used for making parts with high resistance to failure and with a long life. One present-day example of such steels involves Q&P-processed martensitic steels. Their properties are dictated by their treatment, as well as their alloying, particularly by the silicon content. Silicon fundamentally affects microstructure evolution during Q&P processing and, as a result, mechanical properties. With this way it is possible to receive microstructures consinsting of martensite and retained austenite with an ultimate tensile stress of more than 1600 MPa and a uniform elongation of more than 12 %.

1. Introduction

Under usual conditions, heat treatment of steel leads to transformation of austenite to martensite, bainite, pearlite or their mixtures, and to simultaneous precipitation of carbides. Carbides form during thermodynamic processes because the solubility of carbon in austenite and ferrite varies with temperature and with the steel's chemistry. The impact of carbide precipitates on mechanical properties is given by their morphology, size, distribution, crystal structure, and other factors. In many engineering applications, carbides are an undesirable constituent in steel, as they often impair strength properties, most notably fatigue. This adverse effect of carbide precipitation can be, to a great extent, suppressed by silicon addition. Steels with sufficient silicon levels can be unconventionally heat treated to obtain excellent strength and deformation properties [1]. One of the treatment routes that rely on the above-described principle is Q&P processing.

The Q&P process is based on carbon partitioning between martensite and retained austenite so that carbide precipitation is prevented and the excess carbon stabilizes retained austenite [2]. It involves quenching to a temperature below the M_s in order to produce a certain amount of martensite which can be controlled by choosing the quenching temperature below the M_s. This step is followed by partitioning which is beeing executed between the tempereatures of quenching and M_s. During the partitioning step carbon diffuses from martensite to austenite. As austenite becomes saturated with carbon, its stability increases. The resulting microstructure is a mixture of martensite and retained austenite and exhibits excellent mechanical properties [3-7]. However, it is imperative that precipitation of carbide phases is avoided during Q&P processing. If this fails, the microstructure will develop differently: the carbon atoms that diffuse from martensitic laths will not enrich retained austenite but instead create undesirable brittle carbide phases. To prevent the precipitation of carbides the silicon amount of about 1.5 - 2% should be added in steel. The silicon prevents precipitation because of its low solubility in carbides. For the formation of carbides the diffusion of silicon is necessary because they can form only in silicon depleted zones of microstructure. However the temperatures during QP processing of medium carbon steels are set to low value to be enough for diffusion of silicon and the carbides cannot form or are forming only in a limited extent [8] (figure 1).

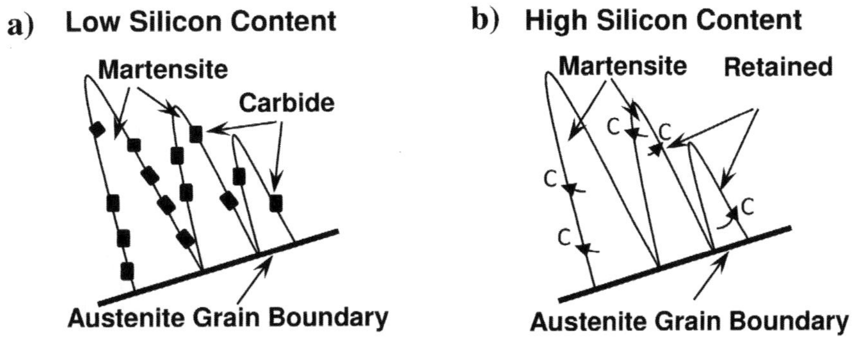

Figure 1. Effect of silicon content on the carbon partitioning mechanism during the Q&P process: a) low-silicon steel, b) high-silicon steel.

2. Experimental programme

The objective of this experiment was to compare the effects of silicon content on the evolution of microstructure and mechanical properties in two low-alloy steels with 0.43 % carbon, 0.56-0.6 % manganese, and 1.33 % chromium upon identical treatments involving the Q&P process (table 1). The only difference between them was the amount of silicon.

Table 1. Chemical composition of the experimental steels.

Steel designation	C	Si	Mn	Cr	Mo	Nb	P	S	M_s [°C]	M_f [°C]
42CrMoS4	0.42	max 0.4	0.6	1.2	0.15	-	0.03	0.04	305	184
42SiCr	0.42	2.03	0.56	1.33	0.16	0.03	0.005	0.003	298	178

To explore the effect of silicon on microstructure evolution, heat treatment routes with Q&P processing were designed (figure 2). The quenching temperatures were chosen to provide approximately 50 % martensite in the microstructure prior to partitioning. The partitioning temperatures were above and below the M_s.

The holding time at the partitioning temperature was 600 s. The specimens had a diameter of 8 mm. Prior to the experimental heat treatment, they were annealed at 1050 °C for 20 minutes. The experiment was carried out in a thermomechanical simulator that used induction-resistance heating.

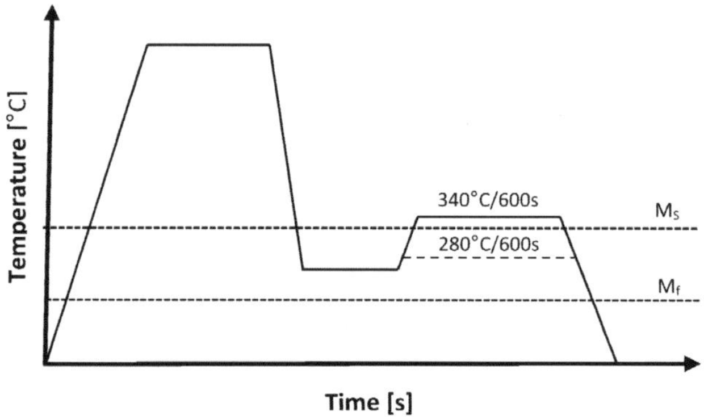

Figure 2. Diagram of the heat treatment route with the Q&P process.

3. Results and discussion

3.1 Microstructure analysis

Microstructures of the specimens were examined using scanning electron microscopy. Therefore samples were etched with 4 % picral (for 30 seconds), and 10 % Na2S2O5 solution (for 180 seconds) after grinding and polishing. Phase composition was examined using TESCAN Vega microscope image analysis For better resolution capability, the micrographs were processed by image transformation.

Using SEM, mixed microstructures of tempered and untempered martensite and bainite were found in the specimens (figures 3 and 4). The microstructure of the 42CrMoS4-QP280 specimen was a mixture of tempered and untempered martensite with lath morphology (figure 3-a). Close examination of the microstructure revealed the presence of fine carbides. The 42CrMoS4-QP340 specimen contained a mixture of tempered lath martensite and lower bainite which were, however, impossible to distinguish with certainty due to their similar appearance (figure 3-b). With regard to the differences in the microstructures of the specimens of the 42CrMoS4 steel, it can be said that austenite stabilization has not occurred in the 42CrMoS4-QP340 specimen. Therefore, the austenite decomposed and due to low silicon content, a bainite-like microstructure formed.

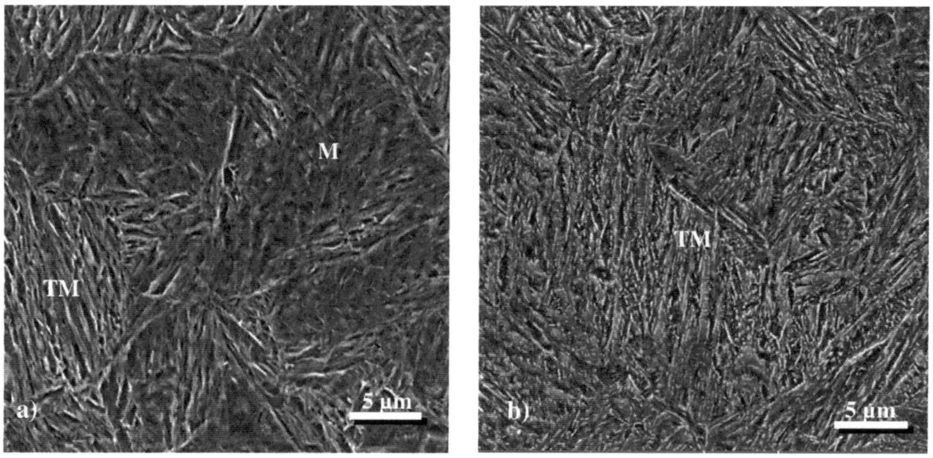

Figure 3. SEM-Micrographs of 42CrMoS4 specimens after Q&P processing obtained using SE mode (phase contrast). a) QP-280 – Mixture of lath tempered martensite (TM) and untempered martensite (M), b) QP-340 – Mixture of lath tempered martensite (TM) (and bainite).

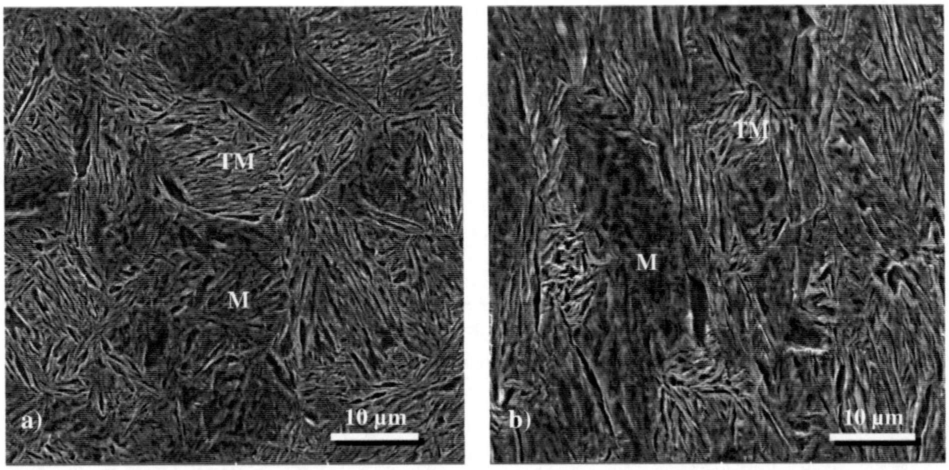

Figure 4. SEM-Micrographs of 42SiCr specimens after Q&P processing obtained using SE mode (phase contrast). Mixture of lath tempered martensite (TM) and untempered martensite (M) in a) QP-280 and b) QP-340.

Both specimens 42SiCr-QP280 and 42SiCr-QP340 contained low-temperature tempered and untempered lath martensite (figure 4). This shows that in this steel, unlike in 42CrMoS4, partitioning at 340 °C has not led to austenite decomposition into bainite. Here, austenite remained present up until the last phase of the treatment. Therefore, it transformed to martensite during quenching (figure 4-b). Both observed martensites had lath morphology. The laths were free from visible carbide precipitates. This confirms the assumption that in order for austenite to stabilize during the Q&P process the steel must contain a sufficient amount of silicon, which prevents carbides from forming.

3.2 Mechanical properties

Mechanical properties were determined by means of standard-based static tensile testing of miniature test pieces [1]. The ultimate strength (UTS) and yield stress (YS) values were comparable in both 42SiCr-QP-340 and 42CrMoS4-QP-340 (table 2). The QP-280 route led to the highest ultimate strength of 1891 MPa in the 42SiCr steel, although the yield stress was lower than that of the 42CrMoS4 steel by approximately 240 MPa (figure 5).

A significant difference between the investigated steels was the value of uniform elongation, which in the case of steel 42SiCr reached values of 9-12% in the case of steel 42CrMoS4 of 4-6%.

Table 2. Mechanical properties of the experimental steels.

Specimen designation	YS [MPa]	UTS [MPa]	A [%]
42CrMoS4-QP-280	1308	1786	4
42CrMoS4-QP-340	1398	1649	6
42SiCr-QP-280	1067	1891	9
42SiCr-QP-340	1394	1665	12

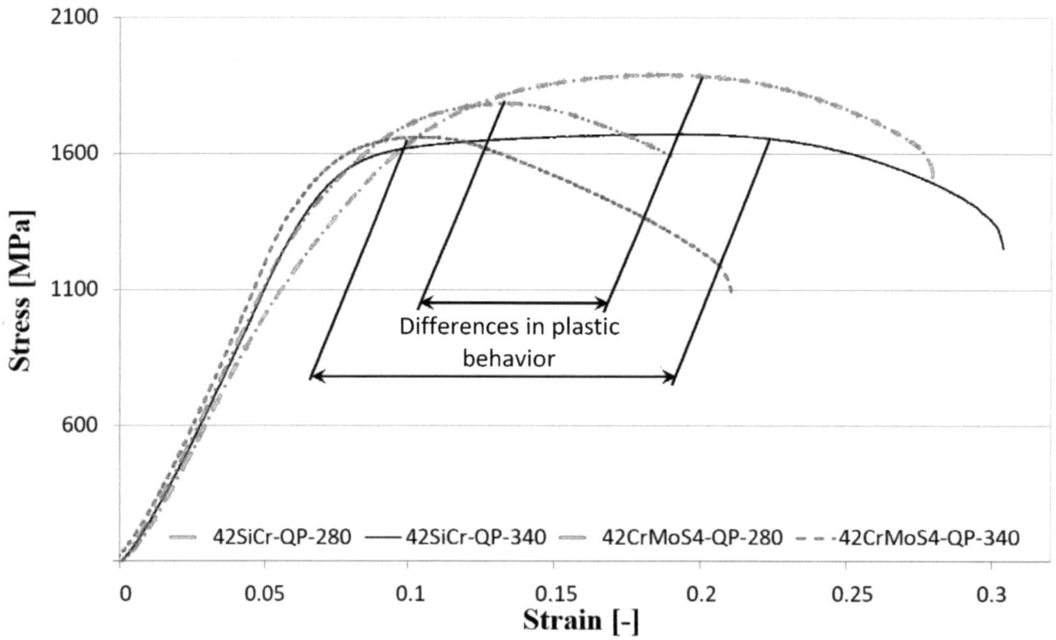

Figure 5. Stress-strain plots for specimens of the 42SiCr and 42CrMoS4 steels upon Q&P processing.

The results of evaluation of microstructure showed the presence of tempered and untempered martensite (and bainite) in the structure of the experimental steels. In the case of steel 42CrMoS4 fine carbides were observed in the martensite, which originated during the step of quenching.

The carbide precipitation also occurred during the decomposition of austenite to bainite structure type. In contrast, in steel 42SiCr most likely occurred due to the higher content of silicon to stabilization the retained austenite as thin films between martensite laths. The presence of austenitic films could be the reason for higher uniform elongation for 42SiCr steel samples. This conclusion, however, will be subject to further research with respect to the detailed description of the structure using TEM and other analytical methods.

Regarding to ultimate tensile strength it can be stated, that in case of steel 42CrMoS4 lower values compared to steel 42SiCr were achieved. This may be caused by carbide precipitation, or a different volume ratio of tempered and untempered martensite or austenite grain size etc. Identifying the true cause will be next step of future research.

4. Summary and conclusion

An experiment that compared two steels, whose alloying only differed in the silicon content, was carried out to explore the hypothesis that in order for Q&P processing to be successful, carbide formation must be prevented. The presence of silicon during Q&P processing enables carbon to diffuse from martensite into austenite, to enrich and therefore stabilize it. As a result, martensitic-austenitic microstructures can form which exhibit high strength and good ductility.

The effects of parameters of the Q&P process on the evolution of microstructure and mechanical properties were examined in two low-alloy steels that contained 0.42 % carbon and very similar levels of manganese and chromium. There were different silicon levels. In 42CrMoS4, the silicon content was 0.4 %, whereas in 42SiCr, it was 2 %.

The Q&P process route was designed to produce approximately 50 % martensite at the end of quenching. The subsequent partitioning steps involved temperatures below and above the M_s.

In those specimens of the 42CrMoS4 and 42SiCr steels that underwent partitioning below the M_s, the resulting microstructures contained a mixture of tempered and untempered lath martensite. Fine carbide precipitates were found in the 42CrMoS4 steel with a low silicon content. The specimens of 42SiCr showed a strength of 1891 MPa and uniform elongation of 9 %. The average ultimate strength and uniform elongation of the specimens of the 42CrMoS4 steel were 1786 MPa and 4 %, respectively.

The Q&P process routes which involved partitioning above the M_s led to different microstructure evolution. In the 42CrMoS4 steel, a mixed microstructure of tempered martensite and, most likely, lower bainite formed. The latter was a product of decomposition of the untransformed austenite during partitioning. By contrast, the microstructure of 42SiCr – which had a high silicon level – consisted of tempered and untempered martensite without any carbide precipitates. Ultimate strengths in both experimental steels were comparable. The average uniform elongation of the 42SiCr steel was 12 %, whereas in 42CrMoS4, it was 6 %.

Acknowledgements
This paper includes results created within the project LO1502 Development of Regional Technological Institute. The project is subsidised by the Ministry of Education, Youth and Sports from specific resources of the state budget for research and development.

References

[1] Vorel I, Jirkova H, Masek B and Kurka P 2015 Production of shaped semi-products from AHS steels by internal pressure *Mater. tehno.* **49** 739–44
[2] Kucerova L 2011 Comparison of microstructures and properties obtained after different heat treatment strategies of high strength low alloyed steel *J. Iron Steel Res. Int.* **18** 427–31
[3] Speer J G, Matlock D K, De Cooman B C and Schroth J G 2003 Carbon partitioning into austenite after martensite transformation *Acta Mater.* **51** 2611–22
[4] Speer J G, Rizzo Assunção F C, Matlock D K and Edmonds D V 2005 The "quenching and partitioning" process: Background and recent progress *Mat. Res.* **8** 417–23
[5] Tsuchiyma T, Tobota J, Tao T, Nakada N and Takaki S 2012 Quenching and partitioning treatment of a low-carbon martensitic stainless steel *Mater. Sci. Eng.* A **532** 585–92
[6] Jirkova H, Kucerova L and Masek B 2012 Effect of quenching and partitioning temperatures in the Q-P process on the properties of AHSS with various amounts of manganese and silicon *Mater. Sci. Forum* **706-709** 2734–9
[7] Masek B, Jirkova H, Kucerova L and Zemko M 2011 Improvement of Mechanical properties of automotive components using hot stamping with integrated Q-P process *J. Iron Steel Res. Int.* **18** 730–4
[8] Kozeschnik E and Bhadeshia H K D H 2008 Influence of silicon on cementite precipitation in steels *Mater. Sci. Tech.* **24** 343–7

IOP Conf. Series: Materials Science and Engineering **181** (2017) 012036 doi:10.1088/1757-899X/181/1/012036

Macromechanical finite-element simulations for predicting microstructures by experimental calibration

T Mehner[1], A Bauer[2], B Awiszus[2] and T Lampke[1]

[1]Institute of Materials Science and Engineering, Chemnitz University of Technology, 09125 Chemnitz, Germany
[2]Professorship Virtual Production Engineering, Institute for Machine Tools and Production Processes, Chemnitz University of Technology, 09107 Chemnitz, Germany

E-Mail: thomas.mehner@mb.tu-chemnitz.de

Abstract. Macromechanical simulations provide excellent opportunities for rapid calculations of forming processes. Geometrical dimensions and residual stresses can be calculated with very good agreement. More complex forming simulations (e.g. crystal dynamic simulations or calculations with representative volume elements) are necessary if microstructural magnitudes, like crystallite sizes and microstrains, have to be included. Using the example of cold-rolling, this paper aims to describe a different approach for connecting macromechanical finite-element simulations with key parameters of the microstructure. By means of X-ray diffraction and confocal microscopy, crystallite sizes, microstrains, texture and roughness values are determined and correlated to the plastic strain. The plastic strain can be simulated easily and the microstructure after forming can be predicted. As a result of this calibration, more complex simulations can be avoided. Nonetheless, these calibrated macromechanical simulations can be used for the estimation of microstructure-related properties, like the corrosion behaviour.

1. Introduction

The macromechanical finite-element analysis (FEA) is a versatile tool for the simulation of forming processes. Especially, fast simulations of large volumes can be performed and geometrical measures, plastic strains and residual stresses can be simulated with good accordance after performing a proper calibration of the simulation [1–2]. In these studies, it was found that the mesh with the associated remeshing parameters as well as the friction model and friction parameters are of great importance concerning this calibration [3].

However, physically relevant magnitudes, like crystallite sizes, microstrains, roughnesses or textures are not an outcome of these simulations. Thus, more complex forming simulations (e.g. crystal dynamic simulations or calculations with representative volume elements (RVE)) are necessary to include them. Hereby, RVE were connected to the laws of crystal plasticity where the deformation mostly takes place, because of dislocations in the lattice structure. Different models provide various mechanisms to implement hardening effects, material damage, diffusion processes or recrystallization [4–7]. Therefore, RVE are often used for modelling of more complex material behaviour with the related micromechanical effects while using reinforced composites [8-10] or dual phase steels [11-12].

In this paper, a different approach is presented using experimental calibration curves: Exemplified by cold-rolling, for different plastic strains (and different rolling-pass sequences), crystallite sizes, microstrains, roughnesses and textures calibration curves are determined experimentally and fitted by proper functions to be correlated to the FEA simulations. For proving the accuracy of the results, rolling-pass sequences that were not used for calibration are simulated and compared with the experimental values.

2. Materials and methods

For the experimental cold flat rolling, a commercial DC04 sheet was cut into samples by water-jet cutting with initial measurements of 100 mm × 40 mm × 2 mm and a composition of 0.042 % C, 0.242 % Mn, 0.035 % Al, 0.011 % Si, 0.013 % S, 0.010 % P (mass fraction). The rolling was performed on a Duo/Quarto EW 105x100 strip rolling mill (Bühler und Co GmbH) with a roll speed of 12.12 rpm and different rolling-pass sequences (Table 1) and plastic strains. After rolling, the width of all samples was reduced to 14 mm by water-jet cutting to receive a homogenous sample state. Each sample was taken from the centre of the sheet with a distance of 13 mm from the edges.

Table 1. Rolling-pass sequences used.

Final sheet thickness (mm)	Rolling passes, sheet thickness (mm)
2.0	none
1.8	2.0–1.8
1.7	2.0–1.7
1.5	2.0–1.7–1.5
1.5	2.0–1.5
1.3	2.0–1.7–1.5–1.3
1.3	2.0–1.5–1.3
1.1	2.0–1.5–1.3–1.1
0.9	2.0–1.5–1.3–1.1–0.9
0.7	2.0–1.5–1.3–1.1–0.9–0.7
0.5	2.0–1.5–1.3–1.1–0.9–0.7–0.5

FEA simulations were performed employing the tool simufact.forming 13.0 with a simulation setup as shown in figure 1. The sample geometry as well as the rolling parameters (size of the rolls, rolling kinematics) are identical with the real process. Furthermore, the numerical calculation was processed as 3D coupled setup with a temperature of 20 °C for workpieces, tools and environment. The friction was determined over a combined friction model (Coulomb and shear friction) with $\mu = 0.15$ and $m = 0.3$. The mesh of the specimen was generated from hexahedral elements with 0.5 mm edge length and 4 elements over the sheet thickness. Therefore, the remeshing parameter was set to start the remeshing at an element deformation of 40 %. The material data for the used DC04 were generated experimentally by flat compression tests (at room temperature with a strain rate of 0.1 s^{-1}) and the calculated values were implemented into the simulation via data table. The plastic strains were determined by averaging over at least 50 nodes in the centre of the sample.

X-ray analyses were performed at the surface of the samples at positions in their centre along the rolling direction. X-ray crystallite sizes and microstrains of the samples were determined by X-ray diffraction (XRD) using a diffractometer D8 Discover (Bruker AXS) with Co K_α radiation (information depth of about 15 µm). The tube parameters were: 40 kV, 40 mA, point focus. For beam focussing and limiting, polycap optics and a 0.5 mm pinhole collimator were employed. The used 1D detector Lynxeye XE with 2.1° aperture angle allows measuring with effective measurement times of 192 s/step at a step width of the diffraction angle (2θ) of 0.01°. The 2θ range was 20°–127°. The evaluations of the line width were done with the programme TOPAS (Bruker AXS) under consideration of device-related line-broadening effects. "Microstrain" denotes the FWHM(2θ) of the Gaussian part of the line profile divided by tan θ and is closely related to the dislocation density. Textures were measured utilising a diffractometer D5000 (Siemens) with Co K_α radiation, 40 kV, 15 mA, point focus, a 1 mm pinhole collimator and a scintillation counter. Three pole figures of α-Fe, {110}, {200} and {211}, were measured at diffraction angles of 52.4°, 77.3° and 99.8°, respectively. Tilt angles of 0°–80° with step sizes of 5° for both rotation and tilt were used with a measurement time of 2 s/step. The evaluations were done with the programme MULTEX (Bruker AXS). At least four measurements were averaged for each XRD result.

The roughness average R_a was determined from measurements with a laser scanning microscope (LSM) VK-X200K (Keyence) with a magnification of 400x and an evaluated length of 4.8 mm. The cut-off wavelengths λ_s and λ_c were set to 2.5 µm and 800 µm, respectively – taking the standard DIN EN ISO 3274 into account. For each sample, 20 roughness profiles in the centre of the sample in rolling direction were evaluated and averaged.

Uncertainties given in the diagrams are confidence intervals with confidence levels of 95 %.

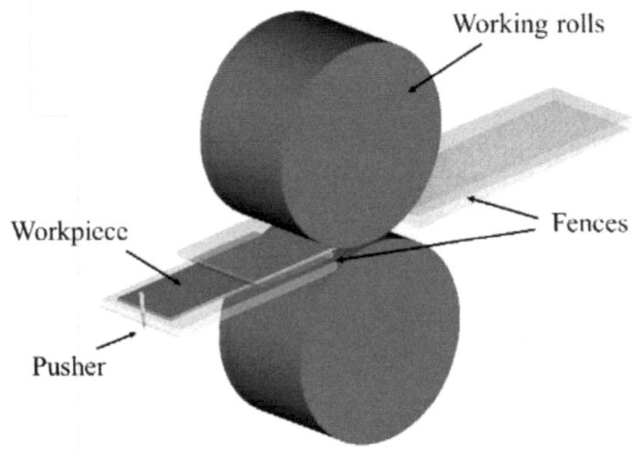

Figure 1. Simulative setup of the cold flat-rolling process.

3. Results and discussions

The novel approach that will be used within this paper is based on the v. Mises equivalent plastic strain φ of the sample. It can be calculated by the individual plastic strains in the x, y and z directions:

$$\varphi = \sqrt{\frac{2}{3} \cdot (\varphi_x^2 + \varphi_y^2 + \varphi_z^2)} \qquad (1)$$

Due to the special sample preparation, the plastic strain is almost homogeneously distributed across the entire sample (figure 2). Small deviation occur at the corners only. The FEA provides plastic strains for the samples, which are shown in figure 3.

Figure 2. FEA simulation of the distribution of the plastic strain (sample 2.0 mm–1.7 mm–1.5 mm).

The FEA results indicate that the plastic strains are independent on the rolling-pass sequence if the final thickness is the same. Along the processing route 2.0–1.5–1.3–1.1–0.9–0.7–0.5, the plastic strain increases quadratically with the number of 0.2 mm thickness-reduction steps.

In the following, the correlation between the plastic strain of the samples and their structural properties is shown. The surface roughness is an important parameter in many applications; it will affect e.g. the corrosion behaviour, friction or sealing properties. R_a is expected to decrease with progressive rolling.

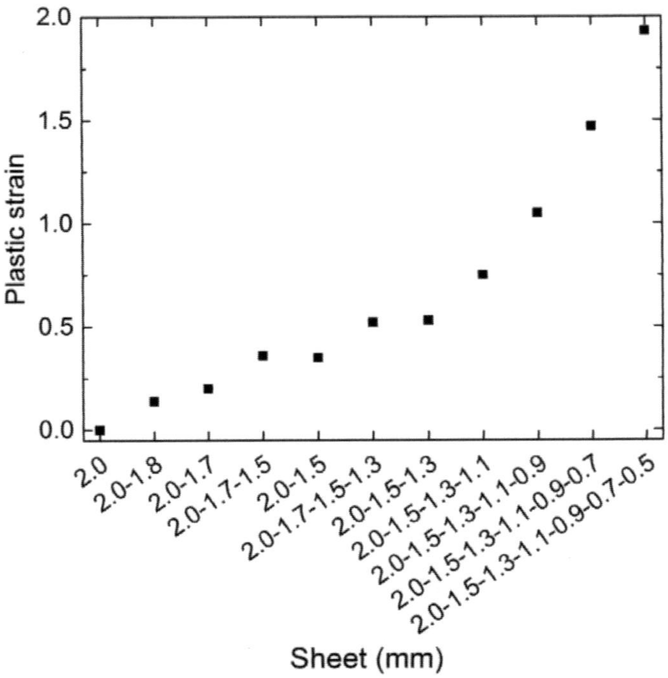

Figure 3. Plastic strains for the considered samples, calculated by FEA.

In figure 4, the measured values of R_a are shown, depending on the plastic strain. The fitting function used is an exponential decay that provides (0.312 ± 0.009) µm as the asymptotical value of R_a. Due to the decay constant of the fitting function (0.102 ± 0.013), an almost constant roughness is reached for plastic strains above 0.4. Therefore, after the rolling pass 2.0–1.5 mm, R_a is less than 0.02 µm above the asymptotical value. The complete fitting function is expressed by:

$$\boldsymbol{R_{a}(\mathrm{fit}) = 0.643\ \mu\mathrm{m} \cdot \exp\left(\frac{-\varphi}{0.102}\right) + 0.312\ \mu\mathrm{m}} \quad (2)$$

In a similar way, the crystallite sizes and microstrains of the samples, measured by XRD, can be evaluated (figure 5). Exponential-decay fitting functions were used. The crystallite sizes show a strong decay with a decay constant of 0.105 ± 0.014 and an asymptotical value of (106 ± 3) nm. Thus, the decay constant is very close to the one of the R_a, which means that the relative decrease of these magnitudes due to cold rolling is very similar. However, it has to be pointed out that the crystallite size of the sheet with $\varphi = 0$ is larger than the maximum value XRD can determine (about 300 nm). Thus, this value (500 nm) is estimated only. Assuming larger values (2000 nm), the decay constant decreases to 0.049 ± 0.009 and the asymptotical value increases slightly to (110 ± 4) nm with $R^2 = 0.904$. Thus, significant differences > 10 % between the two fits occur for low plastic strains ($\varphi < 0.35$). However, the fit quality is better for the fit mentioned first, with the equation:

$$\textbf{crystallite size (fit)} = \textbf{390 nm} \cdot \textbf{exp}\left(\frac{-\varphi}{0.105}\right) + \textbf{106 nm} \tag{3}$$

The experimental microstrains show an almost constant value for plastic strains above 0.2. The exponential decay follows the equation:

$$\textbf{microstrain (fit)} = -\mathbf{0.141°} \cdot \textbf{exp}\left(\frac{-\varphi}{0.086}\right) + \mathbf{0.221°}. \tag{4}$$

Thus, the microstrain reaches a maximum value of 0.221° even for large plastic strains; i.e. above $\varphi \approx 0.4$, the dislocation density is constant.

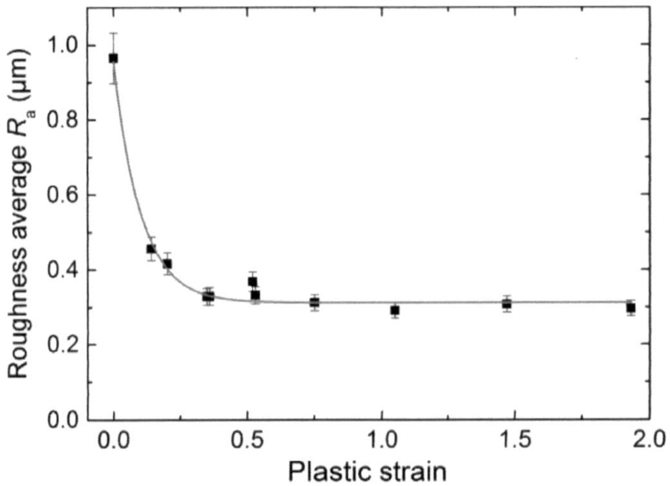

Figure 4. Measured roughness average values dependent on the calculated plastic strains with an exponential decay fitting function ($R^2 = 0.928$).

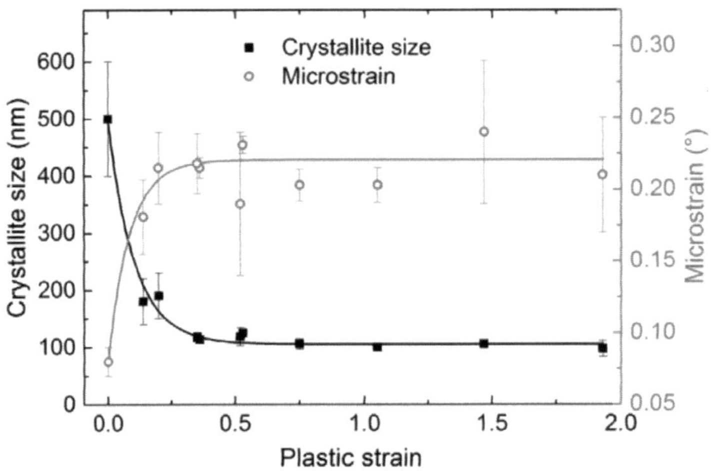

Figure 5. Crystallite sizes and microstrains for the samples with exponential-decay fitting functions ($R^2 = 0.921$ and 0.938, respectively).

The texture can hardly be expressed with one single number. Depending on the application in which the product will be used, different magnitudes may be of interest. In this paper, the fraction of the texture component with {200} lattice planes parallel to the surface is evaluated. Due to the cold rolling of DC04, there are two more texture components: {211} and {111} lattice planes parallel to the surface. In addition, there is an isotropic fraction that declines with progressive rolling passes. The initial state (2.0 mm sheet) was hot rolled, but this kind of rolling texture can still be measured. However, about 40 % of the measured intensity in the pole figures is isotropic for these sheets. In figure 6, the pole figures of a 2.0–1.5 mm sheet are shown.

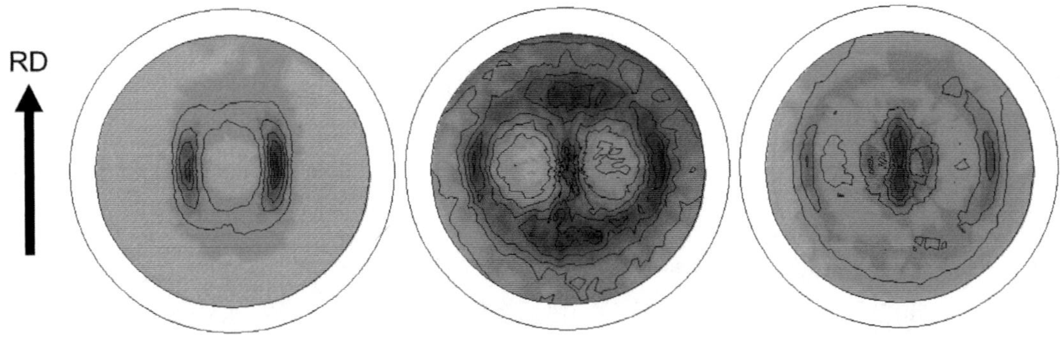

Figure 6. Pole figures (stereographic projection with arbitrary intensity scale) of a cold-rolled 2.0–1.5 mm DC04 sheet, {110} (left), {200} (centre) and {211} (right). RD indicates the rolling direction.

For all of the measured cold-rolled sheets, the most dominant texture component is the {111} component. Its fraction is about 38 % for the 2.0 mm sheet and only increases to about 46 % for the 0.5 mm sheet and thus is hardly affected by cold rolling. A stronger influence is found for the {200} component and the isotropic fraction, figure 7.

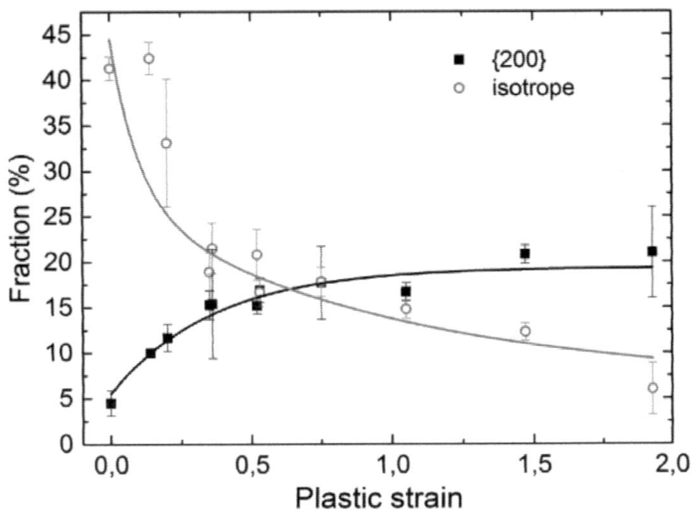

Figure 7. Fraction of the {200} texture component as well as the isotropic fraction for the samples with exponential-decay fitting functions.

IOP Conf. Series: Materials Science and Engineering **181** (2017) 012036 doi:10.1088/1757-899X/181/1/012036

For low plastic strains, the isotropic fraction is decreasing quickly. Thus, an exponential fitting function with two decay constants is used. The {200} component can by described by:

$$\mathbf{fraction}\,\{\mathbf{200}\}\,(\mathbf{fit}) = -\mathbf{14.9}\,\% \cdot \mathbf{exp}\left(\frac{-\varphi}{\mathbf{0.26}}\right) + \mathbf{18.8}\,\%. \quad (5)$$

Using the equations (2)–(5), the structural properties can be predicted. To prove the validity, the values of two additional sheets with different rolling pass sequences are compared with the calculated ones. The rolling pass sequences in table 2 differ significantly from the ones used in the fitting process. Still, the results are in excellent agreement with differences clearly below the experimental uncertainties.

Table 2. Comparison of calculated and measured values.

	2.0–1.4–1.2 mm	2.0–1.9–1.4–0.8 mm
Plastic strain	0.65	1.12
Roughness average (µm)		
experimental/calculated	0.300 ± 0.021/0.313	0.293 ± 0.021/0.312
difference	0.013 (4.3 %)	0.019 (6.5 %)
Crystallite size (nm)		
experimental/calculated	106 ± 9/106	105 ± 11/106
difference	≈ 0 (< 0.1 %)	1 (1.0 %)
Microstrain (°)		
experimental/calculated	0.23 ± 0.04/0.22	0.228 ± 0.024/0.221
difference	-0.01 (4.3 %)	-0.007 (3.1 %)
Fraction {200} texture component (%)		
experimental/calculated	17.6 ± 1.6/17.8	18.5 ± 0.3/18.6
difference	0.2 (1.1 %)	0.1 (0.5 %)

4. Summary and conclusions

It has been shown that the structural properties average roughness, crystallite size, microstrain and texture can be clearly correlated with the plastic strain, which can be calculated by the FEA with low deviations. For the used DC04 cold-rolling process (with $\varphi = 0$ for the hot-rolled sheet with 2 mm thickness), the relations in table 3 were determined. The independence of these properties is most probably a consequence of the low strain-rate sensitivity of the material. If this sensitivity is high, such relations may not be applicable.

Macromechanical FEA simulations allow calculations of plastic strain φ and residual stresses. When including the relations in table 3 into the FEA, it allows predicting of the above-mentioned structural properties, which are not a direct outcome of the simulations. Furthermore, with these calculated magnitudes, it is possible to deduce additional material's behaviour, e.g. corrosion properties. Thus, the usefulness of the FEA can be enhanced significantly without additional calculation time needed.

Table 3. Fitting parameters based on the equation $y(\varphi) = A \cdot \exp(-k/\varphi) + y_0$.

Property	*A*	*k*	y_0
Roughness average (µm)	0.643 µm	0.102	0.312 µm
Crystallite size (nm)	360 nm	0.109	105 nm
Microstrain (°)	-0.141°	0.086	0.221°
Fraction {200} texture component (%)	-14.9 %	0.26	18.8 %

Acknowledgements
The authors gratefully acknowledge the Deutsche Forschungsgemeinschaft (DFG) for funding this work (grant numbers LA 1274/27-1 and AW 6/27-1). The authors thank Marc Pügner for the assistance in XRD measurements and Franz Nozicka for language help.

References

[1] Bauer A, Mehner T, Binotsch C, Sieber M, Awiszus B and Lampke T 2016 *IOP Conf. Series: Mater. Sci. Eng.* **118** 012019
[2] Mehner T, Bauer A, Nickel D, Binotsch C, Awiszus B and Lampke T 2015 *Mater. Today Proc.* **2 Suppl. 1** S141
[3] Bauer A, Mehner T, Awiszus B and Lampke T 2017 *Procedia Eng.* in press
[4] Butz A, Lossau S, Springub B and Roters F 2010 *Int. J. Mater. Forming* **3** 73
[5] Lacaille V, Morel C, Feulvarch E, Kermouche G and Bergheau J M 2014 *Computational Mater. Sci.* **95** 187
[6] Scholtes B, Boulais-Sinou R, Settefrati A, Munoz D P, Poitrault I, Montouchet A, Bozzolo N and Bernacki M 2016 *Computational Mater. Sci.* **112** 57
[7] Stadie-Fröhbus G 2000 *Techn. Mechanik* (in German) **20** 243
[8] Kim B R, Lee H K 2010 *Int. J. of solids and structures* **47** 827
[9] Mohonee V K and Goh K L 2016 *Composites Part B: Eng.* **86** 221
[10] Weidt D, Figiel L 2014 *Computational Mater. Sci.* **82** 298
[11] Wesenjak R, Krempaszky C, Werner E 2016 *Computational Mater. Sci.* **111** 277
[12] Zhou J, Gokhale A M, Gurumurthy A, Bhat S P 2015 *Mater. Sci. Eng.* A **630** 107

IOP Conf. Series: Materials Science and Engineering **181** (2017) 012037 doi:10.1088/1757-899X/181/1/012037

Analogies between continuum dislocation theory, continuum mechanics and fluid mechanics

C B Silbermann and J Ihlemann

Professorship of Solid Mechanics, Chemnitz University of Technology, Reichenhainer Straße 70, 09126 Chemnitz, Germany

E-mail: christian.silbermann@mb.tu-chemnitz.de

Abstract. Continuum Dislocation Theory (CDT) relates gradients of plastic deformation in crystals with the presence of geometrically necessary dislocations. Interestingly, CDT shows striking analogies to other branches of continuum mechanics. The present contribution demonstrates this on two essential kinematical quantities which reflect tensorial dislocation properties: the (resultant) Burgers vector and the dislocation density tensor. First, the limiting process for the (resultant) Burgers vector from an integral to a local quantity is performed analogously to the limiting process from the force vector to the traction vector. By evaluating the balance of forces on a tetrahedral volume element, Cauchy found his famous formula relating traction vector and stress tensor. It is shown how this procedure may be adopted to a continuously dislocated tetrahedron. Here, the conservation of Burger's vector implicates the introduction of the dislocation density tensor. Second, analogies between the plastic flow of a continuously dislocated solid and the liquid flow of a fluid are highlighted: the resultant Burgers vector of a dislocation ensemble plays a similar role as the (resultant) circulation of a vortex tube. Moreover, both vortices within flowing fluids and dislocations within deforming solids induce discontinuities in the velocity field and the plastic distortion field, respectively. Beyond the analogies, some peculiar properties of the dislocation density tensor are presented as well.

1. Introduction

It is well-established that matter of any kind consists of *discrete* particles, such as atoms and molecules. In order to understand the material behavior *microscopically*, *discrete* physical models and theories need to be applied. As an example, plastic deformation of solids is intrinsically tied to the motion of discrete lattice defects such as dislocations [1, 2]. Discrete Dislocation Dynamics (DDD) simulations take into account dislocation lines and their (microscopic) interactions [3]. Though matter is clearly heterogeneous on these small scales, the material body often seems homogeneous on a larger scale. Then, a *continuum* theory is appropriate and effective in order to study the body's *meso- and macroscopical* physical behavior. One prominent example is continuum mechanics, which aims to predict a body's motion/deformation under the action of external loads. Different sub-branches with respect to the state of the body have evolved. In this study, the focus is on continuum mechanics applied to fluids and dislocated solids. In Continuum Dislocation Theory (CDT), the material is considered as *continuously* dislocated. Instead of the (small scale) interaction of individual dislocation lines as in DDD, the collective behavior of dislocation ensembles is of key interest. Surprisingly, CDT shows some striking analogies, both to general continuum mechanics and to fluid mechanics [8, 9]. For simplicity, this is demonstrated on the *geometrically linear* CDT in the present contribution.

First, section 2 recalls basic dislocation properties on the example of a discrete edge dislocation. This is necessary for section 3, where the analogy between Cauchy's stress tensor and Kröner's dislocation density tensor is discussed. Next, section 4 confronts CDT with fluid mechanics and reveals analogous quantities and relations. Finally, section 5 goes beyond the analogies and emphasizes some peculiar properties of the dislocation density tensor.

For the sake of a compact and clear representation, symbolic tensor notation is preferred in this paper. Tensors of n-th order are denoted by a small or capital letter with n underscores, e.g. $\underline{\underline{U}}$ is a second-order tensor. The tensor product is denoted by $\otimes$, the cross product by $\times$ and the contraction of tensors is written symbolically with a dot $(\cdot)$. The coefficients with respect to a certain Cartesian coordinate system $\underline{e}_a$ are $U_{ab} = \underline{e}_a \cdot \underline{\underline{U}} \cdot \underline{e}_b$. Italic indices $a, b, \ldots$ stand for Cartesian coordinates x, y, z, whereas upright indices are labels (e.g. p stands for plastic). The arrangement of second-order tensor coefficients in a quadratic matrix is denoted by $[U_{ab}]$. Differential operators such as divergence and curl are expressed symbolically with the Nabla operator denoted by $\underline{\nabla} = \underline{e}_a \, \partial/\partial x_a$.

2. Preliminary considerations about dislocations

For the following sections some basic dislocation properties are recalled: Dislocations are line-shaped crystal defects. They constitute a discontinuity separating plastically slipped regions from unslipped ones (figure 1). A discrete dislocation within a crystal is characterized by two vectorial quantities: the tangent vector $\underline{t}$ on the dislocation line and the Burgers vector $\underline{b}$. The latter defines both magnitude and direction of the plastic slip carried by the dislocation. Furthermore, $\underline{b}$ is constant along the dislocation line whereas the tangent $\underline{t}$ varies in case of a curved dislocation.

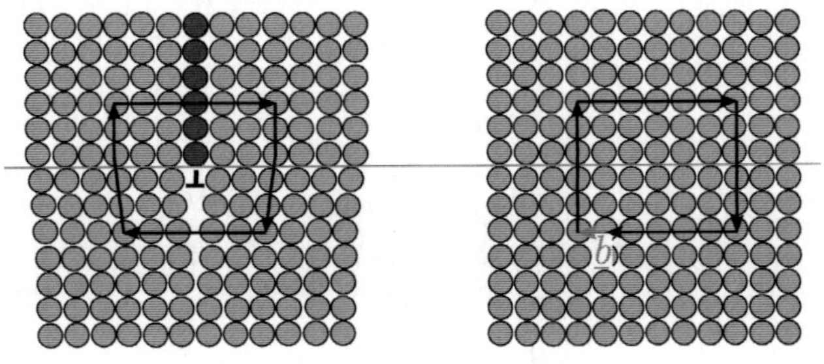

Figure 1. Burgers circuit: a loop around an edge dislocation and the same loop in a perfect simple cubic lattice yield a closure failure – the Burgers vector $\underline{b}$. The red line is the slip plane.

The Burgers vector is obtained by tracing a loop in the dislocated body and evaluating the vector needed to close the loop in the non-dislocated state. This is the well-known Burgers circuit (figure 1). With respect to some sign convention, the obtained closure failure represents the dislocation's elementary Burgers vector $\underline{b}$. For evident geometrical reasons, dislocations cannot end within a crystal, but only at phase boundaries or junctions with other dislocation segments. Considering N (discrete) dislocation lines within some crystal volume, this has the following consequence [3]:

$$\sum_{i=1}^{N} \overset{i}{\underline{b}} = \underline{0} \quad \text{analogous to} \quad \sum_{i=1}^{N} \overset{i}{\underline{F}} = \underline{0} \,, \tag{1}$$

which is Newton's balance of linear momentum for N (discrete) forces $\overset{i}{\underline{F}}$ acting on some solid body. In this sense, the Burgers vector is a conserved quantity. If the length scale is chosen this large that dislocations seem to be distributed continuously, then each material point of the continuum embraces an *ensemble* of dislocations. It may be characterized in an average sense by its *resultant* Burgers vector $\underline{b}_{\mathrm{r}}$.

IOP Conf. Series: Materials Science and Engineering **181** (2017) 012037 doi:10.1088/1757-899X/181/1/012037

3. The analogy between Cauchy's stress tensor and dislocation density tensor

In this section the far reaching analogy between the force vector $\underline{F}$ / traction vector $\underline{s}$ and the resulting Burgers vector $\underline{b}_{\mathrm{r}}$ / dislocation vector $\underline{a}$ is presented. The analogy is further extended by *deriving* the dislocation density tensor in the same way the stress tensor is derived. Therefore, the balance laws (1) are evaluated on an elementary tetrahedron-shaped volume element.

3.1. Definition of local and integral quantities

For everything which follows, assume some finite *continuous* body in space with $\underline{r} = x_a \underline{e}_a$ denoting the position of the material points. Now there are quantities that are *integral* by nature, such as forces or closure failures as explained in section 2. Hence, a corresponding *local* quantity must exist in a continuum theory. Consequently, a limiting process should be possible relating the *local* quantity to the *integral/global* one. However, having in mind the validity of continuum theories (cf. section 1) care must be taken not to violate the continuum hypothesis: Infinitesimal line, surface, volume and mass elements must not become zero. Instead, the elements always have to remain this large that enough of the material's relevant microscopic constituents are included [6]. In table 1 this is demonstrated for a simple solid and a dislocated one.

Table 1. Continuum limit for a simple solid and a dislocated solid.

	Force – Traction	Closure failure – Dislocations
Integral quantity	Force vector $\underline{F}$	Resultant Burgers vector $\underline{b}_{\mathrm{r}}$
Relevant length scale	Atomic lattice: 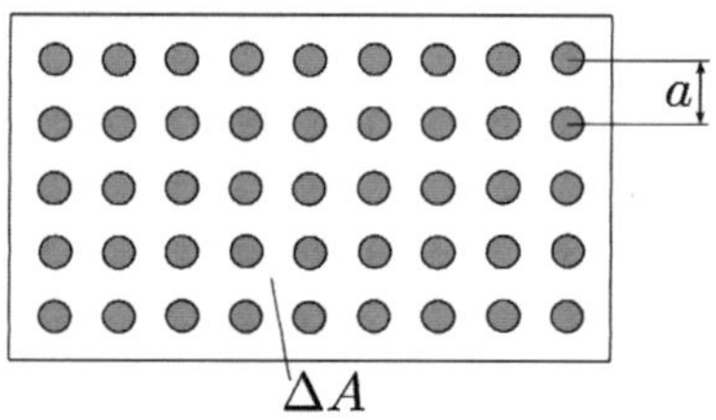	Edge dislocation array: 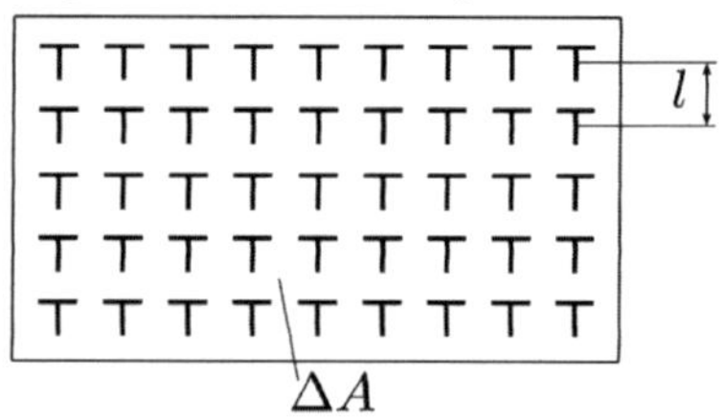
Limiting process	$\underline{s} = \lim\limits_{\Delta A \to 0} \left(\dfrac{\Delta \underline{F}}{\Delta A} \right) \Big\vert_{\Delta A \gg a^2}$	$\underline{a} = \lim\limits_{\Delta A \to 0} \left(\dfrac{\Delta \underline{b}_{\mathrm{r}}}{\Delta A} \right) \Big\vert_{\Delta A \gg l^2}$
Local quantity	$\underline{s} = \dfrac{\mathrm{d}\underline{F}}{\mathrm{d}A}$	$\underline{a} = \dfrac{\mathrm{d}\underline{b}_{\mathrm{r}}}{\mathrm{d}A}$

The definitions of the traction and the dislocation vectors as differential quotients are only valid, if the (infinitesimal) area element is much bigger than the mean atomic spacing – the lattice constant a – and the mean dislocation spacing l, respectively (figures in table 1). The other way round, a material point in the continuum sense averages over many atoms, dislocations and other possible microstructural features [1].

3.2. Evaluation of balance laws on the tetrahedral volume element

The well-known balance of forces $(1)_2$, applied on a infinitesimal, tetrahedron-shaped volume element implicates the introduction of the stress tensor [4, 5]. By exploiting the conservation of Burgers' vector in a similar way, the dislocation (density) tensor can be identified (table 2).

The elementary tetrahedron is obtained by cutting an elementary cube with faces perpendicular to the Cartesian coordinate axes. This results in a new cut face with an area $\mathrm{d}A$ and with an arbitrary outward unit normal $\underline{n} = n_a\,\underline{e}_a$ (figures in table 2), where Einstein's summation convention is to be applied. The other three faces have the area elements $\mathrm{d}A_a = n_a\,\mathrm{d}A$ and the normals $-\underline{e}_x$, $-\underline{e}_y$ and $-\underline{e}_z$. As they point in negative coordinate direction, the corresponding tractions $\underline{s}_a$ and closure failures $\underline{a}_a$ have a minus sign, too [5].

Table 2. Evaluation of balance laws on an elementary tetrahedral volume element.

Tetrahedron element	with tractions:	with closure failures:
	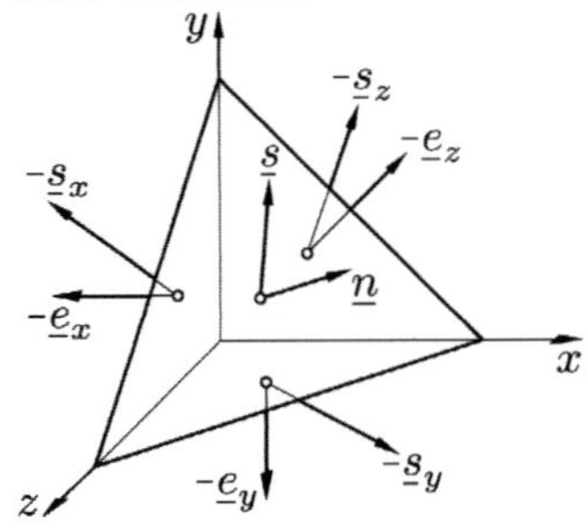	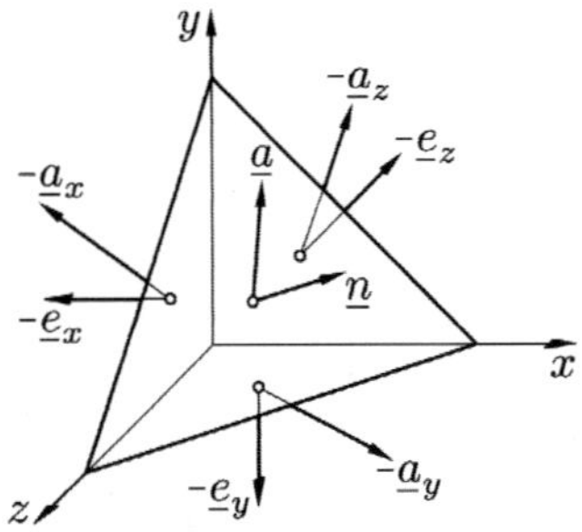
Global balance	$\mathrm{d}\underline{F} + \sum_{a=x}^{z} \mathrm{d}\underline{F}_a = \underline{0}$	$\mathrm{d}\underline{b}_\mathrm{r} + \sum_{a=x}^{z} \mathrm{d}\underline{b}_a = \underline{0}$
Global balance	$\underline{s}\,\mathrm{d}A + (-\underline{s}_a \mathrm{d}A_a) = \underline{0}$	$\underline{a}\,\mathrm{d}A + (-\underline{a}_a\,\mathrm{d}A_a) = \underline{0}$
with local quantities	$\underline{s}\,\mathrm{d}A - \underline{s}_a\,n_a\,\mathrm{d}A = \underline{0}$	$\underline{a}\,\mathrm{d}A - \underline{a}_a\,n_a\,\mathrm{d}A = \underline{0}$
(cf. table 1)	$\underline{s}\,\mathrm{d}A = (\underline{s}_a \otimes \underline{e}_a)\cdot\underline{n}\;\mathrm{d}A$	$\underline{a}\,\mathrm{d}A = (\underline{a}_a \otimes \underline{e}_a)\cdot\underline{n}\;\mathrm{d}A$
Local mapping	$\underline{s}(\underline{r},\underline{n}) = \underline{\underline{\sigma}}(\underline{r})\cdot\underline{n}$	$\underline{a}(\underline{r},\underline{n}) = \underline{\underline{\alpha}}(\underline{r})\cdot\underline{n}$

The transition from the global balances to the local mappings in table 2 is demonstrated on the stresses: the composition $\underline{s}_a\,n_a = \underline{s}_x\,n_x + \underline{s}_y\,n_y + \underline{s}_z\,n_z$ may be re-written into the symbolic expression $(\underline{s}_a \otimes \underline{e}_a)\cdot\underline{n}$, which reveals the second-order tensor

$$\underline{\underline{\sigma}} = \underline{s}_a \otimes \underline{e}_a \quad \text{analogous to} \quad \underline{\underline{\alpha}} = \underline{a}_a \otimes \underline{e}_a\,.^{1} \tag{2}$$

Obviously, it suffices to know the traction/dislocation vectors with respect to three orthogonal faces in order to construct the corresponding second-order tensors [1]. Furthermore, both traction and dislocation *vector* depend on the cutting plane defined by the unit normal $\underline{n}$. In the context of CDT it is natural to choose the crystal's slip planes as cut faces in order to obtain the resultant Burgers vector for the different slip systems. However, as $\underline{\underline{\alpha}}$ is a second-order *tensor*, $\underline{n}$ does not necessarily need to be a slip plane normal but may be related to *any* cut face.

The analogy can even be extended to *local* conservation laws: as will become clear in section 4, the dislocation density tensor is divergence-free, i.e. $\underline{\underline{\alpha}}\cdot\underline{\nabla} = \underline{0}$. In the absence of inertia and body forces the local balance of linear momentum leads to the same condition for the stress tensor, i.e. $\underline{\underline{\sigma}}\cdot\underline{\nabla} = \underline{0}$.[2]

[1] Note that the sequence is not unique here. Likewise, $\underline{s}_a\,n_a = n_a\,\underline{s}_a = \underline{n}\cdot(\underline{e}_a \otimes \underline{s}_a)$ could have been chosen. However, in the context of right differential operators, the normal is to stand right of the second-order tensor.

[2] Here, the right-divergence is applied. In the literature the left-divergence can be found as well [1]. Then, the corresponding stress and dislocation density tensors defined by equations (2) need to be transposed.

4. The analogy between Kelvin's circulation and the resultant Burgers vector

In order to answer the question under which circumstances a solid body (e.g. an aircraft) within some fluid (e.g. air) is able to ascend, Kelvin introduced a scalar quantity called circulation. Consider some fluid flow prescribed by the velocity field $\underline{v}(\underline{r})$ and let C stand for the contour of some vortex tube with the cross-sectional area ΔA (figure 2).[3] Now the circulation is defined as:

$$\Gamma := \oint_C \underline{v} \cdot \mathrm{d}\underline{r} \tag{3}$$

$$= -\int_A (\underline{v} \times \underline{\nabla}) \cdot \underline{n}\, \mathrm{d}A = \int_A (\underline{\nabla} \times \underline{v}) \cdot \underline{n}\, \mathrm{d}A = \int_A \underline{\Omega} \cdot \underline{n}\, \mathrm{d}A\,. \tag{4}$$

Here, Stokes' theorem was used to transform the path integral into the surface integral. Thereby, the derived vectorial quantity $\underline{\Omega} = -\underline{v} \times \underline{\nabla}$ appeared. It is called vorticity (or vortex vector) and represents the curl of the velocity. A solid body within a fluid experiences a *static* lift (Archimedes' Principle). If the fluid (or the solid body) is in motion and the circulation around this body does not vanish, the body additionally experiences a *dynamic* lift [6] (cf. table 3).

Figure 2. Vortex tube within a fluid (left) and dislocation ensemble (right) in a solid body: the lines with tangent $\underline{t}$ depict vortex and dislocation lines, respectively.

In order to answer the question how the internal state (i.e. the dislocation configuration) of a body is related to the plastic deformation of this body, Kröner introduced the *resultant* Burgers vector. It may be thought of as an average value summing up the *elementary* Burgers vectors $\underline{b}$ from all dislocations of some dislocation ensemble (figure 2). Hence, a dislocation ensemble is characterized by a constant *resultant* Burgers vector. Now consider some plastic flow prescribed by the plastic distortion field $\underline{\underline{\beta}}_{\mathrm{p}}(\underline{r})$ and let C stand for the contour of some dislocation ensemble with the cross-sectional area ΔA (figure 2). Then the resultant Burgers vector is defined as [7]:

$$\underline{b}_{\mathrm{r}} := \oint_C \underline{\underline{\beta}}_{\mathrm{p}} \cdot \mathrm{d}\underline{r} \tag{5}$$

$$= -\int_A \left(\underline{\underline{\beta}}_{\mathrm{p}} \times \underline{\nabla}\right) \cdot \underline{n}\, \mathrm{d}A = \int_A (\underline{\nabla} \times \underline{\underline{\beta}}_{\mathrm{p}}^{\mathrm{T}})^{\mathrm{T}} \cdot \underline{n}\, \mathrm{d}A = \int_A \underline{\underline{\alpha}} \cdot \underline{n}\, \mathrm{d}A\,. \tag{6}$$

Here again, Stokes' theorem was used to transform the path integral into the surface integral. Thereby, the derived tensorial quantity $\underline{\underline{\alpha}} = -\underline{\underline{\beta}}_{\mathrm{p}} \times \underline{\nabla}$ appeared. It is called Nye-Kröner-Bilby dislocation density tensor and represents the curl of the plastic distortion [1]. As a consequence of definition (5), $\underline{\underline{\alpha}}$ is an average quantity containing only information about Geometrically Necessary Dislocations (GNDs). Furthermore, as a curl field, $\underline{\underline{\alpha}}$ is divergence-free.

[3] A vortex tube is the surface in the continuum formed by all vortex lines passing through a given closed curve in the continuum [6].

The striking analogy between derivations (4) and (6) is obvious and there are manifold further aspects of it. The correspondence between vortices and dislocations has already been investigated thoroughly for the classical (local) continuum theory of dislocations [8]. Furthermore, a one-to-one relationship between a screw dislocation in gradient elasticity and a smoothed vortex in fluid mechanics has been shown before, including a tabular review of the corresponding field quantities [9]. The present study both reviews some of these results and further enriches the analogy with some more nuances, see table 3. Vortices within a flowing fluid induce discontinuities in the

Table 3. Confrontation of analogous quantities of a flowing fluid and a dislocated solid.

	Flowing fluid [6]	Continuously dislocated solid
Integral quantity	$\varGamma = \oint_C \underline{v} \cdot \mathrm{d}\underline{r} = \int_A \underline{\varOmega} \cdot \underline{n}\ \mathrm{d}A$	$\underline{b}_\mathrm{r} = \oint_C \underline{\underline{\beta}}_\mathrm{p} \cdot \mathrm{d}\underline{r} = \int_A \underline{\underline{\alpha}} \cdot \underline{n}\ \mathrm{d}A$
Significance	$\varGamma = 0$: static lift only	$\underline{b}_\mathrm{r} = \underline{0}$: Compatibility of plastic flow
	$\varGamma \neq 0$: dynamic lift	$\underline{b}_\mathrm{r} \neq \underline{0}$: Incompatibility
Local quantity	$\underline{\varOmega} = \text{-}\underline{v} \times \underline{\nabla}$	$\underline{\underline{\alpha}} = \text{-}\underline{\underline{\beta}}_p \times \underline{\nabla}$
Significance	$\underline{\varOmega} = \underline{0}\ \forall\, \underline{r}$: curl-free flow	$\underline{\underline{\alpha}} = \underline{\underline{0}}\ \forall\, \underline{r}$: GND-free plastic flow
Tangent lines	$\underline{\varOmega} \times \mathrm{d}\underline{t} = \underline{0}$: vortex line $\underline{t}$	$\underline{\underline{\alpha}} \times \mathrm{d}\underline{t} = \underline{\underline{0}}$: dislocation line tangent $\underline{t}$

velocity field and vice versa. Likewise, dislocations within a deforming solid induce discontinuities in the plastic distortion field and vice versa. Furthermore, in both cases the relations are of pure kinematical nature, i.e. they are valid for any admissible motion (including deformation) of a fluid and solid continuum, respectively. Moreover, there is a continuity condition (zero divergence) both for $\underline{\varOmega}$ and $\underline{\underline{\alpha}}$: similar to dislocation lines, vortex lines must form closed loops (or junctions) [8]. In other words: both dislocations and vortices cannot end inside a continuous body [9]. Moreover, vortex line tangents can be post-processed from a given vorticity field $\underline{\varOmega}(\underline{r})$. Likewise, it should be possible to determine dislocation line tangents by a similar mathematical procedure from a given field $\underline{\underline{\alpha}}(\underline{r})$, cf. table 3. Then, the underlying network of dislocation *ensembles* could be revealed as post-processing step. This would be an alternative to tracking the motion of individual dislocation lines, which results in high numerical costs [3].

5. Further properties of the dislocation density tensor

Despite the analogies presented in sections 3 and 4 there are also important differences. As a consequence of the balance of angular momentum the stress tensor is symmetric [4]. The dislocation density tensor is *un*symmetric, which will be illustrated now. In general, according to equation (6), $\underline{\underline{\alpha}}$ has the following coefficients with respect to a Cartesian coordinate system:

$$\underline{\underline{\alpha}} = \text{-}\underline{\underline{\beta}}_\mathrm{p} \times \underline{\nabla} = \text{-}\,\frac{\partial \beta^\mathrm{p}_{ab}}{\partial x_c}\,\epsilon_{bcd}\ \underline{e}_a \otimes \underline{e}_d = \beta^\mathrm{p}_{ab,c}\,\epsilon_{bcd}\ \underline{e}_a \otimes \underline{e}_d = \alpha_{ad}\ \underline{e}_a \otimes \underline{e}_d\ , \tag{7}$$

with the permutation symbol $\epsilon_{bcd} = (\underline{e}_b \times \underline{e}_c) \cdot \underline{e}_d$. The coefficient matrix takes thus the form:

$$[\alpha_{ad}] = \begin{bmatrix} \alpha_{xx} & \alpha_{xy} & \alpha_{xz} \\ \alpha_{yx} & \alpha_{yy} & \alpha_{yz} \\ \alpha_{zx} & \alpha_{zy} & \alpha_{zz} \end{bmatrix} = \begin{bmatrix} \beta^\mathrm{p}_{xz,y}\text{-}\beta^\mathrm{p}_{xy,z} & \beta^\mathrm{p}_{xx,z}\text{-}\beta^\mathrm{p}_{xz,x} & \beta^\mathrm{p}_{xy,x}\text{-}\beta^\mathrm{p}_{xx,y} \\ \beta^\mathrm{p}_{yz,y}\text{-}\beta^\mathrm{p}_{yy,z} & \beta^\mathrm{p}_{yx,z}\text{-}\beta^\mathrm{p}_{yz,x} & \beta^\mathrm{p}_{yy,x}\text{-}\beta^\mathrm{p}_{yx,y} \\ \beta^\mathrm{p}_{zz,y}\text{-}\beta^\mathrm{p}_{zy,z} & \beta^\mathrm{p}_{zx,z}\text{-}\beta^\mathrm{p}_{zz,x} & \beta^\mathrm{p}_{zy,x}\text{-}\beta^\mathrm{p}_{zx,y} \end{bmatrix}\ . \tag{8}$$

The elements on the principal and secondary diagonal represent the screw character ($a = d$) and the edge character ($a \neq d$) of the dislocation ensemble [7].

When the plastic distortion field only depends on a single coordinate x, y or z, the corresponding column in the coefficient matrix of $\underline{\underline{\alpha}}$ is zero (table 4).

Table 4. Special dislocation states for different dependencies of $\underline{\underline{\beta}}_{\mathrm{p}}(\underline{r})$.

	$\underline{\underline{\beta}}_{\mathrm{p}} = \underline{\underline{\beta}}_{\mathrm{p}}(x)$	$\underline{\underline{\beta}}_{\mathrm{p}} = \underline{\underline{\beta}}_{\mathrm{p}}(y)$	$\underline{\underline{\beta}}_{\mathrm{p}} = \underline{\underline{\beta}}_{\mathrm{p}}(z)$
$[\alpha_{ab}] =$	$\begin{bmatrix} 0 & -\beta^{\mathrm{p}}_{xz,x} & \beta^{\mathrm{p}}_{xy,x} \\ 0 & -\beta^{\mathrm{p}}_{yz,x} & \beta^{\mathrm{p}}_{yy,x} \\ 0 & -\beta^{\mathrm{p}}_{zz,x} & \beta^{\mathrm{p}}_{zy,x} \end{bmatrix}$	$\begin{bmatrix} \beta^{\mathrm{p}}_{xz,y} & 0 & -\beta^{\mathrm{p}}_{xx,y} \\ \beta^{\mathrm{p}}_{yz,y} & 0 & -\beta^{\mathrm{p}}_{yx,y} \\ \beta^{\mathrm{p}}_{zz,y} & 0 & -\beta^{\mathrm{p}}_{zx,y} \end{bmatrix}$	$\begin{bmatrix} -\beta^{\mathrm{p}}_{xy,z} & \beta^{\mathrm{p}}_{xx,z} & 0 \\ -\beta^{\mathrm{p}}_{yy,z} & \beta^{\mathrm{p}}_{yx,z} & 0 \\ -\beta^{\mathrm{p}}_{zy,z} & \beta^{\mathrm{p}}_{zx,z} & 0 \end{bmatrix}$

Plane plastic distortions $\underline{\underline{\beta}}_{\mathrm{p}} = \beta^{\mathrm{p}}_{xx}(x,y)\underline{e}_x\otimes\underline{e}_x+\beta^{\mathrm{p}}_{xy}(x,y)\underline{e}_x\otimes\underline{e}_y+\beta^{\mathrm{p}}_{yx}(x,y)\underline{e}_y\otimes\underline{e}_x+\beta^{\mathrm{p}}_{yy}(x,y)\underline{e}_y\otimes\underline{e}_y$ result in $\underline{\underline{\alpha}}(x,y)$ with

$$[\alpha_{ad}] = \begin{bmatrix} 0 & 0 & \beta^{\mathrm{p}}_{xy,x}-\beta^{\mathrm{p}}_{xx,y} \\ 0 & 0 & \beta^{\mathrm{p}}_{yy,x}-\beta^{\mathrm{p}}_{yx,y} \\ 0 & 0 & 0 \end{bmatrix} . \tag{9}$$

Choosing a cut face with normal $\underline{n} = \underline{e}_z$, the corresponding resultant Burgers vector and thus also the dislocation vector have only two (plane) components:

$$\underline{\alpha}(x,y,\underline{n}=\underline{e}_z) = \underline{\underline{\alpha}}(x,y)\cdot\underline{e}_z = (\beta^{\mathrm{p}}_{xy,x}-\beta^{\mathrm{p}}_{xx,y})\,\underline{e}_x + (\beta^{\mathrm{p}}_{yy,x}-\beta^{\mathrm{p}}_{yx,y})\,\underline{e}_y\ . \tag{10}$$

Moreover, certain decompositions known from the stress tensor are not possible. In general, a geometrical dislocation state is neither a symmetric nor a pure spherical tensor. This shall be demonstrated now: Starting from the definition

$$\underline{\underline{\alpha}} = (\underline{\nabla}\times\underline{\underline{\beta}}^{\mathrm{T}}_{\mathrm{p}})^{\mathrm{T}} = -\underline{\underline{\beta}}_{\mathrm{p}}\times\underline{\nabla} = -\underline{\underline{\beta}}_{\mathrm{p}}\cdot\underline{\underline{I}}\times\underline{\underline{I}}\cdot\underline{\nabla} = \underline{\underline{\beta}}_{\mathrm{p}}\cdot\underline{\underline{\underline{\epsilon}}}\cdot\underline{\nabla}\ , \tag{11}$$

and considering that $\underline{\underline{\beta}}_{\mathrm{p}}$ is unsymmetric and $\underline{\underline{\underline{\epsilon}}}$ is totally skew-symmetric, it follows immediately:

$$\underline{\underline{\alpha}} \neq \underline{\underline{\alpha}}^{\mathrm{T}} \quad \text{and} \quad \underline{\underline{\alpha}} \neq \alpha\underline{\underline{I}}\ . \tag{12}$$

This has the following consequence: for an unsymmetric dislocation density tensor, there is no orthonormal coordinate system where $[\alpha_{ad}]$ is a diagonal matrix. Physically speaking, there is *in general* no cut surface where all dislocations piercing the surface appear as screw dislocations. This is evident since otherwise, a dislocation ensemble with screw *and* edge character could be transformed into screw character only by rotating the coordinate system.

However, there is indeed a special case where $\underline{\underline{\alpha}}$ can be symmetric or even spherical. If the plastic distortion tensor is skew-symmetric, a vector may be assigned in the following way:

$$\underline{\underline{\beta}}_{\mathrm{p}} = -\underline{\underline{\beta}}^{\mathrm{T}}_{\mathrm{p}} \quad\rightarrow\quad \underline{\underline{\beta}}_{\mathrm{p}} = \underline{\theta}\cdot\underline{\underline{\underline{\epsilon}}}\ . \tag{13}$$

Physically speaking, there is only a plastic *rotation* now but no plastic *strain*. The rotation vector $\underline{\theta}$ contains both the rotation angle and the axis. Substituting relation (13) into definition (11), Nye's famous formula is obtained:

$$\underline{\underline{\alpha}} = \underline{\theta}\cdot\underline{\underline{\underline{\epsilon}}}\cdot\underline{\underline{\underline{\epsilon}}}\cdot\underline{\nabla} = \underline{\nabla}\otimes\underline{\theta} - (\underline{\nabla}\cdot\underline{\theta})\,\underline{\underline{I}}\ . \tag{14}$$

Obviously, $\underline{\underline{\alpha}}$ is symmetric or even spherical if the rotation gradient $(\underline{\nabla}\otimes\underline{\theta})^{\mathrm{T}}$ is symmetric or spherical, respectively. This case is of practical relevance, because relation (14) is often used to approximate $\underline{\underline{\alpha}}$ from experimental orientation data by methods such as electron backscatter diffraction. Some more properties of $\underline{\underline{\alpha}}$ under special slip conditions can be found elsewhere [10].

6. Conclusions and outlook

In this study, analogies between different branches of continuum mechanics were presented. To be precise, traction vector and dislocation vector can be obtained by a similar physical limiting process such that the continuum hypothesis is preserved. Moreover, similar conservation laws of forces and Burgers vectors may be evaluated on an elementary tetrahedron leading to the introduction of the second-order stress tensor and dislocation density tensor, respectively. Thus, a new way is shown how to motivate the dislocation density tensor. In addition, the quantity vorticity known from fluid mechanics has an analogous significance as the resultant Burgers vector for a continuously dislocated solid. Both quantities may be used to obtain further knowledge about the fluid's flow characteristics and the microstructural state of the solid.

Going beyond the pure mathematical analogy it might be even possible to consider vortices within flowing fluids as a (temporary) microstructure. Indeed, the vorticity vector may be considered as a measure of the defect concentration within a fluid – its vortex density – in total accordance with the dislocation density tensor [8]. It is well-known that macroscopic properties of solids (e.g. hardening) emerge from the collective self-organization of crystal defects during the plastic deformation. Here, dislocation cell patterns are observed often. Analogously, the famous Bénard cells can emerge in the velocity field of fluids under some external thermal load. Furthermore, both dislocations and vortices may form wall structures [8]. Nevertheless, there is an important difference: as the plastic behavior of solids depends on the defect state, the dislocation density tensor enters constitutive equations in CDT. The counterpart in fluid mechanics, the vorticity, cannot enter constitutive equations since it is neither a state nor an objective quantity.

Maybe the presented analogies can help to transfer knowledge between the field theories. There is also a far reaching correspondence between the continuum theory of dislocations and the one of electromagnetism (Maxwell's field theory) [11, 12]. From an even wider perspective, such analogies might be helpful to identify fundamental symmetries leading to unified physical theories.

Acknowledgments

This research was supported by the German National Science Foundation (DFG) within the framework of collaborative research center SFB 692 High-strength aluminum-based light-weight materials for safety components. Furthermore, the help of Prof. Khanh C. Le, Prof. Bernd Platzer, Dr. Ralf Landgraf, Hendrik Donner and Robert Scherzer is gratefully acknowledged.

References

[1] Kröner E 1958 *Kontinuumstheorie der Versetzungen und Eigenspannungen* (Berlin: Springer) p 24
[2] Kröner E 1992 The internal mechanical state of solids with defects *Int. J. Solids Struct.* **29** 1849–57
[3] Cai W and Bulatov V 2006 *Computer Simulations of Dislocations* (New York: Oxford University Press)
[4] Kreißig R and Benedix U 2002 *Höhere Technische Mechanik* (Vienna: Springer) pp 20–27
[5] Ottosen N S and Ristinmaa M 2005 *The mechanics of constitutive modeling* (Amsterdam: Elsevier) pp 52–53
[6] Schade H and Kunz E 2007 *Strömungslehre* ed Kameier F (Berlin: de Gruyter) pp 215–250
[7] Berdichevsky V 2006 Continuum theory of dislocations revisited *Continuum Mech. Thermodyn.* **18** 195–222
[8] Marcinkowski M J 1989 Correspondence between vortices and dislocations *Phys. stat. sol. (b)* **152** 9-31
[9] Lazar M 2004 On the correspondence between a screw dislocation in gradient elasticity and a regularized vortex *Ann. Phys.* **13** 617–22.
[10] Silbermann C B and Ihlemann J 2016 Kinematic assumptions and their consequences on the structure of field equations in continuum dislocation theory *IOP Conf. Ser.: Mater. Sci. Eng.* **118** 012034
[11] Lazar M 2014 On the Fundamentals of the Three-dimensional Translation Gauge Theory of Dislocations *Math. Mech. Solids* **16** 253–64
[12] Lazar M 2002 An elastoplastic theory of dislocations as a physical field theory with torsion *J. Phys. A: Math. Gen.* **35** 1983

IOP Conf. Series: Materials Science and Engineering **181** (2017) 012038 doi:10.1088/1757-899X/181/1/012038

Material selection for climbing hardware using the example of a belay device

E Semenov, S Schwanitz and S Odenwald

Chemnitz University of Technology, Institute of Lightweight Structures, Reichenhainer Str. 70, 09126 Chemnitz, Germany

Abstract. The aim of the research project was to design a novel climbing belay device. The present article describes the details of the therefor performed material selection. Literature research on the materials used in commercially available belay devices revealed a lack of definite information. Thus, a pilot x-ray fluorescence (XRF) test was performed on a small sample of common aluminium belay devices. It revealed the use of a variety of different alloy systems. The selection process continued by compiling a thorough list of constraints and objectives for this safety related piece of sports equipment. Different material options including non-aluminium-materials were discussed. The final material choice was a high strength aluminium alloy with a T6 thermal treatment. The device was designed and calculated by use of CAD and FEM software respectively, aiming to reduce weight. After manufacturing the strength, usability and friction properties of the device have been successfully tested.

1. Introduction

Belay devices are manually operated rope brakes, a central element of climbing safety equipment. Their function is to introduce a braking force to the "life" rope in order to stop a leader fall or allow a controlled rappel. The rope glides through the device against the generated braking force transforming the kinetic energy of the falling climber into thermal energy. The fall comes to a halt as soon as the kinetic energy is fully dissipated. In the process life rope forces can reach values of several kN [1], requiring the belay device to withstand them. This paper describes the process of material selection for a newly developed manual belay device (see figure 1). The main purpose of this particular device is belaying a lead climber directly off the anchor in a multi pitch scenario. It has been designed in accordance to the DIN EN 15151-2:2012 standard.

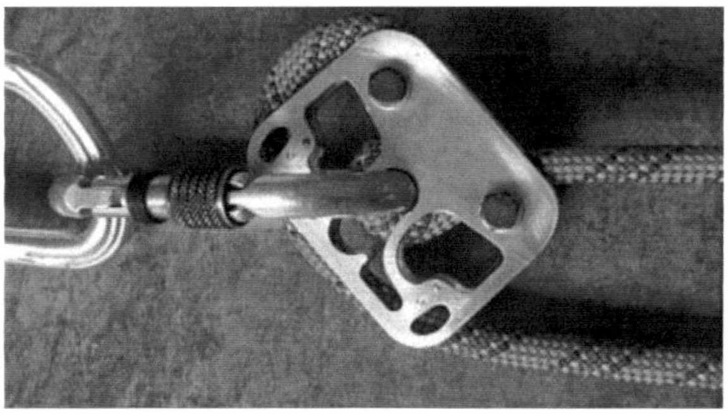

Figure 1. The developed belay device attached to an HMS carabiner and rigged with a climbing rope. The "control" rope can be seen at the bottom and the "life" rope at the top of the image. The control rope force introduced by the operator's hand is amplified by the device, allowing the control of the life rope force.

1. Industrial manufacturing and materials

Due to growing popularity of the sport, mountaineering equipment has become very sophisticated. A part of the development is a growing range of materials in use. Belay devices have traditionally been cast or wrought from aluminium alloys [2]. Typically only this single material was used for the whole device. Modern devices often incorporate a variety of different materials, especially if moving parts are involved. These are most common in belay devices with assisted locking and are typically mounted on steel axles (Grigri by Petzl) and rarely on brass slide bearings (Sirius by Tre, see figure 2). Also, some devices consist of punched sheet metal parts that are held together by steel rivets (Smart by Mammut). Tube-type devices commonly have a retaining steel-wire-sling with a polymer coating. For better ergonomics non-load-bearing polymer parts are incorporated in modern devices (Ergo by Salewa). Yet aluminium alloys remain the most common load bearing material followed by steel, investment cast stainless steel in the case of the Megajul by Edelrid. Even this latter material description is uncommonly precise for an advertisement of climbing equipment. The manufacturers clearly attempt to keep their secrets.

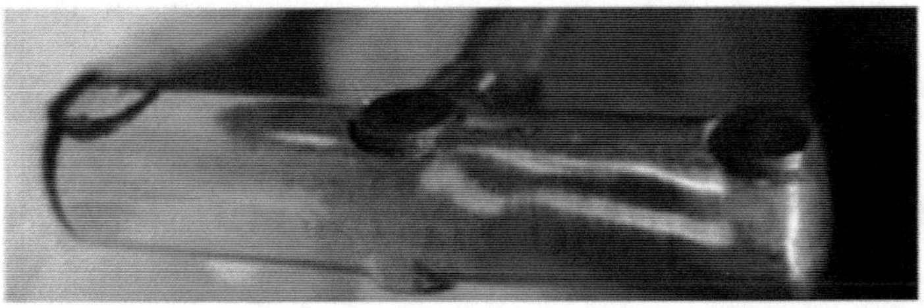

Figure 2. Brass inserts as sliding bearings on a moving component of the Sirius belay device by Tre.

1.1 Literature research

Unfortunately, the scarce information provided by the manufacturers is only marginally supplemented by literature. Blackford [2] merely states that belay devices are made from high strength aluminium alloys, but does not specify further details, even though she provides more precise material information on other types of mountaineering hardware. Two later articles by Fuss and Niegl [3, 4] which deal quite deeply with the matter of belay devices completely ignore the material aspect of the problem. Titt provides interesting ideas in an article on the website of his employer, a mid tier rock anchor manufacturer [5]. In particular it raises awareness of the importance of thermal conductivity to the soundness of the climbing rope. Even though the article does not conform to scientific standards, it has been an important source of inspiration, and therefore deserves credit.

1.2 Pilot x-ray fluorescence test

The lack of information on the alloys in use led to the testing of three different belay devices via x-ray fluorescence analysis (XRF). XRF analysis cannot reveal the exact composition of aluminium alloys due to its low atomic number. While alloying elements can be detected, the aluminium itself remains invisible, thus giving no reference for a calculation of alloying element percentages. Still, by displaying the engaged alloying elements, the method can give an insight on the chosen alloy systems. These in turn allow assumptions about an eventual heat treatment of the devices. The results of the analysis are summarized in table 1. All three devices show the presence of manganese, which is the only alloying element in the Sirius, proving it to be made of a non-heat-treatable 3xxx-series alloy. The other two devices additionally show the presence of copper, one of them also containing zinc, which indicates 2xxx- or 7xxx-series alloys, respectively. All this shows that no consensus on a material choice for belay devices has been found yet among the manufacturers. Even heat treatable alloys are not a standard choice despite their superior structural properties.

Table 1. Alloying elements detected via XRF analysis and possible conclusions.

Belay device	Manufacturer	Alloying elements			Possible alloy series			Impurities
Bat Brake	Edelrid	Mn	Cu		3xxx	2xxx		Fe
ATC Guide	Black Diamond	Mn	Cu	Zn	3xxx	2xxx	7xxx	
Sirius	TRE	Mn			3xxx			Fe
					non-heat-treatable	heat-treatable		

1.3 Assumptions on manufacturing techniques

By examining manufacturing traces on the surface of commercially available climbing equipment, an attempt is made to identify the manufacturing methods of the above-mentioned devices. The Sirius and the Bat Brake were presumably drop forged. They exhibit an appropriate geometry for this forming process and feature circumferential burrs, which indicate a removal of the excess material by stamping (see figure 3). The manufacturing method of the ATC cannot be determined as easily. It has no obvious traces of stamping or similar procedures and a more complex geometry than the Bat Brake or the Sirius. Several undercuts indicate that the ATC might be investment cast similar to the Magajul.

Figure 3. Burrs on the edges of the Bat Brake by Edelrid, where the excess material from drop forging was presumably removed by stamping.

1.4 Conclusion

There is only little to no information on the exact alloys, manufacturing techniques and heat treatment of commercially available belay devices. It is clear that the material class of aluminium alloys is the most popular, being the choice of most manufacturers. Stainless steel is the second, far less popular choice. No other materials are being used as base material for belay devices at the present.

2. Material selection

After the foregoing insight into the material choices in the industry, an own independent effort to systematically select a material is made. In the following, general design requirements for belay devices are rephrased into unique constraints and objectives. The latter are then used to systematically sort out inappropriate material families and classes and single materials leaving behind the most appropriate material choice.

2.1 Design requirements

From the functions a belay device performs, the obstacles it meets and the environmental influences it endures the following requirements are derived:

- Climbing safety equipment is supposed to ensure safety without compromising the mobility of the climber. Thus **low weight at a sufficient breaking strength** is a major requirement.
- For the sake of sustainability the device **should last as long as possible.**
- Obviously, the device **should not damage the rope**.

IOP Conf. Series: Materials Science and Engineering **181** (2017) 012038 doi:10.1088/1757-899X/181/1/012038

2.1.1 Constraints

- The devices are often being used outdoors in bad weather or at the sea side. Thus **sufficient corrosion resistance** is required for longevity.
- **No rope damaging features should be contained within the material**. Their exposure by abrasion or corrosion, no matter how slow, could damage the rope.
- The **melting temperature** of the material should not be reached, as this would damage the rope.
- The material **should be affordable and available.**
- A **sufficient size of pre-product** is required to meet the dimensions of the future device.

2.1.2 Objectives

- **Maximize weight specific yield strength** for low weight at sufficient strength!
- **Maximize abrasion and wear resistance** for longevity!
- A belay device in outdoor use will often have contact to the rock surface, occasionally this will also occur under considerable loads. Experience shows that belay devices accumulate grooves and indentations on their outer surface over the years of use. This superficial damage can lead to high stress concentrations and eventually to failure. Sufficient toughness is required, to reduce and withstand local stress concentrations. Unfortunately a thorough series of tests would be necessary to define the exact level of toughness needed. Without this knowledge it is expedient to **maximise toughness.** This only makes sense within the limits of the previous two objectives, since toughness and strength are interdependent and cannot be raised in equal measure.
- In case of a deep fall (say 10 m), when a large percentage of the energy must be absorbed by friction inside the belay device (factor-2-fall), an 80-kg-climber and an assumed braking duration of 1 s the average heat flow rate to be dissipated almost reaches approximately 8 kW. Due to the low melting temperature of the rope material (typically PA 6.6) and its potential damage, high temperatures and thus the accumulation of thermal energy is to be avoided. To make this possible **thermal conductivity should be maximized**. As in the previous point, for future projects it would make sense to determine a minimum level of thermal conductivity and capacity to ensure safety.

2.2 Choice of material family

The material families taken into consideration are ceramics, composites, polymers and metals.

Ceramics are too brittle and would not only suffer from stress concentrations at superficial indentations, they could also be damaged if dropped from a height. Composites contain fibres that might be exposed by abrasion. Especially glass or metal fibres might damage the rope if exposed. Composites with a polymer matrix feature the same downsides as the polymers themselves: low thermal conductivity [6], low melting temperature (for thermoplastic polymers) and low abrasive resistance. Only metals meet all requirements mentioned above, thus qualifying for the next selection cycle.

2.3 Choice of material class

The material classes taken into consideration are high strength steels, stainless steels, magnesium alloys, titanium alloys and aluminium alloys. High strength steels and magnesium alloys drop out due to their typically low corrosion resistance, which makes them impractical in the outdoors. Titanium alloys and stainless steels exhibit a very low thermal conductivity compared to aluminium (see figure 4), which might lead to overheating and rope damage. By the method of elimination the aluminium alloys remain to be the only acceptable material class. It is important to note that aluminium is chosen over stainless steel just as high thermal conductivity is chosen over wear resistance for a reason: the first influences the safety of the device, while the second only prolongs the life span. Picking safety over life span is an easy choice.

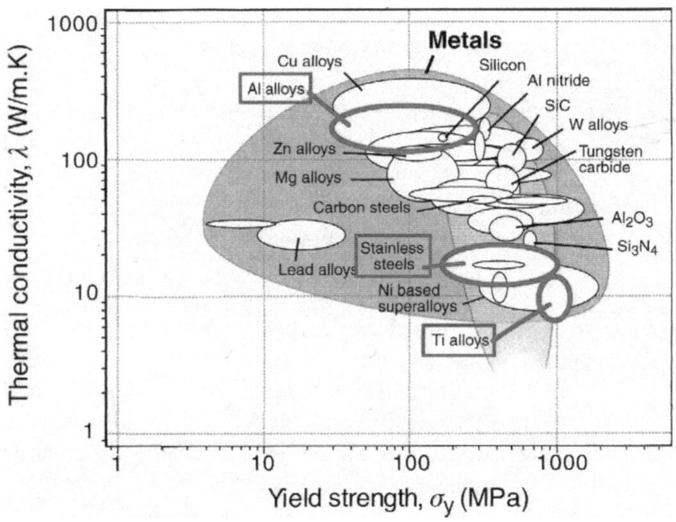

Figure 4. Material property chart of yield strain vs. thermal conductivity, an extract from [6, p. 252].

2.4 Choice of single material

To make a choice within the material class of aluminium alloys it is expedient to check which ones of the constraints and objectives remain unfulfilled and can be influenced by the imminent choice. Three remaining goals result from this: to maximize weight specific yield strength, to maximize abrasion and wear resistance and to ensure sufficient toughness.

The overall wear resistance of aluminium alloys is relatively low. It increases with rising hardness and yield strength, but this is only one of the influencing parameters [6]. The presence of above-average hard grains in a relatively soft surrounding grain structure can lead to considerable elevation of wear resistance [7], without necessarily changing the macroscopic hardness. At the same time such grains or precipitates could lead to a faster abrasion of the rope, thus conflicting with one of the above formulated constraints: No rope damaging features should be contained within the material. Another issue is the continuous formation of an oxide layer on the surface of aluminium alloys and the option of anodising the surface intentionally, both having a yet unpredictable effect on wear behaviour in combination with a textile surface. Due to all these difficulties only the positive effect of increasing yield strength shall henceforth be taken into account for wear resistance. This agrees with the objective to maximise the weight specific yield strength. The densities of the different alloys are fairly similar, so the alloy with the highest yield strength can be chosen. Also this simple approach doesn't seem to affect the toughness strongly [6], which makes it acceptable within the formulated objectives. Thus the strongest accessible aluminium alloy is chosen: EN AW-7075.

2.5 Material treatment

The T6-heat-treatment has been applied, the resulting material being EN AW-7075-T6. The treatment involves three steps: dissolving the alloying elements in the solid aluminium at high temperatures; quenching to provoke supersaturation at low temperatures; and aging at medium temperatures to receive finely dispersed precipitates that maximise yield strength.

3. Further development

3.1 Construction

The conception of the created belay device was structured according to the VDI 2221 guideline. Handling, rope friction and strength were taken into account. The belay device was modelled with CREO Parametric (Parametric Technology GmbH, Unterschleißheim, Germany). The effects of the 7 kN force on the life rope required by the DIN EN 15151-2 standard were simulated via CREO

Simulate and the bolted joint additionally calculated according to the VDI 2230 guideline. The geometry was iteratively altered to save weight and to meet the required safety margins according to the FKM-guideline. The braking force generated by the device was adjusted to match that of the ATC (a popular tube-type device) and the Munter hitch, being two widely accepted references.

3.2 Manufacturing
The belay device prototype has been CNC-milled and hand-finished. The two symmetrical halves were bolted together using a torque wrench. The elongation of the bolts was controlled with a micrometer to meet the required pretension.

3.3 Testing
Usability, rope friction and strength of the prototype were tested. The usability was tested with volunteers in a safe environment. The resulting belay device has minor usability issues but has otherwise proven to be fit for use. The rope friction test involved a mechanical ("simulated") hand that substituted the force of the operator's hand. The friction generated by the device meets expectations. Being the only material-related test the strength test is described below. The test was performed in accordance to the requirements of the DIN EN 15151-2 standard (see figure 5). The device was connected to the lower attachment point of a tension testing machine with a climbing carabiner as it would be in the intended application. The "life" rope was connected to the upper attachment point while the "control" rope was directed to an additional attachment point at a 30° angle from the pulling axis. The arrangement was then loaded with 7 kN life rope force. The device withstood the applied forces without major deformations. The only plastic deformation was an indentation of the contact area between the attachment carabiner and the device body. This indentation being a typical deformation with most belay devices and being located in an area of compressive stress is considered harmless. Hence the device passed the test.

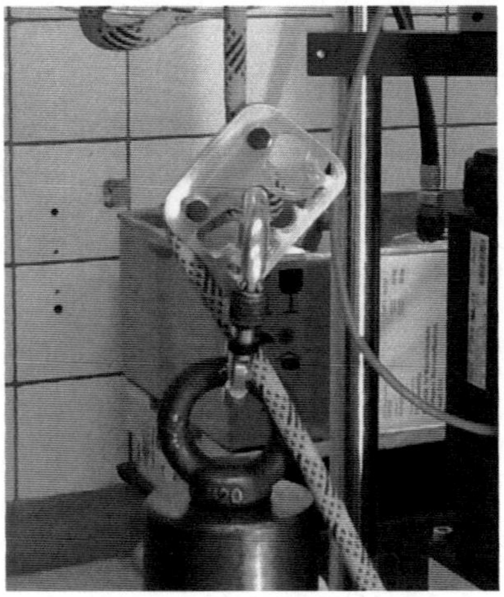

Figure 5. The belay device as it was tensioned during the pull test: The device attached to the tension testing machine with a carabiner, the life rope going upward and the control rope leaving the device at an angle of 30°.

4. Conclusion
Though the systematic approach towards a material choice for a climbing belay device results in the selection of a very typical material, the high strength aluminium alloy EN AW-7075, it shows that some constraints and objectives need a more precise formulation. The call for toughness, wear resistance and thermal conductivity is clearly motivated but not properly quantified. This results in the inefficient maximisation of quantities, that most probably only need to exceed a discrete value. Defining these values should be the topic of future studies.
For the present project the material selection depicted above has been a success. A prototype has been built from the chosen material and its functionality and strength have been successfully tested.

Acknowledgements

The authors gratefully acknowledge funding by the German Research Foundation (Deutsche Forschungsgemeinschaft, DFG) within the framework of SFB 692.

The authors would also like to thank Mr. S. Fritsch for imparting the XRF and the final strength test of the device as well as for providing the construction material alongside its thermal treatment and Mr. S. Müller for precious technical support in designing and machining the prototypes.

References

[1] Blackford J R 2003 Materials in mountaineering *Materials in sports equipment* ed M Jenkins (Boca Raton, Cambridge: CRC Press) chapter 11 pp 279–325

[2] Semmel C 2002 Die Handbremse *DAV Panorama* **4** 61–3

[3] Fuss F K and Niegl G 2010 Design and mechanics of belay devices and rope brakes *Sports Technology* **3** 68–87

[4] Fuss F K and Niegl G 2014 Design and mechanics of mountaineering equipment *Routledge handbook of sports technology and engineering* ed F K Fuss, A Subic, M Strangwood and R Metha (London: Routledge) chapter 17 pp 277–92

[5] Titt J 2009 Belay Device Theory, Testing and Practice *online* 22 august 2016 http://www.bolt-products.com/Glue-inBoltDesign.htm

[6] Ashby M, Shercliff H and Cebon D 2008 *Materials. Engineering, science, processing and design.* (Amsterdam: Elsevier)

[7] Drossel G and Kammer C 2010 *Aluminium-Taschenbuch* (Duesseldorf: Aluminium-Verlag)

IOP Conf. Series: Materials Science and Engineering **181** (2017) 012039 doi:10.1088/1757-899X/181/1/012039

Influence of ECAP temperature on the formability of a particle reinforced 2017 aluminum alloy

S Wagner, M Härtel, P Frint and M F-X Wagner

Chemnitz University of Technology, Institute of Materials Science and Engineering, 09125 Chemnitz, Germany

E-mail: swetlana.wagner@mb.tu-chemnitz.de

Abstract. Severe plastic deformation methods are commonly used to increase the strength of materials by generating ultrafine-grained microstructures. The application of these methods to Al-Cu alloys is, however, difficult because of their poor formability at room temperature. An additional reduction of formability of such alloys occurs when ceramic particles are added as reinforcement: this often triggers shear localization and crack initiation during ECAP. This is the main reason why equal-channel angular pressing (ECAP) of aluminum matrix composites (AMCs) can generally only be performed at elevated temperatures and using ECAP dies with a channel angle larger than 90° (e.g. 120°). In this study we present a brief first report on an alternative approach for the improvement of the formability of an AMC (AA2017, 10 % SiC): ECAP at low temperatures. We show that, using a temperature of -60 °C and a channel angle of 90° (corresponding to an equivalent strain of 1.1), ECAP of the AMC can be successfully performed without material failure. The mechanical properties of the strongly deformed AMC are analyzed by tensile testing. Our results indicate that the increased formability of the AMC at low temperatures can be attributed to the suppression of unstable plastic flow that affects formability at room temperature.

1. Introduction

Compared to conventional aluminum alloys, aluminum matrix composites (AMCs) combine advanced mechanical properties, such as high specific strength or high wear resistance, with enhanced creep resistance and thermal stability [1, 2]. Equal-channel angular pressing (ECAP) of AMCs also in principle offers the possibility of strain-hardening and induces an ultrafine-grained microstructure [3, 4]. Most relevant publications on ECAP, however, report on pure metals or their alloys [5-8]. Only a limited number of studies have been focused on ECAP of AMCs [3, 9-11], mainly because AMCs are prone to shear localization or the formation of cracks during processing. Most of these earlier investigations have considered AMCs produced by melting routes [4]. There are almost no publications about ECAP of powder metallurgical, SiC-reinforced AA2017, because such difficult-to-work materials are hard to deform at lower temperatures [11-13]. A homogeneous, crack-free deformation is typically only possible at significantly elevated temperatures, larger channel angles of the ECAP die (> 90°) and using a sufficiently large amount of back pressure.

In the present study, we demonstrate that an AMC can be successfully performed by ECAP at a low temperature (-60 °C), even using an ECAP die with a channel angle of 90°, and without any surface cracks. The purpose of this brief report is to document that low temperatures can improve the formability of the AMC, and reduce its tendency for unstable plastic deformation, which enables processing in a 90°-ECAP die and thus to introduce sufficient amounts of severe plastic deformation to considerably increase the material's strength.

2. Experimental

The chemical composition (wt.-%) of the 2017 alloy matrix material was 4.1 % Cu, 0.7 % Mg, 0.8 % Mn, 0.1 % Si, 0.2 % Fe, and balance Al. It was obtained as commercial, gas-atomized, spherical powder with a particle size below 100 µm. The base material and 10 vol.-% of SiC particle powders (particle sizes less than 2 µm) were milled for about 4 h in a high energy ball mill (Simoloyer1 CM08 by Zoz). Hot isostatic pressing (HIP) and extrusion of the mixed powders was performed at Powder Light Metals GmbH in Gladbeck (Germany) to produce semi-finished AMC products. Further details on the fabrication procedures are given in [11].

(a) (b)

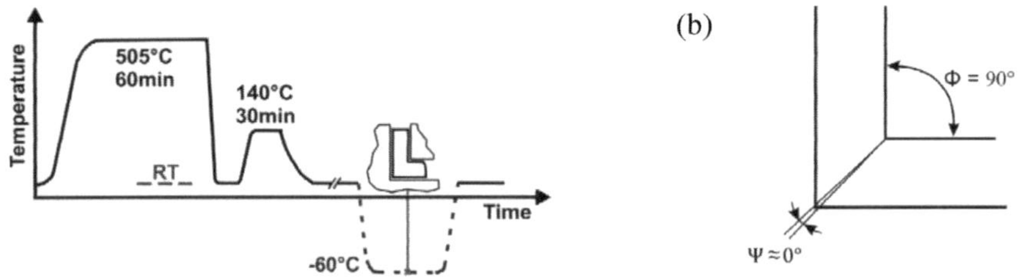

Figure 1. Schematic illustration of (a) the thermo-mechanical route associated with low temperature ECAP processing: solution annealing at 505 °C for 60 min and water quenching, short time under-aging heat treatment at 140 °C for 30 min, and one pass of ECAP at -60 °C; in an ECAP die (b) with a channel angle of Φ = 90°. The shear deformation is introduced in the shear zone (characterized by the corner angle Ψ, which is quite small when an appropriate back pressure is applied).

Figure 1a and table 1 summarize the processing route used in this study: Prior to ECAP, the billets with a gauge length of 80 mm were heat treated at 140 °C for 30 min after solution annealing and water quenching. This short under-aging (ua) treatment increases the workability of the reinforced material at room temperature (RT) and at elevated temperatures [11]. ECAP was carried out in a custom-built ECAP die that allows for cooling by use of an external cryostat, with a channel angle of Φ = 90° (Ψ ≈ 0°, see figure 1b); this corresponds to an effective strain of 1.1 per pass [15]. The ECAP die design includes moveable walls to minimize friction [14]; further details on this new approach to low temperature ECAP have been published in [16]. To further reduce friction, all billets were lubricated using MoS_2 [17] prior to forming. The material was first cooled in liquid nitrogen and then placed in the ECAP die (which was already at a temperature of -60 °C). After keeping the billet in the ECAP tool for 5 minutes, the ECAP process was conducted with a pressing speed of 5mm/min and a back pressure of 110 MPa. The two different conditions of thermo-mechanical treatment of the material (i.e., the ua material condition prior to ECAP, and the E1 condition after 1 pass of ECAP) used in this study are summarized in table 1.

Table 1. Material conditions of AA2017 + 10 vol.-% SiC and the corresponding thermo-mechanical processing parameters.

Condition	Processing
ua	solution annealing at 505 °C, 60 min; water quenching to RT; short time aging at 140 °C, 30 min
E1(-60 °C)	ECAP (N=1) of the ua condition at -60 °C; ECAP tool with a channel angle of Φ = 90° and an outer corner angle of Ψ ≈ 0°; effective strain of φ = 1.1 [15]; pressing speed 5 mm/min; back pressure 110 MPa

For the characterization of strength and ductility, cylindrical tensile samples (with a length of 10.5 mm and a diameter of 3.5 mm) were machined from the billets in extrusion direction after ua, and after one pass of ECAP, respectively. Quasi-static tensile tests were performed in a conventional testing machine (Zwick-Roell) with a constant cross-head speed corresponding to an initial strain rate of 10^{-3} s^{-1}. While the E1 condition was only tested at RT to document the increase of strength after ECAP, the ua condition was tested at different temperatures (RT, -60 °C, -196 °C). Three samples for each temperature were tested. Low temperature testing was realized using two different experimental methods: For testing at -196 °C, liquid nitrogen was filled in a special double-ring cooling device placed around the specimen. For tensile testing at -60 °C, a climate chamber was used.

3. Results and Discussion

As reported in our earlier publications [11-13], ECAP of powder metallurgically produced AA2017 + 10 % SiC was only possible using an ECAP tool with a channel angle of 120°. The equivalent plastic strain per ECAP pass was 0.67. Our first results presented below demonstrate that ECAP of this material can indeed be performed with a tool with a channel angle of 90° when the temperature is decreased to -60°C.

Prior to the single pass of low temperature ECAP considered in this study, the AMC was ua in order to increase the workability of the material. Figure 2a shows (true) stress-strain curves of the initial (prior to ECAP) ua material condition at different testing temperatures. The ua condition exhibits serrated flow at room temperature, which is directly related to unstable plastic flow and the formation of deformation bands. It is generally well known that such unstable deformation processes often depend on temperature and/ or strain rate, e.g. when they result from a Portevin-Le-Châtelier effect, but further work is required to analyze the microstructural origins in the AMC considered here. Tensile tests were also performed at much lower temperatures (-60 °C and -196 °C) in order to observe the effect of temperature on the unstable plastic flow. There is still some serrated flow at -60 °C (but less than at RT), whereas the stress-strain curve measured at -196 °C shows no traces of unstable plastic flow. This indicates that a reduction of temperature can be beneficial in suppressing localized plastic deformation in this AMC.

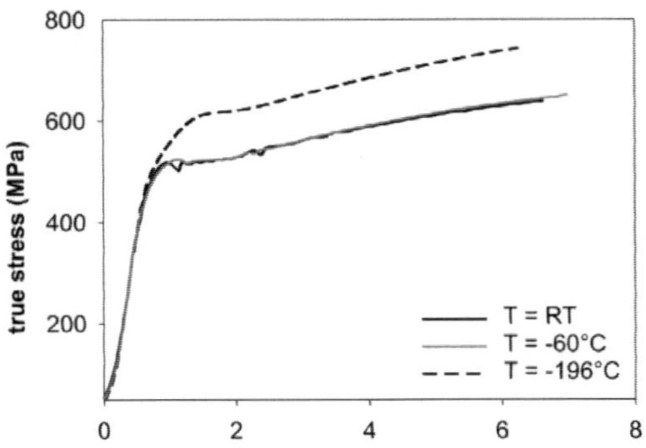

a)

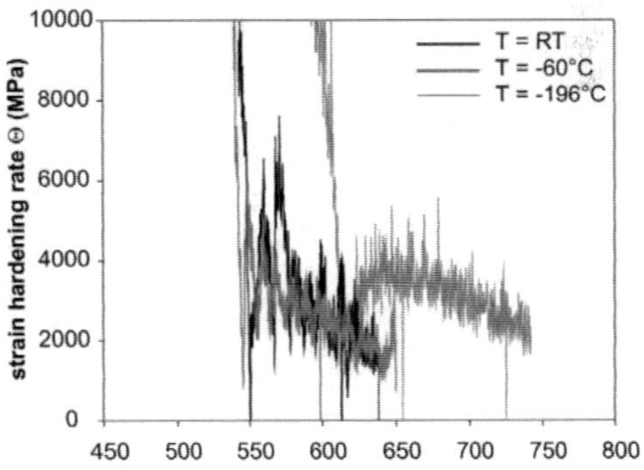

b)

Figure 2. a) True stress-strain curves and b) smoothed data of strain hardening rate curves as a function of true stress of AA2017 + 10 % SiC in the ua condition, measured at RT, -60 °C and -196 °C.

In our previous work on low temperature ECAP of an AA7075 alloy, we observed an increase of ductility with decreasing testing temperatures [18]. Moreover, tensile strength of this alloy monotonously increases from RT to -196 °C. At low temperatures, thermally activated cross slip of screw dislocations is strongly suppressed, which leads to an increase of strain hardening and therefore decreases the alloy's tendency for localization of plastic strain. The material behavior of the AMC considered in the present study is different: first, strength hardly increases when the testing temperature is reduced from RT to -60°C, and a somewhat higher initial yield stress is only observed at -196 °C. Second, ductility is not increased at lower temperatures. It is likely that this effect is related to the presence of the SiC particles, which may place a limit on the maximum admissible total strain before cracks are initiated in the AMC and fracture occurs; an analysis of fracture surfaces to determine microstructural fracture mechanisms is currently under way.

Figure 2b shows the strain hardening rate Θ (which in this study we simply define as the first derivative of true stress with respect to true strain), obtained from the true stress-strain data shown in figure 2a, plotted against true stress for the three investigated temperatures. Prior to numerically determining Θ values, the data was numerically smoothed in order to reduce fluctuations in the Θ-σ plots. The strain hardening rate at -60 °C does not differ from the one measured at RT. At -196 °C, the strain hardening rate is significantly higher. A separation of the data into strain hardening stages III (dynamic recovery processes) and IV (interaction of the increasing number of dislocations with grain boundaries) cannot be performed. This indicates that the microstructural processes related to hardening stages III and IV take place simultaneously [18]; the presence of the SiC particles may further obscure the individual contributions of the different microstructural processes to the macroscopic stress-strain behavior. We finally note that all tensile samples failed prior to the onset of necking, which may again be explained by brittle failure caused by the ceramic particles under tensile loading. It is important to note that this failure mechanism cannot directly be transferred to the compressive/ shear loading conditions during ECAP.

Figure 3 shows billets of the investigated AMC after ECAP processing at two different pressing temperatures: As expected, processing at RT in the 90°-tool leads to a pronounced formation of cracks in regular intervals (figure 3a). Despite of the fact that at -60 °C no improvement of ductility was observed during tensile testing, the AMC could be deformed almost homogeneously at -60 °C in the ECAP die with a channel angle of $\Phi = 90°$ (figure 3b). This first result shows for the first time that homogeneous severe plastic deformation of AMCs via ECAP is indeed possible at low temperatures. This positive effect of decreased processing temperatures on formability can be related to an increase of work hardening. Moreover, lower temperatures seem to reduce the material's tendency for serrated flow, which may also be beneficial in suppressing early cracking. Furthermore it should be pointed out that, especially in the case of particle reinforced materials, their mechanical behavior under tensile loading is not suitable to fully describe the deformation behavior under compressive loading; compression or shear testing is likely to provide more relevant data on ductility that can be more readily used to predict formability in severe plastic deformation processes like ECAP.

We finally provide some information on the effect of low temperature ECAP on the mechanical behavior of the ECAP-processed AMC. Figure 4 shows an overview of engineering stress strain curves of the investigated conditions, comparing the pre-ECAP (ua) and the post-ECAP behavior. Obviously, low temperature ECAP of the ua condition results in an increase of yield strength by about 19 %, but it simultaneously decreases ductility. This loss of ductility is typical for SPD-processed materials that are not yet ultrafine-grained or nanocrystalline [3, 11, 19], and may be partly reversed by subsequent heat treatments.

a) b)

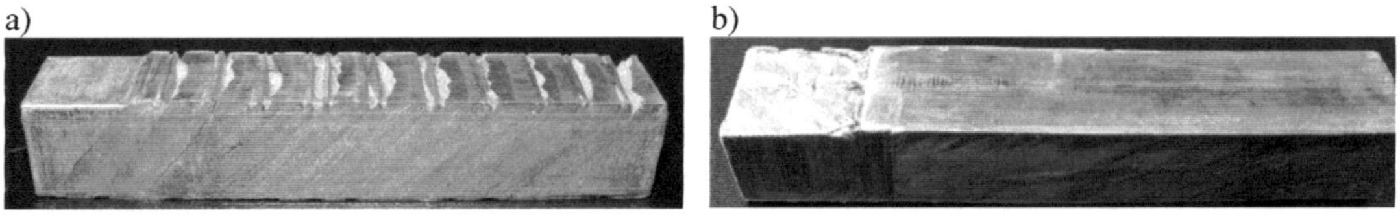

Figure 3. ECAP billets: a) heterogeneously deformed and cracked sample after RT deformation in an ECAP die with an internal angle of $\Phi = 90°$; b) homogeneously deformed and crack-free sample after deformation at -60 °C in the same ECAP die.

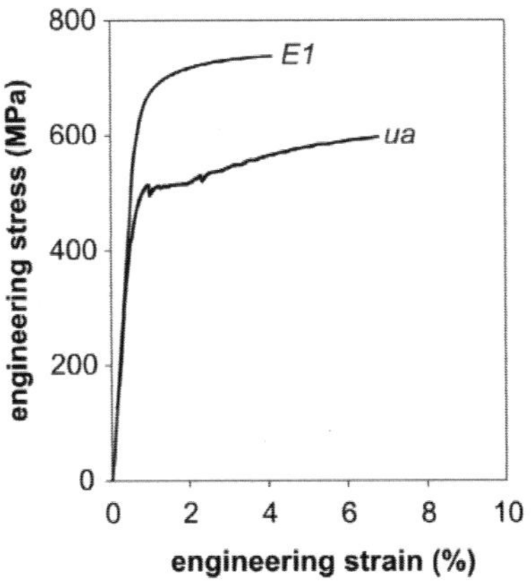

Figure 4. Engineering stress-strain curves of AA2017 + 10 % SiC after one pass of ECAP (E1), compared to the ua (pre-ECAP) condition.

4. Summary and conclusions

ECAP of an aluminum matrix composite (AA2017 + 10 % SiC) at a low temperature (-60 °C) was successfully performed in a die with a channel angle of 90°, resulting in a homogeneous deformation and no cracking. Our results can be summarized as follows:

1. The under-aged condition of the AMC (AA2017 + 10 % SiC) exhibits serrations in the stress-strain curve at room temperature. Testing at lower temperatures indicates a reduction of unstable plastic flow. At -60 °C, some serrations occur, while at -196 °C this effect is completely suppressed.
2. Reducing the temperature from RT to -60 °C does not result in an increase of ductility under tensile loading. This can be explained by the presence of SiC particles, which generally lead to a more brittle deformation behavior in tension. All samples failed prior to the onset of necking. The strain hardening rates determined from tensile testing are almost equal at RT and at -60 °C. Only at -196 °C the strain hardening rate is significantly higher.
3. Despite of the only minor changes observed when changing the testing temperature from RT to -60 °C in tensile testing, ECAP of the AMC can be successfully performed at a pressing temperature of -60 °C. Severe plastic deformation by low temperature ECAP results in an increase of the tensile strength of about 25 % compared to the initial under-aged condition, but simultaneously decreases ductility.

Acknowledgments

The authors gratefully acknowledge funding by the German Research Foundation (Deutsche Forschungsgemeinschaft, DFG) within the framework of SFB 692.

References

[1] Mani B and Paydar M H 2010 *J. Alloys Comp.* **492** 116
[2] Hort N and Kainer K U 2004 *Metallische Verbundwerkstoffe* (Weinheim: Wiley–VCH)
[3] Valiev R Z, Islamgaliev R K and Kuzmina N F 1999 *Scr. Mater.* **40** 117
[4] Sabirov I, Kolednik O, Valiev R Z and Pippan R 2005 *Act. Mater.* **53** 4919
[5] Kim J K, Jeong H G, Hong S I, Kim Y S and Kim W J 2001 *Scr. Mater.* **45** 901
[6] Kim W J, Kim J K, Kim H K, Park J W and Jeong J H 2008 *J. Alloys Comp.* **450** 222
[7] Valiev R Z, Islamgaliev R K and Alexandrov I V 2000 *Prog. Mater. Sci.* **45** 103
[8] Kim W J, Chung C S, Ma D S, Hong S I and Kim H K 2003 *Scr. Mater.* **49** 333
[9] Ramu G and Bauri R 2009 *Mater. Design* **30** 3554
[10] Chen L J, Ma C Y, Stoica G M, Liaw P K, Xu C and Langdon T G 2005 *Mater. Sci. Eng.* A **410-411** 472

[11] Wagner S, Siebeck S, Hockauf M, Nestler D, Podlesak H, Wielage B and Wagner M F-X 2012 *Adv. Eng. Mater.* **14** 388
[12] Wagner S, Podlesak H, Siebek S, Nestler D, Wagner M F-X, Wielage B and Hockauf M 2010 *Materialwiss. Werkstofftechn.* **41** 704
[13] Härtel M, Wagner S, Frint P, Wagner M F-X 2014 *IOP Conf. Ser.: Mater. Sci. Eng.* **63** 012080
[14] Frint S, Hockauf M, Frint P and Wagner M F-X 2016 *Mater. Design* **97** 502
[15] Iwahashi Y, Wang J T, Horita Z, Nemoto M and Langdon T G 1996 *Scr. Mater.* **35** 143
[16] Fritsch S, Scholze M and Wagner M F-X 2012 *Materialwiss. Werkstofftechn.* **43** 561
[17] Frint P, Wagner M F-X, Weber S, Seipp S, Frint S and Lampke T 2017 *J. Mater. Process. Tech.* **239** 222
[18] Fritsch S, Scholze M and Wagner M F-X 2016 *IOP Conf. Ser.: Mater. Sci. Eng.* **118** 012030
[19] Muñoz-Morris M A, Calderón N Gutierrez-Urrutia I and Morris D G 2006 *Mater. Sci. Eng.* A **425** 131

IOP Conf. Series: Materials Science and Engineering **181** (2017) 012040 doi:10.1088/1757-899X/181/1/012040

Composition of highly concentrated silicate electrolytes and ultrasound influencing the plasma electrolytic oxidation of magnesium

F Simchen*, L-M Rymer, M Sieber and T Lampke

Institute of Materials Science and Engineering, Chemnitz University of Technology, 09125 Chemnitz, Germany
*corresponding author: frank.simchen@mb.tu-chemnitz.de, +49371 531 30115

Abstract. Magnesium and its alloys are increasingly in use as lightweight construction materials. However, their inappropriate corrosion and wear resistance often prevent their direct practical use. The plasma electrolytic oxidation (PEO) is a promising, environmentally friendly method to improve the surface characteristics of magnesium materials by the formation of oxide coatings. These PEO layers contain components of the applied electrolyte and can be shifted in their composition by increasing the concentration of the electrolyte constituents. Therefore, in contrast to the use of conventional low concentrated electrolytes, the process results in more stable protective coatings, in which electrolyte species are the dominating constitutes. In the present work, the influence of the composition of highly concentrated alkaline silicate electrolytes with additives of phosphate and glycerol on the quality of PEO layers on the magnesium alloy AZ31 was examined. The effect of ultrasound coupled into the electrolyte bath was also considered. The process was monitored by recording the electrical process variables with a transient recorder and by observation of the discharge phenomena on the sample surface with a camera. The study was conducted on the basis of a design of experiments. The effects of the process parameter variation are considered with regard to the coatings thickness, hardness and corrosion resistance. Information about the statistical significance of the effects of the parameters on the considered properties is obtained by an analysis of variance (ANOVA).

1. Introduction

Magnesium alloys are increasingly in use as lightweight construction materials. Their low density, high specific strength and good damping behaviour against mechanic vibrations and electromagnetic oscillations make these alloys interesting for applications in automotive and aerospace industries as well as housing materials for high-quality electronics. However, the insufficient resistance against corrosion and wear often prevent the direct practical use of magnesium materials. [1–2] The plasma electrolytic oxidation is a promising environmentally friendly surface treatment method to handle these challenges. During the process, the workpiece is immersed in an aqueous electrolyte, opposed to a cathode and receives an increasing anodic polarisation. This procedure leads to the formation of an anodic film on the material surface and to the development of a gas-steam shell enveloping the workpiece geometry. The voltage drop over the cell arrangement is concentrated between the metal substrate and the gas-steam shell/electrolyte interface. Afterwards, this equipotential area called quasi cathode becomes the starting point of spark discharges from the electrolyte through the gaseous phase and the anodic film into the substrate. The discharges leave oxidized spots and produce protective oxide ceramic PEO layers by further oxide formation, melting and remelting processes subsequently. [3] This method is well understood and commercially used for aluminium materials. The excellent technical characteristics of the developing alumina allow for the formation of hard and resistive coatings. In contrast, the PEO layers on magnesium materials consist mainly of magnesia, which has less protective properties. In addition to oxide compounds of the substrate material, the PEO layer contains varying amounts of electrolyte constitutes in dependency of its composition and concentration. For the plasma electrolytic oxidation of magnesium materials, substances with the anions AlO_2^-, SiO_3^{2-}, PO_4^{3-}, F^- and SnO_3^- are common electrolyte components. Thus, the formed oxide ceramic coatings can contain significant amounts of Al, Si, P, F and Sn [4–7].

Ultrasound is a common process parameter to improve the results of wet-chemical surface treatment methods. On the field of the anodic surface modification, a few publications describe the positive influence of ultrasound on the process characteristics and treatment results. According to these studies, ultrasound coupled into the electrolyte bath during the conventional anodic oxidation can increase the growth rate of anodized layers on aluminium in oxalic acid [8] or support the formation of a regular cell structure while anodizing titanium in nitric acid [9]. For plasma electrolytic oxidation of aluminium it has been reported that ultrasound can increase the compactness and the α-Al_2O_3 amount of the PEO layers [10] and decreases the formation voltage during the process [11]. Concerning the influence of ultrasound on the PEO process on magnesium, there are hardly any data, but first studies for the alloy ZK61 also indicate a positive influence on the resulting corrosion resistance [12].

In our present work the magnesium alloy AZ31 was used. The study follow the research approach of focussing on the incorporation of electrolyte constitutes into the PEO layer to shift its composition in favour of more stable phases than MgO. The main component of the electrolyte systems used is silicate in combination with the additives phosphate and glycerol. Figure 1 gives a graphical representation of the composition of common low concentrated electrolytes for the PEO process on magnesium, based on the mentioned constitutes (a) and in direct comparison to the electrolyte mixtures used in the present study (b).

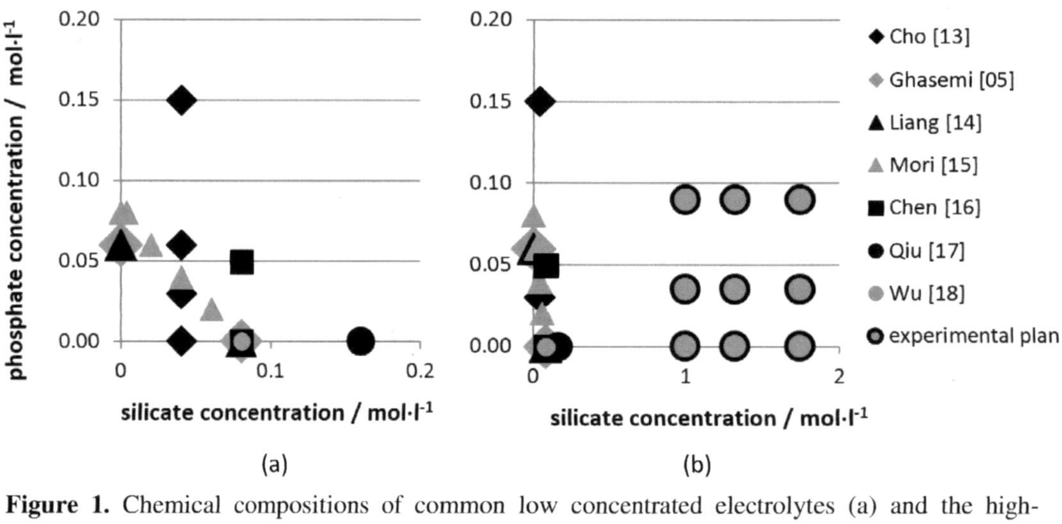

Figure 1. Chemical compositions of common low concentrated electrolytes (a) and the high-concentrated electrolytes of this study (b) for the PEO treatment of magnesium.

To ensure the environmental compatibility the use of fluoride was renounced. To investigate a possible influence of ultrasound on the process, ultrasound is coupled into the electrolyte bath by pulse mode.

2. Experimental Methods

The magnesium alloy AZ31, the composition of which is given in table 1, was used as the sample material for all tests.

Table 1. Chemical composition of the AZ31 alloy.

Material	Alloying elements (wt%)						
	Mg	Al	Zn	Ti	V	Fe	Mn
AZ31	Balance	2.5–3.5	0.6–1.4	≤ 0.05	≤ 0.005	≤ 0.005	≤ 0.2

The single samples had a cylindrical geometry with a diameter of $d_s = 20$ mm and a height of $h_s = 10$ mm, and were included into the electrical circuit by a magnesium wire with a diameter of $d_c = 3.2$ mm laterally pressed in (this electrical contact is visible in figure 7 on the right side of the AZ31 samples). Thus, the active sample surface results as $A_s = 1248.6$ mm². The magnesium wire by itself was insulated electrically against the electrolyte. The current-regulated electrical regime was applied by a rectifier (Plating Electronics, Germany) providing effectively 550 V and 50 A at its maximum. The current pattern was predefined as a symmetric pulse of $I_1 = I_2 = 2$ A with durations of $t_1 = t_2 = 5$ ms. The resulting current densities were $i_1 = i_2 = 16$ A/dm².

The electrolytes were prepared with deionised water, different amounts of $Na_2SiO_3 \cdot 5\ H_2O$, Na_2HPO_4, and $C_3H_5(OH)_3$. The pH value was between 13.1 and 13.5. Therefore, the phosphate ions were present as PO_4^{3-} rather than HPO_4^{2-}. The ultrasound was coupled into the electrolyte bath by a sonotrode (Weber Ultrasonics, Germany) with frequency of 30 kHz and a power output of 78 W by pulse mode. The concentrations of the three electrolyte components and the duration of the ultrasound pulse were varied on three different levels with a logarithmically oriented graduation. A design of experiments (DOE) according to the Box-Behnken design [19] was used. Table 2 gives an overview of the parameters, their notations, and level values. Figure 2 shows a graphical representation of the experimental plan.

Table 2. Overview of the varied process parameters.

Parameter		Notation	Levels		
			1	2	3
Ultrasound pulse duration	($s \cdot min^{-1}$)	U	1	8	60
Silicate concentration	($mol \cdot l^{-1}$)	S	1.00	1.32	1.75
Phosphate concentration	($mol \cdot l^{-1}$)	P	0	0.035	0.09
Glycerol concentration	($mol \cdot l^{-1}$)	G	0	0.02	0.04

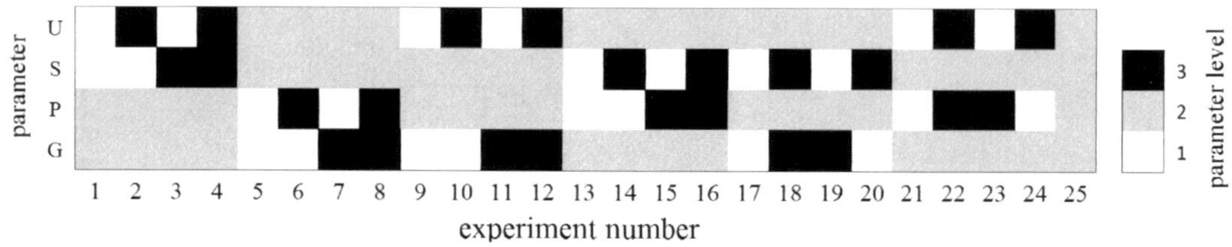

Figure 2. Graphical representation of the experimental plan used in this study showing the designation of parameters and parameter levels (1–3) according to table 2.

In the further article the capital letters U, S, P and G in combination with the corresponding parameter level are used as short nomenclature to describe the parameter set of a certain experiment. For example the parameter set of experiment 1 with an ultrasound pulse of 1 s, followed by pulse pause of 59 s and an electrolyte composition of 1.00 mol/l silicate, 0.035 mol/l phosphate, and 0.02 mol/l glycerol is abbreviated as U=1, S=1, P=2 and G=2.

To enable a statistical support of the obtained results, every experiment was carried out three times. Furthermore, a randomisation of the experiment order is desirable to minimize the influence of experimentally not determinable disturbances. Therefore and to limit the experimental effort, the electrolytes were prepared by systematic up-concentration. Within a certain electrolyte the experiments with different ultrasound pulse modes were carried out in randomised order. The electrolytes were prepared with a volume of 5 liters and filled in a cuboidal acrylic glass container, which was surrounded by cooling water with a temperature of 5 °C. All chemicals used: sodium metasilicate pentahydrate (Sigma-Aldrich, Germany), disodium phosphate (Merck, Germany) and glycerol (Roth, Germany) were of an analytical grade.

To observe the development of the electrical process variables, a transient recorder DL850 (Yokogawa, Japan) was connected to the electrical circuit. The voltage signal was acquired directly from the cell electrodes. The current flow was calculated based on Ohm's law and the recorded voltage drop over a measuring resistor of $R_m = 1\ \Omega$, which was included in the cathode strand. The electric measurements had a measuring interval of 1 min, a measuring duration of 200 ms and a sampling rate of 100 kS/s. The measured values of the electrical process observation were processed by a low-pass filter with a cut-off frequency of 15 kHz. The resulting data give a statement about the development of the maximum anodic and cathodic voltage during the processes. In addition, they enable to check whether the predefined current form has been correctly transferred to the experimental setup by the rectifier. To monitor the discharge behaviour on the sample, a camera GoPro Hero 3 (GoPro, USA), was immersed into the cooling water and oriented to the sample surface. Figure 3 shows a schematic representation of the complete experimental setup.

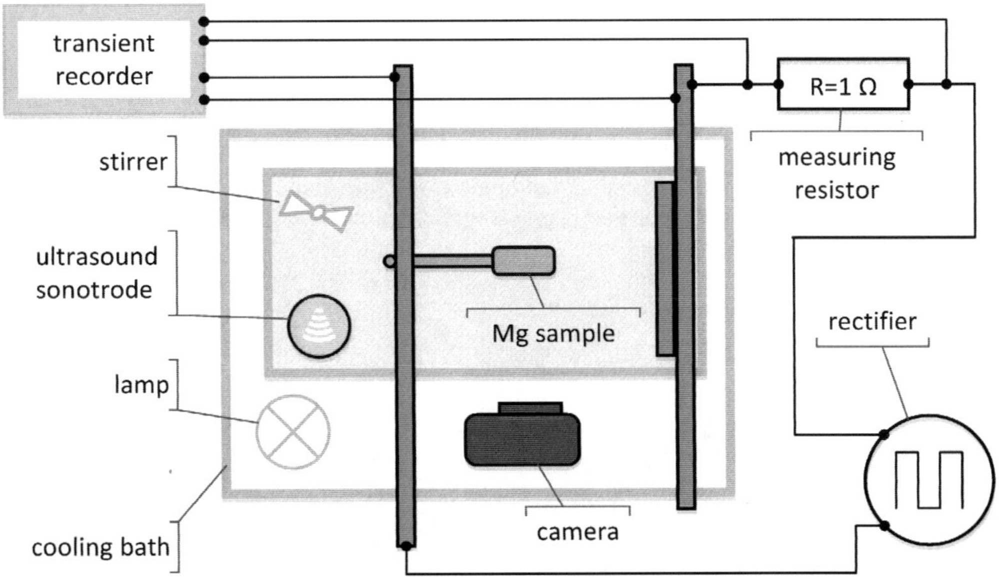

Figure 3. Schematic representation of the experimental setup.

The evaluation of the recorded videos allows determining to which extent the variation of the investigated process parameters influences the occurrence of certain discharge stages as well as their temporal transition to one another.

After the PEO treatment, the three samples of every experiment were cut into two halves. Two different samples of every experiment were prepared to cross-sections. Afterwards, these cross-sections were used for measurements of the layers thickness and Martens hardness at an optical microscope and a hardness measuring instrument Fischerscope HM2000 XYm (Fischer, Germany). The layer thickness *d* was determined by 10 measurements at each cross-sections. The Martens hardness *HM* was measured to examine a possible influence of the process parameter variation on the mechanical characteristics of the PEO-layer material. The measure force of 50 mN was applied in 30 s and kept constant for further 30 s. A Vickers diamond was used as indenter. The measurements were conducted 5 times at each cross-sections. To minimize a falsifying influence of the PEO-layers' morphology, the indents were placed in compact regions of the coatings. To investigate the corrosion behaviour of the PEO layers, current density potential measurements were carried out for each repetition of the single experiments. The curves were recorded in a three-electrode arrangement, in which the test sample is the working electrode with an active surface area of 0.785 cm^2. A platinum sheet was used as counter electrode and a Ag/AgCl electrode as reference electrode. The used test solution was 0.1 mol/l NaCl, had a volume of 250 ml and a temperature of 25 °C. The potential was varied from -1800 mV up to -600 mV with a potential feed of 1 mV/s. Afterwards, the Butler-Volmer equation, which is shown in equation 1, was fitted manually with a MATLAB tool on the measured data by varying i_0 and α.

$$i(i_0, \alpha) = i_0 \cdot \left[\exp\left(\frac{\alpha \cdot z \cdot F \cdot (\varphi - \varphi_0)}{R \cdot T}\right) - \exp\left(-\frac{(1-\alpha) \cdot z \cdot F \cdot (\varphi - \varphi_0)}{R \cdot T}\right)\right] \quad (1)$$

In this equation, F and R are the Faraday and the universal gas constant, respectively. T is the absolute measuring temperature and $z = 2$ is the charge number according to the oxidation state of magnesium ions. The quantity α describes the transfer coefficient of the anodic and cathodic corrosion reaction. The corrosion current density i_0 quantifies the net current flow at the open circuit potential and can be obtained from the fit. It provides an integral and well comparable statement about the corrosion properties and defect densities of the produced coatings, and is used for the evaluation of the experiments in this study. Figure 4 shows a diagram with a measured and fitted current-density potential curve. The dashed lines show the corresponding Tafel plots, which intersection represents the corrosion current density.

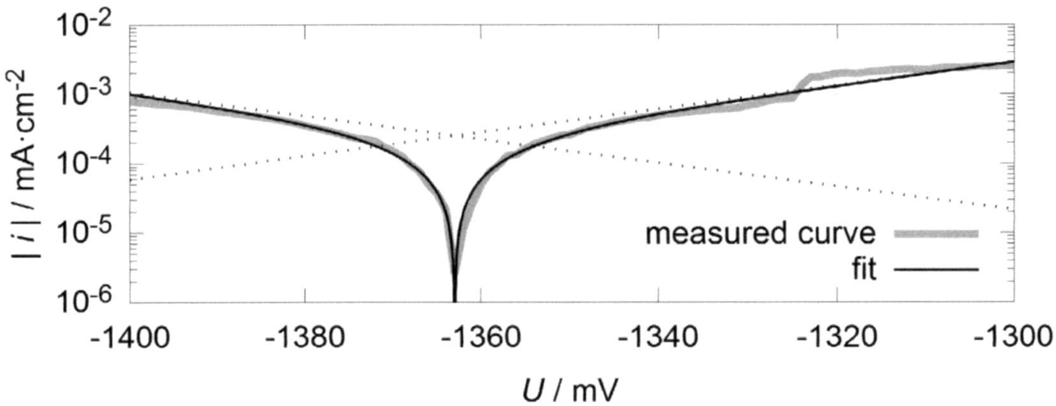

Figure 4. Measured $i(U)$ curve and fitted $i(U)$ curve according to the Butler-Volmer equation.

Information about the statistical significance of the effects of the parameter variations on the maintained layer properties are generated by an ANOVA. Further investigations respecting the layers morphology, chemical composition and phase structure were conducted at selected samples. The layers morphology was examined at a scanning electron microscope (SEM) 1455VP (LEO, Germany) using an excitation voltage of 25 kV based on back-scattered electron (BSE) images, which show an element specific contrast. The chemical compositions of the investigated coatings were measured in dependency of the substrate distance $d_{substr.}$ by an energy-dispersive X-ray spectroscopy (EDX) system included in the scanning electron microscope. X-ray diffraction (XRD) measurements were carried out to determine the coating's phase structure. Therefore, an X-ray diffractometer D8 Discover (Bruker AXS, USA) was used. The samples were measured with Co-K_α radiation in a line focus with Bragg-Brentano geometry and a Lynxeye-XE detector (Bruker AXS, USA) on a length of 10 mm for 3.5 h. The phase allocation was based on the PDF-2 data base of the International Centre for fiffraction Data (ICDD).

IOP Conf. Series: Materials Science and Engineering **181** (2017) 012040 doi:10.1088/1757-899X/181/1/012040

3. Results and Discussion

Figure 5 shows representative current density and voltage curves of the applied bipolar pulse regime. The black dashed line describes the predefined current pattern, the solid line displays the measured curve. The grey graph indicates the corresponding measured voltage signal.

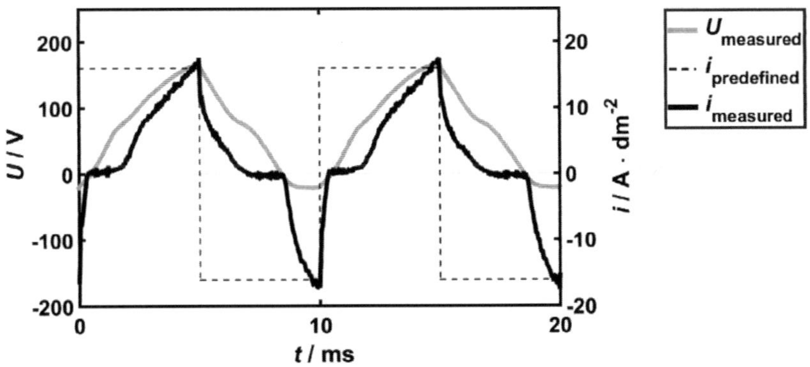

Figure 5. Predefined current-density curve (dashed black line) and representative measured current-density (solid black line) and voltage (dashed black line) curves of the applied bipolar pulse regime.

The measured current-density course shows a significant deviation from its predefined pattern. This can be explained by the fact that the absolute current values of these processes are at the lowest part of the working range of the used rectifier, where its working accuracy is restricted. As a result, the applied current pattern in this study was a kind of transition form between triangular and rectangular pulse, which has to be taken into account for reproduction or upscaling of these processes. Figure 6 shows envelopes of the course of the maximum anodic and cathodic voltages of the three repetitions of the experiment couples 9/10 (a), 15/16 (b), 7/8 (c), and 9/11 (d). At these experiment couples, three experimental parameters were kept constant whereas one was shifted from level one up to level three (see graphical representation of the DOE at figure 2).

The influence of the parameter variations on the resulting process voltages is of interest for economic considerations respecting the total consumption of electrical energy. Obviously, the parameters ultrasound and phosphate concentration do not influence the voltage courses during the PEO process (figure 6 (a) and (c)). A higher silicate concentration lowers the anodic process voltage by approximately 50 V (figure 6 (b)). An elevated glycerol concentration (figure 6 (d)) increases the anodic process voltage by 20–50 V. It also prevents the appearance anodic voltage sags like shown in figure 6 (d) in the process time between 300 and 400 s. This observation can be made also for the experiment couples 6/8, 17/19, and 20/18, which also differ only with respect to the shift of the glycerol concentration from level one up to level three. This suggests that the electrolyte additive glycerol has a stabilizing effect on the process in addition to its anodic voltage increase. Whether there is indeed a positive influence on the resulting layer properties can be determined by the investigations below.

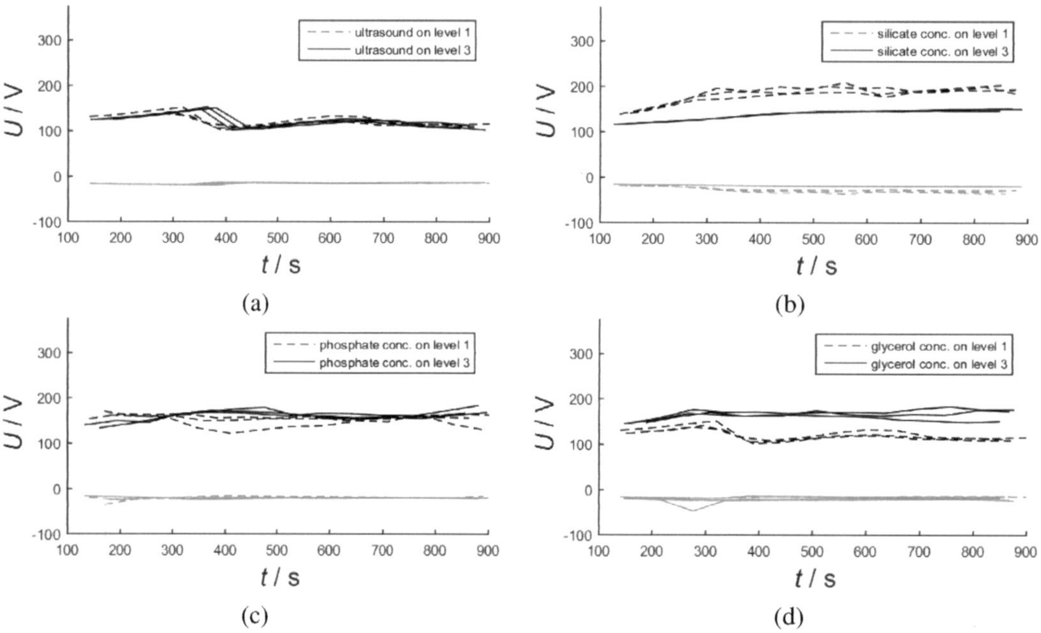

Figure 6. Envelopes showing the course of the maximum anodic and cathodic voltages over the three repetitions of the experiment couples 9/10 (b), 15/16 (b), 7/8 (c) and 9/11 (d); the experimental parameters of these couples were similar except for one process parameter, which was varied from level one up to three.

The evaluation of the recorded video data allows for a categorization of the discharge behaviour during the process into three different stages, which are depicted representatively for the example of experiment 23 in figure 7 (a–f). Stage I shows high-frequency, uniformly distributed, bright little discharges (figure 7 (a)–(b)). Stage II shows high-frequency, uniformly distributed, weakly luminous little discharges accompanied by rare low-frequency, single, bright big discharges (figure 7 (c)–(d)). Stage III shows low-frequency, localized, bright big discharges (figure 7 (e)–(f)).

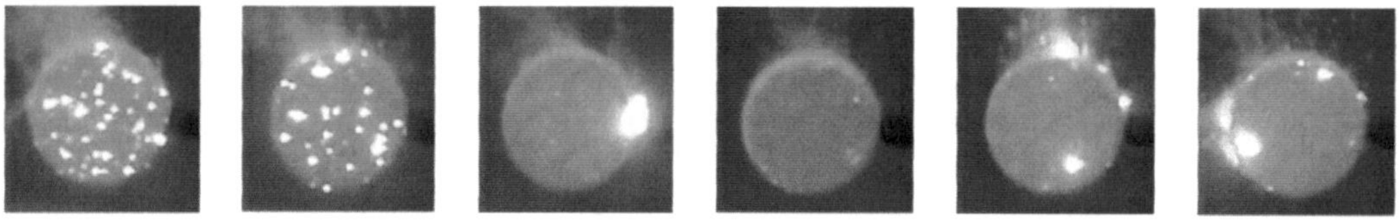

(a) 02:00 (b) 04:00 (c) 06:00 (d) 08:00 (e) 12:00 (f) 13:00

Figure 7. Exemplary photographs with process time (mm:ss) for discharge stages I (a)–(b), II (c)–(d), and II (e)–(f), recorded from experiment 23, obtained by the optical process monitoring.

Figure 8 (a) shows an effect diagram, which represents the parameters influence on an investigated quantity, in this case the transition time of discharge stage I into discharge stage II $t_{I \to II}$. The bar plots show the effect of every single investigated process parameter. The error bar includes the process parameter variation over the complete experimental plan. The value of the error probability p_{error} is obtained by the ANOVA and provides a statement whether the observed effect is statistically significant or not. In this study, a confidence interval of 5 % is chosen. Thus, parameters with a p_{error} of lower than 5 % are considered as statistically significant. The ultrasound as well as the silicate and phosphate concentration accelerate the transition by raising the parameter level, while an increased glycerol concentration delays it. All effects are statistically significant. Figure 8 (c)–(d) shows effect diagrams for the influence of the investigated parameters on the layer properties thickness, martens hardness and corrosion current density.

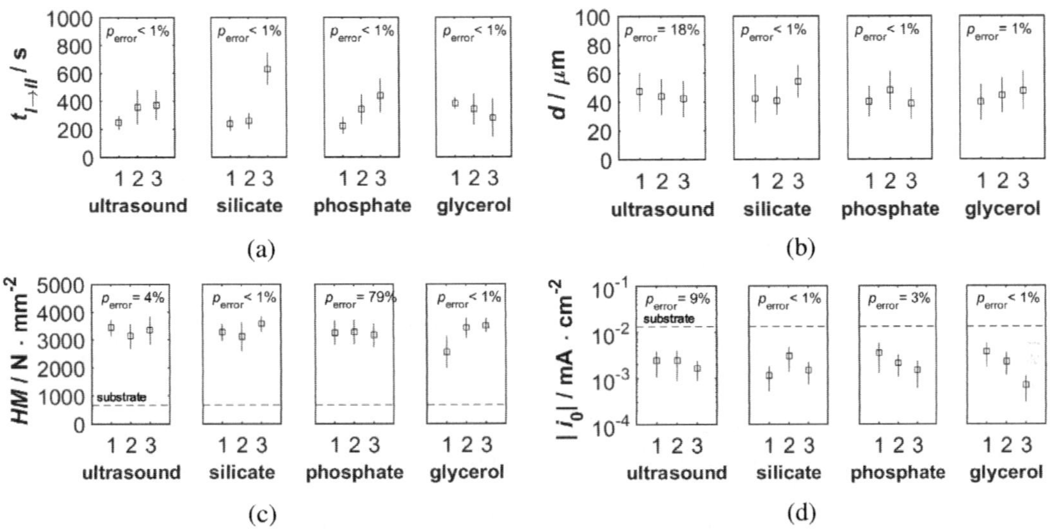

Figure 8. Effect of the investigated parameters on the transition time between the discharge stage I and II (a) and the layer thickness (b), hardness (c) and corrosion current density (d).

A higher layer thickness is supported by increasing the silicate and glycerol concentration, while the phosphate concentration results in a maximum at level 2. The ultrasound coupling had no statistically significant effect (figure 8 (b)). The hardness of the PEO layers is increased by raising the glycerol concentration level, and shows maximum values for ultrasound pulses on level one and silicate concentrations on level three. The phosphate concentration has no statistically significant influence (figure 8 (c)). In comparison to the untreated substrate material, represented by the dashed line, the hardness clearly increases by the PEO treatment. Figure 8 (d) shows the obtained data for i_0 where a low value is desirable. The corrosion current density is minimized by increasing the phosphate and glycerol concentration. The silicate concentration leads to reduced values at level 1 and 3 while ultrasound has no statistically significant influence. In comparison to the untreated substrate material, represented by the dashed line, the corrosion current densities are clearly lowered by the PEO treatment. The PEO layer of the experiments 11, 12 and 16 show the lowest corrosion current densities, which indicates that they possess the lowest defect densities. Additionally, they show high values for layer thickness and hardness. Table 3 gives a quantitative overview of this optimal experimental parameters and the resulting layer properties. Figure 9 shows backscatter electron micrographs of the corresponding cross-sections.

Table 3. Quantitative overview of the optimal parameters obtained by this study and the resulting layer properties.

Experi-ment	Parameters ultrasound (s·min^{-1})	$c_{silicate}$	$c_{phosphate}$ (mol·l^{-1})	$c_{glycerol}$	Layer properties i_0 (10^{-4} mA · dm^{-2})	HM (N · mm^{-2})	d (µm)
11	1	1.32	0.035	0.04	3.5 ± 1.3	3600 ± 200	43 ± 18
12	60	1.32	0.035	0.04	3.1 ± 1.2	3700 ± 300	58 ± 21
16	8	1.75	0.090	0.02	2.5 ± 0.2	3700 ± 200	64 ± 14

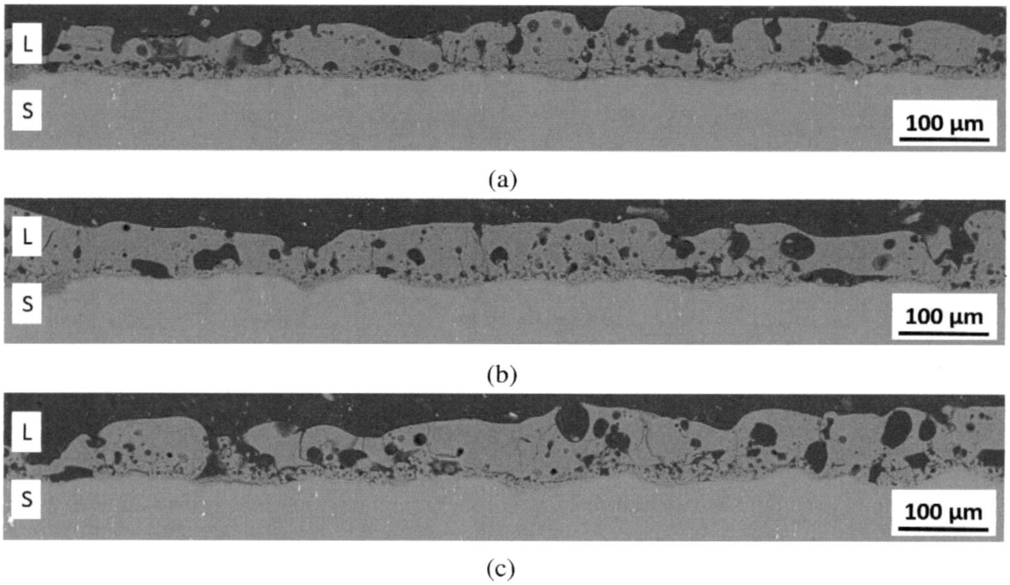

Figure 9. BSE micrographs of the substrate (S) and PEO layers (L) of the experiments 11 (U=1, S=2, P=1, G=3), 12 (U=3, S=2, P=2, G=3) and 16 (U=2, S=3, P=3, G=2).

The micrographs show that the PEO layers consist of an inner zone with a fine porosity and an outer zone with areas with a relative compact structure combined with big porosities. The results of EDX measurements with increasing substrate distance $d_{substr.}$ are depicted in table 4.

Table 4. Chemical composition in dependency of the distance from the substrate for the PEO layers of the experiments 11 (US=1, S=2, P=2, G=3), 12 (U=3, S=2, P=2, G=3) and 16 (U=2, S=3, P=3, G=2).

Experiment 11					Experiment 12					Experiment 16				
$d_{substr.}$ (µm)	Element content (at%)				$d_{substr.}$ (µm)	Element content (at%)				$d_{substr.}$ (µm)	Element content (at%)			
	O	Si	Mg	Na		O	Si	Mg	Na		O	Si	Mg	Na
5	56	16	25	2	12	6	18	17	3	4	58	15	23	3
24	60	21	13	4	35	61	21	15	3	20	60	22	14	4
46	58	19	16	6	45	61	20	15	3	40	60	22	14	4
56	59	21	14	6	55	59	20	14	6	60	59	21	15	5

Al, Zn < 1%

The measurements show that the electrolyte species silicon and sodium dominate the oxide coating composition. Only close to the substrate/layer interface there is more magnesium than silicon. Figure 10 shows a diffraction diagram of exemplary XRD measurements of untreated magnesium substrate and selected PEO coatings of the experimental plan. The diffractograms of the PEO layers show the same peaks like the untreated substrate and in addition a broad slightly peak between $2\Theta = 19–47°$. This indicates that the peaks in the diffractograms originate from the substrate and the PEO layers by themselves have an amorphous structure. These observation are confirmed by experiments with the electrolytes of this experimental plan which were not part of the present study.

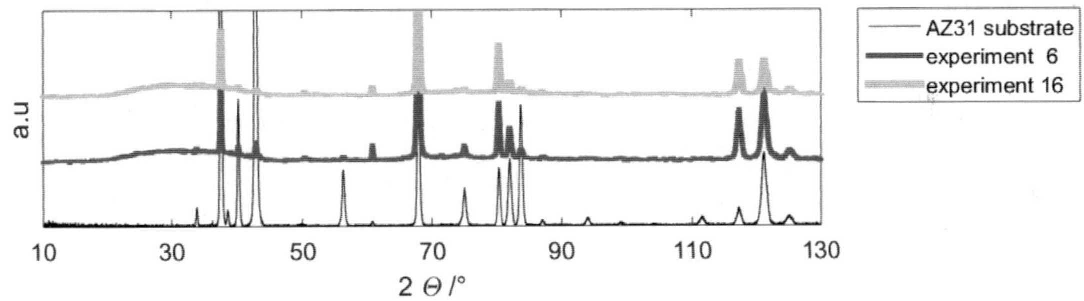

Figure 10. Diffractograms of the XRD analyses for the untreated AZ31 substrate and the PEO layers of experiment 6 (US=2, S=2, P=3, G=1) and 16 (US=2, S=3, P=3, G=2).

4. Summary and Conclusions

The influences of electrolyte composition and ultrasound pulses on the plasma electrolytic oxidation of the magnesium alloy AZ31 were investigated based on three-dimensional experimental plan according to the Box-Behnken design. Therefore, the electrical and optical process characteristics were observed and the PEO layer's properties were examined. Within the investigated parameter range, the following findings were obtained.

The ultrasound pulses and phosphate concentration have no influence on the process voltage. The maximum anodic voltage was decreased by approximately 50 V by increasing the silicate concentration by 0.75 mol/l, which enables a lowering of the electrical energy consumption. Increased glycerol concentrations lead to a higher process voltage but also prevent the appearance of anodic voltage sags. The PEO coatings obtained in this study have an amorphous structure and consist of oxygen, silicon and magnesium mainly. Additionally, low amounts of sodium, zinc and aluminium are detectable. The atomic content of oxygen is around 60 % and does not vary in dependency of the substrate distance. Close to the substrate ($d_{substr.} \leq 5$ µm) magnesium is in majority within the oxide structure with around 24 % while in the outer parts of the coatings the magnesium silicon ratio is around 2:3. The high amount of electrolyte constitutes in the layer can be explained by the high concentrations of the used electrolytes. The coating method presented in this study is therefore a transitional form between a conversation and a deposition process. The modification of the electrolyte composition leads to different effects. Increased glycerol concentrations have a positive, statistically significant influence on all investigated layer characteristics. This could be explained by its process-stabilizing effect, which was suggested by the previously described influence on the course of the maximum anodic voltage. In general, glycerol can be identified as a beneficial electrolyte additive. The phosphate concentration has no statistically significant influence on the hardness, shows an increased layer thickness at level 2, and leads to a lowered corrosion current density at increased levels. Thus, phosphate is a statistically proven beneficial electrolyte additive for the plasma electrolytic oxidation of magnesium. However, the data obtained by this study give no clear explanation for its mode of action. There was no phosphorous detectable during the EDX analysis of the coating compositions. This suggests that the positive effect of increased phosphate concentrations can be justified by a modified layer morphology with a lower defect density. Increased silicate concentrations lead to thicker and harder coatings. This can be explained by the significant incorporation of silicon into the PEO layer. On average, the lowest and therefore best values for i_0 were obtained at the first and third silicate level. But two of the three experiments with the absolute lowest corrosion current density were obtained by experiments with a silicate concentration on level two, see table 3. This suggests that there are strong interactions between the electrolyte's constitutes during the PEO-layer formation process. The assumption is supported by the resulting data of the optical process observation. The increase of the phosphate and glycerol concentrations lead to lower values for i_0 but has directly contradicting effects on the transition of discharge stage I into discharge stage II. Ultrasound has a strong influence on the duration of discharge stage I but no statistically significant effect on the layer thickness or corrosion current density. Its influence on the layer hardness is statistically significant but not in a technologically relevant magnitude.

A better understanding of the process parameter interactions could lead to an explanation of these partly contradictory observations. For a systematic statistical evaluation of such interactions the DOE used in this study is not appropriate. Therefore, selected areas of the parameter space should be investigated by full-factorial experimental plans. Additionally, the process monitoring used in this study could be improved: The measurements of the anodic and cathodic voltage maxima should be more high frequent. For a more effective evaluation of the data obtained by the optical process observation, automatized algorithms, which quantify discharge characteristics like duration, brightness and frequency are necessary.

Acknowledgment
The financial support of this work by the Deutsche Forschungsgemeinschaft (DFG) is gratefully acknowledged (project number La-1274/34-1).

References

[1] Ling G and Atrens A 1999 *Adv. Energy Mater.* **1** 11–33
[2] Gray J E and Luan b 2002 *J. Alloys Compd.* **336** 88–113
[3] Yerokhin A L, Nie X, Leyland A and Matthews A 2000 *Surf. Coat. Technol.* **122** 73–93
[4] Sieber M, Simchen F, Scharf I and Lampke T 2016 *Surf. Coat. Technol.* **25** 1157–62
[5] Ghasemi Raja V S, Blawert C, Dietzel W and Kainer K U 2010 *Surf. Coat. Technol.* **204** 1469-78
[6] Barchiche C E, Rocca E and Hazan J 2008 *Surf. Coat. Technolog.* **202** 4145–52
[7] Liang J, Guo B, Tian J, Liu H, Zhou J and Xu T 2005 *Appl. Surf. Sci.* **199** 121–6
[8] Zhang R, Jiang K, Zhu Y, Qi H and Ding G 2011 *Appl. Surf. Sci.* **258** 586–9

[9] Enomoto N, Kurakazau M, Inada M, Kamada K, Hojo J and Lee W 2009 *J. Ceram. Soc. Jpn.* **117**, 369–372
[10] Shen D, Donglei H, Liu F, Guo C, Cai J, Li G and Ma H 2014 *Ultrasonics* **54** 1065–70
[11] Dejiu S, Jingrui C, Guolong L, Donglei H, Lailei W, Haojie M, Yonghong X, He C and Yaqian Y 2014 *Vacuum* **99** 143–148
[12] Qu L, Li M, Liu M, Miao L, Zhang E and Ma C 2013 *J. Adv. Ceram.* **2** 227–34
[13] Cho J Y, Hwang D Y, Lee D H, Yoo B and Shin D H 2009 *Trans. Nonferrous Met. Soc. China* **19** 824–8
[14] Liang J, Hu L and Hao J 2007 *Appl. Surf. Sci.* **253** 4490–96
[15] Mori Y, Koshi A, Liao J, Asoh H and Ono A 2014 *Corros. Sci.* **88** 254–62
[16] Chen H, Lv G, Zhang G, Pang H, Wang X, Lee H and Yang S 2010 *Surf. Coat. Technol.* **25** 32-5
[17] Qiu Z, Zhang Y, Li Y, Sun J, Wang R and Wu X 2015 *RSC Adv.* **5** 63738–44
[18] Wu D, Liu X, Lu K, Zhang Y and Wang H 2009 *Appl. Surf. Sci.* **255** 7115–20
[19] Box G E P and Behnken D W 1960 *Technometrics* **2** 455–75

IOP Conf. Series: Materials Science and Engineering **181** (2017) 012041 doi:10.1088/1757-899X/181/1/012041

Electrochemical deposition of iridium and iridium-nickel-alloys

J Näther[1*], F Köster[1], R Freudenberger[2], C Schöberl[2] and T Lampke[3]

[1]HS Mittweida University of Applied Sciences Fakultät Ingenieurwissenschaften Fachbereich Fertigungs- und Werkstofftechnik, Technikumplatz 17, 09648 Mittweida, Germany
[2]Forschungsinstitut für Edelmetalle und Metallchemie FEM, Katharinenstraße 17, 73525 Schwäbisch Gmünd, Germany
[3]Institute of Materials Science and Engineering, Chemnitz University of Technology, 09125 Chemnitz, Germany
* mailto: naether@hs-mittweida.de

Abstract. To develop durable and reliable electronic contacts, precious metals are still very important for finish plating of contact surfaces. The lesser-known iridium might be an interesting alternative to substitute gold alloys, platinum or rhodium for applications with highest demands to wear and corrosion resistance such as sliding and plug contacts. As matters stand there is no commercial electrolyte for iridium plating. Initial investigation screened the parameter range for different iridium compounds when an iridium layer occurred on the substrates. This approach showed that the oxidation state of iridium is crucial to reach contenting deposits. Best results came from Ir(IV) electrolyte with high bromine concentration coming from the starting compound, while electrolytes made from Ir(III) compounds gave very poor deposits. In subsequent experiments different organic compounds were added to the electrolytes to improve plating efficiency and stability of the solutions. So found electrolytes gave crack-free deposits up to two microns with a micro-hardness of 600 HV. To reduce the iridium content in the layer, iridium-nickel-alloys were investigated, finding that a nickel-content of 10 wt% raised the layer hardness to more than 900 HV.

1. Introduction

Despite iridium having some remarkable physical and chemical properties regarding the needs of electronic or chemical industries, the platinum metal is only used for niche applications. Insoluble anodes made by thermal oxidation of iridium compounds are used for electrolysis or electroplating processes. [1] Furthermore there is a wide field of scientific uses in sensory applications. Iridium oxide is known to be pH-sensitive, so it became a preferred material for measurements in biological media. Thin iridium layers can be used as planar micro sensors for pH-measurement or as stimulating electrode for medical purposes. Such iridium layers are commonly made by physical vapour deposition or thermal oxidation of iridium compounds.[2] Vacuum-metallized iridium-platinum-layers are used as wear protection for hardware with direct contact to molten glass.[3]

There are some uses as alloying constituent in platinum alloys to improve hardness and corrosion resistance. Such materials can be found in jewellery or premium pen tips as well as electronic contacts and spark plug tips. Electronic contacts are implemented by tight fittings or welding contacts. There is a much wider field of application for iridium as a contact material, due to its good electrical conductivity, the high hardness and the outstanding corrosions resistance. The galvanic deposition of iridium layers could unlock these potentials, if an efficient electrochemical process can be developed.

Incipient investigations concerning iridium plating were carried out by Tyrell and Conn. [4-5] The present research project pursues the improvement of the electrodeposition process for iridium and the identification of coating characteristics. The project aims to utilize iridium as an option for finish plating of contact surfaces and to search for new alloying elements to improve the properties of such surfaces.

2. State of the art

To date there is no commercial electrolyte for the deposition of iridium. There were some investigation to reduce iridium from fused salt electrolytes, which gave good adherent and ductile deposits up to 100 µm. The plating bath was based on a mixture of sodium- and potassium cyanide, but the process required high operating temperatures combined with water and oxygen free conditions, which where ensured by an argon atmosphere. The high expenditure for equipment and process control as well as the high failure proneness caused by smaller contamination made the process not very beneficial for contact plating. [6]

First investigations of aqueous iridium electrolytes were realised in the 1960s. Comprehensive studies of different electrolytes were carried out by Tyrell. These electrolytes were made by dissolving hydrated IrO_2 in various acids. He found that an electrolyte based on hydrobromic acid gave best results. [4] Conn used an electrolyte based on iridium(III)chloride and sulphamic acid, but he needed unusual high anodic current densities to sustain the deposition [5]. MacNanamara investigated different iridium-halogen complexes and concluded, that the tetravalent state is crucial for the deposition [7]. All found electrolytes had low cathode efficiencies, which decreased further when the solution was worked. A more recent paper by Sheela verified Tyrells experiments, but used another preparation route for the iridium bromine solution [8]. All mentioned works were concerned with the deposition of pure iridium layers and focussed on the behaviour of the electrolyte itself. There is no detailed information about the characteristics of the iridium layers.

Very few publications dealt with the co-deposition of other metals to create an alloyed iridium layer. Tyrell and Sheela investigated the deposition from iridium-platinum solutions. [9,10] Sheela implemented iridium-platinum alloys with amounts of 30 % to 70 % iridium depending on the used current densities and the iridium contents of the solutions.

3. Experimental

The iridium salts used in the present work are outlined in table 1. To allow a big amount of experiments, the electrolyte volumes were limited to 30 ml in beakers of 50 ml. The iridium concentration was set to 5 g/l iridium for the experiments. To reduce the evaporation loss and to get a constant distance between anode and cathode, the electrodes were mounted in a lid. The substrates were made from printed circuit board material with 30µm copper and etched to samples with a size of 1 cm². To ensure that the strong acidic electrolyte with pH=2 won`t attack the substrates a coating system of 15 µm nickel and 1µm gold was applied before the iridium plating to protect the copper. The pure nickel layer was applied by galvanic deposition from the electrolyte Nicorsit EPG supplied by BECE. The gold layer was made by galvanic deposition from the Dodurex electrolyte Puramet 202 to get pure gold layers.

Table 1. Iridium salts used for the experiments.

chemical designation	valence	chemical purity	supplier
sodium hexabromoiridate	(IV)	99.9%	Metakem GER
hydrogen hexachloroiridate	(IV)	99.9%	Metakem GER
iridium acetate	(III)	99.9%	Heraeus GER
iridium chloride	(III)	99.9%	Alfa Aesar GER

IOP Conf. Series: Materials Science and Engineering **181** (2017) 012041 doi:10.1088/1757-899X/181/1/012041

Sections of the substrates which did not dip into the electrolyte where protected by an adhesive tape, to avoid any contamination caused by the aggressive vapour. Platinum sheets of 4 cm² or insoluble iridium oxide electrodes were used as anode. So the anode-cathode-ratio amounted 4:1. The convection of the electrolyte was ensured by a magnetic mixer in combination with a controlled heating. The temperature varied from room temperature up to 80 °C with constant steps of 20 K. Since very low but constant currents where required – especially for long term experiments – a potentiostat was used as source of electricity.

First of all some screening experiments were made to get an overview of the electrochemical properties of the dissolved iridium salts and their behaviour in different conditions. The main parameters were temperature, pH-level und current densities. The plating time was set to 15 min. The parameter range is summarised in table 2.

Table 2. Parameter range for screening experiments.

parameter	lowest value	highest value	intervals
temperature	20°C	80°C	20K
pH-value	pH=2	pH=8	pH=2
current density	3mA/cm²	21mA/cm²	9mA/cm²

Next attempt was to improve the deposition characteristics by different additives. As the mode of action for these additives wasn`t known in detail, a practical selection was made from compounds used in similar platinum metal electrolytes. The investigated additives and their concentrations are listed in table 3. To reduce the number of experiments, the temperature was set to 80°C while the pH-value was set to pH=2. Electrolytes, which gave an iridium layer within the first 15 min were determined for longer periods to study the efficiency behaviour during the process. Therefore the layer thickness was determined after 15min of plating, bevor the process continued on the same substrate.

Table 3. Additives used during experiments.

chemical designation	chemical formula	investigated concetrations [mmol]		
sulphamic acid	H_2NSO_3H	52	206	412
oxalic acid	$C_2H_2O_4$	5.5	28	55
ascorbic acid	$C_6H_8O_6$	5.6	28	56
nitrite acid	HNO_3	3.2	16	32
phosphonic acid	H_3PO_3	2.4	12	24
sulphuric acid	H_2SO_4	2	10	20

The layer thickness was determined by x-ray fluorescence (Oxford X-Met 7500, acceleration voltage 45 kV, 30 s measurement time). The reference thickness for the calibration was ascertained by measuring the polish cross section. Due to the small cathode area there were no quantifiable weight differences. So the cathodic efficiencies were calculated from the layer thickness determined by x-ray fluorescence. The measured layer thickness divided by the theoretical thickness calculated from Faraday`s law gave a good approximation of the efficiency. All efficiencies given in this paper were calculated from a minimum of three experiments. The quality of the layers was rated by visual impression. For layers of more than 1µm thickness the micro hardness was measured by nanoindentation with the Fischerscope H100c (Fischer, Germany) The test force was adjusted to 3 mN for a period of 10 s until the force was retracted. An optical analysis of the formed layers by microscopy gave information about cracks and pores.

Latest experiments were concerned with the development of iridium-alloy deposits. Different elements were considered to give an alloy with iridium to reduce the noble metal consumption and these were phosphor, silver, copper and nickel. To determine the contents of the alloys x-ray fluorescence analysis and energy dispersive x-ray analysis (Bruker Quantax 200, energy 25 keV , 2 min measurement time, area 2.5 x 1.8 mm²) were used. To avoid the distortion of the results by measuring the nickel signal coming from the substrate, the substrates were directly coated with 2µm gold from the Puramet electrolyte, without any nickel interlayer.

4. Results

Iridium chloride
The aqueous solution showed an orange colour and pH 1,8 after dissolving the salt. It was not possible to get any deposit from this solution by itself. According to reference 6, there was a very thin iridescent film when sulphamic acid was added with concentrations of more than 200 mmol/l and high temperatures of 80 °C occurred. In the investigated range of current densities, no further deposit was observed. All other additives did not tend to result in any advance. Raising the pH-value led to black precipitate formed on the anode. This effect was increasing for higher current densities.

Iridium acetat
The dissolved salt had a very low electrolytic conductivity, so inorganic acids were added to reduce the electrolytic resistance. The cyan coloured solution did not give any deposit. The addition of oxalic and sulphamic acid led to very thin layers in conditions of 80 °C and pH=2, similar to the layers formed in the iridium chloride electrolyte. The precipitate formation for higher pH-values was although very similar to that observed in the iridium chloride electrolyte.

Iridium hexachloroiridate
This solution gave black-brown powdery deposits for current densities of more than 10 mA/cm² and temperatures higher than 60 °C, while there was not any layer formation for lower current densities. The addition of nitrite acid as well as sulphuric acid led onto metallic depositions, with very low cathode efficiencies of less than 10 %. After 30 min the deposition ceased. There were no possibilities found to restart the process.

Sodium hexabromoiridate
The bromoiridate gives the best results of all investigated iridium compounds. Adherent bright layers are formed from 3 to 12 mA/cm² in conditions of pH=2 and temperature range from 60 °C to 80 °C. The cathode efficiency averages 40% at the beginning of the plating process for 3mA/cm² and 80 °C, but decreases with increasing current densities. Plating with 60 °C gives poor efficiencies of less than 5%. During the first 15min the electrolyte changes colour from green to dark blue. If the electrolyte is worked with current densities between 3 and 12 mA/cm² the efficiency slowly decreases. For current densities of more than 30 mA/cm² the deposition stops after 45 min. If the anodic current density exceeds 20 mA/cm² a black powder covers the anode. After filtering off this precipitate, the iridium concentration halved.

The most interesting additives are nitrite and oxalic acid. The nitrite acid raises the cathodic efficiency to 50 % while a current density of 3 mA/cm² is applied. The improved efficiency is stable for 3 h. Because of the low pH-level, the substrates are attacked through some defects in the gold layer, caused by delamination. For that reason this electrolyte was no longer regarded. The oxalic acid has a comparable positive effect on the deposition rate with 3 mA/cm² and raised the cathodic efficiency to more than 60 %. The colour change from green to blue is inhibited during the first 30 min of the process. Similar to the behaviour of the pure bromoirdate solution, the efficiency is not stable for longer terms and relapses to 30 % after 3 h. Thereafter the efficiency decreases over process time. This behaviour is depicted in figure 1. After 6 h the experiments were stopped, while the efficiency was reduced to less than 15%. Refreshing the electrolyte by adding oxalic acid during the process did not show any positive effect.

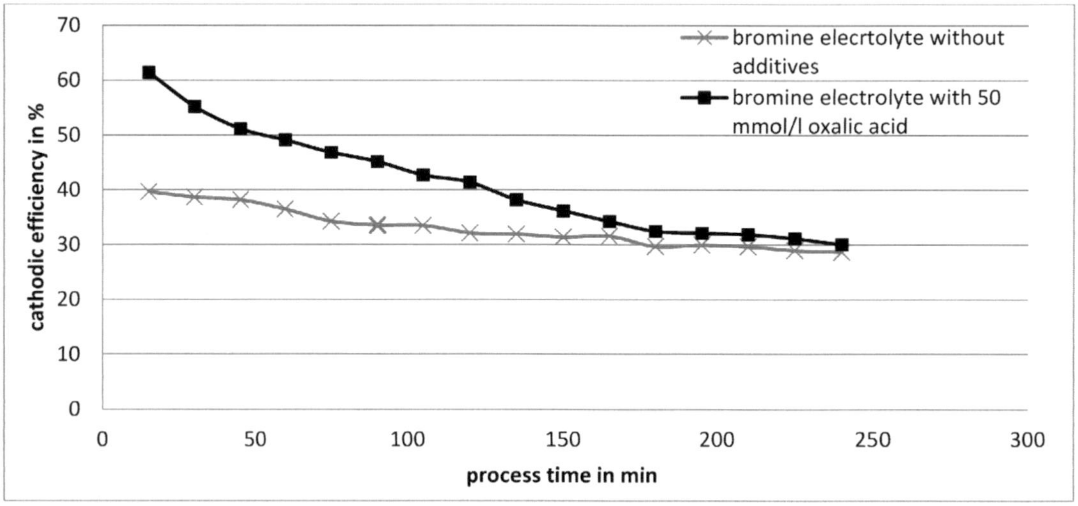

Figure 1. Electrolytes worked with 3 mA/cm² and 80 °C.

The layer thickness is limited to 2 µm. Higher thicknesses caused cracked deposits and delamination due to internal stress. A clear correlation between current density and propensity for cracked layers was found. The cracking occurred much earlier when the electrolyte was worked with high current densities (see also figure 2 and figure 3). The maximum thickness of 2 µm was reached with 3 mA/cm² starting from fresh electrolyte. Worked electrolytes tend to give cracked deposits with increasing age. Furthermore the electrolytes with oxalic acid tend to crack earlier, but there was no quantified limit for layer thicknesses from oxalic electrolytes. The hardness was measured to 600 HV for pure iridium deposits with small quantities of bromine and oxygen incorporated. There is no sign, that oxalic acid had any impact to the layer hardness.

Figure 2. Cracked layer, 21mA/cm², 1 µm.

Figure 3. Crackfree layer, 3mA/cm², 1 µm.

Iridium alloys

As the best results came from the hexabromoiridate solution, this electrolyte should be used for the alloying experiments. For that reason silver could not be investigated. Free bromine would react with silver ions to form the low-soluble silver bromide. Phosphor added as phosphonic or phosphoric acid did not have any impact to the deposition, but there was never any incorporated phosphor measured. The addition of sodium hypophosphite resulted in a colour change to orange when the solution was heated. No deposition is possible from this electrolyte. The copper solved as copper sulphate hindered the deposition, so that the efficiency decreased very fast until the reduction stopped. This effect was already observed, when the copper on the substrates was attacked by bromine vapour. Very low concentrations reduced the efficiency to less than 5 %.

From the suggested elements, only nickel could be co-deposited by dissolving nickel sulphate in the bromine electrolyte reaching a nickel concentration of 1 g/l. The temperature behaviour followed the base electrolyte made from hexabromoiridate. Thin deposits are obtained from 60 °C to 90 °C with an efficiency optimum at 80 °C. The nickel content raises the efficiency for low current densities to more than 80 % for initial deposition, but working this electrolyte decreased the efficiency over time to 30 %, which conforms to the values of the pure iridium electrolytes. For higher current densities the efficiency never exceeded 40 % while a fast decline to 30 % was observed after 1 h.

The current density has the strongest impact to the nickel content of the obtained layers. The relation for an electrolyte with 1g/l nickel is shown in figure 4. Low current densities gave also low nickel contents in the coating, while raising the current density to more than 10mA/cm² leads to nickel contents of 16%. This content is very stable when higher current densities up to 20 mA/cm² are applied. Nickel concentrations from 0.1 g/l also give incorporations of 10 % for plating with 6 mA/cm².

To date of this publication there are no indications that higher nickel concentration would lead to higher incorporation rates than the maximum of 17 % that was already reached with 1 g/l nickel.

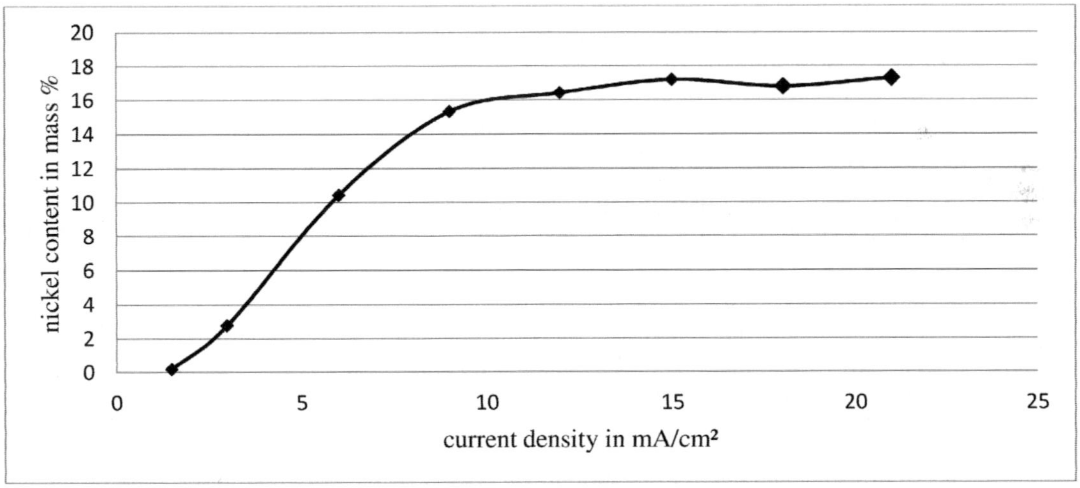

Figure 4. Nickel content as a function of cathodic current density (with 1g/l nickel).

The characterisation of the iridium-nickel-deposits showed crack-free layers for thicknesses smaller than 2 μm made from fresh solution with 3 mA/cm². Deposits from worked electrolytes or higher current densities resulted in cracked layers and delamination as already observed for the pure iridium electrolyte. Crack-free iridium-nickel-layers of 1.5μm thickness possessed a hardness of 900HV measured by nanoindentation.

5. Conclusion

The experiments confirm the assumption made by Macnamara and Tyrell [4,7], that the tetravalent state is crucial for the reduction of iridium from aqueous solutions, due to the fact that there was no adequate deposition from trivalent iridium salts. However, there should be additional studies to explain the mechanism of the reduction process for iridium.

The only working electrolyte is still based on bromine solutions. High temperatures are identified as primary parameter to form an appropriate complex to reduce iridium from aqueous solutions. This formation starts at 60 °C and has an optimum at 80 °C. Second to temperature a low pH-level is vital for the deposition. The current efficiency raises with decreasing pH-level. For practical aspects a pH-level lower than pH 1.5 is not useful, due to massive substrate attack.

The current efficiencies that were obtained in previous experiments by Tyrell and Sheela [4,8] can be improved by the addition of oxalic acid to a maximum of 60 %. Further investigations should be concerned with the maintenance of this efficiency level.

The electrolyte made from hexabromoiridate gives bright, adherent and crack-free deposits up to 2µm thickness with a plating rate of 0,8µm/h for optimum parameter values, which is an adequate form for practical applications. The layer hardness of 600HV is very encouraging to force the development of iridium plating for electronic contacts with abrasive strain. Furthermore the co-deposition of nickel in crack-free layers with 2µm thickness heightened the hardness to 900HV. This is a promising result to develop contact plating with excellent wear resistance compared to other platinum metals or hard-gold. For coming experiments contact and corrosions resistance of such layers should be determined to get a more detailed characterisation.

Acknowledgment

The present work is a co-operation between the "University of Applied Sciences Mittweida" and the "Institute for precious metal chemistry FEM" in Schwäbisch Gmünd. The research project "Development and characterisation of iridium- and iridium-alloy-deposition for electrical contact applications" is publicly sponsored by budgets of German Federation of Industrial Research Associations (AiF) with the identification number 18543BG.

References

[1] Patent specification DE102004015633A1 2005
[2] Kakooei S, Ismail M C and Wahjoedi B A 2013 Electrochemical Study of Iridium Oxide Coating on Stainless Steel Substrate *Int. J. Electrochem. Sci*, **8** pp 3290-3301
[3] Bobzin K, Bagcivan N, Brögelmann T and Münstermann T 2013 Development and qualification of a MSIP PVD iridium coating for precision glass moulding *Materialwissenschaft und Werkstofftechnik* **44** (8) pp 673-678
[4] Tyrrell C J 1965 The Electrodeposition of Iridium *Transaction of the Institute of Metal Finishing* **43** pp 161-166
[5] Conn G A 1965 Iridium Plating *Plating* pp 1258-1261
[6] Harding W B 1978 Platinum Surface *Finishing* **65** (2) pp 30-35
[7] MacNamara E L 1962 The Electrodeposition of Iridium *Journal of the Electrochemical Society* **109** pp 61-63
[8] Sheela G and Pushpavanam M 1999 Electrodeposition of Iridium *Bulletin of Electrochemistry* **15** (5-6) pp 208-210
[9] Sheela G, Pushpavanam M and Pushpavanam S 2005 A bromide electrolyte for the electrodeposition of platinum and platinum-iridium alloy *Transactions of the Institute of Metal Finishing* **83** (2) pp 77-81
[10] Tyrell C J 1967 *Transcactions of the Institute of Metal Finishing* **45** p 53

Localised anodic oxidation of aluminium material using a continuous electrolyte jet

D Kuhn[1], A Martin[1], C Eckart[1], M Sieber[2], R Morgenstern[2], M Hackert-Oschätzchen[1], T Lampke[2] and A Schubert[1]

[1]Professorship Micromanufacturing Technology, Faculty of Mechanical Engineering, Chemnitz University of Technology, 09107 Chemnitz, Germany
[2]Institute of Materials Science and Engineering, Chemnitz University of Technology, 09125 Chemnitz, Germany

E-mail: danny.kuhn@mb.tu-chemnitz.de

Abstract. Anodic oxidation of aluminium and its alloys is often used as protection against material wearout and corrosion. Therefore, anodic oxidation of aluminium is applied to produce functional oxide layers. The structure and properties of the oxide layers can be influenced by various factors. These factors include for example the properties of the substrate material, like alloy elements and heat treatment or process parameters, like operating temperature, electric parameters or the type of the used electrolyte. In order to avoid damage to the work-piece surface caused by covering materials in masking applications, to minimize the use of resources and to modify the surface in a targeted manner, the anodic oxidation has to be localised to partial areas. Within this study a proper alternative without preparing the substrate by a mask is investigated for generating locally limited anodic oxidation by using a continuous electrolyte jet. Therefore aluminium material EN AW 7075 is machined by applying a continuous electrolyte jet of oxalic acid. Experiments were carried out by varying process parameters like voltage or processing time. The realised oxide spots on the aluminium surface were investigated by optical microscopy, SEM and EDX line scanning. Furthermore, the dependencies of the oxide layer properties from the process parameters are shown.

1. Introduction

Anodic oxidation of aluminium and many aluminium alloys is well established and most commonly used as prevention against wearout and corrosion. During anodic oxidation, the aluminium substrate material is connected to the anodic potential and immersed into the electrolyte bath. This electrochemical process is used to create a layer of aluminium oxide on the whole surface of the substrate material. The oxide layer grows into the material as well as to the outside and is characterised by its high joint strength with the substrate material.

In order to localise the anodic oxidation to confined sections of the surface, covering materials, such as hoses, plastic foils and insulation tapes, are currently used. These covering materials need to be inert during the anodic oxidation to avoid undercutting. In addition, the covering materials need to be deposited on the material's surface and removed after the anodisation without damage. Applications of localised anodic oxidation are found on components and component groups, which consist of multiple materials. In these cases the surface characteristics need to be modified sequentially, hence, the anodisation of all sections is technically unnecessary or even damaging. Moreover, in most cases the tribological load, the demand of an adhesive grounding or the decorative effect is limited to segments of the component. In these cases, the anodisation of the whole immersed component surface is an unnecessary waste of the resource electrical energy. In addition, the localised anodisation offers an approach to the modification of electrical, optical and corrosion-related characteristics.

The application of a continuous electrolyte jet is an adequate alternative for the creation of locally confined oxide spots. Electrochemical machining (ECM) applying a continuous electrolyte jet (Jet-ECM) is a non-conventional technology used to localise the electrochemical machining process at high current densities. So far, it is utilized to remove electrically conductive materials [1, 2]. Figure 1 shows the principle of Jet-ECM.

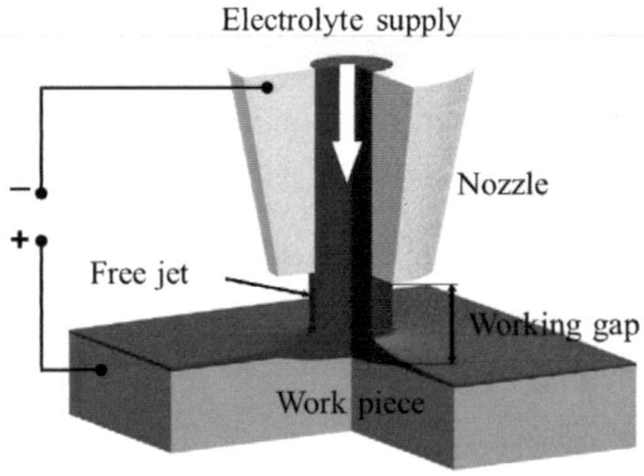

Figure 1. Principle of Jet-ECM [2, 3].

The basic principle of electrochemical machining is the anodic dissolution of work-piece material through electric charge transport. The dissolution takes place at the interface between the work-piece surface and the electrolyte, which is a liquid ion conductor. The special characteristic of Jet-ECM is the supply of fresh electrolyte through a micro nozzle at a mean jet velocity of approximately 20 m/s [2, 3].

The electrolyte is ejected perpendicularly towards the work-piece surface in the form of a continuous free jet surrounded by atmospheric air. This leads to a highly localised machining area, because the distribution of the current density is locally confined by the impinging jet.

Extremely high current densities of several hundred A/cm² can be realised [4]. Thus, excellent surface qualities can be machined. In addition, the high jet velocity leads to a very good supply with fresh electrolyte. Hence, there is no need for interrupting the process for flushing phases.

In this paper the application of a free jet for anodic oxidation is investigated. By using an acidic electrolyte the application of this method for the realisation of locally confined oxide spots is analysed.

2. Experimental Setup

For the experimental investigations in localised anodic oxidation, a modular test rig is used [2, 5]. The machine frame in gantry design and the working table are made of granite, which guarantees the required mechanical and thermal stiffness for micro-machining applications.

The electrolyte is delivered by a pulsation-free pump and ejected from the nozzle in Z-direction to the work-piece. Used up electrolyte is collected in a disposal tank. A power supply is used to provide the process voltage up to 300 V. All electrical and kinematic operations are controlled by a personal computer. Therefore, customized control software based on National Instruments LabVIEW was developed [2, 3, 7]. The scheme in Figure 2 illustrates the experimental setup [2, 3].

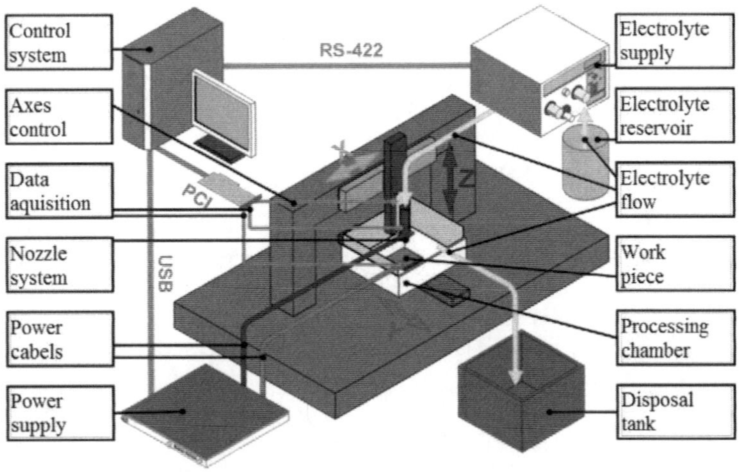

Figure 2. Scheme of the modular test rig [2, 6].

Figure 3 shows a photograph of the granite table and the processing chamber of the existing test rig.

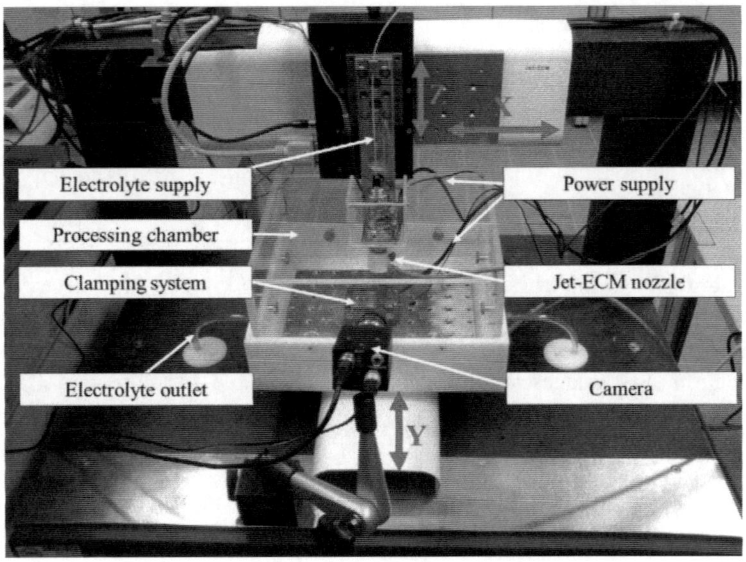

Figure 3. Photograph of the existing test rig [7].

The relative movement between nozzle and work-piece is carried out by a three-axis positioning system of linear stages. With this setup travel distances of 150 mm in the direction of X and Y and 50 mm in Z-direction are possible. The maximal speed of the linear stages is limited to 5 mm/s by the control software. The nozzle is adapted to the Z-stage, which is mounted to the X-stage. A processing chamber with a lateral dimension of 200 x 200 mm^2 is adapted to the Y-stage. The work-piece clamping system is integrated into the processing chamber [7].

3. Design of Experiments and Evaluation

The anodic oxidation of aluminium material using a continuous electrolyte jet was investigated by applying oxalic acid on an EN AW-7075 work-piece in peak-aged condition T6. Therefore, localised anodic oxidation without nozzle movement was carried out under variation of processing time and processing voltage. The three-axis setup was applied to realise the infeed movements from one machining point to another.

The experiments were carried out on a work-piece with lateral dimensions of 250 x 200 mm² and a thickness of 3 mm.

In order to remove the natural oxide layer, bating of the work-pieces was carried out before the anodic oxidation process. At first, the work-piece was cleaned with ethanol. Then it was immersed into an aqueous solution with a mass fraction of 10 % NaOH at a temperature between 50 °C and 60 °C for two minutes. Finally, pickling was carried out with nitric acid aqueous solution with a mass fraction of 10 % for another two minutes. The processing parameters for the investigations in localised anodic oxidation are summarised in Table 1.

Table 1. Process parameters.

Parameter	Value
Work-piece material	AlZn5,5MgCu (EN AW 7075 T6)
Electrolyte type , mass fraction	Oxalic acid, 4.3 %
Electrolyte conductivity	52 mS/cm
Electrolyte temperature	22°C
Pump delivery rate	10 ml/min
Nozzle diameter	100 µm
Working gap	2500 µm
Voltage	10 V, 20 V, 40 V, 80 V, 160 V
Processing time	300 s, 600 s

Besides the constant processing parameters given by the work-piece, the electrolyte, the ambient conditions and the setup, the voltage was varied between 10 V to 160 V and the processing time was varied between 300 s and 600 s.

Secondary electron (SE) images were obtained at the sample surface using a scanning electron microscope (SEM) Zeiss Leo 1455 VP. Energy dispersive X-ray spectroscopy (EDX) line scans were conducted across the diameter of anodised spots using an excitation voltage of 5 kV. The oxide thickness was estimated according to the calculated and measured Al/O ratios for different oxide thicknesses on EN AW-7075 alloy in [8].

4. Experimental Results

The results of the experiments in localised anodic oxidation were investigated optically. The oxide spots were arranged with a lateral distance of 4 mm from one another. The experiments with 10 V processing voltage are positioned on the left-hand side of the work-piece. The voltage increases up to 160 V at the right-hand side. In the upper part of the work-piece the experiments with a processing time of 300 s are arranged while in the lower section the same voltages were used at a processing time of 600 s. Figure 4 shows a microscope image of the work-piece surface.

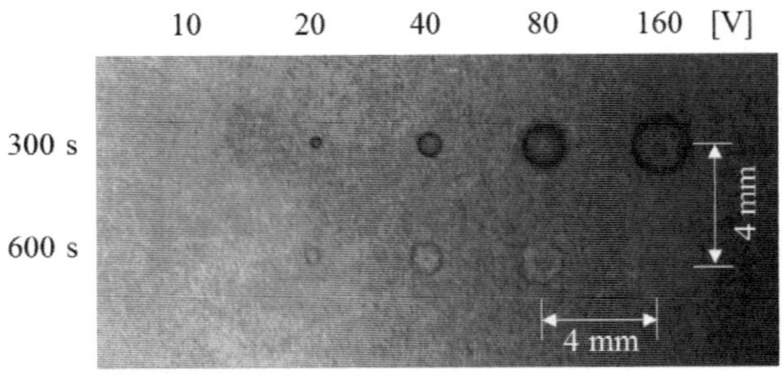

Figure 4. Microscope image of the work-piece with local oxide spots.

From the microscope image it can be recognized that the surface of the work-piece contains a visible change in colour for both of the applied processing times. The circular interference colours represent areas with approximately the same oxide thickness. From the colour gradient within the spots it can be concluded that the oxide thickness varies across the spot diameter. Furthermore, it can be seen that the spot diameter increases with increasing voltage.

In order to detect the content of oxygen and aluminium and to evaluate the oxide thickness on the work-piece surface inside and outside the oxide spots, the work-piece was investigated by SEM and EDX line scanning. Figure 5 and Figure 6 show SEM images of the experiments with processing voltages of 10 V and 160 V after a processing time of 300 s. From the SEM image in Figure **6**, it can be derived that the process influences the surface of the work-piece. The machined area shows a dark circle in its centre with a lighter ring around it. At the dark areas on the right side of Figure 6 contaminants on the surface of the work-piece can be seen. In Figure 5, no influence on the work-piece surface was recognized.

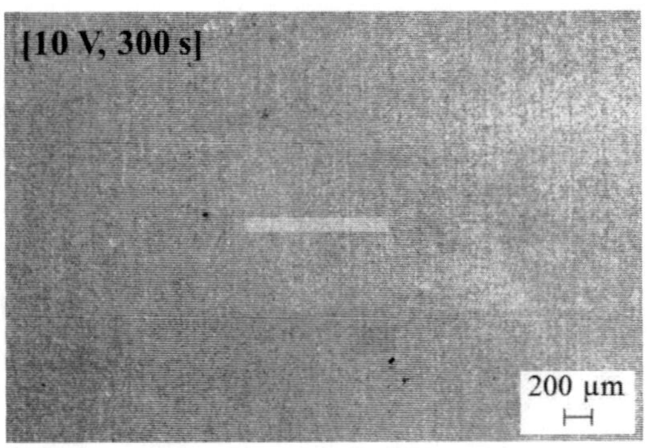

Figure 5. SEM image of the area machined at 10 V and 300 s.

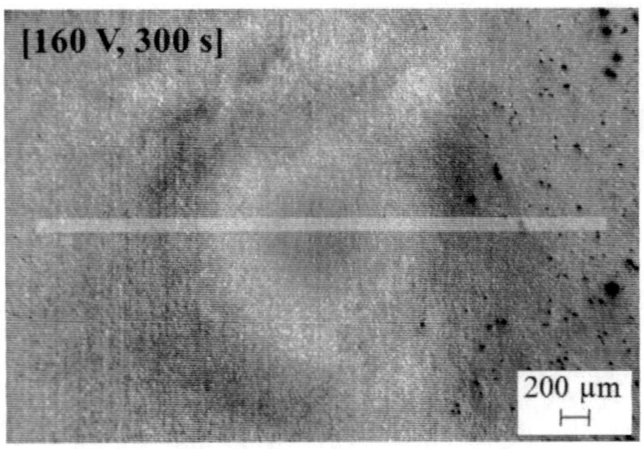

Figure 6. SEM image of the area machined at 160 V and 300 s.

Figure 7 and Figure 8 show the results of EDX line scans. The EDX line scans were captured along the yellow lines highlighted in the SEM images figure 5 and figure 6.

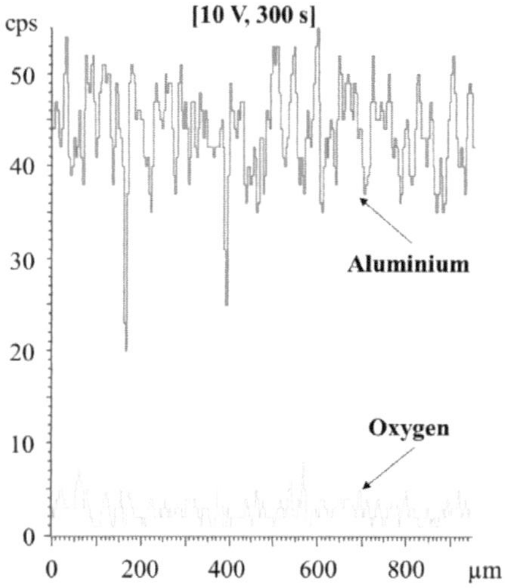

Figure 7. EDX line scan of the area machined at 10 V and 300 s.

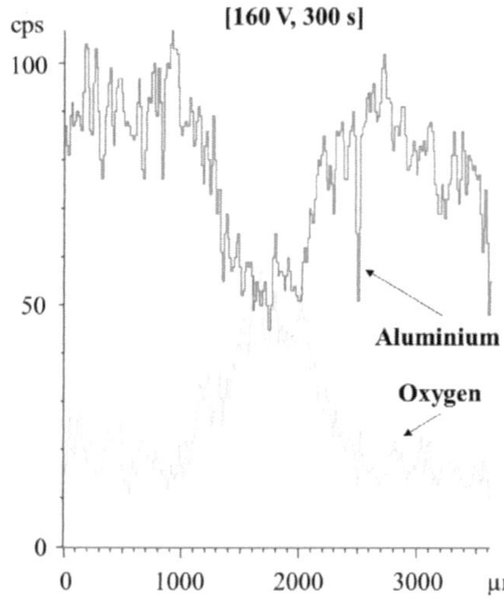

Figure 8. EDX line scan of the area machined at 160 V and 300 s.

Figure 7 represents the analysis of the oxide spot machined with 10 V and 300 s. It can be seen that the content of oxygen is low in comparison to the aluminium content, which is between 40 counts per second (cps) and 50 cps. Counts per second in EDX analyses are a measurement unit for the quantity of interactions of electrons with the atoms of the displayed element. The resulting absolute values differ in dependence of the measurement parameters, which are defined by the measurement software. Therefore, the ratio between the cps of the single elements is significant for the investigation of the oxide evolution.

Figure **8** shows the analysis of the oxide spot with the parameters 160 V and 300 s. A significant increase of oxygen content can be seen in the middle of the detected line. In this area, the content of aluminium drops from nearly 100 cps to 50 cps while the content of oxygen increases from 20 cps to 50 cps. This can be attributed to an increasing oxide thickness towards the centre of the spot.

Figure 9 and Figure 10 show SEM images of the experiments with voltages of 10 V and 160 V at a processing time of 600 s

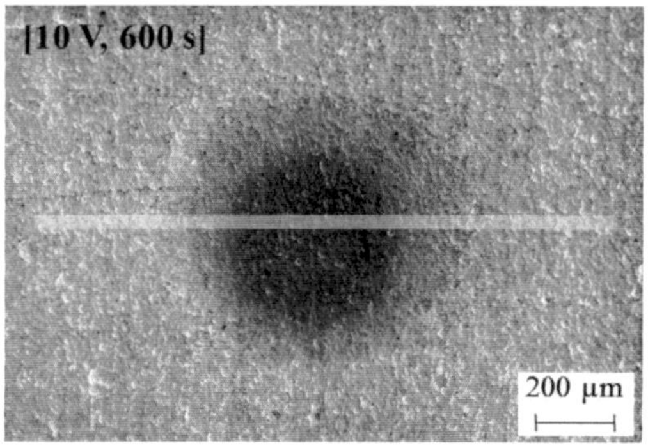

Figure 9. SEM image of the area machined at 10 V and 600 s.

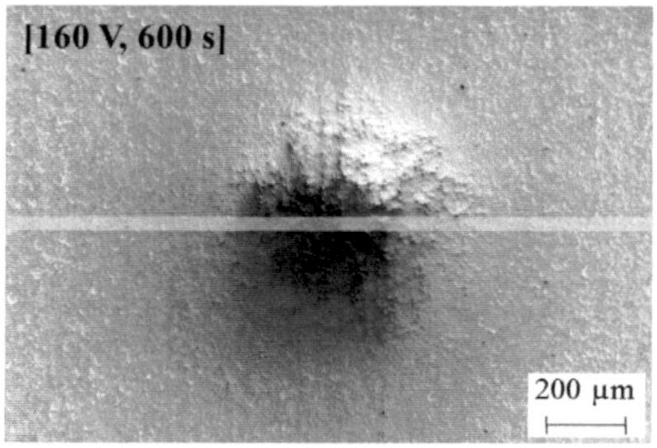

Figure 10. SEM image of the area machined at 160 V and 600 s.

The SEM images show that the work-piece surfaces of both machined areas were influenced by the process. In Figure 9, which shows the experiment with the parameters 10 V and 600 s, a circular area with a dark centre and an increasing brightness to the outside can be recognized. In comparison to the SEM image of the oxide spot in Figure 5, which was machined at shorter processing time of 300 s, the centre has a significantly darker appearance. The SEM image in Figure 10, with the experimental parameters of 160 V and 600 s, shows a dark area in the centre and a significant change in the work-piece surface compared with the SEM image in Figure 6. The differing darkness represents the differing intensity of the interactions of the electrons in SEM imaging caused by differing electric conductivities. From the differing darkness, the hypothesis can be derived, that a larger oxide layer than before was developed. From the comparison of all SEM images, it can be derived that the processing time as well as the voltage have significant influences on the machined areas.

Figure 11 and Figure 12 show the EDX line scans of the SEM images in Figure 9 and Figure 10.

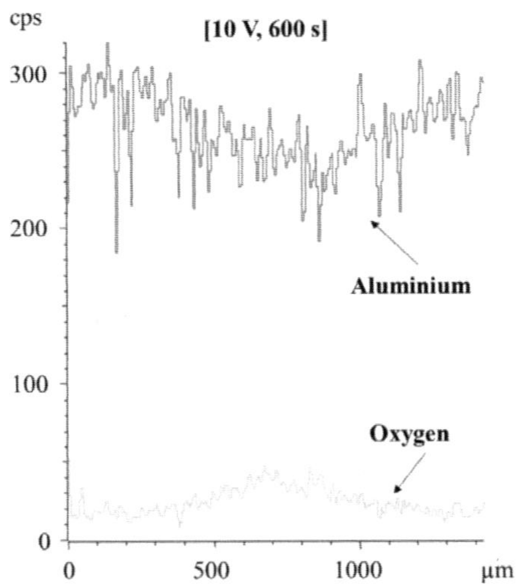

Figure 11. EDX line scan of the area machined at 10 V and 600 s.

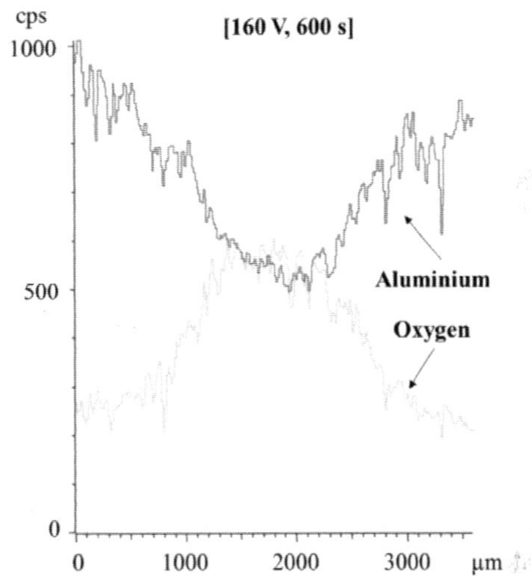

Figure 12. EDX line scan of the area machined at 160 V and 600 s.

In Figure 11 it can be seen that the content of oxygen increases slightly to 30 cps in the centre of the detected line while the content of aluminium decreases slightly from 300 cps to 250 cps. Consequently, an oxide layer was developed which is significantly thicker than the oxide layer, which was created after half of the processing time as shown in Figure 7. In the SEM image of Figure 9 a significant influence with a diameter of approximately 500 µm could be detected, but the EDX analysis in Figure 11 hardly represents an amount of oxygen.

In comparison to Figure 11, the content of oxygen in Figure 12 with 500 cps to 600 cps is significantly higher. Furthermore, it can be seen that the content of the aluminium drops from nearly 1000 cps to 500 cps. This indicates that a comparatively thick layer of aluminium oxide was developed during the experiment. In addition, a comparatively wide distribution of oxygen can be derived. Thus, it can be asserted that a higher voltage leads to an increase of the machined area.

Stoichiometric alumina exhibits an O/Al ratio of 1.5. However, slightly higher O/Al ratios may be measured by EDX analysis at the oxide surface due to additional OH-groups. If the excitation depth of the electron beam exceeds the oxide thickness, the EDX measurement will be influenced by the subjacent aluminium alloy. Consequently, the measured O/Al ratio decreases with decreasing oxide thickness.

Figure 13 displays a graph of the oxide thickness as function of the applied voltage. The oxide thickness d_o was estimated from the calculated ratios of the measured O/Al ratios as already published in [8]. The given values represent the maxima of the thicknesses, which are situated in the centre of the machined oxide spots. Maximal thicknesses between 40 nm and 290 nm were realized at a processing time of 300 s while at a processing time of 600 s thicknesses up to 500 nm were achieved. By comparing the two different processing times, it can be seen, that the longer processing time leads to a larger oxide thickness.

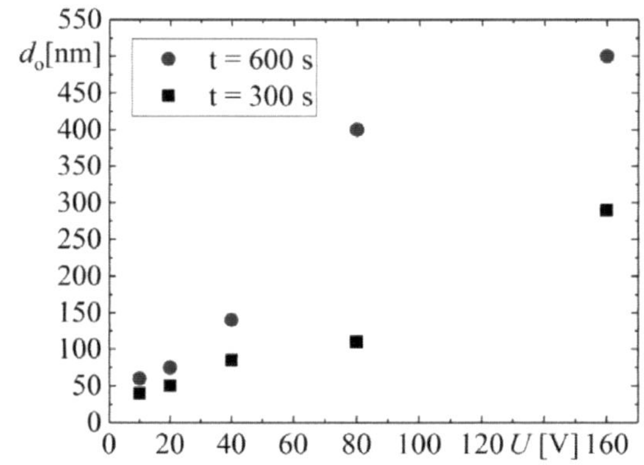

Figure 13. Thickness d_o of the realized oxide spots as function of the applied voltage *U*.

Furthermore, it can be seen, that an increase in oxide thickness is achieved through increasing the processing voltage. In addition, a significantly wider distribution of the local oxide spot was detected at higher voltage. Thus, it can be derived, that the diameter and the thickness of the local oxide spots can be adjusted by adjusting the voltage. The thickness of the local oxide layer can additionally be adjusted by adjusting the processing time.

5. Summary

In this paper an adequate alternative for generating locally confined anodic oxide spots without preparing the substrate by a mask was investigated applying a continuous electrolyte jet. In the experiments a work-piece specimen of EN AW 7075 was machined using oxalic acid. Since aluminium develops a natural oxide layer, bating of the work-piece was carried out before starting the experiments. The machining experiments were carried out at processing voltages between 10 V and 160 V and at processing times of 300 s and 600 s.

The machined areas were investigated optically, by SEM and EDX line scanning. In the optical investigation a colour change of the work-piece surface was recognized. The SEM images showed changes on the surface in the machined areas as well. In both analyses it was asserted, that the voltage as well as the processing time influence the size of the machined areas. The EDX line scanning results showed, that at increasing voltage and processing time the content of oxygen increases and the content of aluminium decreases in the centre of the machined areas. This proves, that oxide layers were developed during the process. Furthermore, it can be seen, that the thickness of the oxide layer increases at increasing voltage and processing time. Local oxide spots with thicknesses between 40 nm and 500 nm were achieved. The diameter and the thickness of the local oxide spots can be adjusted by the voltage and the processing time.

Acknowledgments
The authors thank the DFG (German Research Foundation) for supporting this work carried out within the framework of project SCHU 1484/20-1 and LA 1274/40-1.

References

[1] Natsu W, Ikeda T and Kunieda M 2007 Generating Complicated Surface with Electrolyte Jet Machining *Precis. Eng.* **31** **1** 33-9
[2] Hackert-Oschätzchen M, Meichsner G, Zinecker M, Martin A and Schubert A 2012 Micro Machining with Continuous Electrolytic Free Jet *Precis. Eng.* **36** **4** 612-19
[3] Hackert-Oschätzchen M, Martin A, Meichsner G, Zinecker M and Schubert A 2013 Microstructuring of Carbide Metals Applying Jet Electrochemical Machining *Precis. Eng.* **37** **3** 621-34
[4] Hackert M 2010 Entwicklung und Simulation eines Verfahrens zum elektrochemischen Abtragen von Mikrogeometrien mit geschlossenem elektrolytischen Freistrahl (Dissertation) *Scripts Precision and Microproduction Engineering* Vol 2 A Schubert Verlag Wissenschaftliche Scripten
[5] Schubert A, Zeidler H, Martin A, Hackert-Oschätzchen M, Kempen K and Kruth J P 2014 Jet-Electrochemical Machining of Selective Laser Melted Aluminum and Steel Alloys for Micro Injection Moulds *Proc. of the 14th euspen Int. Conf.* **2** 367-70
[6] Schubert A, Hackert-Oschätzchen M, Meichsner G, Zinecker M and Martin A 2011 Evaluation of the Influence of the Electric Potential in Jet Electrochemical Machining *In: Proc. of the 7th Int. Symp. on Electrochemical Machining Technology* pp 47-54
[7] Martin A, Kuhn D, Hackert-Oschätzchen M and Schubert A 2016 Design and Realisation of a Specific Device for Jet Electrochemical Machining of Rotating Work Pieces *Proc. of the 12th Int. Symp. on Electrochemical Machining Technology 2016*
[8] Nickel D, Dietrich D, Morgenstern R, Scharf I, Podlesak H and Lampke T 2016 Anodisation of Aluminium Alloys by Micro-Capillary Technique as a Tool for Reliable, Cost-Efficient, and Quick Process Parameter Determination *Adv. Mater. Sci. Eng.* Article number 1374897

IOP Conf. Series: Materials Science and Engineering **181** (2017) 012043 doi:10.1088/1757-899X/181/1/012043

Influence of the heat treatment condition of alloy AlCu4Mg1 on the microstructure and properties of anodic oxide layers

R Morgenstern, D Dietrich, M Sieber and T Lampke

Institute of Materials Science and Engineering, Chemnitz University of Technology, 09125 Chemnitz, Germany

Abstract. Due to their outstanding specific mechanical properties, high-strength, age-hardenable aluminum alloys offer a high potential for lightweight security-related applications. However, the use of copper-alloyed aluminum is limited because of their susceptibility to selective corrosion and their low wear resistance. These restrictions can be overcome and new applications can be opened up by the generation of protective anodic aluminum oxide layers. In contrast to the anodic oxidation of unalloyed aluminum, oxide layers produced on copper-rich alloys exhibit a significantly more complex pore structure. It is the aim of the investigation to identify the influence of microstructural parameters such as size and distribution of the strengthening precipitations on the coating microstructure. The aluminum alloy EN AW-2024 (AlCu4Mg1) in different heat treatment conditions serves as substrate material. The influence of the strengthening precipitations' size and distribution on the development of the pore structure is investigated by the use of high-resolution scanning electron microscopy. Integral coating properties are characterized by non-destructive and light-microscopic thickness measurements and instrumented indentation tests.

1. Introduction

Due to their high specific strength, aluminum alloys are widely used as lightweight construction materials. The outstanding mechanical properties of the age-hardenable aerospace alloy EN AW-2024 (AlCu4Mg1) are related to the precipitation of microstructural heterogeneities. According to Sha et al. [1], copper- and magnesium-rich atomic clusters and Guinier-Preston-Bagaryatsky zones (GPB zones) crucially influence the mechanical strength. Furthermore, rod-shaped Al_2CuMg precipitates (S-phase), Al_2Cu precipitates with different degree of coherence and lattice structures as well as different nanoscale iron-, silicon- and manganese-containing precipitates develop during artificial aging at higher temperatures [2]. In general, the open circuit potential (OCP) of intermetallic phases (IMPs) and the surrounding aluminum solid solution (Alss) are different. In the presence of an electrolyte, the potential difference leads to the selective electrochemical corrosion of the less noble phases.

Aluminum alloys can be efficiently protected from corrosion by the electrochemical conversion of the aluminum substrate surface into an anodic aluminum oxide (AAO) layer. The development of an ordered hexagonal pore-structure during the anodic oxidation process has already been described by Keller et al. [3] in 1953. Recent studies describe the effect of alloying elements and impurities on the anodic oxidation of high-strength aluminum alloys. For copper-containing alloys, the influences of highly-dispersed copper atoms and copper-rich intermetallic phases can be differentiated. Copper atoms in solid solution accumulate to copper-rich particles at the substrate-oxide interface due to the preferential oxidation of aluminum atoms. The anodic overpotential at the interface increases with the depletion of aluminum atoms enabling the oxidation of copper-rich particles. In contrast to aluminum oxide, copper oxide is an electronic conductor which allows for the evolution of oxygen gas bubbles at the anode. As copper enrichment and oxidation repeat periodically, the ordered pore-structure is superimposed by repetitive gas-voids [4]. According to Ma et al. [5], microscale copper-rich IMPs such as Al_2CuMg are rapidly anodically dissolved as soon as they get in contact with the electrolyte, because the copper-rich oxide formed at the IMP's surface is not electrochemically stable in the electrolyte. After rapid dissolution of the IMPs, microscale cavities remain within the AAO layer.

In naturally and artificially aged conditions, which are of high technical importance, the degree of copper segregations lies in between the highly dispersed state (copper in solid solution) and the thermodynamic equilibrium state (microscale S-phase precipitates). It is the aim of this work to clarify, at which point of copper segregation a different microstructure of AAO layers is observed and to what extent the macroscopic coating properties are influenced.

2. Material and methods

Sheets of the aluminum alloy EN AW-2024 (chemical composition in wt%: 3.8–4.9 Cu, 1.2–1.8 Mg, 0.3–0.9 Mn, ≤0.5 Si, ≤0.5 Fe, ≤0.1 Cr, ≤0.25 Zn, ≤0.15 Ti) with dimensions of 50x25x1,5 mm^3 were used as substrate material. All sheets were solution annealed (505 °C, 1 h) and quenched in water, immediately followed by pretreatment and anodization (condition W), natural aging (room temperature, > 14 days, condition T4), peak aging (180 °C, 16 h, condition T6), overaging (220 °C, 20 h, condition T7).

Sample pretreatment included etching in 3 wt% sodium hydroxide at 50 °C for 5 min and pickling in 1:1 nitric acid at room temperature for 30 s. Each step was followed by rinsing under deionised water. The anodic oxidation was conducted at 22 °C (room temperature) and 5 °C in 20 vol% sulphuric acid (Merck, Germany) at a constant current density of 2 A/dm^2 for a duration of 1 hour. The coating thickness was measured by non-destructive eddy-current testing (Fischerscope MMS) as well as by optical microscopy on cross-sections.

Metallographic cross-sections were prepared by diamond grinding accomplished by a finish using a silica oxide polishing suspension. The microstructure was investigated by scanning electron microscopy (SEM, Zeiss, LEO 1455VP) and field emission scanning electron microscopy (FE-SEM, Zeiss, NEON40EsB). Both secondary electron (SE, topography contrast) and backscattered electron (BSD, element contrast) detectors were applied. Hardness profiles of the AAO cross-sections were measured by nanoindentation (UNAT) using a Berkovich indenter and a normal load of 5 mN.

3. Results and discussion

3.1 Substrate and coating microstructure

During solidification of EN AW-2024 alloy, iron and silicon rich intermetallic phases precipitate from the liquid phase. Because of the low solubility of iron and silicon in the Alss even at high temperatures, these primary phases remain largely unchanged after solution annealing and quenching. Also, microscale Al_2CuMg precipitates, which formed due to slow cooling, are not completely solved during one hour of solution annealing. The complete dissolution of this phase would require a significantly longer annealing time, which leads to undesirable grain coarsening. However, this phase is partly dissolved and formed to a spherical or ellipsoidal shape.

As shown in figure 1, the iron- and silicon-rich primary precipitates mostly exhibit an irregular elongated shape due to previous sheet processing. After anodizing, these precipitates leave highly porous areas within the AAO layer. As described above, the spherical Al_2CuMg precipitates are rapidly dissolved as soon as they get in contact with the electrolyte. Thus, they leave microscale spherical cavities within the conversion layer. The different coating properties described below are not caused by these microscale defects as they can be observed in the AAO layers of all heat treatment conditions.

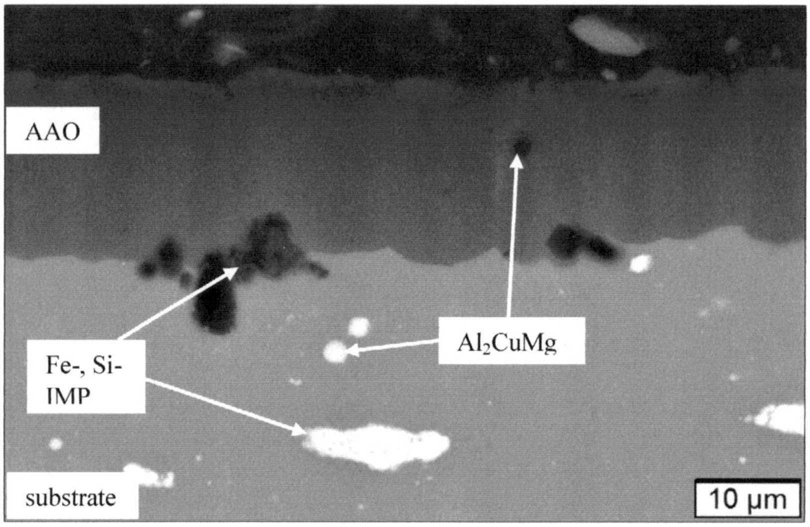

Figure 1. Microscale precipitations in the substrate leading to microscale defects within the AAO layer (SEM, BSD picture).

Clusters of copper and magnesium atoms already precipitate some hours after solution annealing and water quenching from the supersaturated Alss [1]. The microstructures of the heat treatment conditions W and T4 differ in respect of the size and number of clusters as well as of the amount of copper and magnesium remaining in solid solution. However, as atomic clusters cannot be recognized at FE-SEM pictures because of their small size, the substrate microstructure of the condition T4 is also representative for the condition W. As shown in the BSD picture figure 2a, the condition T4 exhibits bright spheroidal precipitates, which contain heavier elements than aluminum, e.g. iron, silicon or manganese. After the anodic oxidation, these precipitates leave spheroidal pores or highly porous areas within the AAO layer. However, the microstructure of the AAO layer is determined by cyclic enrichment of highly dispersed copper at the substrate-coating interface and the oxidation of copper rich particles being accompanied by oxygen evolution. As described above, this leads to the formation of fine gas voids along the ordered pore channels (figure 3a).

Artificial aging at 180 °C (peak-aged condition T6) leads to the development of a network of bundled GPB zones in the substrate alloy (figure 2b). Due to the higher copper content of these zones, a more porous and less stable oxide is formed leading to a stronger dissolution. At the SE image of the AAO cross-section (figure 3b), these porous areas appear as wells. Between the wells, a porous microstructure similar to the conditions T4 and W can be observed. Consequently, a sufficient amount of highly dispersed copper atoms or clusters exists between the GPB zones leading to copper enrichment at the substrate-coating interface and the cyclic generation of gas voids.

During artificial aging at 220 °C (overaged condition T7), the precipitation of the S-phase at grain boundaries is more pronounced. Some grain boundaries as well as two triple junctions appear as bright "dotted" lines in figure 2c (element contrast) due to the copper rich precipitates. Within close proximity of the S-phase, e.g. around grain boundaries, areas with a dark appearance indicate a local copper depletion. This observation is confirmed by the findings of Sha et al. [1]. In contrast to the GPB zones in condition T6, the rod-like S-phase precipitates in the grain interior exhibit a higher material contrast and a stricter orientation to one another. Some of the numerous bright spheroidal phases in figure 2c might be assignable to the Ω-phase being another modification of Al_2CuMg, which typically forms at temperatures above 200 °C [2].

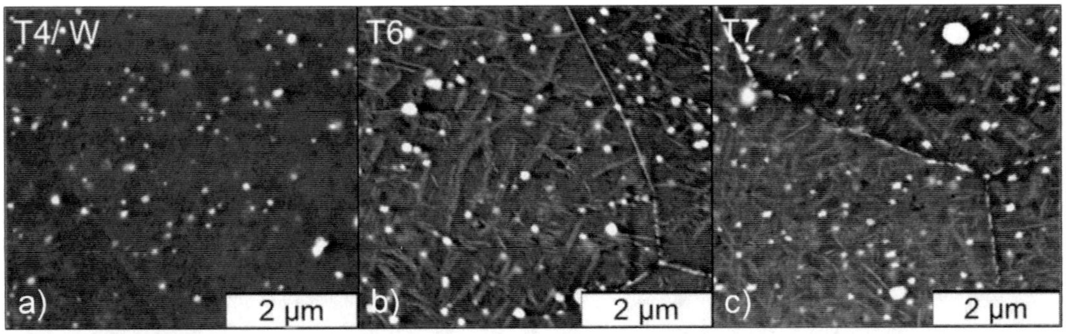

Figure 2. Microstructure of alloy EN AW-2024 in heat treatment conditions T4/W (a), T6 (b) and T7 (c) (FE-SEM, BSD).

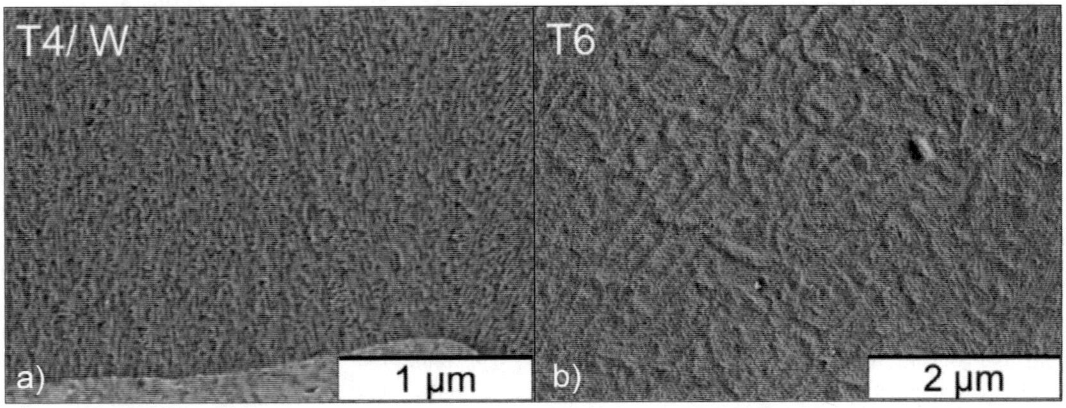

Figure 3. Microstructures of AAO coatings on alloy EN AW-2024 in heat treatment conditions T4/W (a), T6 (b) after anodizing at 22 °C (FE-SEM, SE).

Similar to microscale S-phase precipitates, the fine S-phase rods in the overaged condition leave cavities in the AAO layer after room temperature anodizing due to their rapid dissolution in the electrolyte. In the coating cross section (figure 4a), these cavities appear as elongated wells within the former grain interior and along the former grain boundaries. Areas of the AAO in close vicinity of the elongated wells appear to be less porous in comparison with the AAO microstructure of the conditions W and T4. The formation of additional gas voids along the ordered pore channels cannot be observed there. This finding correlates with the copper depletion in the surrounding of S-phase precipitates. During anodic oxidation of these depleted areas, no significant copper enrichment takes place at the substrate-coating interface and no significant oxygen evolution takes place at the alumina barrier layer. Again, some ellipsoid pores are visible within the coating. They might be attributable to the dissolution of the Ω-phase.

Similar to room temperature anodizing, hard anodizing of the conditions T4 and W at 5 °C is also characterized by the cyclic copper enrichment and oxidation at the substrate-coating interface leading to the cyclic formation of small gas voids. However, the influence of the heat treatment condition on the coating microstructure is less pronounced. As shown in figure 4b, elongated areas containing fine gas voids can be observed next to more compact areas with an ordered pore structure. In contrast to room temperature anodizing of condition T7 (figure 4a), the porous areas exhibit less topography contrast.

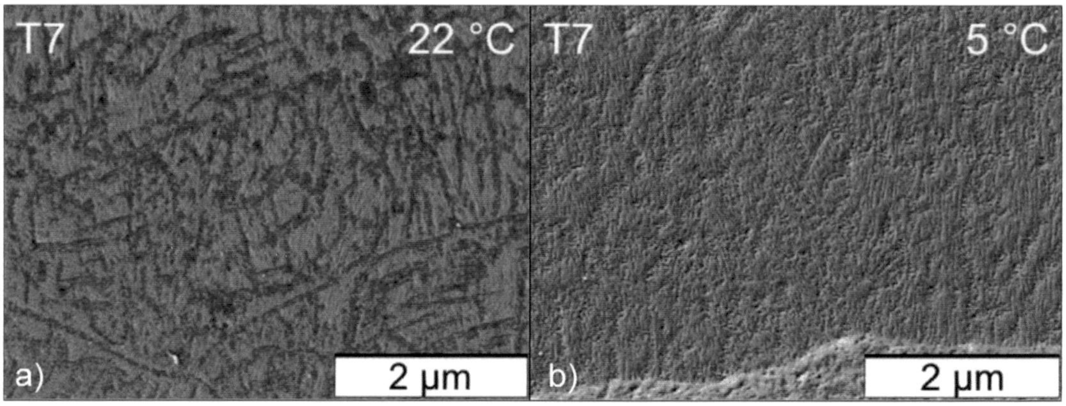

Figure 4. Microstructures of AAO coatings on alloy EN AW-2024 T7 after anodizing at 22 °C (a) and at 5 °C (b) (FE-SEM, SE).

3.2 Coating properties

It can be seen from figure 5a that for room temperature anodizing the coating thickness strongly depends on the heat treatment condition. Although an identical charge quantity of 2 Ah/dm² (and also approx. the same potential of about 13 V) was applied to all samples, conditions W and T4 exhibit the lowest coating thicknesses of about 18 to 19 µm. This might be due to the oxygen evolution, which occurs at the substrate-coating interface as a result of the cyclic enrichment and oxidation of copper. As electrical charge is consumed by this side-reaction, the current efficiency of the oxide formation and thus the coating thickness decrease. With increasing segregation of the alloying elements (clusters → GPB zones → S-phase) the thickness of the AAO layer increases. Possibly, chemical dissolution takes part during the rapid dissolution of the S-phase to a greater extent as the copper-rich oxide is not stable in the sulfuric acid electrolyte. Consequently, oxygen evolution is reduced due to the smaller amount of electrochemical dissolution. As no significant oxygen evolution takes place during the oxidation of copper-depleted regions, the highest current efficiency and coating thickness can be observed for condition T7. The anodizing behavior of condition T6 can be seen as an intermediate between T4 and T7, therefore it is consistent that the coating thickness also lies in between.

In general, higher coating thicknesses are achieved at 5 °C as the chemical dissolution of the AAO is smaller. However, the chemical dissolution of copper-rich regions such as the S-phase also decreases significantly. Consequently, electrochemical processes dominate the dissolution of the S-phase and the oxygen evolution is more pronounced. In contrast, less oxygen evolution takes place during the oxidation of copper-depleted areas. In total, the different extents of oxygen evolution seem to compensate one another. Consequently, the anodization of heat treatment conditions with both high copper segregation and high copper dispersion take place with a similar current efficiency. For all heat treatment conditions the coating thickness amounts to about 40 µm.

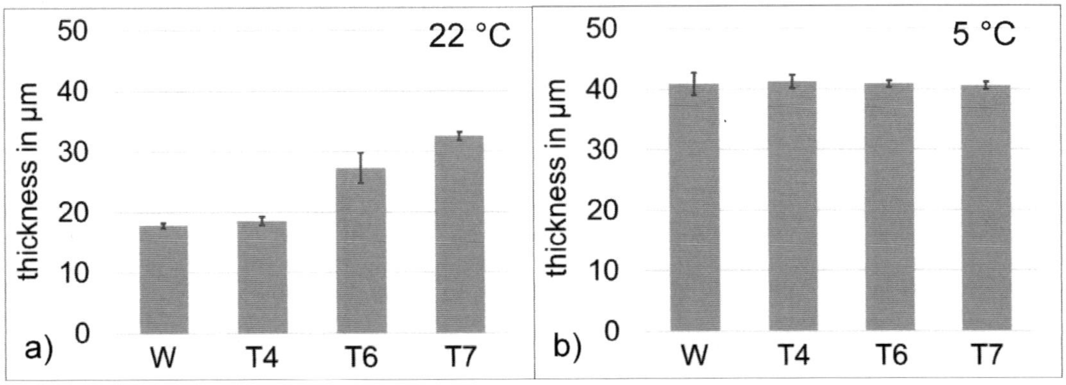

Figure 5. Coating thicknesses after anodizing of different heat treatment conditions at a) 22 °C, b) 5 °C.

The measured hardness values include the hardness of compact AAO and pores and therefore correlate with the average coating porosity of the examined volume. As the pore channel diameter enlarges with increasing exposition time in the electrolyte due to chemical dissolution, the hardness generally decreases with increasing distance from the substrate. This especially applies to room temperature anodizing because of the increasing reactivity of the sulphuric acid electrolyte at higher temperatures. In order to illustrate the typical deviation of hardness values for room temperature anodization, figure 6a shows all measuring points for the condition T7. It can be seen that the trendlines for conditions W and T4 are in the same range correlating with the similar coating microstructure. Compared with this, the trendline for condition T7 lies at a higher level which implies that the heterogeneous microstructure consisting of compact areas and elongated cavities (see figure 4a) exhibits a lower global porosity and thus a higher global hardness than the homogeneous coating microstructure of the conditions W and T4 being characterized by numerous small gas voids (see figure 3a). The hardness profile of AAO coatings on condition T6 lies in between those of conditions W/ T4 and T7, which corresponds with the intermediate heterogeneity of the coating microstructure. For hard anodizing, only a slight influence of the heat treatment condition on the AAO hardness can be observed. As can be seen from figure 6b, the single hardness points for conditions W and T7 partly overlap one another and the trendlines are close together. This again corresponds with the minor microstructural differences between AAO coatings on the conditions W/ T4 and T7 after hard anodizing.

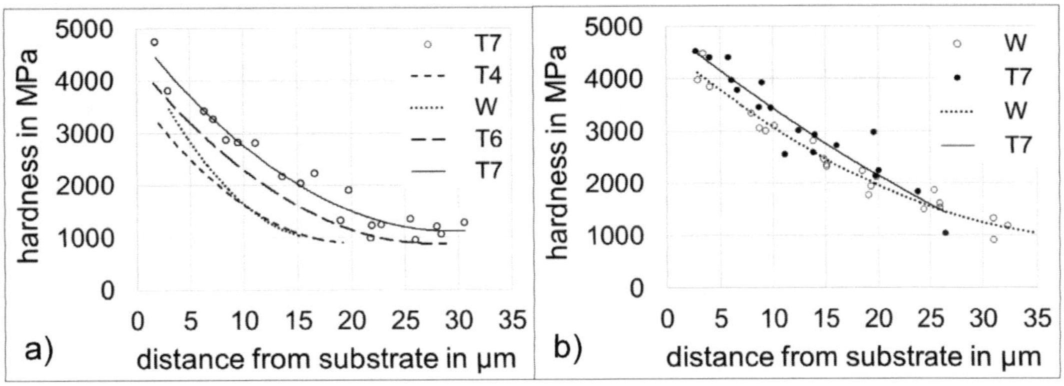

Figure 6. Hardness profiles of AAO coatings on different heat treatment conditions after anodizing at a) 22 °C and b) 5 °C.

4. Conclusions

It was shown that for room temperature anodizing the microstructure and the macroscopic properties of AAO coatings significantly depend on the heat treatment condition of the substrate alloy AlCu4Mg1. In case that alloying elements are highly dispersed within the substrate (e.g. in Alss or atomic clusters), oxygen evolution takes place at the substrate-coating interface due to the cyclic enrichment and oxidation of copper. This leads to the formation of fine gas voids along the ordered pore channels as well as to the reduction of the current efficiency of oxide formation and thus to a reduction of the coating thickness. After artificial aging, the segregation of copper increases and copper-depleted regions develop next to copper-rich precipitates such as the S-phase. As no significant copper enrichment takes place during the anodization of copper depleted regions, no significant oxygen evolution and void formation occurs. The S-phase rapidly dissolves during anodizing due to electrochemical dissolution and the chemical attack of the electrolyte leaving elongated cavities in the AAO layer. Therefore, less oxygen evolution occurs during room temperature anodization of the condition T7 resulting in higher current efficiency and coating thickness. The heterogeneous AAO microstructure consisting of compact regions and elongated cavities exhibits a higher hardness than the homogeneous AAO microstructures of conditions W and T4.

Basically, the same mechanisms of AAO formation apply to room temperature and hard anodizing at 5 °C for the conditions W and T4. However regarding the condition T7, the chemical dissolution of copper-rich regions such as the S-phase is less emphasized as the reactivity of the electrolyte decreases with lower temperature. As a result, the oxygen evolution is not reduced significantly and the current efficiencies for the anodization of different heat treatment conditions are similar. Consequently, coating properties like thickness and hardness are not significantly influenced by the heat treatment condition for hard anodizing.

It can be concluded from these results that for industrial room temperature anodizing of alloy AlCu4Mg1, process conditions have to be adjusted for different heat treatment conditions in order to fulfill coating specifications economically, e.g. a specific coating thickness can be achieved in a shorter process time on condition T7 compared with condition T4. Regarding hard anodizing, different heat treatment conditions might be anodized within the same process with the objective of optimizing the throughput as thickness and hardness of AAO coatings do not vary significantly.

Acknowledgement

The authors gratefully acknowledge funding by the German Research Foundation (Deutsche Forschungsgemeinschaft, DFG) within the framework of the Collaborative Research Centre 692 (SFB HALS 692, subproject B2). The technical assistance of J. Xu, D. Spieler, K. Muhr and C. Gläser is also gratefully acknowledged.

References

[1] Sha G, Marceau R K W, Gao X, Muddle B C and Ringer S P 2011 *Acta Mater.* **59** 1659
[2] Zhang F, Levine L E, Allen A J, Campbell C E, Creuziger A A, Kazantseva and Ilavsky J 2016 *Acta Mater.* **111** 385
[3] Keller F, Hunter M S and Robinson D L 1953 *J. Electrochem. Soc.* **100** 411
[4] Curioni M, Roeth F, Garcia-Vergara S J, Hashimoto T, Skeldon P, Thompson G E and Ferguson J 2010 *Surf. Interface Anal.* **42** 234
[5] Ma Y, Zhou X, Thompson G E, Curioni M, Zhong X, Koroleva E, Skeldon P, Thomson P and Fowles M 2011 *Corros. Sci.* **53** 4141

IOP Conf. Series: Materials Science and Engineering **181** (2017) 012044 doi:10.1088/1757-899X/181/1/012044

Downscaled anodic oxidation process for aluminium in oxalic acid

M Sieber[1], R Morgenstern[1,3], D Kuhn[2], M Hackert-Oschätzchen[2], A Schubert[2] and T Lampke[1]

[1]Institute of Materials Science and Engineering, Chemnitz University of Technology, 09125 Chemnitz, Germany
[2]Professorship Micromanufacturing Technology, Chemnitz University of Technology, 09107 Chemnitz, Germany
[3]corresponding author, email address: roy.morgenstern@mb.tu-chemnitz.de

Abstract. The increasing multi-functionality of parts and assemblies in several fields of engineering demands, amongst others, highly functionalised surfaces. For the different applications, on the one hand, there is a need to scale up surface modification processes originating in the nano- and micro-scale. On the other hand, conventional macro-scale surface refinement methods offer a huge potential for application in the said nano- and micro-scale. The anodic oxidation process, which is established especially for aluminium and its alloys, allows the formation of oxide ceramic layers on the surface. The build-up of an oxide ceramic coating comes along with altered chemical, tribological and electrical surface properties. As a basis for further investigations regarding the use of the anodic oxidation process for micro-scale-manufacturing, the scale effects of oxalic acid anodising on commercially pure aluminium as well as on the AlZn5.5MgCu alloy are addressed in the present work. The focus is on the amount of oxide formed during a potentiostatic process in relation to the exchanged amount of charge. Further, the hardness of the coating as an integral measure to assess the porous oxide structure is approached by nano-indentation technique.

1. Introduction

Service properties of technical products are, amongst others, determined by surface properties. The demand for highly functionalised surfaces arises in all fields of engineering. Not only because of the material and process efficiency, but also because of the trend to miniaturisation of assemblies and products, a scaling-down of conventional manufacturing processes is required. The anodic oxidation process is a suitable means to produce oxide coatings on a plethora of electrically conductive materials, amongst them established light metals like aluminium, magnesium and titanium and also substrates like graphite or silicon. For the latter ones, the formation of very thin, yet functional, e.g. electrically isolating oxide coatings is possible by a strong local anodic polarisation. This polarisation can be realised utilising devices as small as an AFM tip, for instance, and using the ambient humidity as the electrolyte. The application of the polarisation and the associated electric field limits the geometrical extent of the oxidised surface region to the micrometre or even the nanometre range [1-6]. To form functional oxides on metals like aluminium, titanium or magnesium, the ambient humidity is not sufficient because of the low thickness of the oxide layers, which are formed under ambient conditions [7]. Therefore, alkaline or acidic media have to be used, which entails an increased expenditure both with regard to the technical realisation of the process and its costs. The following chemical reactions occur during the oxidation of the aluminium and describe the oxide formation (1) and the oxide dissolution in the acidic electrolyte (2):

$$2Al + 3H_2O \rightarrow Al_2O_3 + 6H^+ + 6e^- \quad (1)$$

$$Al_2O_3 + 6H^+ \rightarrow 2Al^{3+} + 3H_2O \quad (2)$$

As pointed out in [8], the distribution of the electric field is vital for the formation of a thick oxide layer, since the barrier layer, which is formed at the beginning of the process, impedes the mass transfer and thus further oxide formation. Therefore, a sufficiently high electric field has to be applied to ensure oxide layer growth.

In the present work, scaling effects of the anodic oxidation of aluminium in an oxalic acid electrolyte are investigated in the centimetre, millimetre and micrometre scale. To exclude the influence of inhomogeneous electric field formation, the geometrical extent of the surface modification is limited by restricting the active surface area.

2. Methods and Materials

Two different substrate alloys were used, namely the commercially pure aluminium alloy Al99.5 (EN AW-1050, nominal composition in wt%: Si ≤ 0.25, Fe ≤ 0.4, Cu ≤ 0.05, Mn ≤ 0.05, Mg ≤ 0.05, Zn ≤ 0.07, Ti ≤ 0.05) and the frequently used high-strength alloy AlZn5.5MgCu (EN AW-7075, nominal composition in wt%: Si ≤ 0.4, Fe ≤ 0.5, Cu 1.2–2.0, Mn ≤ 0.3, Mg 2.1–2.9, Cr 0.18–0.28, Zn 5.1–6.1, Ti ≤ 0.2). Both substrates were supplied as sheet metal with a thickness of 1.5 mm and 3 mm for Al99.5 and AlZn5.5MgCu, respectively. Sample pretreatment included etching in 3 wt% NaOH solution (60 °C, 3 min) and in the case of AlZn5.5MgCu pickling in 1:1 diluted nitric acid (room temperature, 30 s). The electrolyte for the anodic oxidation was comprised of 0.8 mol/l oxalic acid (supplied as oxalic acid dehydrate, Merck, Germany). The electrolyte temperature was 25 °C. All the used chemicals were of analytical grade.

The anodic oxidation was carried out with a constant voltage of 60 V for 15 min. Three different scales of the active surface area were considered, namely the centimetre, millimetre and micrometre scale. The power supply for the experiments in the centimetre scale was a pe1028 (Plating Electronic, Germany) and the active surface of the sample was approx. 50 cm². The power supply for the millimetre and micrometre scale was a potentiostat Zahner IM6 with voltage booster CVB120 (Zahner, Germany) and the active surface was approx. 80 mm² and 500 µm², respectively. For both of the power supplies, the values of current and voltage were logged internally during the process with an acquisition rate of 1 sample/s. A saturated Ag/AgCl electrode was used as reference in both the millimetre and the micrometre scale. A platinum sheet and a platinum wire served as the counter electrode in the millimetre and the micrometre scale, respectively. No reference was used in the centimetre scale and lead sheets served as the counter electrode in this case. All potential values acquired in three-electrode-arrangement are given in relation to the standard hydrogen electrode. In the centimetre and the millimetre scale, the electrolyte was constantly stirred, while in the micrometre range, the electrolyte movement was realised via a volume flow through the cell of 4 µl/s. A schematic representation of the different experimental setups for the considered scales is given in figure 1. The coating thickness was determined by light microscopy in the cross sections of the samples. Being a conversion coating process, a part of the coating thickness lies inside of the original substrate dimensions, while the other part lies outside. The relation of these dimensions was considered by partially shielding the substrate by a polyurethane lacquer in the centimetre scale and by the given experimental setup in the other scales. Thus, it was possible to recognise the original substrate surface in the cross section and to determine the parts of the oxide coating, which apparently grew in inward and in outward direction. The hardness of the coatings was obtained from instrumentated indentation tests in the cross sections of the coatings with a Berkovich indenter (Asmec Unat, Germany). A load of 5 mN was applied (load time 10 s, hold time 5 s, unload time 4 s). The indents were placed at different locations in the coatings. Because of the rather low coating thicknesses, the typical decay of the hardness with increasing distance from the substrate does not occur. Therefore, the mean hardness values are given for the considered coatings.

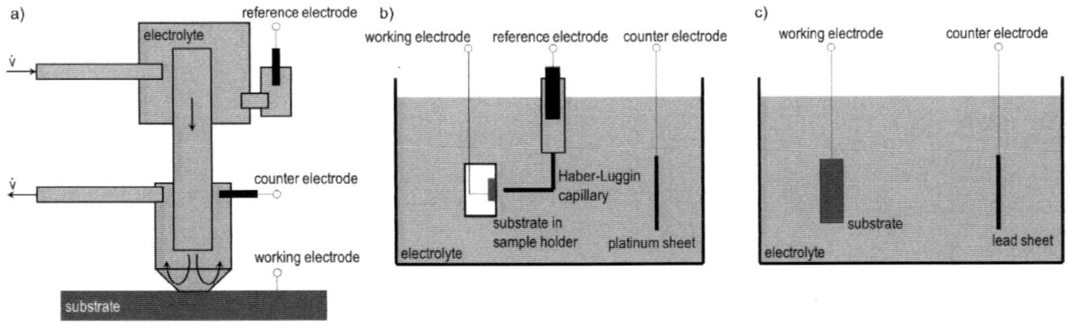

Figure 1. Schematic representation of the experimental setup for the three different scales – a) micrometre scale: a capillary filled with electrolyte is placed on the substrate and the inner capillary diameter defines the active surface area; b) millimetre scale: the sample is placed in a sample holder, which limits the active surface area; c) centimetre scale: the sample is placed in an electrolyte basin, the active surface area is defined by the sample size.

3. Results and Discussion

A representative curve for the evolution of the current density over the process time for the potentiostatic regime at 60 V for 15 min is given in figure 2. Initially, the barrier oxide layer formation takes place, which leads to a decrease of the current flow within the first few seconds of the process. The reason for this is the character of the barrier layer, which acts as a resistance against current flow. Afterwards, the formation of the typical pore structure takes place, which is accompanied by an increase of the current density. With the growth of the porous layer, the resistance of the oxide layer increases slightly with the time and therefore the current density slowly decreases in the course of the process. The transferred amount of charge, which is calculated by integration of the current density over the time, is shown in figure 3 for the different substrate alloys and scales. It is obvious, that the amount of charge transferred in the process is significantly lower for the AlZn5.5MgCu alloy compared to the commercially pure aluminium substrate Al99.5. The reason for this behaviour lies in the alloying elements, mainly Zn and Cu, which impede the substrate conversion process and therefore lead to a lower current density and hence charge transfer. Also, it has to be noted that the charge transfer in the centimetre and the millimetre scale is in a similar range, but there is significantly less charge transfer in the micrometre range. Two effects are responsible for this behaviour: For the AlZn5.5MgCu alloy, excessive oxygen evolution occurs during the anodic oxidation. While in the millimetre and the centimetre range, the oxygen can freely flow off through the electrolyte, it reduces the effective cross section of the electrolyte-containing capillary, and therefore impedes the current flow. This leads to a premature end of the process after approximately 700 s due to complete blocking of the surface. Additionally, the electrolyte-containing capillary itself acts as a significant resistance due to its small cross-section so that also for the Al99.5 substrate, for which practically no gas evolution occurs, the charge transfer is significantly lower in the micrometre range. Thus, the influence of the experimental setup is, although undesirable, not negligible in this case.

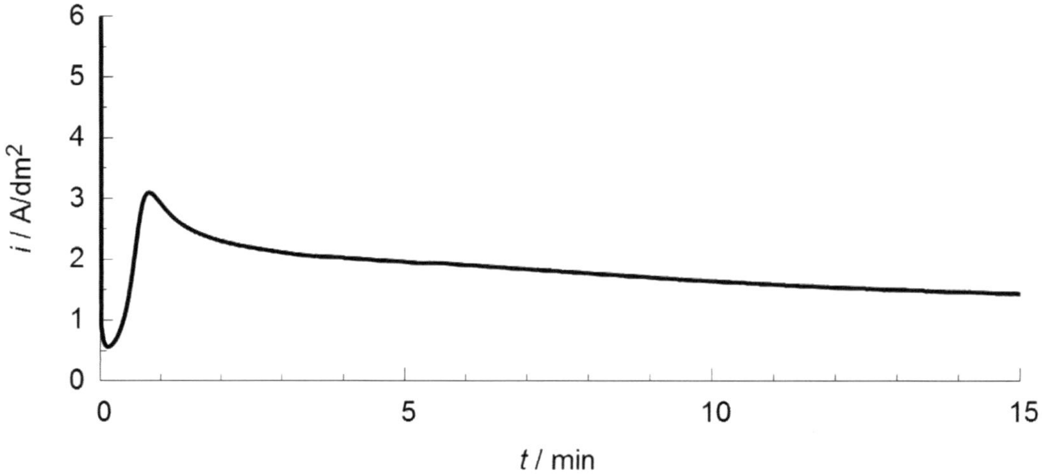

Figure 2. Representative evolution of the current density over the process time for a potentiostatic anodic oxidation process of Al99.5 at 60 V in 0.8 mol/l oxalic acid at room temperature in the micrometre scale. Qualitatively, the same curve appears for the centimetre and the millimetre scale for both alloys with different absolute values of the current density.

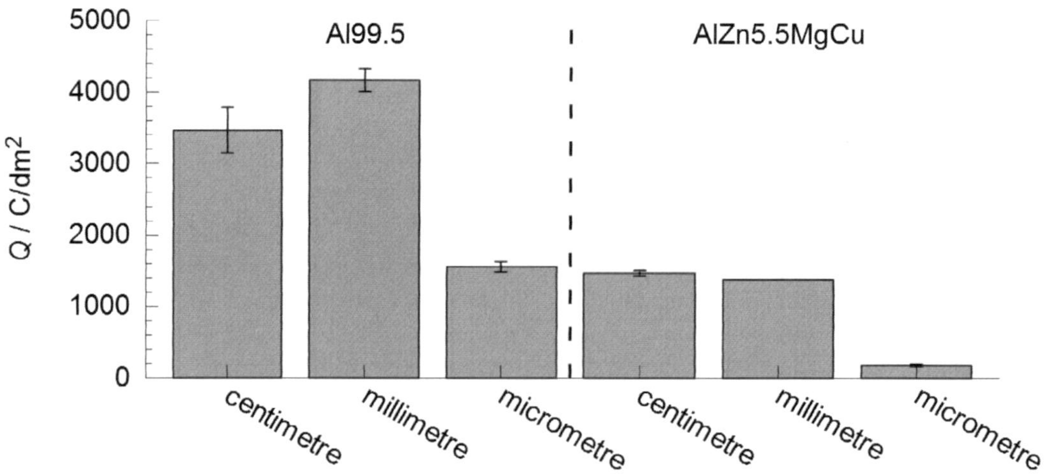

Figure 3. Charge transfer in the potentiostatic anodic oxidation process (60 V in 0.8 mol/l oxalic acid at room temperature, 15 min) for the two different alloys in the considered scales of the active surface area.

Figure 4 shows a representative cross section of an oxide coating with adjacent active and inactive surface areas, which allow the determination of the "inward" and "outward" coating growth. The resulting thickness values of the formed oxide coatings for the considered alloys and scales of the active surface are shown in figure 5. In accordance with the impediment of the anodic oxidation process by the alloying elements and the charge consumed by the evolution of oxygen, the coating thickness formed on AlZn5.5MgCu is generally lower than on Al99.5. Considering the millimetre and the centimetre scale, there is no significant change of the coating thickness. The coating thickness is much lower in the micrometre scale because of the geometrical conditions in the micrometre scale setup and, consequently, the effect of the electrolyte resistance. It is noticeable that the relative amount of "inward" coating growth becomes higher with the reduction of the scales. However, no single effect can be made responsible for this behaviour. Except for the experimental setup and the occurring resistances therein, the conditions of the electrolyte flow in the system as well as the temperature distribution within the substrate (local or global temperature rise and relevance of the heated volume compared to the volume of the sample) have to be considered. Further investigation focussing on both abstraction and definition as well as isolation of experimental conditions is necessary to clarify the found behaviour.

An overview of the hardness of the oxide coatings in the cross section is given in figure 6. It is noticeable that the hardness of coatings on the Al99.5 substrate is significantly higher than on the AlZn5.5MgCu substrate, which is caused by a higher density of flaws in the coatings on the alloy. These originate from the disturbance of the aluminium conversion process by the alloying elements, which are either dissolved or incorporated into the oxide. Both of the mentioned processes are accompanied by oxygen evolution and as a result, the regular pore structure of the coatings becomes disturbed and larger flaws arise. This consequently leads to a lower coating hardness in accordance with the literature [9, 10]. There is no significant effect of the scale of the active surface area on the hardness. This inevitably implies that the temperature distribution at the interface substrate/oxide/electrolyte is roughly identical for all the scales, which otherwise would exert an effect on the hardness according to the literature [11]. The hardness of the oxide coatings on AlZn5.5MgCu produced in the micrometre scale is not considered, since the low coating thickness did hardly allow for an appropriate placement of the hardness indents.

Figure 4. Representative cross section of an oxide coating with adjacent active and inactive surface areas (E – embedding resin, O – oxide coating, S – substrate).

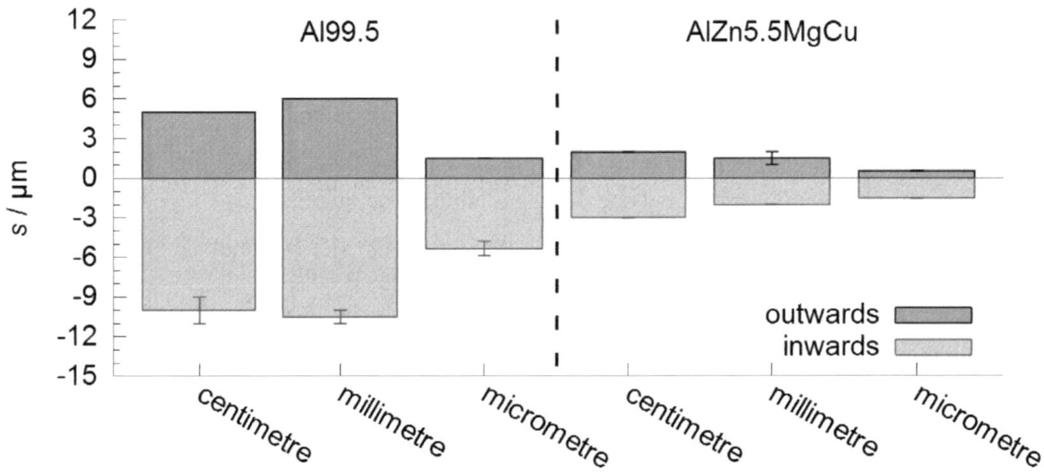

Figure 5. Thickness of the oxide coating formed in the potentiostatic anodic oxidation process (60 V in 0.8 mol/l oxalic acid at room temperature, 15 min) for the two different alloys in the considered scales of the active surface area. The combined height of the boxes represents the coating thickness, the single boxes represent the amount of oxide grown outwards and inwards with reference to the original dimensions of the substrate (0).

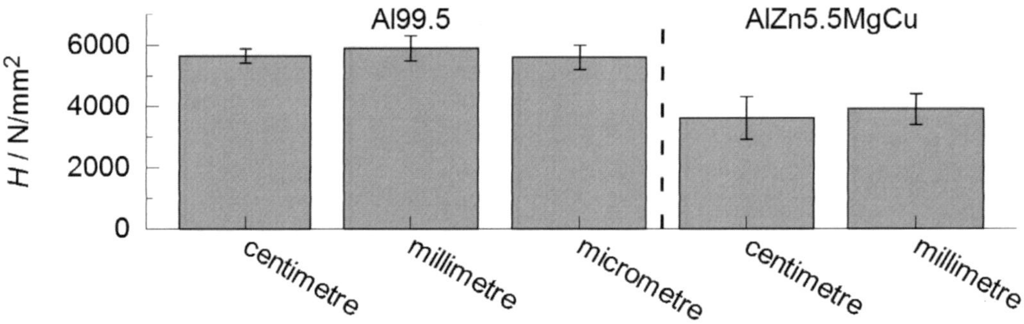

Figure 6. Hardness measured in the cross section of the oxide coatings produced on the two different alloys in the different scales of the active surface area.

4. Summary

The effect of different scales of the active surface area on the anodic oxidation of Al99.5 and AlZn5.5MgCu in oxalic acid was considered with regard to the amount of charge transferred during the anodic oxidation, the resulting coating thickness and hardness. The presence of the alloying elements, especially zinc and copper, in the AlZn5.5MgCu alloy impedes the oxide formation reaction and therefore decreases both the charge transfer and the resulting coating thickness in comparison to Al99.5. The hardness in the cross section of the coating is lower for the AlZn5.5MgCu alloy, which is also attributed to the alloying elements, which led to an increased number and size of flaws in the coatings in accordance with the literature. The different scales of the active surface area do not exert an effect on the hardness. The experimental arrangement led to a higher system resistance in the micrometre scale so that the charge transfer and hence the coating thickness were reduced in comparison to the millimetre and the centimetre scale.

Acknowledgments
The authors gratefully acknowledge funding of this work by the German Research Foundation (Deutsche Forschungsgemeinschaft, DFG, project LA1274/40-1). The support by Dagobert Spieler, Elke Benedix, Gabriele Tauchmann, Sigrid Kaufmann, Christel Gläser, Karla Muhr and Franz Nozicka (all from the Institute of Materials Science and Engineering) is gratefully acknowledged.

References

[1] Bartosik M, Kolibal M, Cechal J, Mach J and Sikola T 2009 *J Nanosci Nanotechnol* **9** 5887
[2] Hu K, Wang Q and Wangyang P 2013 *Int J Nanomanuf* **9** 496
[3] Kolanek K, Gotszalk T, Zielony M and Grabiec P 2008 *Materials Science Poland* **26** 271
[4] Li Y F, Sugiyama M and Toshiyoshi H 2015 *J Microelectromech Syst* **24** 1471
[5] Mori G, Lazzarino M, Ercolani D, Biasiol G, Sorba L, Heun S and Locatelli A 2005 *J Appl Phys* **98** 114303
[6] Suehisa Y, Aoki T, Shimizu J, Zhou L and Eda H 2009 *Key Eng Mater* **389-390** 424
[7] Davis Z J, Abadal G, Hansen O, Borise X, Barniol N, Perez-Murano F and Boisen A 2003 *Ultramicroscopy* **97** 467
[8] Jee W Y, Kim J G and Chu C N 2015 *Int. J. Precis. Eng. Manuf.* **16** 2623
[9] Runge J M and Hossain T 2015 *Mater Today Proc* **2** 5055
[10] Zhou X, Habazaki H, Shimizu K, Skeldon P, Thompson G E and Wood G C 1996 *Corros Sci* **38** 1563
[11] Aerts T, Jorcin J B, De Graeve I and Terryn H 2010 *Electrochim Acta* **5** 3957

IOP Conf. Series: Materials Science and Engineering **181** (2017) 012045 doi:10.1088/1757-899X/181/1/012045

Gibbs energy calculation of electrolytic plasma channel with inclusions of copper and copper oxide with Al-base

V M Posuvailo[1], M D Klapkiv[1], M M Student[1], Y Y Sirak[1]and H V Pokhmurska[2]

[1]Karpenko Physico-Mechanical Institute of NAS of Ukraine
5, Naukova Str., Lviv, 79060, Ukraine.
[2]Institute of Materials Science and Engineering, Chemnitz University of Technology, 09125 Chemnitz, Germany

Abstract. The oxide ceramic coating with copper inclusions was synthesized by the method of plasma electrolytic oxidation (PEO). Calculations of the Gibbs energies of reactions between the plasma channel elements with inclusions of copper and copper oxide were carried out. Two methods of forming the oxide-ceramic coatings on aluminum base in electrolytic plasma with copper inclusions were established. The first method – consist in the introduction of copper into the aluminum matrix, the second - copper oxide. During the synthesis of oxide ceramic coatings plasma channel does not react with copper and copper oxide-ceramic included in the coating. In the second case is reduction of copper oxide in interaction with elements of the plasma channel. The content of oxide-ceramic layer was investigated by X-ray and X-ray microelement analysis. The inclusions of copper, $CuAl_2$, Cu_9Al_4 in the oxide-ceramic coatings were found. It was established that in the spark plasma channels alongside with the oxidation reaction occurs also the reaction aluminothermic reduction of the metal that allows us to dope the oxide-ceramic coating by metal the isobaric-isothermal potential oxidation of which is less negative than the potential of the aluminum oxide.

1. Introduction

Plasma electrolytic oxidation is used to form oxide-ceramic coatings on aluminum alloys. These coatings have a high micro-hardness, wear and corrosion resistance, dielectric properties and are applied in automotive, instrument engineering, radioelectronics, in manufacturing casings of computers and mobile phones and so on. Their synthesis, phase composition and properties have been widely studied in [1-4]. However, such coatings have certain disadvantages, in particular, high brittleness.

However it is found that oxide-ceramic coating on the copper–doped alloy D16 have better physical and mechanical characteristics [5-7]. Analysis of the microstructure of these coatings revealed the presence of copper-containing inclusions of nanometer size [5]. Conventional models, representing the processes of plasma electrolytic processing, claim that in spark channels only the reactions of oxidation of the alloying elements, available in the initial alloy, take place.

It is assumed that, alongside with the reactions of all alloy elements oxidation in the plasma channel, the secondary reactions of aluminothermic reduction of copper oxides also occur, the isobaric-isothermal oxidation potential of which is less negative than the oxidation of aluminum. Aluminothermic copper oxide reduction leads to the formation of a composite oxide-ceramic layer based on aluminium and precipitations of pure metal. Such pure metal precipitations improve the mechanical and tribological characteristics of the oxide-ceramic layers.

IOP Conf. Series: Materials Science and Engineering **181** (2017) 012045 doi:10.1088/1757-899X/181/1/012045

The purpose of this work is to substantiate by the thermodynamic calculations the possibility of the secondary aluminothermic reactions and confirm the possibility of the formation of composite oxide-ceramic layers with metal inclusions.

2. Materials and Methods

The relative stability of metal oxides at different temperatures is determined by the change in the Gibbs energy:

$$\Delta G = \Delta H - T \cdot \Delta S \quad (1)$$

where ΔG is Gibbs energy, ΔH is enthalpy, ΔS is entropy, T is absolute temperature. Enthalpy ΔH and entropy ΔS of these substances are calculated from the equations:

$$\Delta H_T = \Delta H_0 + \int_0^T \Delta c_p dT \quad (2)$$

$$\Delta S_T = \Delta S_0 + \int_0^T (\Delta c_p / T) dT \quad (3)$$

where ΔH_0 is standard enthalpy of substance, Δc_p is isobaric heat capacity, ΔS_0 is standard entropy substance.

Consider the chemical reaction

$$v_1A_1+v_2A_2+\ldots \leftrightarrow v_1'A_1'+v_2'A_2'+\ldots \quad (4)$$

in which v_i is the number of moles of substance A_i, involved in reaction (4) and v_i' is, accordingly, the number of moles of reaction products A_i'.

The ΔG_r – the Gibbs energy of chemical reaction (4) is associated with the values of Gibbs energy substances ΔG, participating in the reaction, described by equation [8]:

$$\Delta G_r= (v_1'\Delta G(A_1', T)+ v_2'\Delta G (A_2', T)+\ldots) - (v_1\Delta G(A_1, T)+ v_2\Delta G (A_2, T)+\ldots) \quad (5)$$

Enthalpy of chemical reactions ΔH_r was calculated from enthalpies of reactants and reaction products ΔH from standard elements for all substances involved in the reactions.

$$\Delta H_r= (v_1'\Delta H(A_1', T)+ v_2'\Delta H (A_2', T)+\ldots) - (v_1\Delta H(A_1, T)+ v_2\Delta H (A_2, T)+\ldots) \quad (6)$$

For calculations the tabulated values of Gibbs energy and enthalpy of substances formation were used [9].

3. The method of obtaining the oxide ceramic coatings

The samples of two types were manufactured for investigations. The first consisted of the copper and aluminum powders; the second – of copper oxide (CuO) and aluminum. All powders were technical pure with grains size 50 – 100 μm. The weight ratio of the components in the samples was 50 to 50 wt.%. The powders were mixed in a ball mill Retsh PM 100 for 40 min with velocity 100 rpm. They were compacted under pressure of 980 MPa during 12 hour. The sample size was 20 mm diameter by 4 mm height. The oxide-ceramic coating was synthesized in the electrolyte with a composition of 1 g/l KOH and 1 g/l of water glass. The current density was 30 A/dm^2 for 120 min. The density ratio of the cathode and anode currents was *Ic/Ia* = 1. The durations of anode and cathode pulses were 6 ms. Transverse sections of the oxide-ceramic coating were prepared by the diamond polishing paper and a diamond paste. The samples were washed with distilled water and ethanol in the ultrasonic bath before examining. Their surface was examined by the scanning electron microscope EVO 40XVP with a microanalysis system INCA Energy.

The X-ray phase analysis was performed using the diffractometer DRON 3.0m (Cu-Kα radiation) in a step-wise mode (0.05 ° step, exposition at a point 10-15 seconds, 2Θ=20-90 °, the voltage at the cathode – 30 kV, current – 10 mA). The obtained diffraction data were clarified by the Rietveld method [10] using the program package GSAS [11].

4. Results and discussion

It was found previously that in the discharge channel there were 4 regions that differed in both temperature and composition [12-16]. The temperature in the central zone of the discharge channel is 7.000 – 10.000 K and decreases to 600 K in the adjacent areas. In the discharge channel there are Al, Al_2O_3 in different states of aggregation and gaseous H_2O, OH, H, H_2, O_2, O, AlO, Al_2O and Al_2O_2. The final product of the synthesis is the oxide-ceramic coating which is mainly composed of Al_2O_3. The interactions of CuO and Cu with electrons, ions, H_2O, OH, H, H_2, and components of the electrolyte are not considered here.

Let us determine the most probable interaction of inclusions of copper and its oxide with the substances of discharge channels. We calculate the free energy (Gibbs energy) and heat (enthalpy) reactions of copper oxide CuO with aluminum using equation (5, 6):

$3CuO_{cr\,l}+2Al_{cr\,l} \rightarrow \alpha\text{-}Al_2O_{3cr\,l} + 3Cu_{cr\,l}$ (7);

$3CuO_g+2Al_g \rightarrow \alpha\text{-}Al_2O_{3cr\,l} + 3Cu_l$ (8);

$3CuO_g+2Al_g \rightarrow \alpha\text{-}Al_2O_{3cr\,l} + 3Cu_g$ (9);

$3CuO_g+2Al_l \rightarrow \alpha\text{-}Al_2O_{3cr\,l} + 3Cu_g$ (10);

$3CuO_g+2Al_l \rightarrow \alpha\text{-}Al_2O_{3cr\,l} + 3Cu_{cr\,l}$ (11).

The obtained dependences are shown in figure. 1 (a, b).
The subscripts $_g$ and $_{cr\,l}$ indicate the reagents in gaseous, crystalline and liquid states.

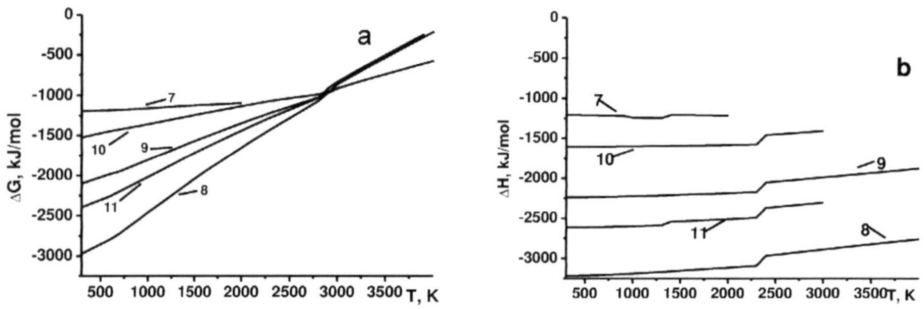

Figure 1. Dependence of Gibbs energy (a) and enthalpy (b) reactions on the process temperature: $_l$ (7); (8); (9); (10); (11).

Calculate the Gibbs energy and heat reactions of copper oxide with aluminum oxide gas (figure 2a, b):

$CuO_g+2AlO \rightarrow \alpha\text{-}Al_2O_{3cr\,l}+Cu_{cr\,l}$ (12);

$CuO_{cr\,l}+2AlO \rightarrow \alpha\text{-}Al_2O_{3cr\,l}+Cu_{cr\,l}$ (13);

$CuO_g+2AlO \rightarrow \alpha\text{-}Al_2O_{3cr\,l}+Cu_g$ (14);

$CuO_{cr\,l}+2AlO \rightarrow \alpha\text{-}Al_2O_{3cr\,l}+Cu_g$ (15).

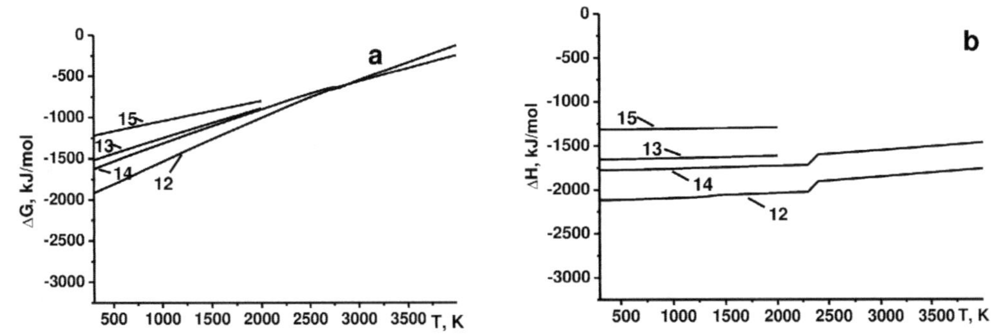

Figure 2. Dependences of Gibbs energies (a) and enthalpy (b) of reactions on the process temperature: (12); (13); (14); (15).

The calculated free energies (Gibbs energy) and heat of reactions of oxide copper with dialuminum oxide Al_2O are shown in figure 3 (a, b).

$2CuO_{cr\,l}+Al_2O \rightarrow \alpha\text{-}Al_2O_{3cr\,l}+2Cu_g$ (16);
$2CuO_g+Al_2O \rightarrow \alpha\text{-}Al_2O_{3cr\,l}+2Cu_g$ (17);
$2CuO_{cr\,l}+Al_2O \rightarrow \alpha\text{-}Al_2O_{3cr\,l}+2Cu_{cr\,l}$ (18);
$2CuO_g+Al_2O \rightarrow \alpha\text{-}Al_2O_{3cr\,l}+2Cu_{cr\,l}$ (19).

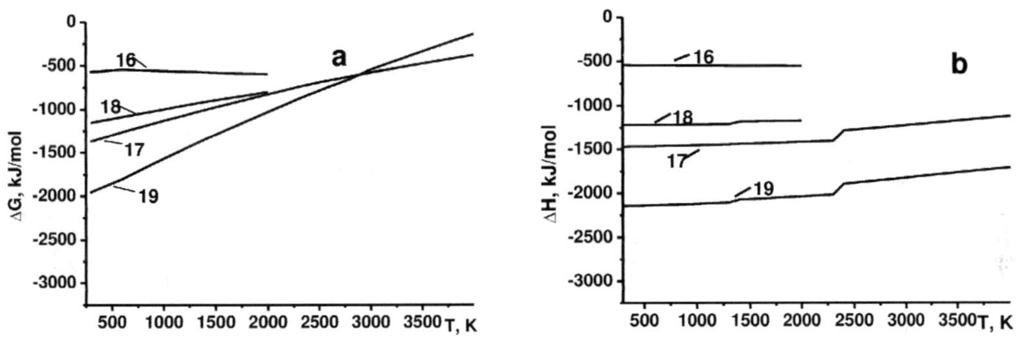

Figure 3. Dependence of Gibbs energy ΔG_r (a) and enthalpy ΔH_r (b) of reactions on the process temperature: (16); (17); (18); (19).

Determination of the free energy and the heat reactions of oxide copper with dialuminum oxide Al_2O in the presence of atomic oxygen figure 4 (a, b):

$CuO_{cr\,l}+ Al_2O +O \rightarrow \alpha\text{-}Al_2O_{3cr\,l}+Cu_{cr\,l}$ (20);
$CuO_g+Al_2O +O \rightarrow \alpha\text{-}Al_2O_{3cr\,l}+Cu_{cr\,l}$ (21);
$CuO_{cr\,l}+Al_2O +O \rightarrow \alpha\text{-}Al_2O_{3cr\,l}+Cu_g$ (22);
$CuO_g+Al_2O +O \rightarrow \alpha\text{-}Al_2O_{3cr\,l}+Cu_g$ (23).

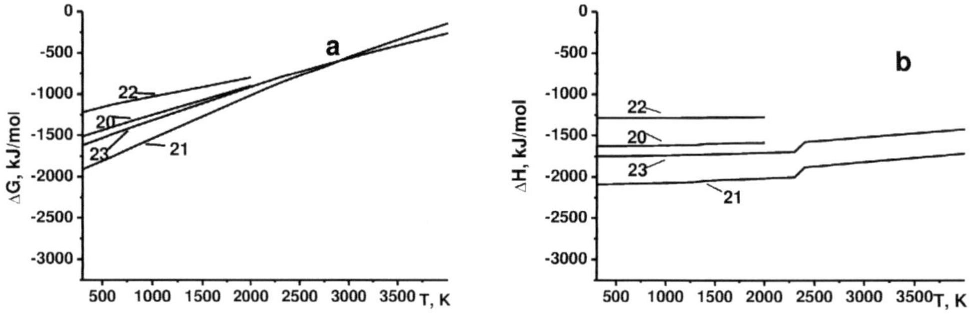

Figure.4. Dependence of Gibbs energy (a) and enthalpy (b) of reactions on the process temperature: (20); (21); (22); (23).

The calculated free energies and heat reactions of copper oxide with dialuminium dioxide (figure. 5 (a, b)):

$CuO_{g}+Al_2O_2 \rightarrow \alpha\text{-}Al_2O_{3cr\,l}+Cu_{cr\,l}$ (24);

$CuO_{cr\,l}+ Al_2O_2 \rightarrow \alpha\text{-}Al_2O_{3cr\,l}+Cu_{g}$ (25);

$CuO_{g}+ Al_2O_2 \rightarrow \alpha\text{-}Al_2O_{3cr\,l}+Cu_{g}$ (26);

$CuO_{cr\,l}+ Al_2O_2 \rightarrow \alpha\text{-}Al_2O_{3cr\,l}+Cu_{cr\,l}$ (27).

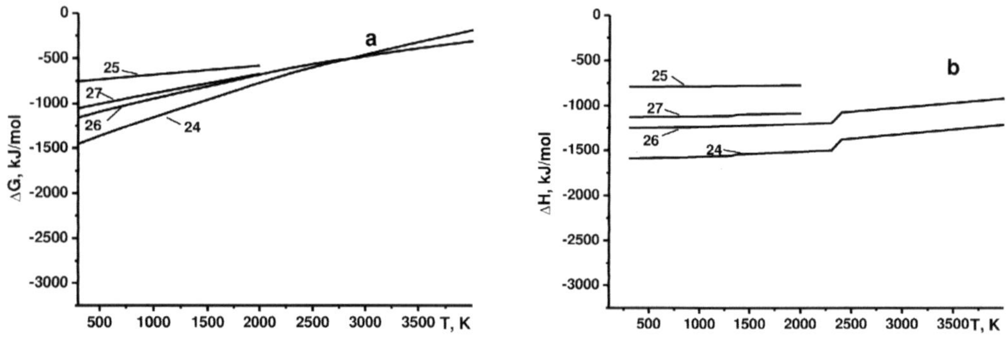

Figure 5. Dependences of Gibbs energy (a) and enthalpy (b) of reactions on the process temperature: (24); (25); (26); (27).

The temperature dependence of the calculated free energies and heat reactions of copper oxide with aluminum in the molecular oxygen is illustrated in figure 6 (a, b).

$CuO_{cr\,l}+ 2Al_{cr\,l}+O_2 \rightarrow \alpha\text{-}Al_2O_{3cr\,l}+Cu_{cr\,l}$ (28);

$CuO_{g}+ 2Al_{g}+ O_2 \rightarrow \alpha\text{-}Al_2O_{3cr\,l}+Cu_{g}$ (29);

$CuO_{cr\,l}+ 2Al_{g}+ O_2 \rightarrow \alpha\text{-}Al_2O_{3cr\,l}+Cu_{g}$ (30);

$CuO_{g}+ 2Al_{cr\,l}+ O_2 \rightarrow \alpha\text{-}Al_2O_{3cr\,l}+Cu_{g}$ (31);

$CuO_{cr\,l}+2Al_{cr\,l}+ O_2 \rightarrow \alpha\text{-}Al_2O_{3cr\,l}+Cu_{g}$ (32);

$CuO_{cr\,l}+2Al_{g}+ O_2 \rightarrow \alpha\text{-}Al_2O_{3cr\,l}+Cu_{cr\,l}$ (33);

$CuO_{g}+ 2Al_{cr\,l}+ O_2 \rightarrow \alpha\text{-}Al_2O_{3cr\,l}+Cu_{cr\,l}$ (34).

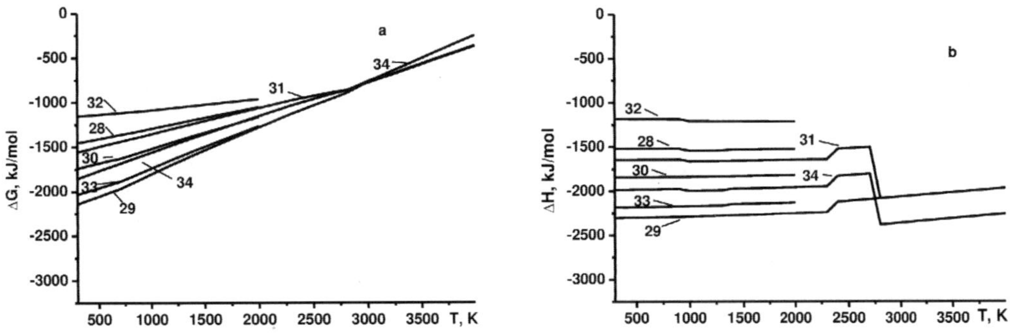

Figure 6. Dependences of Gibbs energy (a) and enthalpy (b) of reactions on the process temperature: (28); (29); (30); (31) (32); (33); (34).

The value of the free energy and enthalpy of reactions that occur with the reagents in the gaseous state is more negative than the reaction of reagent in the solid state, and is represented correspondingly, in release of a greater amount of heat. Besides Gibbs energy for such reactions:

$3CuO_g+2Al_g \rightarrow \alpha\text{-}Al_2O_{3cr\,l} +3\ Cu_g$ (9); $3CuO_g+2Al_l \rightarrow \alpha\text{-}Al_2O_{3cr\,l} + 3Cu_{cr\,l}$ (11);

$3CuO_g+2Al_l \rightarrow \alpha\text{-}Al_2O_{3cr\,l} + 3Cu_g$ (10); $CuO_g+ 2Al_g+ O_2 \rightarrow \alpha\text{-}Al_2O_{3cr\,l}+Cu_g$ (29).

is the lowest, what determines the priority of these reactions. Reactions involving AlO, Al_2O, Al_2O_2 take place in the final period of the plasma channel existence.

Separately the copper oxide will be reduced to copper by aluminium and will precipitate in the oxide-ceramic coating. The reverse reactions (reduction of aluminum oxide by copper) at 5000 K will not occur because of the positives values of the free energy in this temperature range.

Since in the central part of the discharge channel the temperature reaches 7000 - 10000 K, all the components are in the atomized and ionized states, and oxidizing of Al and Cu takes place in it. Reduction of copper oxide by aluminum will take place in the low-temperature areas of the discharge channel and the adjoining area.

It is possible to predict the behavior of copper and copper oxide inclusions in the aluminum matrix based on the above calculations of Gibbs energy and heat effects of interaction between the copper oxide with elements of the plasma channel and reverse reactions (interaction of copper with aluminum oxide Al_2O_3).

Contrary to the valve metals, which are characterized by the rapid oxide films formation, on the copper surface it is formed more slowly and consists of the oxides Cu (I) and Cu (II), which are semiconductors.

The breakdown of the film leads to anodic dissolution of inclusions that come out on the sample surface. This is proved by the absence of discharge channels on the surface of the compacted powder samples of aluminum and copper at first minutes of synthesis.

After dissolution of copper particles on the specimen surface spark discharges appear. Copper does not react with aluminium oxide and enters the oxide-ceramic coating in the form of metallic inclusions in figure 8 a, b.

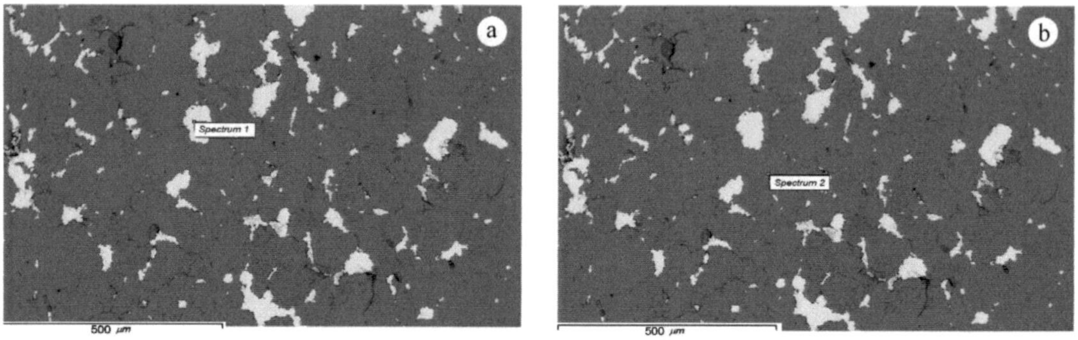

Figure 7. SEM image of the cross-section of compacted powders of copper and aluminum. The bright areas correspond to copper (a), gray - to aluminium (b).

Spectral analysis of point 1 in figure 7a indicates that this is Cu, and point 2 in figure 7b corresponds to Al.

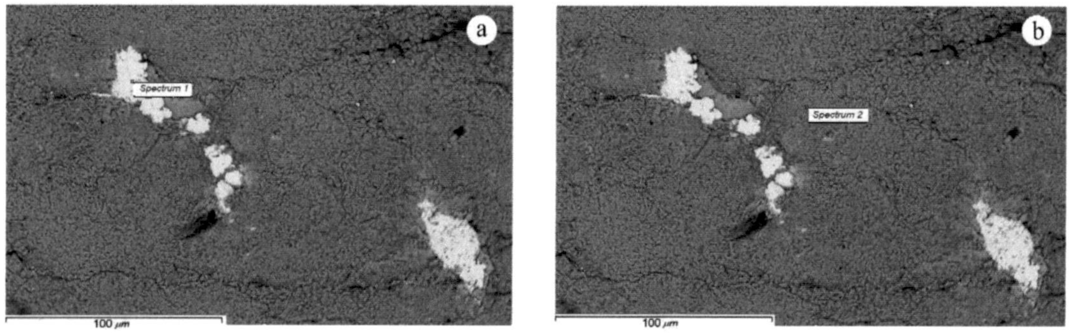

Figure 8. SEM image of the oxide-ceramic coating cross-section synthesized on the specimen with a mixture of compacted powders of copper and aluminium, (a, point 1)- copper, (b, point 2) – aluminium oxide.

After synthesizing the oxide ceramic coating on compacted powders of Al and Cu the coating was examined by electron microscopy. Figure 8 presents a photomicrograph of the cross-section of the oxide ceramic coating formed on a sample of compacted powders of copper and aluminum. Point 1 in figure 8 a , according to the results of spectral analysis, indicates the inclusion of Cu – 93.66 atomic% and O - 6.34 atomic%, and point 2 in Fig. 8 b Al_2O_3 with a small admixture of Cu (Al - 35.60 atomic%, O - 63.55 atomic%, Cu - 0.85 atomic%).

Unlike copper, copper oxide reacts with the active elements of the discharge channel. The aluminothermic reduction reactions of copper oxide by aluminum and incomplete oxide take place. The reaction runs with the release of large amounts of heat.

The oxide ceramic coating composition synthesized in compacted powders of copper and aluminum oxide was studied by X-ray analysis. Figure 9 shows the diffraction patterns of the initial mixture of CuO-Al (after powders milling for 40 min).

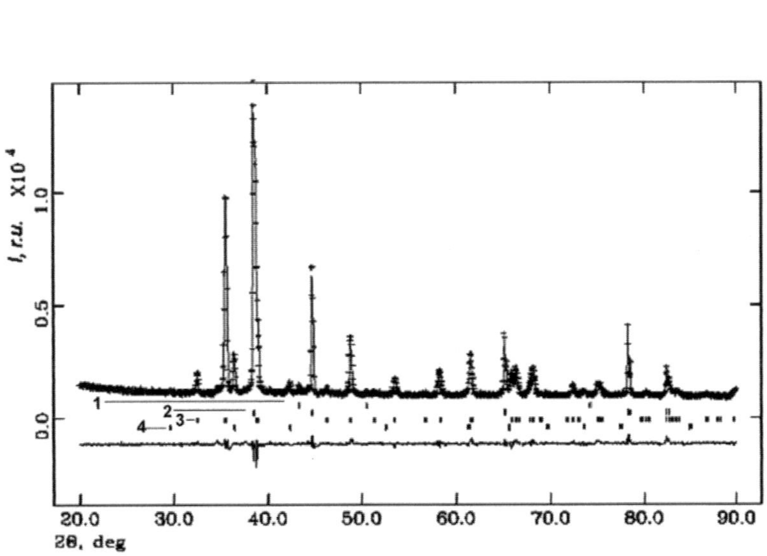

Figure 9. Experimental (+), calculated (solid line -) and the difference (bottom line -) X-ray diffraction profiles.

Vertical bars correspond to the location of the Bragg reflections (peaks) of constituting phases (top-down):

1 - Cu - 0.5(1) mass.% (FCC, *a*= 3.6153(8) Å); 2 - **Al** - 44.4(2) mass.% (FCC, *a*= 4.0493(1) Å);
3 - CuO - 51.4(2) mass.% (monoclinic, *a*= 4.6857(3) Å, *b*= 3.4256(2) Å, *c*= 5.1328(3) Å, *β*=99.341(3)°); 4 - **Cu_2O** - 3.7(1) mass.% (cubic, *a*= 4.2692(3) Å)
R-factors (factors of discrepancy): R_p=3.21%; R_{wp}=4.15%.

As has been described above, reactions of copper oxide reduction by aluminum occur with the release of a large amount of heat, which leads to the formation of the uneven surface of oxide-ceramic coating. For phase analysis the powdered oxide-ceramic coating was used. Figure 10 shows its diffraction pattern.

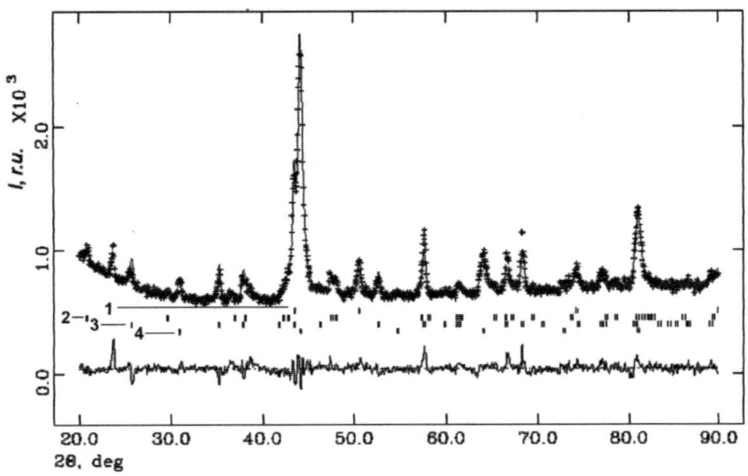

Figure 10. Experimental (+) calculated (solid line -), and the difference (bottom line -) X-ray diffraction profiles.

Vertical bars correspond to the locations of the Bragg reflections (peaks) of constituting phases (top-down):

1 - Cu - 7.4(4) mass.% (FCC, *a*= 3.622(1) Å)
2 - $CuAl_2$ - 8.7(9) mass.% (tetragonal, *a*= 6.067(5) Å, *b*= 4.877(6) Å)
3 - α-Al_2O_3 - 47.3(9) mass.% (rhombohedron, *a*= 4.765(1) Å, *c*=13.009(5) Å)
4 - Cu_9Al_4 - 36.6(5) mass.% (cubic, *a*= 2.9108(6) Å) – disordered phase (
R-factors (factors of discrepancy): R_p=3.47%; R_{wp}=4.61%.

Based on the phase composition analysis it is found that the oxide ceramic coating is composed of Cu, intermetallic compounds $CuAl_2$, Cu_9Al_4 and Al_2O_3. Most likely, the intermetallic compounds are formed due to the excessive amount of Al in the discharge channels at the boundary of the oxide-ceramic coating and the compressed powder material. Unidentified peak at an angle of $2\theta = 23.75$ ° corresponds to spinel $AlSiO_2$. It is also found that the oxide-ceramic coating formed on the compacted powders of copper oxide and aluminum does not contain copper oxides. This proves our assumption about copper oxide reduction by aluminum in the discharge channels. As noted above, the reaction of copper oxide reduction by aluminum takes place with the release of a large amount of heat, which leads to the formation of the uneven surface of the oxide-ceramic coating. To form a more uniform coating it is proposed to reduce the content of CuO or introduce Cu_2O in the powder.

5. Conclusions

Calculations of Gibbs energy and enthalpy of chemical reactions between components of discharge channels and copper or copper oxide have shown the possibility of oxide-ceramic coatings alloying with copper inclusions in two ways.

1 - Introduction of copper particles into the powder material, which will not interact with plasma channel and remain as inclusions.
2 - Introduction of copper oxides into the powder material that will reduce in the discharge channels. The thermodynamic calculations were checked by analyzing the transverse sections of the oxide-ceramic Cu - Al coating and by X-ray analysis of CuO - Al.

The investigations of the phase composition of the oxide-ceramic coatings show the presence of copper and intermetallic materials $CuAl_2$, Cu_9Al_4 in them, thus indicating an excess of aluminum in the discharge channel.

It is established that in the spark plasma channels alongside with the oxidation reactions, the alumothermic metal reduction reactions take place and allows add in the oxide-ceramic coating with metal inclusions, the Gibbs energy oxidation of which is less negative than the aluminum oxidation.

The results this investigations can be used for the further development of the PEO process in order to obtain new aluminium alloys for the preliminary generation of plasma spraying.

References

[1] Chernenko V I, Snezhko L A and Papanov I I 1991 *Preparation of coatings by anode-spark electrolysis* Leningrad 127
[2] Wenbin Xue, Zhiwei Deng, Yonchun Lai and Ruyi Chen 1998 *J. of American Ceramic Society* **81** 1365
[3] Yerokhin A L, Nie X, Leyland A, Matthews A and Dowey S J 1999 *Surf. Coat. Tech.* **122** 73
[4] Nie X, Meletis E I, Jiang J C, Leyland A, Yerokhin A L and Matthews A 2002 *Surf. Coat. Tech.* **149** 245
[5] Student M M, Dovhunyk V M, Klapkiv M D, Posuvailo V M, Shmyrko V V and Kytsya A P 2012 *Mater. science.* **48** (2) 180
[6] Student M, Dovhunyk V, Klapkiv M, Posuvailo V and Shmyrko V. The effect structure of PEO layers at aluminum arc sprayed coating on changes a mixed electrode potential and friction coefficient in a couple with steel at the stage of grinding *Proceedings of the XI International Conference and Exhibition "Problems of corrosion and corrosion protection of structural materials" June 4–6, 2012* Lviv, Ukraine 622-625
[7] Student M M, Shmyrko V V, Klapkiv M D, Lyasota I M, and Dobrovolska L N 2014 *Mater. science.* **50** (2) P. 290
[8] Gurvich L V Veyts I V Medvedev V A Khachkuruzov G A Yungman V S and others *Thermodynamic properties of individual substances*. Vol 1. Book 1. Moskva 1978 495
[9] http://kinetics.nist.gov/janaf
[10] *The Rietveld Method* Ed. by R A Young, Oxford University Press, 2000.
[11] Larson A C and Von Dreele R B *GSAS–General Structure Analysis System*, LANSCE, MS-805, Los Alamos National Laboratory, Los Alamos, NM 87545 USA, 1994.
[12] Klapkiv M D, Nykyforchyn H M. and Posuvailo V M 1994 *Mater. science.* **30**(3) 333
[13] Posuvailo V M and Klapkiv M D 1997*Mater. science.* **33**(3) 383
[14] Curran J A, Clyne T W 2005 *Surf. Coat. Tech.* **199** 168
[15] Klapkiv M D 1999 *Mater. science*, **35**(2) 279
[16] Stojadinović S, Perić M, Petković M, Vasilić R, Kasalicaa B, Belća I and Radić-Perić J 2011 *Electrochim Acta.* **56** 10122

Institute of Physics
Dirac House, Temple Back
Bristol BS1 6BE UK

ISSN: 1757-8981
ISBN 978-1-5108-3772-0